Springer
Proceedings in Physics 21

Springer Proceedings in Physics

Managing Editor: H. K. V. Lotsch

Volume 1 *Fluctuations and Sensitivity in Nonequilibrium Systems*
Editors: W. Horsthemke and D. K. Kondepudi

Volume 2 *EXAFS and Near Edge Structure III*
Editors: K. O. Hodgson, B. Hedman, and J. E. Penner-Hahn

Volume 3 *Nonlinear Phenomena in Physics* Editor: F. Claro

Volume 4 *Time-Resolved Vibrational Spectroscopy*
Editors: A. Laubereau and M. Stockburger

Volume 5 *Physics of Finely Divided Matter* Editors: N. Boccara and M. Daoud

Volume 6 *Aerogels* Editor: J. Fricke

Volume 7 *Nonlinear Optics: Materials and Devices* Editors: C. Flytzanis and J. L. Oudar

Volume 8 *Optical Bistability III* Editors: H. M. Gibbs, P. Mandel, N. Peyghambarian, and S. D. Smith

Volume 9 *Ion Formation from Organic Solids (IFOS III)* Editor: A. Benninghoven

Volume 10 *Atomic Transport and Defects in Metals by Neutron Scattering*
Editors: C. Janot, W. Petry, D. Richter, and T. Springer

Volume 11 *Biophysical Effects of Steady Magnetic Fields*
Editors: G. Maret, J. Kiepenheuer, and N. Boccara

Volume 12 *Quantum Optics IV* Editors: J. D. Harvey and D. F. Walls

Volume 13 *The Physics and Fabrication of Microstructures and Microdevices*
Editors: M. J. Kelly and C. Weisbuch

Volume 14 *Magnetic Properties of Low-Dimensional Systems*
Editors: L. M. Falicov and J. L. Morán-López

Volume 15 *Gas Flow and Chemical Lasers* Editor: S. Rosenwaks

Volume 16 *Photons and Continuum States of Atoms and Molecules*
Editors: N. K. Rahman, C. Guidotti, and M. Allegrini

Volume 17 *Quantum Aspects of Molecular Motions in Solids*
Editors: A. Heidemann, A. Magerl, M. Prager, D. Richter, and T. Springer

Volume 18 *Electro-optic and Photorefractive Materials* Editor: P. Günter

Volume 19 *Lasers and Synergetics* Editors: R. Graham and A. Wunderlin

Volume 20 *Primary Processes in Photobiology* Editor: T. Kobayashi

Volume 21 *Physics of Amphiphilic Layers* Editors: J. Meunier, D. Langevin, and N. Boccara

Volume 22 *Semiconductor Interfaces: Formation and Properties*
Editors: G. Le Lay, J. Derrien, and N. Boccara

Springer Proceedings in Physics is a new series dedicated to the publication of conference proceedings. Each volume is produced on the basis of camera-ready manuscripts prepared by conference contributors. In this way, publication can be achieved very soon after the conference and costs are kept low; the quality of visual presentation is, nevertheless, very high. We believe that such a series is preferable to the method of publishing conference proceedings in journals, where the typesetting requires time and considerable expense, and results in a longer publication period. Springer Proceedings in Physics can be considered as a journal in every other way: it should be cited in publications of research papers as *Springer Proc. Phys.*, followed by the respective volume number, page number and year.

Physics of Amphiphilic Layers

Proceedings of the Workshop,
Les Houches, France
February 10–19, 1987

Editors: J. Meunier,
D. Langevin, and N. Boccara

With 200 Figures

Springer-Verlag Berlin Heidelberg New York
London Paris Tokyo

Dr. Jacques Meunier
Dr. Dominique Langevin
CNRS, Lab. de Spectroscopie Hertzienne de l'ENS, 24 rue Lhomond,
F-75231 Paris Cedex 05, France

Professor Nino Boccara
Centre de Physique, Université Scient. et Médic., Côte des Chavants,
F-74310 Les Houches, France

ISBN-13: 978-3-642-83204-8 e-ISBN-13: 978-3-642-83202-4
DOI: 10.1007/978-3-642-83202-4

Softcover reprint of the hardcover 1st edition 1987

2153/3150-543210

Preface

This volume is the proceedings of the International Winter School on the Physics of Amphiphilic Layers, which was held at the "Centre de Physique des Houches", February 10–19, 1987.

Amphiphilic layers play an essential role in the behavior of a great variety of disperse systems such as micelles, microemulsions and vesicles. They can also exist as isolated mono- or bilayers, or constitute extended liquid crystalline structures. Although the properties of these different systems may at first sight seem unrelated, their theoretical interpretations are based on only a small number of common concepts, for example, curvature energy and thermal roughening. This was the reason for bringing together about 65 scientists from this research area – both theoreticians and experimentalists – from universities as well as from industrial laboratories, with different backgrounds: physicists, chemists, biologists.

The topics treated are: mono- and bilayers; interactive forces between layers, with special emphasis on steric forces; ordered structures, in particular swollen lamellar phases and defects; vesicles; micelles, including polymer-like systems; microemulsions, especially random bicontinuous structures; and porous media.

The topic of this workshop was proposed by N. Boccara, Director of the Centre de Physique des Houches, to the members of the council, who accepted the project. We wish to thank them for the confidence they had in us. The program of the Winter School was put together by the scientific committee, and we wish to thank its members for their very constructive work.

The conference was sponsored by:

Centre National de la Recherche Scientifique (MPB-Chimie)
Commissariat à l'Energie Atomique (IRF)
Ministère de la Défense (DRET)
Ministère de l'Education Nationale (DCRI)
Institut National de la Santé et de la Recherche Médicale
National Science Foundation

Further financial support came from the following companies:

ENIRICHERCHE
EXXON Research and Engineering Company
Institut Français du Petrole
L'OREAL
Standard Oil Company

Paris, France
February 1987

J. Meunier
D. Langevin

Contents

Part I Mono- and Bilayers

Part II Surface Thermal Fluctuations

Part VI Microemulsions

Part VII **Porous Media**

Part I

Mono- and Bilayers

Molecular Theory for Amphiphile Packing and Elastic Properties of Monolayers and Bilayers

A. Ben-Shaul[1], *L. Szleifer*[1], *and W.M. Gelbart*[2]

[1]Dept. of Physical Chemistry and the Fritz Haber Molecular Dynamics Center, The Hebrew University, Jerusalem 91904, Israel
[2]Dept. of Chemistry, The University of California, Los Angeles, CA 90024, USA

1. Introduction

In a number of recent papers we have presented and developed a statistical thermodynamic theory for amphiphile chain packing in micellar aggregates and membrane bilayers[1]. The central quantity in the theory is the (singlet) probability distribution function (pdf) of chain conformations, with the aid of which one can calculate conformational properties (e.g. bond order parameters and chain segment spatial distributions), as well as relevant thermodynamic functions such as the average internal (gauche/trans) energy, conformational entropy and free energy. The conformational pdf accounts explicitly for the packing constraints imposed on a given chain by (excluded volume interactions with) its neighbors. These constraints, and consequently the pdf depend on the geometrical characteristics of the microenvironment in which the chains are packed; namely, the curvature of the hydrocarbon-water interface of the micelle, membrane or monolayer and the surface density of polar heads.

Conformational and thermodynamic properties calculated by our mean-field type theory show very good agreement with corresponding results from large-scale molecular dynamics simulations[2] and available experimental data. Similarities and differences with respect to other theoretical approaches (e.g. [2-4]) have been discussed in detail elsewhere[1]. We have applied the theory to study a variety of issues pertaining to the conformational and thermodynamic properties of amphiphilic systems. These include, for example, the relative importance of (hydrocarbon) 'tail' vs. (hydrophilic) 'head' free energies in determining the preferred micellar geometry, energetic vs. entropic contributions to the packing free energy, the extent and importance of the hydrocarbon-water surface roughness and the conformational statistics and thermodynamics of mixed amphiphilic aggregates. Another important and natural extension of the theory is its application to calculating elastic properties of membrane bilayers and interfacial films (monolayers). In section 3 we briefly present a few preliminary results concerning this application. Before doing so, we briefly outline, in section 2, the formal and physical basis necessary for such calculations.

2. Packing Constraints and Conformational Statistics

Many and various experiments, as well as theory, reveal that the density of hydrocarbon chain segments (monomers) in the hydrophobic interior of amphiphilic bilayers is uniform throughout, and its value is similar to that in the bulk liquid hydrocarbon[5-7]. (We consider here only the liquid or so-called 'liquid-crystalline' state of the chains. Also, although the discussion below is general our calculations correspond to amphiphiles with simple alkyl chains). Uniform segment density does not imply random configurational chain statistics. In fact, apart from the boundary condition associated with the confinement of chains to one side of the water-hydrocarbon interface, the uniform density condition is the (only) relevant constraint on chain-packing statistics in our theory, as explained below. The uniform monomer density (i.e. no 'holes') is ensured by the attractive long-range (van der Waals) interactions responsible for the cohesiveness of the hydrocarbon core. In the case of surfactant monolayers at oil-water interfaces, uniform density does not require that the hydrocarbon chains form a 'compact' film of constant thickness. This is because the oil molecules (in the case of a good solvent) can fill up the 'bays' characterizing a rough chain-oil interface, Fig. 1. The extent of this roughness, which allows more conforma-

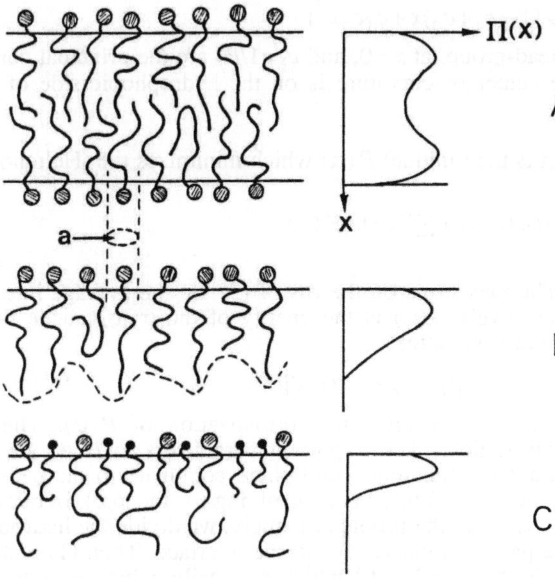

Fig.1 Schematic pictures of a bilayer (A), a monolayer (B), and a mixed monolayer (C)

tional freedom to the chains, decreases with chain length and surface density of head-groups. It should also be noted that bilayers, although often treated as such, are not a 'sandwich of two compact monolayers' either. Rather, the two monolayers interdigitate each other, often significantly, thereby gaining configurational free energy, Fig. 1. All these qualitative notions regarding chain packing can be expressed in simple mathematical terms,as shown below.

Let $P(\alpha)$ denote the probability of finding a chain in conformation α. The conformation of a $-(CH_2)_n-CH_3$ chain, in the rotational isomeric state (RIS) model, is fully specified by the sequence of trans/gauche bonds along the backbone and the three Euler angles describing the overall orientation of the chain relative to the interface. Let $\phi(x;\alpha)dx$ denote the number of (centers of) chain segments which, for a chain in conformation α are found in the interval $x, x+dx$, where x is the normal distance from the water-oil interface, Fig. 1. Equivalently, $v\phi(x;\alpha)dx$ is the volume taken up by the chain in $x, x+dx$ where v is the segment's volume ($\approx 27Å^3$ for CH_2). Because of the nonzero volume (area) of the chains it is clear that the total area occupied by N equivalent chains originating from the interface can not exceed $A(x)$, the cross sectional area of a surface parallel to the interface at distance x. This implies that for all x

$$<\phi(x)> = \sum_{\alpha} P(\alpha)\phi(x;\alpha) \leq a(x), \qquad (1)$$

where $a(x)=A(x)/N$ is the average area available per chain at distance x from the interface. This inequality is the appropriate packing constraint on $P(\alpha)$ for a monolayer. If there is some a priori reason to assume that the monolayer is compact and of width L then (1) should be used with the strict equality for all $x \leq L$. More generally, the equality applies when the chains are packed at uniform density. The extension of (1) to mixed monolayers and to bilayers (pure or mixed, symmetrical or asymmetrical) is straightforward[1]. Note that the packing constraints on $P(\alpha)$ describe the average (mean-field) effect of excluded volume interactions of a given chain with its neighbors. The attractive interactions do not affect $P(\alpha)$ because the density is uniform for all chain configurations.

The geometry of an amphiphile layer (or aggregate) enters the packing constraints via $a(x)$. Explicitly, for an interface with principal radii of curvature R_1 and R_2,

$$a(x)=a[1-(c_1+c_2)x+c_1c_2x^2] \quad (2)$$

where $a=a(0)$ is the average area per head-group, at $x=0$, and $c_i=1/R_i$ are the principal curvatures. c_i is assumed positive when the center of curvature is on the hydrophobic side of the interface.

The conformational pdf of the system is the (unique) $P(\alpha)$ which minimizes the (Helmholtz) free energy per chain,

$$f_c=\sum_\alpha P(\alpha)\varepsilon(\alpha)+kT\sum_\alpha P(\alpha)\ln P(\alpha) \quad (3)$$

subject to the packing constraints (1). The two terms on the rhs of (3) are the average internal energy and the entropy, per chain, respectively. $\varepsilon(\alpha)$ is the energy of the trans/gauche bond sequence corresponding to α. The minimization yields

$$P(\alpha)=\exp[-\beta\varepsilon(\alpha)-\beta\int\pi(x)\phi(x;\alpha)dx]/z \quad (4)$$

where $\beta=1/kT$ and the partition function, z, ensures the normalization of $P(\alpha)$. The x-integration extends over the hydrophobic region. The important parameters in $P(\alpha)$ are the lateral pressures (tensions) $\pi(x)$ which are the Lagrange multipliers conjugate to the packing constraints. They are determined by substituting (4) into (1) and solving for the $\pi(x)$. In our calculations we simplify the solution by discretizing the problem, namely, we divide the hydrophobic region into, say L, imaginary layers parallel (concentric) to the interface. Then (1) and (4) yield a set of L coupled equations for the π_i's (i=1,...,L) which are easily solved by standard methods[1].

As noted above, for 'compact' films or aggregates in which the hydrophobic region is uniformly packed only by segments of the chains, the constraints in (1) are strict equalities. For non-compact films (1) is a real constraint only for some regions of x, while for others the inequality is trivially satisfied. More explicitly, for a given x, (1) is a relevant constraint only if $\langle\phi(x)\rangle_f$, the lateral extension of a free chain (i.e. a single chain with no neighbors around) exceeds $a(x)$, so that the packed chain must be squeezed in order to fit (on the average) into the area available to it. Consistent with this notion $\pi(x)=0$ for all x where (1) is an 'irrelevant' constraint. On the other hand $\pi(x)>0$ for all x where (1) is a real constraint ensuring $\langle\phi(x)\rangle=a(x)$. In general, large values of $\pi(x)$ indicate strong chain confinement at the corresponding x. Typical lateral pressure profiles are shown schematically in Fig. 1. It is seen for example that for a monolayer $\pi(x)$ decreases from finite values at small x (where chains push each other), to zero towards the rough region where both chain and oil segments are present. It is interesting to note that the lateral pressure profile in a bilayer is very nearly a superposition of the $\pi(x)$ for two, non-compact, monolayers. Consistent with this observation we find that many other bilayer properties (e.g. chain free energies) are, to a good approximation, a superposition of the corresponding monolayer properties.

After evaluating the $\pi(x)$ we can use $P(\alpha)$ to calculate any desired chain conformational or thermodynamic property, e.g. the conformational free energy (3). Formally, substituting (4) into (3), and noting that $\langle\phi(x)\rangle=a(x)$ for all $\pi(x)\neq 0$, we find $f_c=-kT\ln z-\int dx\,\pi(x)a(x)$. Noting also that $\langle\phi(x)\rangle=-kT\partial\ln z/\partial\pi(x)$ it follows that the free energy changes associated with changing the geometry of the system, i.e. changing $a(x)$, are given by

$$\delta f_c=-\int dx\,\pi(x)\delta a(x)+\delta^{(2)}f_c+\cdots \quad (5)$$

with $\pi(x)=-\partial f_c/\partial a(x)$. [For compact, 'volume-preserving' systems such as bilayers the rhs includes in addition the term $(\partial f_c/\partial L)\delta L$, accounting for changes in the thickness of the chains' layer].

The first order terms in (5) determine the equilibrium (minimal f_c) state of the chains. $\delta^{(2)}f_c$ stands for second order variations involving second order derivatives of f_c (first derivatives of π). When evaluated at the equilibrium state these derivatives give the chain contribution to the elastic constants[8-10]. Based on our theory we can calculate f_c for any desired geometry, and thus evaluate its derivatives with respect to geometrical parameters such as the area per head group, a, and the curvatures c_i, cf (2). However, f_c is not the only geometry dependent term in the free energy of the amphiphilic layer, f. In particular f includes an important contribution from the head-group region, f_h. This quantity is usually modeled in terms of the opposing forces[6,7]: namely, the effective attraction between molecules resulting from the tendency to minimize oil-water contact on the one hand, and the (electrostatic or excluded volume) repulsion between head-groups on the other. The attraction is usually described by a phenomenological surface tension term γa, with $\gamma \sim 0.1kT/\mathring{A}^2 \sim 50 dyn/cm$. The repulsion is often expressed as K/a' where a' is the area per head-group at the surface of repulsion[7,11]. If this surface is located at a distance d from the oil-water interface then $a'=a[1+d(c_1+c_2)+d^2c_1c_2]$. Thus, for a monolayer $f_h=\gamma a+K/a'$ becomes (up to quadratic terms in dc_i),

$$f_h=\gamma a+(\gamma a_h^2/a)[1-d(c_1+c_2)+d^2(c_1^2+c_2^2+c_1c_2)] \tag{6}$$

where $a_h=(K/\gamma)^{1/2}$ is the value of a which minimizes f_h for a planar layer, $c_1=c_2=0$. It should be noted that although a_h is often referred to as the 'optimal area per head group'[7], this statement ignores the a-dependence of f_c. Namely, the real optimum, a_0, is the value of a which minimizes the sum $f=f_c+f_h$. This minimization yields a_0 which is intermediate between a_h and a_c where a_c is the optimal area for chain packing. In some cases identification of a_0 with the experimental equilibrium area, and using the calculated a_c yields a_h which is much smaller than a_0, sometimes even smaller than $a_t \sim 21\mathring{A}^2$ which is the minimal (all-trans) area allowed for chain packing. This implies that the dominant repulsion balancing the interfacial attraction γa is due to the chains and not to the heads. Consequently, the form of the head-group repulsive term, as well as the exact value of the parameter a_h, are of secondary importance.

3. Elastic Constants

Writing

$$\delta f=\delta^{(1)}f_c+\delta^{(1)}f_h+\delta^{(2)}f_c+\delta^{(2)}f_h+\cdots \tag{7}$$

we can determine the equilibrium geometry of a layer from the first order variations and its elastic moduli from the second order terms. As a first example let us consider the stretching elasticity of a planar bilayer. Fig. 2 shows f_c, calculated by our theory, for a bilayer of $-(CH_2)_{11}-CH_3$ chains as a function of a, revealing a minimum for f_c at $a_c \approx 45\mathring{A}^2$. Also shown

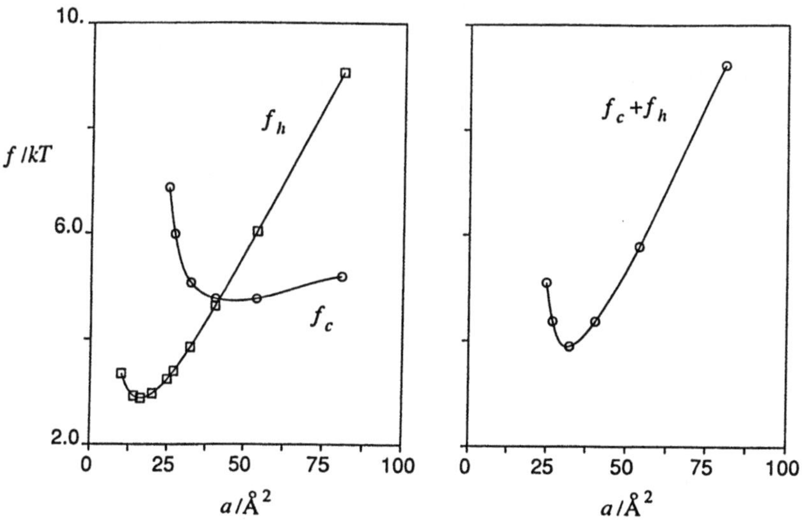

Fig.2 Free energies of a planar bilayer of 12-carbon chains as a function of the area per molecule. The separate contributions of heads and chains are shown on the left and their sum on the right

is f_h, with the parameters $\gamma=0.12kT/\text{Å}^2$ and $a_h\approx16\text{Å}^2$, chosen such that the minimum in $f=f_c+f_h$ occurs at an experimentally typical value, $a_0\approx33\text{Å}^2$. The figure reveals quite clearly that the optimal area is due primarily to the balance between the attractive surface term γa in f_h and the strong chain repulsion operative at $a<a_c$. Formally, the equilibrium condition is $\delta^{(1)}f=\delta^{(1)}f_c+\delta^{(1)}f_h=0$ with $\delta a(x)=\delta a=const$ and $c_1=c_2=0$ in (2). Using (5) and (6) this yields $\int\pi(x)dx=\gamma[1-(a_h/a_0)^2]$.

From the behavior of f around $a=a_0$ one can derive a stretching elastic constant k_s, defined via[8]

$$\delta f/a_0=(1/2)k_s(\delta a/a_0)^2 \tag{8}$$

For the example in Fig. 2 we obtain $k_s\approx0.70kT/\text{Å}^2\approx300erg/cm^2$. δf and correspondingly k_s is a sum of chain and surface (head) terms. From (6) one finds that the surface contribution to k_s is $2\gamma(a_h/a_0)^2$ which for the chosen parameters gives $\sim0.05kT/\text{Å}^2$, i.e. considerably smaller than the chain term.

Curvature elasticity is usually modeled in terms of Helfrich formula for the free energy per unit area[8-10]

$$f/a=(1/2)k(c_1+c_2-c_0)^2+\bar{k}c_1c_2 \tag{9a}$$

$$=(2k+\bar{k})c^2-2kc_0c+(1/2)kc_0^2 \qquad (c_1=c_2) \tag{9b}$$

where k and $\bar{k}$ are the bending constants corresponding to splay and saddle splay deformations, respectively. The second equality corresponds to spherical deformations ($c_1=c_2$), and c_0 is the spontaneous curvature related to the equilibrium curvature $c_{eq}=(k/2K)c_0$. In analogy to k_s the phenomenological parameters in (9) can be determined by calculating $\delta f=\delta f_c+\delta f_h$ for different values of c_1 and c_2 (i.e. different deformations) and evaluating the appropriate coefficients. (Combinations of the parameters can be related to moments of the lateral tensions[8-10]). Again, the bending constants are sums of chain and surface terms. In Fig. 3 we

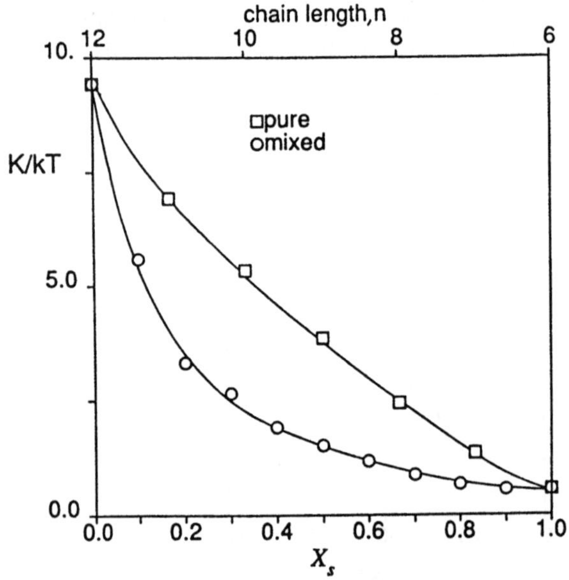

Fig.3 Bending constants, $K\equiv k+\bar{k}/2$, of a mixed, C_{12}/C_6, monolayer as a function of short chain fraction, (lower scale, circles), and of a pure monolayer as a function of chain length, (upper scale, squares)

show K_c, the chain contribution to $K \equiv k + \bar{k}/2$ for a mixed monolayer of 12 and 6 carbon chains [$-(CH_2)_{n-1}-CH_3$ with n=12 and 6] as a function of the short chain fraction, X_s. Also shown is K_c of a pure monolayer as a function of chain length n. It should be stressed that the results in Fig. 3 correspond to the same area per chain for all cases, $a = a_0 = 32\AA^2$ regardless of n or X_s. This would be the case if a_0 is dictated solely by the balance between the opposing surface forces, and if the head groups are the same for all chains. Otherwise one must take into account the variation of a_0 with chain length and composition. Another assumption employed in the calculation of K_c is that the $a = a_0$ is kept constant upon bending, i.e. the neutral surface is assumed to coincide with the water-oil interface. The values of K_c obtained in this way are approximate (upper bounds) since the location of the neutral surface generally depends on chain length as well as on the specific parameters of head group interactions. In the same approximation the head group contribution to K can be estimated from the quadratic terms in (6), namely, $K_h \sim 6\gamma(a_h/a_0)^2 d^2 \sim \gamma d^2$. Using $\gamma \sim 0.1kT/\AA^2$ and, e.g. $d \sim 3\AA$ this yields $K_h \sim kT \sim 5\times10^{-14}$ erg which is considerably smaller than K_c for the long chain film. For mixed monolayers, particularly surfactant-alcohol films, K_c is expected to decrease significantly. But K_h will also decrease (approximately linearly) due to the screening of electrostatic repulsion resulting from the addition of alcohol[12].

The most significant qualitative trend apparent from the results in Fig. 3 is the very efficient lowering of the bending constant of a ('long chain') surfactant film by dilution with short chain molecules. A similar behavior has been observed experimentally[13]. Qualitatively, this follows from the fact that addition, even of a small amount, of short chains relieves much of the conformational strain associated with confining the longer chains to small cross sectional areas, as schematically shown in Fig. 1. In terms of the packing constraints: a small amount of short chains suffices to make (1) an irrelevant packing constraint for $x > l_s$ where l_s is the average length of a short chain in the film. The fast decrease of K_c with X_s should be contrasted with its nearly linear variation with chain length. [A faster decrease of K_c with n is expected when the n dependence of a_0 is taken into account. This behavior can be explained by simple scaling arguments]. Extensive and more accurate calculations of elastic constants for pure and mixed bilayers and monolayers are now in progress[14].

Acknowledgment We thank D. Kramer for his help in some of the calculations and D. Roux, J. L. Viovy and S. Safran for helpful discussions. The financial support of the US-Israel Binational Science Foundation and the US National Science Foundation is gratefully acknowledged. The Fritz Haber Molecular Dynamics Center is supported by the Minerva Gesellschaft für die Forschung, Munich, FRG.

References

1. A. Ben-Shaul, I. Szleifer and W.M. Gelbart, Proc. Natl. Acad. Sci., U.S.A. **81** , 4601 (1984), J. Chem. Phys. **83** , 3597 (1985). I. Szleifer, A. Ben-Shaul and W.M. Gelbart, J. Chem. Phys. **85** , 5345 (1986) and a study of mixed aggregates, J. Chem. Phys. (in press). A. Ben-Shaul and W.M. Gelbart, Ann. Rev. Phys. Chem. **36** , 179 (1985). J. L. Viovy, W. M. Gelbart and A. Ben-Shaul, J. Chem. Phys. (submitted).
2. P. van der Ploeg and H. J. C. Berendsen. Mol. Phys. **49,** 233 (1983). see also, B. Jonsson, O. Edholm and O. Teleman, J. Chem. Phys. **85** 2259 (1986).
3. D. W. R. Gruen. J. Phys. Chem. **89,** 146 (1985), Prog. Polymer Sci.**70,** 6, (1985).
4. K. A. Dill and P. J. Flory. Proc. Natl. Acad. Sci. USA. **77,** 3115 (1980), **78,** 676 (1980), K. A. Dill and R. S. Cantor. Macromolecules **17,** 380 (1984).
5. H. Wennerstrom and B. Lindman, Phys. Rep. **52,** 1, (1980).
6. C. Tanford, "The Hydrophobic Effect" 2nd ed. (Wiley- Interscience, New York, 1980).
7. J. N. Israelachvili, D. J. Mitchell and B. W. Ninham, J. Chem. Soc. Faraday Trans. II **72,** 1525 (1976).

8. W. Helfrich, Z. Naturforsch. **28c** , 693 (1973), **29c** , 510 (1974), and Les Houches Session XXXV, 1980, "Physics of Defects", R. Balian et al. eds., North-Holland Publ. Comp. (1981), p. 716.

9. E. A. Evans and R. Skalak, CRC Critical Rev. Bioeng., **3** , 181 (1979).

10. A. G. Petrov and I. Bivas, Progr. in Surf. Sci. **16** , 389 (1984).

11. D. J. Mitchell and B. W. Ninham, J. Chem. Soc. Faraday Trans. II **77,** 609 (1981).

12. See e.g. A. Ben-Shaul, D. H. Rorman, G. V. Hartland and W. M. Gelbart, J. Phys. Chem **90** , 5277 (1986).

13. J. M. Di Meglio, M. Dvolaitzky and C. Taupin, J. Phys. Chem. **89** , 871 (1985). see also the papers by J. M. Di Meglio and D. Roux in this volume.

14. I. Szleifer, D. Kramer, A. Ben-Shaul, W. M. Gelbart and D. Roux, to be published.

Chain Packing and the Compressional Elasticity of Surfactant Films

W.M. Gelbart[1] *and A. Ben-Shaul*[2]

[1]Dept. of Chemistry, The University of California, Los Angeles, CA 90024, USA

[2]Dept. of Physical Chemistry and the Fritz Haber Molecular Dynamics Center, The Hebrew University, Jerusalem 91904, Israel

1. INTRODUCTION

We summarize briefly our work on the effect of excluded volume on the conformational statistics and thermodynamics of semi-flexible chain packing in surfactant layers. Our aim is to establish the important extent to which "tails" compete with "heads" in determining the equilibrium state and compressional properties of mono- and bi-layer amphiphilic films. Using the mean-field theory developed originally for micellar systems [1] we present first some typical results for chain free energies in bilayers as a function of area per molecule. Ensuing estimates for the optimum area per molecule and compression (stretching) force constant are compared with those arising from head-group interactions alone. Then, in order to treat the dependence of these properties on chain contour length (e.g. alkyl carbon number) we discuss a diffusion-equation approach to conformational statistics which takes into account the constraint of uniform segment density. This description provides a natural justification for a scaling of layer thickness which leads to approximately universal compressional behavior. It also allows new insights into the way in which lateral pressure profiles serve to "squeeze" "free" chains into uniformly packed ones. We then treat the important case of mixed-chain systems and derive the corresponding surface-volume relations and their consequences for relative stabilities of planar and curved layers. Finally, we present a simple but accurate model for calculating bending elastic properties of surfactant films in terms of their compressional force constants. In particular we account for how decreasing chain length (carbon number) can lower the curvature energy from 10's of kT to kT.

2. HEADS VS. TAILS: CHAIN ELASTICITY EFFECTS

In the prevailing theories [2] of micellar self-assembly, the surfactant chains are assumed to be "passive." That is, the free energy per molecule, g(a), is taken to depend on head-group area, a, only through the surface contributions. Modeling these latter terms via the notion of opposing forces [3], for example, one writes $g(a) = \gamma a + \frac{C}{a} + g_c$ where γ is an interfacial tension, C characterizes the (steric, electrostatic, or hydration) repulsion between head groups, and g_c is the chain contribution (assumed to be independent of a). The optimum area per molecule is then given by $a = (C/\gamma)^{1/2} = a_0$. At the same time, from surface/volume constraints one has the requirement that $a = k\nu/\ell$ where ν is the chain volume, ℓ the micellar dimension, and k=1,2,3 for bilayers, cylinders, and spheres, respectively. It follows that the preferred aggregate geometry will be the highest-curvature (hence smallest size) one which allows $a_0 = k\nu/\ell$ with $\ell \leq \ell_{max}$. Here ℓ_{max} is the maximum micellar dimension, corresponding to the fully stretched "all-trans" chain length.

Using the mean-field theory developed with Szleifer and described elsewhere [1,4], however, it is possible to calculate directly how g_c depends on area per molecule. This simple formulation has been shown to be highly accurate in determining the details of conformational statistics (e.g. bond order-parameter profiles and segment distributions) and thermodynamics, as compared with full molecular dynamics simulations. In the present context, for say a 12-carbon tail, we find that $g_c(a)$ has a minimum at $a^*=45Å^2$ and that stretching the (uniform monomer density) packed chains down to $a=25Å^2$ costs as much as 3kT. Note that, for reasonable values of γ and a_0, the corresponding change in $g_h(a)$ is only half as large. For this reason, the optimum value of a is no longer $a_0=(C/\gamma)^{1/2}$ but rather something significantly bigger: $a_0<a_{opt}<a^*$. That is, the a-dependence of g_c can not be neglected--the chains are not "passive" and do not simply "go along for the ride" as the area per molecule is optimized.

A further result of the chain contributions is to enhance dramatically the compressional elasticity of a surfactant layer. Writing $g_c(a) \to g_c(\ell) = \kappa(\ell-\ell^*)^2$, for example (using $a \to \frac{v}{\ell}$), one can inquire about the contour length (carbon number, n) dependence of both κ and ℓ^*. Suppose g_c depended on the micellar dimension ℓ (bilayer half-thickness) only through the scaled variable $\tilde{\ell} = \ell/bn^{1/2}$ where b is the monomer size. Then one would have $\ell^* \sim n^{1/2}$ and $\kappa \sim n^{-1}$. This indeed turns out to be approximately the case, as discussed in the sections which follow.

3. ARBITRARY CHAIN-LENGTH APPROACH

In recent work with Viovy [5] we have generalized our earlier mean-field theory to the case of arbitrarily large n. This requires that one give up a description which explicitly enumerates chain conformations (whose number increases as 3^n in the usual rotational isomer state model), switching instead to a random walk language. Indeed it is possible to show that the monomer propagator [6] obeys a quasi-one-dimensional diffusion equation in an external field. Unlike in the classic excluded volume problem, however, the external field is not the local segment concentration (to be determined self-consistently in terms of the propagator) but rather the lateral pressure profile which arises from the constraint of uniform monomer density. Again one is able to solve directly for either the chain conformational statistics or thermodynamics with no adjustable parameters.

Plots of g_c vs. ℓ for increasing n show that both the horizontal and vertical positions of the minima can be accounted for in terms of "free" chain properties. (Here and henceforth we speak primarily of bilayer, systems.) A "free" chain is one which does not "feel" its neighbors but rather only the constraint due to the impenetrable interfaces. Thus g_c vs. ℓ for a free chain is monotonic decreasing: at small ℓ ($<bn^{1/2}$) the free energy is high because of all the conformations denied by the presence of the interfaces; for larger ℓ, g_c^{free} levels off to a constant. For each n this asymptote turns out to correspond closely with the minimum in g_c vs. ℓ for the packed chain. Indeed, both the horizontal ($\ell^* \sim n^{1/2}$) and vertical ($g^* \sim \ln n$) positions of the minima in $g_c^{packed}(\ell)$ are well described by the onsets of the free-chain asymptotes.

The lateral pressure profiles, $\pi(Z)$, which emerge from the diffusion equation approach also serve to point up the crucial role played by free chains in understanding the statistics and thermodynmics of uniformly-packed ones. First we find again (as in our earlier treatments [1,4]) that $\pi(Z)$ is smallest for $\ell \approx \ell^*$, consistent with the fact that the lateral pressures tell us the extent to which "free" chains must be "squeezed" in order to transform them into uniformly packed ones. (Recall that ℓ^* is the optimum bilayer half-thickness, the value which minimizes $g_c(\ell)$ and allows the packed chains to most closely resemble "free" ones.) Furthermore, the "shape" of $\pi(Z)$--the dependence of lateral pressure on distance Z from the interfaces--follows closely that of $\rho^{free}(Z)$, the monomer density of a free chain with the same ℓ: one must squeeze hardest in high-concentration regions in order to achieve uniform segment density.

4. MIXED-CHAIN SYSTEMS

Our original mean-field theory has also been extended to the important case of mixed- (e.g. short- and long-) chain systems [7]. These are of great interest for understanding microemulsions which are stabilized by addition of short-chain alcohols, say, and biological membranes which have incorporated cholesterol and other "solutes." Before proceding to a discussion of our results for mixed-chain conformational statistics and thermodynamics, we explain first new features of the surface/volume geometric constraints for these systems.

Recall that, for pure (i.e. single component) micelles, the relation $a=k\nu/\ell$ follows trivially from requiring that the aggregate's surface is covered by heads (area per molecule a) and its interior volume filled by chains (ν). For mixture of short (s) and long (ℓ) chains with mole fraction X_s, this becomes $a=(k/\ell)(X_s\nu_s+(1-X_s)\nu_\ell)$ with a the average area per molecule. Now, however, the maximum micellar dimension ℓ_{max} is no longer simply $\ell_{all-trans,\ell}$, the fully stretched length of the longest chain. Rather, for a given composition X_s, one must guarantee that ℓ is small enough that there is not too much "interior" volume (i.e. space inaccessible to the shorter chain) for the longer chain to fill at constant segment density. Defining this latter bound by $\ell_>$, we have that ℓ_{max} is equal to the smaller of $\ell_{all-trans,\ell}$ and $\ell_>$. Since $\ell_>$ is dependent on X_s (as well as on the different contour lengths), it follows that the minimum value of a ($a=a_{min}\leftrightarrow \ell=\ell_{max}$) is a function of composition. That is, instead of $a_{min}^{pure} = k\nu/\ell_{all-trans,\ell}$ as in the pure surfactant case, one has $a_{min} = f^{(k)}(X_s)\ a_{min}^{pure}$ where $f^{(k)}(X_s)\leq 1$. Note that $a_{min}^{pure} = k\nu/\ell_{all-trans,\ell}$ is independent of n since both chain volume (ν) and fully stretched length ($\ell_{all-trans,\ell}$) are linear functions of n. Thus, in the limits $X_s \to 0$ or 1, $a_{min} \to a_{min}^{pure} \approx (22\ \text{Å}^2)k$, while for intermediate values it is smaller: for spheres (k=3), for example, $f^{(3)}(X_s)$ is as small as $\frac{2}{3}$ in the case of $\nu_\ell/\nu_s \approx 2$.

The fact that mixed spherical micelles can have average areas per molecule which are significantly smaller than in the pure case means that addition of short chains should preferentially stabilize higher curvature. Indeed we find for the $\nu_\ell/\nu_s \approx 2$ system that the average free energy per chain becomes lower for a sphere (vs. a bilayer) when the molefraction is sufficiently different from 0 or 1. That is, for the pure "long" or "short" chain systems the lamellae are stable, whereas spherical curvature becomes preferable upon the addition of sufficient amounts of the other chain. This phenomenon accounts for the break up of lamellar phases observed by Charvolin and Mely [8] as they move from $X_s=0$ or 1 in aqueous solutions of mixed fatty acids.

We have also used our mean field theory to treat explicitly the conformational statistics in mixed-chain systems. One simply needs to generalize the constraints of constant segment density to include molefraction-weighted contributions from both chains. Minimizing the average free-energy per chain with respect to the probability distribution function for each chain, subject to uniform density, leads to a lateral pressure profile (dependent on area and composition) which again determines the conformational statistics. We find in general that the longer chains occupy smaller areas per molecule at the interface: they are stretched there in order to "save themselves" for filling up the "interiors." Again these predictions are consistent with the order-parameter measurements reported by Charvolin and Mely [9] for mixed fatty acids.

5. COMPRESSION MODEL FOR BENDING ELASTICITY

In this final section we outline briefly work with Roux [10] relating curvature elasticity to the compressional properties of a planar film. Consider the bending at constant area per molecule, of a planar layer of thickness ℓ. For a spherical deformation, for example, $R = \infty \to R < \infty$ implies $\ell \to \ell'(a,R)$ in order to maintain the constant head group area a. The bent layer has a higher free energy than the original planar one (assuming, say, a symmetric bilayer) because the constituent molecules have been both stretched and splayed. In the preceding paper [4] we described how this free energy change--and the corresponding bending constants--can be calculated from a full mean-field treatment of chain packing. Here we consider instead an approximation to this approach in which splay contributions are explicitly neglected. That is, we take into account only the free energy change associated with the "compression" $\ell \to \ell'(R)$: $\delta g_{c,\text{bending}} \approx \delta g_{c,\text{compression}} = \kappa(\ell'(R) - \ell^*)^2$. Here κ and ℓ^* are the force constant and optimum thickness for the planar layer. Expanding $\ell'(R)$ in powers of $\frac{1}{R}$ then gives the following result for the bending constant: $k \approx \kappa v^4/a^5$. Note the strong (and inverse) dependence on area per molecule. Also, from our planar-layer mean-field calculations we find $\kappa \sim n^{-1.1}$ (to be compared with n^{-1} from scaling), implying $k \sim n^3$ since $v \sim n$.

Taking into account the slow variation of a ($=a^*_{\text{bilayer}}$) with n gives rise to a somewhat weaker dependence of k on n ($k \sim n^{2.5}$), in agreement with preliminary experimental data by Roux et al. [11]. Furthermore, we obtain absolute magnitudes for 15-carbon chains (10's of kT) which suggest that indeed the measured bending energies are due predominantly to the chains rather than to the head groups. When the area per molecule is allowed to respond to the bending, i.e. when a is optimized for each R, the values of k estimated from the compression model are more than twice as small, consistent with the full mean field calculations [4]. The variation of k with carbon number remains aboutthe same ($\sim n^{2.5}$), accounting for bending energies as small as kT for $n \approx 4$ or 5.

We gratefully acknowledge our collaborations with Igal Szleifer, Jean-Louis Viovy, and Didier Roux, as well as financial support from the U. S. National Science Foundation, the U. S.-Israel Bi-national Science Foundation, and the Minerva Gesellschaft für die Forschung.

REFERENCES

1. A. Ben-Shaul, I. Szleifer, W. M. Gelbart: Proc. Nat. Acad. Sci. (USA) 81, 4601 (1984), and J. Chem. Phys. 83, 3597 and 3612 (1985); A. Ben-Shaul, W. M. Gelbart: Ann. Rev. Phys. Chem. 36, 179 (1985); I. Szleifer, A. Ben-Shaul, W. M. Gelbart: J. Chem. Phys. 85, 5345 (1986).
2. J. N. Israelachvili, D. J. Mitchell, B. W. Ninham: J. Chem. Soc. Faraday Trans. II72, 1525 (1976).
3. C. Tanford: The Hydrophobic Effect, 2nd ed. (Wiley, New York, 1980).
4. A. Ben-Shaul, W. M. Gelbart: preceding paper in these proceedings.
5. J.-L. Viovy, W. M. Gelbart, A. Ben-Shaul: J. Chem. Phys., submitted.
6. S. F. Edwards: Proc. Phys. Soc. 85, 613 (1965); P.-G. de Gennes: Rep. Prog. Phys. 32, 187 (1969); H. Yamakawa: Modern Theory of Polymer Solutions (Harper and Row, New York, 1971).
7. I. Szleifer, A. Ben-Shaul, W. M. Gelbart: J. Chem. Phys., submitted.
8. J. Charvolin, B. Mely: Mol. Cryst. Liq. Cryst. 41, 209 (1978).
9. B. Mely, J. Charvolin: In Physicochimie des Composes Amphiphiles, ed. by P. Perron (CNRS, Paris, 1979).
10. D. Roux, unpublished.
11. D. Roux, C. Safinya, E. Sirota, G. Smith, unpublished.

Dynamics of Phase Transitions in Langmuir Monolayers of Polar Molecules

F. Brochard[1], *J.F. Joanny*[2], *and D. Andelman*[3]

[1]Université Pierre et Marie Curie, Laboratoire "Structure et Réactivité aux Interfaces", 11 Rue Pierre et Marie Curie, F-75231 Paris Cedex 05, France

[2]Dept. de Physique des Matériaux, Université Claude Bernard Lyon I, 43 Bd. du 11 Novembre 1918, F-69622 Villeurbanne Cedex, France

[3]C.R.S.L., Exxon Research & Engineering Co., Route 22 East, Annandale, NJ 08801, USA

Insoluble monolayers in the presence of dipolar forces have been predicted to form lamellar and hexagonal supercrystal phases at the liquid-gas or liquid-expanded - liquid-condensed transition . We study here the dynamics of the initial stages of the formation of these supercrystal phases including the flow induced in the supporting liquid in three different geometries a) free films b) films on a solid c) film on a bulk liquid phase.

INTRODUCTION

Electrostatic dipolar interactions have recently been shown to be of major importance in Langmuir monolayers of polar molecules /1/2/3/ : they may change drastically the phase diagram of the monolayer. In insoluble monolayers of amphiphilic molecules, dipolar forces may have two origins :

i) neutral molecules carry a permanent electrostatic dipole of the order of a few millidebye with a preferential orientation perpendicular to the interface ;

ii) charged monolayers build up an electrical double layer which can be considered as a layer of dipoles at sufficiently high ionic strength.

At the liquid-gas or liquid-expanded - liquid-condensed phase transition, the repulsive long-range interactions between the dipole ($\sim 1/r^3$) give rise to spatially periodic arrangement of the molecules (sometimes called supercrystals) in a significant part of the coexistence region /1/2/.

Our aim here is to study the dynamics of formation of the supercrystal phases of a monolayer initially in a one-phase isotropic region of the phase diagram, which has been rapidly quenched into the coexistence region either in a two-phase region or in a one-phase region corresponding to a spatially periodic, supercrystal phase, by changing either the temperature of the surface pressure.

If in the final state the homogeneous isotropic phase is metastable, the new equilibrium phase grows first by a nucleation process, which requires an activation energy to initiate the phase separation. If the isotropic phase is unstable after the quench, the equilibrium state forms by amplification of the local fluctuations of concentration which in monolayers corresponds to a divergence of the longitudinal Lucassen modes.

We study here this process known as spinodal decomposition. The nucleation process and the kinetics of domain growth (Ostwald Ripening) will be addressed in a separate study. In addition to the role of long-range forces, spinodal decomposition in Langmuir monolayers is interesting and different from usual spinodal decomposition in three-dimensional liquid systems /9/ for the following reasons :

a. as in any two-dimensional system, gravity plays no role : this is thus an example of phase separation in zero-gravity ;

b. although the static phase diagram is purely two-dimensional, motions in the monolayer induce a backflow in the water substrate. We will show that dissipation is dominated by the viscous dissipation in the substrate and has thus a three-dimentional character ;

c. the first unstable mode (the spinodal line) does not occur at zero-wave vector as in systems with short-range forces but at a finite wave vector q^*. This situation is also found in other systems such as block copolymers /4/ (in three dimensions) or ferrofluids /5/ (in two dimensions) ;

d. contrary to usual liquid-liquid phase separations where the soft mode (critical mode) is diffusive, thus overdamped, outside the spinodal region, the "LUCASSEN" modes /6/ are propagating ; we show that in the unstable spinodal regions, the propagation disappears and the unstable fluctuations grow exponentially without any oscillations.

We first quickly recall the complex diagram of Langmuir monolayers of polar molecules. The amplification of the Lucassen mode is then studied in different geometries : films on a thin liquid slab above a solid substrate for which viscous dissipation is the dominant effect and the more common case of a monolayer on an infinite bulk water phase for which both inertia and viscous dissipation play a role.

1. SUPERCRYSTAL PHASES of POLAR MONOLAYERS

We call μ the mean value (vertically oriented) of the permanent dipoles of the amphiphilic molecules. If Σ is the area per polar head, the dipole $P = \mu\,\Sigma^{-1}$ is derived from the contact potential V

$$V = \frac{P}{\varepsilon} = \frac{\mu}{\varepsilon\Sigma_0}\,\phi, \tag{1}$$

where Σ_0 is the area corresponding to close packing, ϕ the surface fraction occupied by molecules and ε the dielectric constant seen by the dipoles at the interface (which might be different from the water bulk value). Equation (1) is valid both for neutral and charged monolayers at high ionic strength, the equivalent dipole in this case is $\mu = e\,\kappa^{-1}$, where e is the elementary charge and κ^{-1} the Debye-Hückel screening length.

In the supercrystal phases the surface fraction varies periodically on the surface. Close to a liquid-gas or liquid-expanded - liquid-condensed critical point, it is sufficient to describe this variation by a one-mode expansion

$$\Phi(r) = \phi_0 + \Phi_q(\vec{r}),$$

the undulation Φ_q being periodic with a period $2\pi/q$.

The electrostatic free energy associated with the undulation ϕ_q is

$$F_D = -\frac{1}{2}\,|q|\,\frac{\mu^2}{\Sigma_0^{\,2}}\,\phi_q^2\,\frac{\varepsilon_0}{(\varepsilon+\varepsilon_0)}\,. \tag{2}$$

The phase diagram of the monolayer may thus be deduced from a Landau-Ginzburg expansion of the free energy :

$$\frac{F_s - F_0}{kT} = \frac{1}{2}\,\Delta\psi^2 + \frac{1}{4}\,B\psi^4 + \Sigma_0\Phi_c^{\,2}\sum_q\left[-K\,|q| + \frac{1}{2}\,L\,q^2\right]\psi_q^{\,2}, \tag{3}$$

where $\psi = \frac{\Phi - \Phi_c}{\Phi_c}$ is the order parameter, $\Delta = \alpha\frac{T - T_0}{T_0}$ vanishes at T_0. L is related to the attractive short-range interaction and K to the long-range repulsion $\left[K = \frac{1}{2kT}\frac{\mu^2}{\Sigma_0^2}\frac{\varepsilon_0}{(\varepsilon + \varepsilon_0)} = \frac{\lambda}{\Sigma_0^{1/2}}\right.$, where λ is a small parameter measuring the strength of dipolar interactions $\left.(\lambda \sim 10^{-3})\right]$.

The functional (3) is optimal for a wave vector q^*

$$q^* = \frac{K}{L} \tag{4}$$

of the order of 1000 Å for $\lambda \sim 10^{-3}$. Setting $q = q^*$ in (3) leads to a renormalization of the mean field critical temperature ($T_c = T_0 \left[1 + \frac{1}{\alpha}\frac{K^2}{L}\Sigma_0\,\Phi_c^2\right]$).

The phase diagram deduced from (3) is shown on fig.1.

a) Near the critical point, the monolayer builds up a supercrystalline phase spatially modulated, of period $2\pi/q^*$. The order may be smectic or triangular. Near the axis $\psi = 0$, the smectic phase is always stable. Near the isotropic phase coexistence line, one finds a triangular phase made of a network of circular islands /2/.

b) At lower temperatures, all wavevectors contribute to the surface fraction Φ and the one-mode approximation breaks down : the period is not any more related to q^* and increases exponentially /2/.

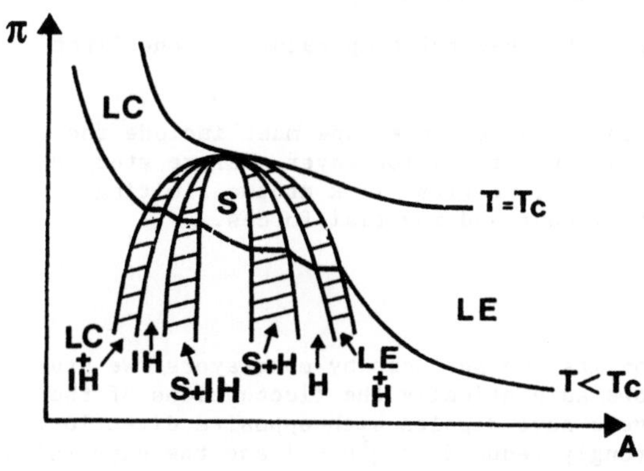

Fig. 1 : Phase diagram and π - A curves of a Langmuir monolayer of polar molecules : the two isotropic phases liquid-condensed - liquid expanded are separated by the hexagonal (H), stripe (S) and inverted hexagonal (IH) phases.

2. SPINODAL DECOMPOSITION of POLAR MONOLAYERS

The monolayer, in an isotropic homogeneous phase, is suddenly quenched into the coexistence region (see fig. 1). The critical mode is the longitudinal mode (fig. 2) first discussed by LUCASSEN /6/, and in more details by KRAMER /7/. The monolayer acts as an elastic membrane : we call $u_q = u_0\, e^{iqx}\, e^{iwt}$ the displacement of the monolayer in the direction x of the wavevector q. The fluctuation of the surface density ψ_q is related to u_q by a conservation equation $\psi_q = -\partial u_q / \partial x$.

For any concentration fluctuations ψ_q, there is an elastic force F_{el} opposing the displacement u_q which can be calculated from the free energy(3)

$$F_{el} = E(q)\,\frac{\partial^2 u}{\partial x^2}\,. \tag{5}$$

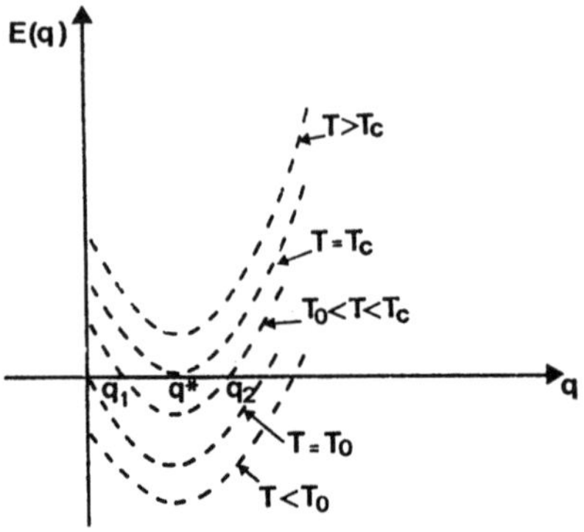

Fig. 2 : Elastic modulus E of a polar Langmuir monolayer as a function of wave vector q for several temperatures ; because of the dipolar repulsions, the sign of E is reversed at finite q value ($q = q^*$) corresponding to the period ($2\pi/q^*$) of the supercrystal.

E(q) is the monolayer membrane elasticity :

$$\frac{E(q)}{kT} = \frac{\Delta}{\Sigma_0} + 2\Phi_c^2 \left[-K\,|q| + \frac{1}{2} L\, q^2 \right] .$$

Whenever the elasticity is positive, the elastic force is a restoring force and the monolayer is stable with respect to the fluctuation ψ_q.

If the elasticity E(q) is negative, the elastic force increases the fluctuation and the membrane is unstable.

The condition E(q) = 0 thus defines the spinodal line.

The variation of E(q) is sketched on fig. 2 for several temperatures ; the first unstable mode arises for $T = T_c$, $q = q^*$.

In order to describe the dynamics of the Lucassen mode, one must include the motion of the liquid driven by the fluctuations of the monolayer. Before studying the case of a monolayer on a bulk liquid phase, we first look at two limiting cases in order to show the relative role of viscous and inertial forces.

a. "Free" Film

A water film of thickness d is bounded on its two surfaces by monolayers. We study the limit $qd \ll 1$ where the liquid follows adiabatically the fluctuations of the membrane. In soap films, the two monolayers have dipoles with opposite direction. The dipolar interactions may thus be strongly reduced if $q^*d < 1$ and the expression of E(q) may be different from (6). The problem is purely inertial. The equation of motion is :

$$2E(q) \frac{\partial^2 u}{\partial x^2} = d\rho \frac{\partial^2 u}{\partial t^2} , \qquad (6)$$

ρ is the density of the film.
The dispersion relation of the longitudinal modes is then

$$w^2 = \frac{2E}{d\rho} q^2 . \qquad (7)$$

When the elasticity E is positive, we find two sound waves ($w = \pm \sqrt{\frac{2E}{d\rho}}\, q$). The sound velocity goes to zero for E = 0. For negative E, the two modes become non propagative ($iw = \frac{1}{\tau} = \pm \sqrt{\frac{2|E|}{\rho d}}\, q$), one diverges and the other one is overdamped. We shall now look at the opposite limit of viscous dissipation in the liquid.

b. Thin Liquid Slab on a Solid

The monolayer bounds a very thin water film deposited on a solid plane. Any fluctuation in the monolayer induces a shear in the film. In the limit $qd \ll 1$, the lubrication approximation /8/ holds and the equation of motion, given by a balance between the elastic and the viscous forces, can be written as :

$$E \frac{\partial^2 u}{\partial x^2} = \frac{\eta}{d} \frac{\partial u}{\partial t} , \quad (8)$$

η is the film viscosity.
This leads to a diffusive mode with a relaxation time

$$\frac{1}{\tau} = E \frac{d}{\eta} q^2 . \quad (9)$$

The fluctuation is amplified for negative membrane elasticity E. If the quenching temperature is close to T_c, the amplified modes are the modes around q^* $[q_1 < q < q_2]$ as shown on fig.3. As T decreases, the threshold q_1 of unstability goes to zero and the wave vector q_M of the fastest mode increases. q_M varies from q^* at T_c to a large value at low temperature :

$$q_M^2 \simeq \Delta\Sigma_0^{-1} \Phi_c^{-2} L^{-1} . \quad (10)$$

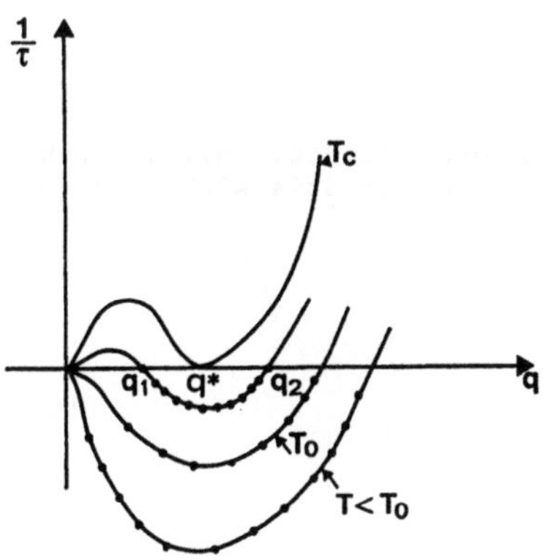

Fig. 3 : Relaxation time of the longitudinal mode of a polar Langmuir monolayer deposited on a thin liquid film. The first instability ($1/\tau < 0$) arises for a mode of wave length equal to the period of the supercrystal.

c. Monolayer on a Bulk Liquid

The longitudinal modes of a monolayer on a bulk liquid have been studied in detail by LUCASSEN /6/. Taking into account the coupling between longitudinal and transversal modes (capillary waves), the dispersion relations are given by the roots of the following equation :

$$(x'^2 + 2x' + Q_\gamma)\,(x'^2 + 2x' + Q_E \overset{+}{\sqrt{1+x'}}) = (2x' + Q_E)\,(2x' \overset{+}{\sqrt{1+x'}} + Q_\gamma) , \quad (11)$$

where $x' = \frac{i\omega\rho}{\eta q^2}$, $Q_\gamma = \frac{\gamma\rho}{\eta^2 |q|}$ and $Q_E = \frac{E(q)\rho}{\eta^2 |q|}$.

$\overset{+}{\sqrt{\ }}$ means that the real part of the square root is positive ; γ is the surface tension ($\gamma = \gamma_0 - \pi$, where γ_0 is the pure water surface tension ($\gamma_0 \simeq 70$ dynes cm^{-1}) and π the surface pressure). In (11), we have neglected gravity ($q\kappa^{-1} > 1$, where κ^{-1} is the Laplace capillary length). We study the roots of (11) as a function of the quenching temperature T.

1° Shallow Quench ($T \lesssim T_c$)

If T is close to T_c, E(q) is very small compared to γ. In this limit, one can assume that the water surface remains flat as the monolayer fluctuates.

The dispersion relation for the longitudinal mode can be deduced from (11) by taking the limit $C_\gamma/Q_E \to \infty$:

$$\frac{x'}{\sqrt[+]{1+x'} - 1} = - \frac{E\rho}{\eta^2|q|} = -Q_E \ . \tag{12}$$

The roots of (12) depend upon the sign and the amplitude of Q_E.

a) $|Q_E| >> 1$

If Q_E is large and positive, the roots of (12) are

$$x' = e^{\pm 2i\pi/3} \, |Q_E|^{2/3} \tag{13}$$

$$\text{or} \quad \frac{1}{\tau} = \frac{E(q)^{2/3}}{\eta^{1/3}\rho^{1/3}} q^{4/3} e^{\pm 2i\pi/3} . \tag{13'}$$

The modes are propagative and involve both inertia (ρ) and viscosity (η). One can check that the elastic energy is balanced by the kinetic energy of a slab of liquid of thickness l ($l^{-1} = \sqrt[+]{q^2+i\omega\rho/\eta}$).

If Q_E is negative, (12) has only one root :

$$x' = |Q_E|^{2/3} \tag{14}$$

or a negative relaxation time :

$$\frac{1}{\tau} = - |E(q)|^{2/3} q^{4/3} / \eta^{1/3} \rho^{1/3} . \tag{14'}$$

This mode is amplified exponentially, without any oscillation.

b) $|Q_E| << 1$

In this limit, (12) reduces to

$$x' = - \frac{E\rho}{2\eta^2|q|} . \tag{15}$$

The relaxation time is :

$$\frac{1}{\tau} = \frac{E(q)}{2\eta} |q| . \tag{15'}$$

If E is positive, the mode is overdamped.
If E is negative, the mode is amplified exponentially.

For a clear shallow quench near T_c, the most unstable wave vector q_M is very close to q^* and $Q_E << 1$. The characteristic time for the divergence of the fluctuation is then given by (15') with $q = q^*$

$$\frac{1}{\tau} = - |E(q^*)| \frac{q^*}{\eta} = - \left|\frac{\Delta}{\Sigma_0} - \Phi_c \frac{K^2}{L}\right| \frac{q^* kT}{2\eta} . \tag{16}$$

The amplified mode is purely viscous. This regime is similar to the case of a liquid slab of thickness $d = q^{-1}/2$ deposited on a solid (see (9)).

2° Deep Quench $(T << T_c)$

Far from T_c, the elastic modulus E is usually much larger than the surface tension γ. We assume now $Q_E >> Q_\gamma$ in (11).

If $|Q_E| >> 1$, we find here also that $x'^{3/2} = -Q_E$, and the dispersion relation is given by (13') if $E > 0$ and by (14') for negative E.

If $|Q_E| << 1$, one finds $x' = - \frac{Q_E}{2}$.

For a deep quench, $|E|$ is large, and for the most unstable mode, $|Q_E| >> 1$; the dispersion relation is then given by (14'). For a deep quench, we have a viscous inertial behaviour. The wave vector q_M is given by (10). It is much larger than q^* and dipolar interactions are negligible in the small time regime of the spinodal decomposition.

3 . CONCLUSIONS

We have studied spinodal decomposition in a Langmuir monolayer of polar molecules in the coexistence region between the liquid and gas or liquid-expanded and liquid-condensed phases where periodic supercrystal phases form. We summarize here the most important results :

- The unstable critical mode of the monolayer is the longitudinal Lucassen mode.

- The spinodal line where the elastic modulus vanishes does not occur at zero wave-vector but at the equilibrium wave-vector at the critical point q^* resulting from a competition between dipolar and short-range forces.

- A deep quench at a low temperature leads to a quite classical spinodal behaviour, long-range forces play no role, and the only important feature is the flow induced in the water substrate.

- After a shallow quench, close to the critical point, the diverging fluctuation has the equilibrium wave-vector q^*. At longer time, the growth of the equilibrium periodic phase will require only a variation in the amplitude of the modulation of the surface concentration.

REFERENCES

1. D. Andelman, F. Brochard, P.G. de Gennes, J.F. Joanny : C.R.A.S. 301, 675 (1985)
2. D. Andelman, F. Brochard, J.F. Joanny : J. Chem. Phys., in press
3. A. Fischer, M. Lösche, H. Möhwald, E. Sackmann : J. Phys. Lettres 45, L-785 (1984)
4. L. Leibler : Macromolecules 13, 1602 (1980)
5. R.E. Rosensweig : Ferrohydrodynamics (Cambridge University Press, New York 1985)
6. J. Lucassen : Transactions of the Faraday Society 64, 2221 (1968)
7. L. Kramer : J. Chem. Phys. 55, 2097 (1971)
8. G.K. Batchelor : An Introduction to Fluid Dynamics (Cambridge University Press, 1967)
9. J.D. Gunton, J.M. San Miguel, P.S. Sahni : in Phase Transitions and Critical Phenomena, Vol. 8 (C. Domb, J. Lebowitz editors, Academic Press 1983)

Investigation of Phase Transitions in Fatty Acid Monolayers by Fluorescence Microscopy

F. Rondelez

Université Pierre et Marie Curie, Laboratoire "Structure et Réactivité aux Interfaces", 4 place Jussieu, F-75231 Paris Cedex 05, France

This paper is a brief summary of several recent experiments performed on monolayers of fatty acids adsorbed on water and which constitute nearly perfect two-dimensional systems. Fluorescence microscopy allows one to observe directly the structures formed i) during expansion of the monolayer from the liquid expanded phase into the liquid expanded-gas coexistence region, ii) during compression from the liquid expanded phase into the liquid expanded - liquid condensed coexistence region. In the former case, gas bubbles and metastable foams are observed. In the latter case, the structures of the growing liquid condensed domains range from circular platelets to self-similar fingering patterns, depending on the temperature of the experiment relative to the critical temperature for the phase transition.

INTRODUCTION

Monolayers of amphiphilic molecules, e.g. fatty acids and phospholipids, spread at the air-water interface, form quasi-perfect two-dimensional systems. Depending on the molecular surface density, the monolayer may be solid, liquid or gaseous-like, much as in three-dimensional systems /1/. The phase transitions between these various physical states have been extensively studied by surface manometry, an expression first coined by PETHICA /2/ but which describes a technique introduced by LANGMUIR /3/ more than fifty years ago. Plots of surface pressure isotherms reveal well-defined plateaux and/or kinks which have been associated with the liquid expanded-gas transition (in the low pressure range, < 1 dyne cm^{-1}), the liquid expanded - liquid condensed transition (in the intermediate pressure range, $\simeq 1 - 10$ dyne cm^{-1}) and the liquid condensed - solid transition (in the high pressure range, > 10 dyne cm^{-1}). Depending on the temperature, some of these transitions may or may not exist : there are critical temperatures T_{c1} and T_{c2} respectively, above which the liquid expanded - gas phases or the liquid expanded - liquid condensed phases become indistinguishable. Over the years, important questions however have remained unsettled, even in the very simple case of long-chain fatty acids which constitute convenient archetypes for these so-called Langmuir monolayers. For instance, the liquid expanded - liquid condensed phase transition is described sometimes as first-order and sometimes as second order because the corresponding plateau in the isotherm is generally not truly horizontal /4/. Similarly, the boundaries of the gas - liquid expanded coexistence region for a given fatty acid, such as pentadecanoic acid, vary by as much as a factor 2 on the surface density axis and 40°C on the temperature axis, depending on the authors /5/.

This rather unsatisfactory situation is due to the fact that surface pressure measurements cannot give information on the spatial structures formed during the various phase transitions. Moreover the data are prone to considerable error in the gas phase, because of the very low values involved. Therefore surface pressure should rather be used in conjonction with other techniques, both macroscopic and microscopic. In that respect, the demonstration by VON TSCHARNER and McCONNELL /6/, and LÖSCHE, MOHWALD and SACKMANN /7/, that fluorescence microscopy is sensitive enough to observe the structure of monomolecular films a few Angströms thick deposited on solid or liquid substrates, has represented a considerable break-through. In this technique, also called epifluorescence microscopy, a small amount of fluorescent dye is dissolved in the monolayer. If the monomolecular film is heterogeneous, for instance composed of liquid condensed domains floating in a continuum of the liquid expanded phase, the dye will partition differently between the various phases, depending on its relative solubility. This preferential distribution of the dye will provide good optical contrast. If the liquid condensed phase expels the dye, the liquid condensed domains will appear black over a white background, corresponding to the liquid expanded phase. Very recently, it has been discovered that the quantum yield of some fluorescent dyes is also a strong function of the molecular orientation /8/. For instance, when using 4-(hexadecyl amino)-7-nitrobenz-2-oxa-1,3-diazole dissolved in fatty acids, the gas phase appears black relative to the liquid expanded phase because in the former state, the dye molecules are flat on the interface and their fluorescence is quenched by the water subphase. On the whole, fluorescence microscopy provides a way to discriminate between the liquid expanded, liquid condensed and gas phases, and to investigate the structures in the biphasic regions with a lateral spatial resolution of a few microns.

RESULTS

We have performed detailed investigations of monolayers formed with three fatty acids, namely myristic acid (MA), pentadecanoic acid (PA) and stearic acid (SA). They all show common features which we will only summarize in the following since the results have already been published (or submitted) elsewhere.

1. Transition from the Liquid Expanded Phase to the Gas Phase /8,9/

Starting from the homogeneous all-white liquid expanded phase, gas bubbles are nucleated as the monolayer is expanded and the liquid-gas coexistence line is crossed. This occurs for a molecular surface density of about 70 Å^2 per molecule for PA at 20°C and for 49 Å^2 per molecule for MA at 20°C. As the expansion is continued, the gas bubbles grow in size and the monolayer takes the appearance of Swiss cheese. Eventually the bubbles become so numerous that they start to interact with each other. At this stage, the monolayer takes a structure which is highly reminiscent of a foam, with thin bright lines of liquids separating large gaseous domains. The lines are interconnected into a two-dimensional network as shown in fig. 1. This network is metastable and

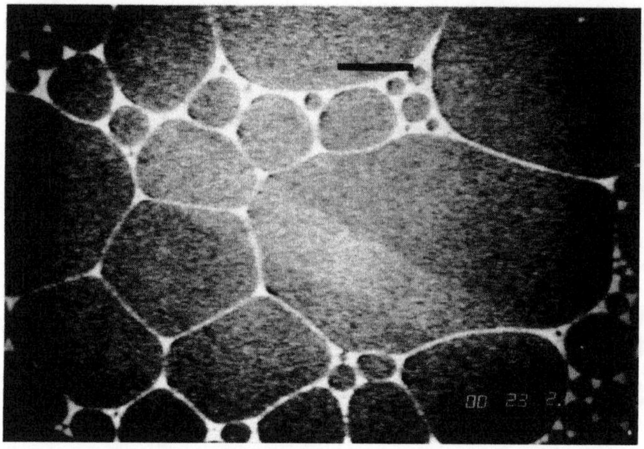

Fig. 1 : Fluorescence image for a pentadecanoic acid monolayer in the liquid expanded-gas coexistence region. Surface density is 400 Å^2 per molecule. T = 20°C. The bar length is 120 µm. The bright lines are thin threads of the liquid expanded phase separating gaseous domains which appear as a dark background.

the thin lines progressively breaks to form white liquid bubbles floating in a continuous dark gaseous phase. These white domains then disappear as the low density branch of the coexistence curve is crossed. This occurs at ≃ 1600 Å^2 per molecule for PA at 20°C and ≃ 1500 Å^2 per molecule for MA at 20°C. The values obtained by fluorescence microscopy for PDA are in good agreement with those proposed by PALLAS from surface pressure data /5-c/. On the other hand, they cast some doubt on the data published previously by the two other experimental groups of ref.5.a-b.

2. Transition from the Liquid Expanded Phase to the Liquid Condensed Phase

2.a) Isothermal Compression at Temperatures Well Below the Critical Temperature /9,10/

When the monolayer is gradually compressed from its homogeneous liquid expanded phase, isolated domains of the liquid condensed phase start to appear at random positions as soon as the liquid expanded - liquid condensed coexistence region is entered. This allows us to locate its right hand side boundary region with an accuracy of ± 0.5 Å^2 per molecule. For PDA at 20°C, we find 45 Å^2 per molecule while for MA at 20°C, we obtain 30 Å^2 per molecule. The liquid condensed domains appear as black dots upon a white background since the dye solubility is lower in the denser phase. Upon further compression, these dots grow in size but their number stays fixed, which is typical of a nucleation and growth process /11/. The coexistence of liquid expanded and liquid condensed regions within the same monolayer is in strong support of a first-order phase transition. The hypothesis made by some authors that the non-zero slope of the plateau region in the surface pressure isotherms could signal a second-order phase transition is therefore not borne out by our experiments.

The shape of the liquid condensed domains is generally circular, as expected for a fluid droplet with finite surface tension. There is however evidence that, at high surface density, geometrical distortions can be induced by long range, probably electrostatic /12/, interactions between the black domains. Fig. 2 shows a typical example where the droplets have become facetted and where long whiskers have started to grow, connecting neighbouring liquid condensed domains in their direction of closest

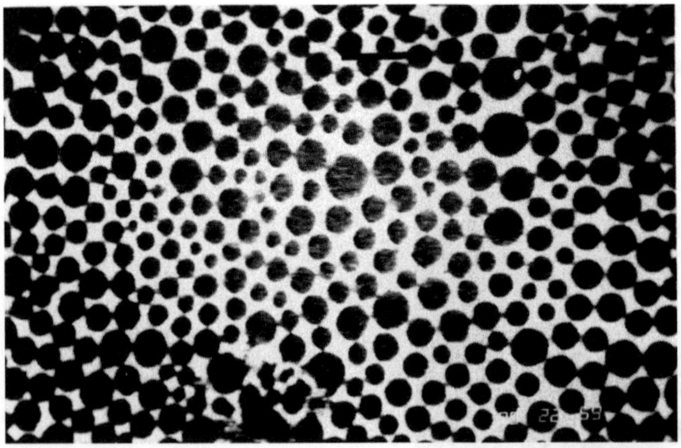

Fig. 2 : Fluorescence image for a pentadecanoic acid monolayer in the liquid expanded - liquid condensed coexistence region. Surface density is 28 Å^2 per molecule. T = 20°C. The bar length is 120 μm. The dark circles and polygons are domains of the liquid condensed phase floating in the continuum liquid expanded phase (white background). Note the bridges, or whiskers, connecting the liquid condensed domains.

approach. This phenomenon, which is particularly evident with pentadecanoic acid monolayers, is still unexplained. The left-hand side of the coexistence curve is reached when all the liquid expanded phase have been transformed into the liquid condensed phase. This corresponds to a surface density around 20-22 Å^2 per molecule, depending on the fatty acid. This boundary is difficult to locate accurately because the liquid condensed phase can easily be confused with the solid phase which is still denser but has practically the same optical appearance under fluorescent light. Judging from a small, but nevertheless well-defined, kink in the surface pressure isotherm, the transformation to the solid phase occurs at 20.5 Å^2 per molecule in the case of myristic acid.

2.b) Isothermal Compression in the Vicinity of the Critical Temperature /10/

The width of the liquid expanded - liquid condensed coexistence region is a strong function of temperature. As shown in the pioneering experiments of ADAM and JESSOP and HARKINS and BOYD /13/, it narrows down as the temperature increases and there is a critical temperature T_{c2} above which the liquid expanded phase becomes indistinguishable from the liquid condensed phase. This is the analog of the critical point for the ordinary gas-liquid transition in three dimensions or of the consolute point for binary mixtures. In the vicinity of T_c , the properties of the system become very peculiar. For instance, the divergence of the isothermal compressibility is at the origin of the well-known phenomenon of critical opalescence in binary mixtures. In the case of monolayers, we will be interested by the fact that the line tension between the liquid expanded and the liquid condensed goes to zero as $|T - T_c|^{\mu}$ where μ is a critical exponent of order unity /14/. Because of the reduction in the interfacial energy near T_c , the growing liquid condensed domains will markedly deviate from the circular shape. We have recently observed /10/ the formation of self-similar, fractal, structures in myristic acid. These arborescent structures are highly reminiscent of the viscous fingers which form when a fluid of low viscosity is pushed through a fluid of high viscosity in a two-dimensional Hele-Shaw cell. In our experiments, the instability of the interfacial front is apparently due to the same diffusion-limited growth mechanism as first discussed by MULLINS and SEKERKA for the case of a solid spherical nucleus growing from its pure melt /15/.

REFERENCES

1. G.L. Gaines : Insoluble Monolayers at Liquid-Gas Interfaces, Interscience Publishers (Wiley, New York 1966)
2. S.R. Middleton, M. Iwahashi, N.R. Pallas, B.A. Pethica : Proc. Roy. Soc. Lond. A396, 143 (1984)
3. I. Langmuir : J. Am. Chem. Soc. 39, 1848 (1917)
4. See for example J.F. Baret : Prog. Surface and Membrane Sci. 14, 291 (1981)
5. a) G.W. Hawkins and G.B. Benedek : Phys. Rev. Lett. 32, 524 (1974)
   b) M.W. Kim, D.S. Cannell : Phys. Rev. A13, 411 (1976)
   c) N.R. Pallas : Ph. D. Dissertation, Clarkson University (Potsdam, New York 1983)
6. V. von Tscharner, H.M. McConnell : Biophys. J. 36, 409, 421 (1981)
7. M. Lösche, E. Sackmann, H. Möhwald : Ber. Bunsenges Phys. Chem. 87, 848 (1983)
8. B. Moore, C.M. Knobler, D. Broseta, F. Rondelez : J. Chem. Soc. Faraday Trans. 2, 82, 1753 (1986)
9. F. Rondelez, J.F. Baret, K.A. Suresh, C.M. Knobler : Proceedings of the 2nd International Conference on Physico-Chemical Hydrodynamics, Huelva, Spain, July 1-15, 1986 (Plenum Press, New York 1987)
10. K.A. Suresh, J. Nittmann, F. Rondelez : to be published (1987)
11. J.D. Gunton, M. San Miguel, P.S. Sahni : in Phase Transitions and Critical Phenomena, vol. 8, C. Domb and J. Lebowitz eds. (Academic Press, London 1983)
12. H.M. McConnell, L.K. Tamm, R.M. Weis : Proc. Natl. Acad. Sci. (USA) 81, 3249 (1984)

13. W.D. Harkins, E. Boyd : J. Phys. Chem. 45, 20 (1941)
References to the earlier work by Adam and Jessop are also included.
14. B. Widom : in Phase Transitions and Critical Phenomena, vol. 2, C. Domb and M.S. Green eds. (Academic Press, London 1976)
15. W.W. Mullins, R.F. Sekerka : J. Appl. Phys. 34, 323 (1963)

Polymorphism of Monolayers of Monomeric and Macromolecular Lipids: On the Defect Structure of Crystalline Phases and the Possibility of Hexatic Order Formation

E. Sackmann, A. Fischer, and W. Frey

Physik Department, Biophysics Group, Technical University Munich, D-8046 Garching, Fed. Rep. of Germany

Phase transitions of monolayers of one-, two- and multichain amphiphiles are compared. The nature and defect structure of the crystalline phases exhibiting long-range orientational order but short-range spatial order are discussed and the question of hexatic phase formation is addressed.
Electron diffraction studies show that fatty acids go over from a tilted (smectic B_c) to a non-tilted (sm B) at a weak first order transition.
The continuous crystal-crystal transition of phospholipids can be understood in terms of a sintering process, but according to X-ray scattering appears to be associated with a change in the crystal structure. The low pressure solid phase has a small but finite shear elastic modulus ($\mu \approx 1$ mN·m^{-1}) and is thus not clearly hexatic. The μ-value of the high pressure phase is a factor of 100 higher. The low shearing rigidity of the former phase is explained in terms of a high defect density. Owing to the rigid mutual coupling of the two chains and the head-group exhibiting a dipolar shape a frustration arises since the chains form a triangular and the heads an orthogonal lattice. This leads to a coupling of dislocations of the former lattice and disclinations (e. g. vortex states) of the liquid crystal-like arrangement of the collectively oriented head-groups.

I. INTRODUCTION

Lipid monolayers exhibit a rich polymorphism. By classical film balance experiments, three (phase) transitions of phospholipid monolayers are established [1, 2]:

(1) a second-order-like transition at very low pressures, which is associated with a discontinuity of the viscosity and is most probably associated with the transition from the two-dimensional foam to the liquid (smectic A) state, [3],

(2) a phase change (at π_m) with characteristic features of a first-order-transition (such as phase coexistence) and a large change in molecular area by about 30 %, which is associated with the melting of the hydrocarbon chains,

(3) a second order-like high-pressure transition between two condensed states at a pressure π_c.

First of all, the monolayer polymorphism is of practical interest since it provides a possibility for the controlled preparation of Langmuir-Blodgett films with heterogeneous lateral organization. On the other hand lipid monolayers appear to behave as close approximations of two-dimensional systems and thus become of growing interest in testing recent theories of dislocation mediated melting and two-dimensional liquid crystal formation. Figure 1 shows for comparison the isotherms of a one chain lipid (namely arachidic acid) and a two-chain phospholipid (namely DMPE). The main transition at π_m corresponds to the smectic A to smectic B transition of lipid bilayers. A remarkable difference between the two lipids is observed at high pressures: while the fatty acid exhibits a clearly visible weak first-order transition the phospholipid shows only a break in the π-A-curves.

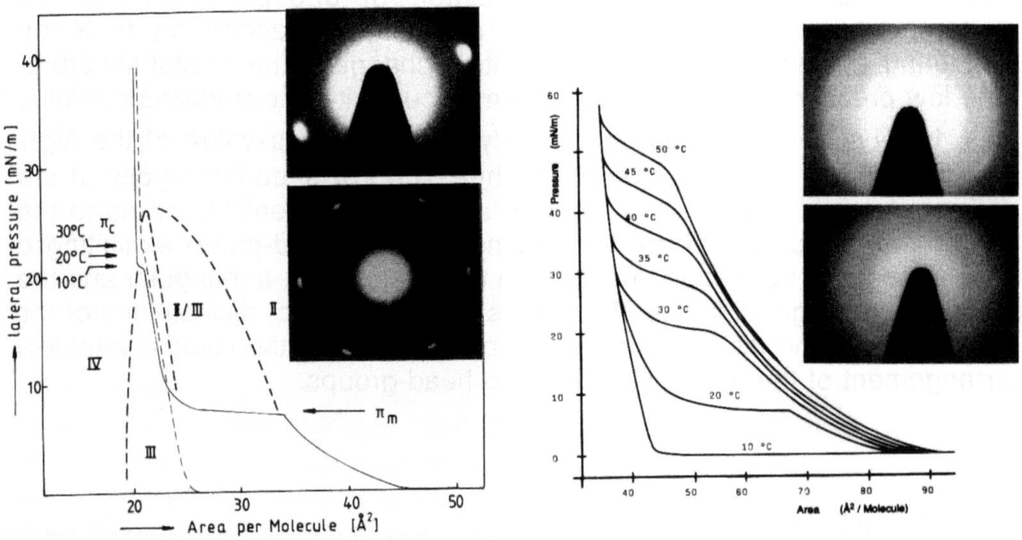

Fig. 1: Comparison of isotherms of fatty acid (arachidic acid; right side) and phospholipid (dimyristoyl phosphatidyl-ethanolamine = DMPE; left side). Note in particular the difference in the behaviour of the transition at π_c: while the two-chain lipid exhibits only a break, the fatty acid shows a weak horizontal deflection that is first-order behaviour. Note furthermore the absence of a sharp edge at the π_m transition for DMPE above 30°C, that is, the chain melting transition becomes continuous. The inserts show the electron diffraction pattern taken from crystalline phase at $\pi > \pi_c$ and $\pi < \pi_c$.

Another noteworthy feature of DMPE is that the main transition is very sharp at low temperatures (T<30°C) whereas the deflection of the isotherm becomes continuous at T>35°C. Such behaviour is indeed found also for other phospholipids [1]. It is very likely that this change in sharpness is due to the fact that at high temperatures where the transition occurs at higher lateral packing densities the transition is determined by the head-group.

A completely different type of main transition was discovered for macromolecular lipids consisting of copolymers of amphiphiles which are interconnected by hydrophilic chains. Monolayers of these lipids show a continuous transition from an expanded state where the hydrophilic chains are adsorbed to the air-water interface exhibiting a pan-cake-like configuration into a condensed phase in which the hydrophilic chains escape into the third dimension forming a brush-like configuration. The transition exhibits a negative Clausius-Clapeyron coefficient: $d\pi / dT < 0$ and is thus entropy driven (cf. Frey et al. [5]). The condensed phase at $\pi > \pi_b$ exhibits also a hexagonal diffraction pattern. In striking contrast to normal lipids, the foam-to-liquid transition occurs at very large areas, which is a consequence of the adsorption of the hydrophilic chain to the air/water interface. This behaviour has been explained in terms of the scaling laws of adsorbed macromolecules [5].

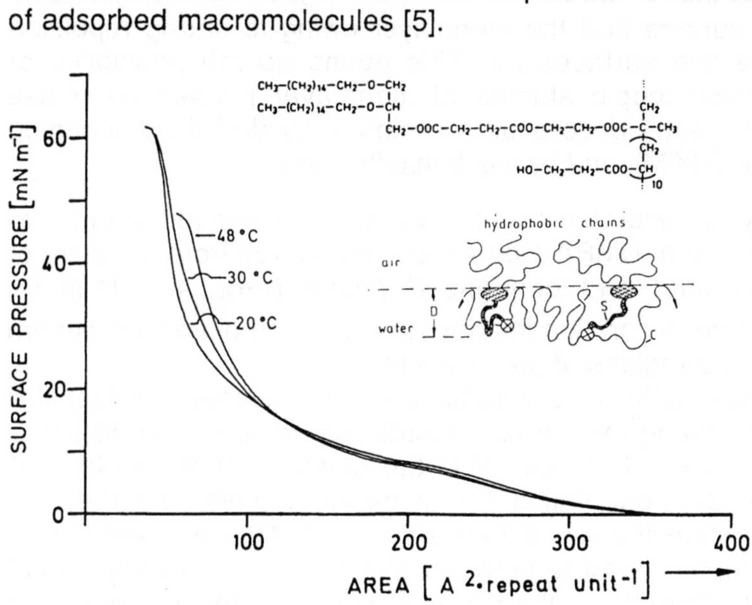

Fig. 2: Isotherm of a macrolipid consisting of (two-chain) amphiphiles interconnected by hydrophilic chains as indicated. Note that the transition at π_b exhibits a weakly negative Clausius-Clapeyron coefficient $d\pi_b / dT < 0$, indicating that it is an entropy-driven process.

II. MICROSTRUCTURE AND ORDER OF MONOLAYER PHASES

Three recent new techniques have drastically improved our understanding of the microscopic structure of lipid monolayer phases: fluorescence microscopy [3], charge decoration electron microscopy in combination with low dose electron diffraction [2], and the recently developed synchrotron X-ray scattering at glancing incidence [4]. First insight into the microscopic structure of the monolayer phases was gained by the microfluorescence technique which demonstrated that at zero lateral pressure, the monolayer forms a two-dimensional foam consisting of fluid and gas-like domains. Surprisingly, the lateral packing density of the gas phase appears to be only a factor of three smaller than that of the fluid phase [2b]. The most important result was, however, that at $\pi_m < \pi < \pi_c$, the fluid and the condensed phase coexist. This follows from the finding that the fluorescent probe molecules are squeezed out from the condensed phase by zone refining (cf. Fig.3) which already points to a crystalline structure.
All monolayer phases can be deposited in a perturbationless way onto electron transparent solid substrates with silicone oxide surfaces. Film disrupture is avoided if the ionic strength of the sub-phase is low. This is explained in terms of the formation of a thin water layer (about 100Å thick) between the glass surface and the monolayer owing to strong repulsion forces between the two surfaces [6]. This opens up the possibility of detailed electron microscopic studies of monolayer phases by phase contrast EM (called charge decoration technique), by dark field scanning transmission EM (DF-STEM), and by electron diffraction.

Figure 4 shows how the electron microscopy can be exploited in order to study the microstructure of DMPA monolayers transferred from the state of coexistence of the fluid and condensed phase ($\pi=\pi_m+1\,mN\cdot m^{-1}$). Simultaneously, charge decoration images and electron diffraction pattern are taken from selected regions of the platelets.
Hexagonal diffraction patterns are obtained in each case. In fact the orientation does not change when the electron beam is moved across a single platelet. Moreover, it is seen that the crystal planes of different platelets exhibit the same orientation if they are merged whereas they are oriented differently when the two-dimensional crystallites are separated. This shows that a repulsive force between platelets arises if their crystal planes are oriented differently in the region of contact. Below, this will be explained in terms of the line tension due to grain boundaries which would have to form in the case of merging.
The electron diffraction studies demonstrate that the dark platelets exhibit a long-range orientational order at all pressures above π_m. We conclude that the same holds for the air-water interface since it is hardly possible that

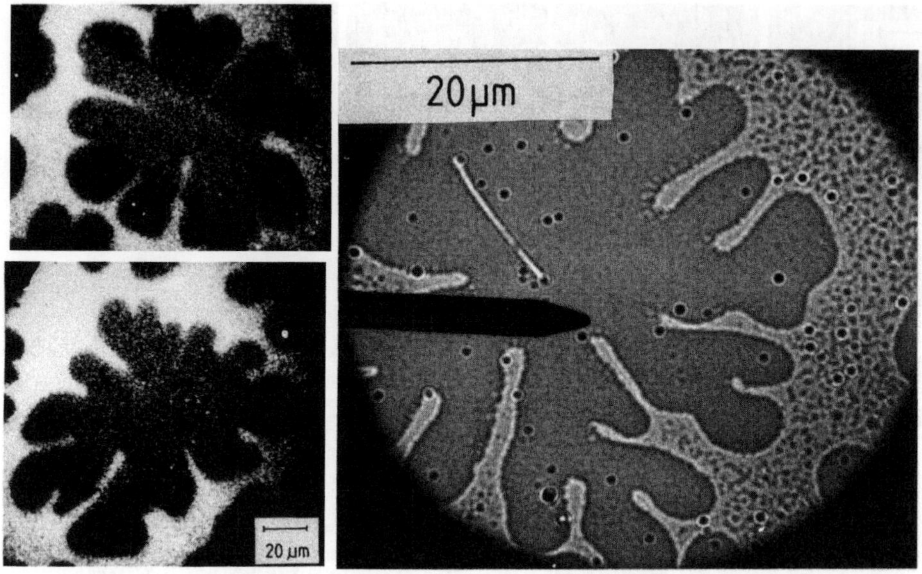

Fig. 3: Demonstration of a perturbationless transfer of lipid monolayer in state of fluid-solid coexistence from air/water interface to solid substrate.
Left side: Fluorescence micrograph of dimyristoyl phosphatidic acid (DMPA) monolayer (containing 0.3 mole% of fluorescent dye) just below chain-melting transition ($\pi = \pi_m + 1$ mN·m^{-1}).
Right side: Electron micrograph taken by (charge decoration) phase contrast electron microscopy [2] of DMPA monolayer deposited from state of coexistence onto electron transparent substrate (Formvar-carbon-SiO-sandwich).

crystalline order over dimensions of some 10 µm is formed during the film transfer. The zone refining of the fluorescent probes also points to a crystalline order of the low pressure condensed phase. Moreover, long-range orientational order of the dark patches is also verified by fluorescence polarization experiments [7,8].
For the phospholipids the electron diffraction pattern and the radial widths of the diffraction spots are the same above and below π_c. The synchrotron experiments [4], however, have revealed a dramatic reduction in the radial width of the diffraction pattern of DMPA above π_c compared to $\pi < \pi_c$. Provided radiation damage can be excluded [9] this shows that the crystalline platelets of the phospholipids undergo a conformational change at π_c.

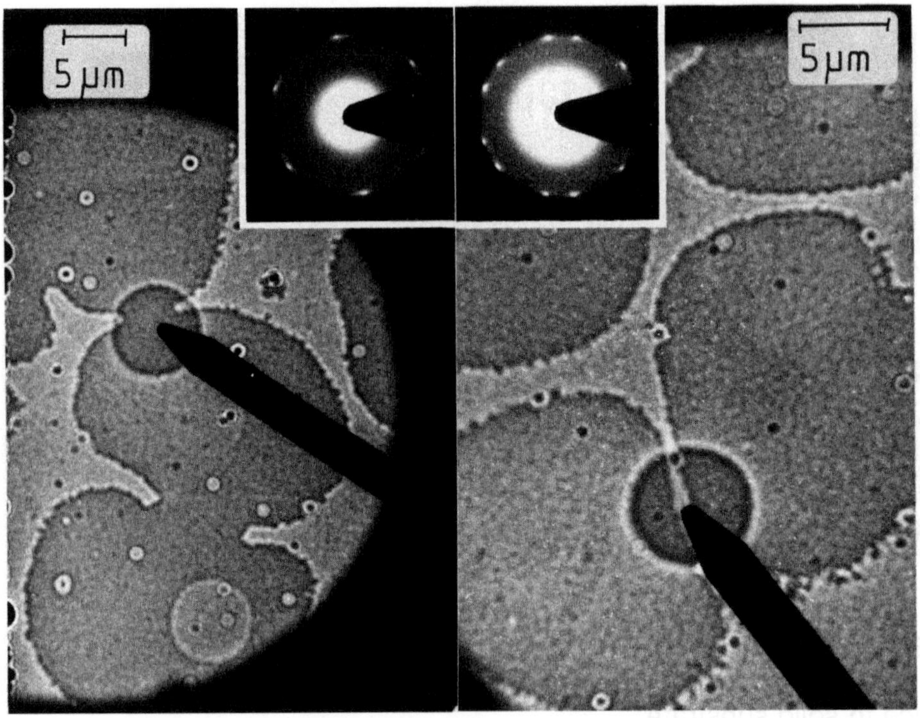

Fig. 4: Charge decoration electron micrograph of DMPA monolayer in the coexistence region and low dose electron diffraction pattern taken from the circular patches within the platelets. The change in brightness of these patches is a consequence of the charging up of the substrate, which helps to locate the region from which the diffraction pattern is taken. The dark pointers visible in both the phase contrast and in the diffraction micrographs allow the crystal plane orientations to be determined.
Right side: Case of separated platelets. The hexagonal diffraction patterns of the two platelets are mutually rotated, which shows that the crystal planes of the domains differ by about 30 degrees.
Left side: Diffraction pattern taken at different sites of merged platelets. The orientations of the crystal planes of the interconnected platelets are completely in register.

In the case of the fatty acid, the electron diffraction pattern of the high- and the low-pressure solid phases are clearly different (cf. Fig. 1). The latter is most probably a tilted (smectic C) phase as follows from the C2-symmetry of the pattern (cf. Fig. 1 and reference [2 b]).

III. DISLOCATIONS AND DISCLINATIONS OF CRYSTALLINE MONOLAYER PHASES

Depending on the size of the head-groups, the chains may be oriented normal (such as for DMPE and monovalent DMPA) or may be tilted (such as for phosphatidylcholines [2 a]) with respect to the monolayer normal. Even in the former case it must be realized that:

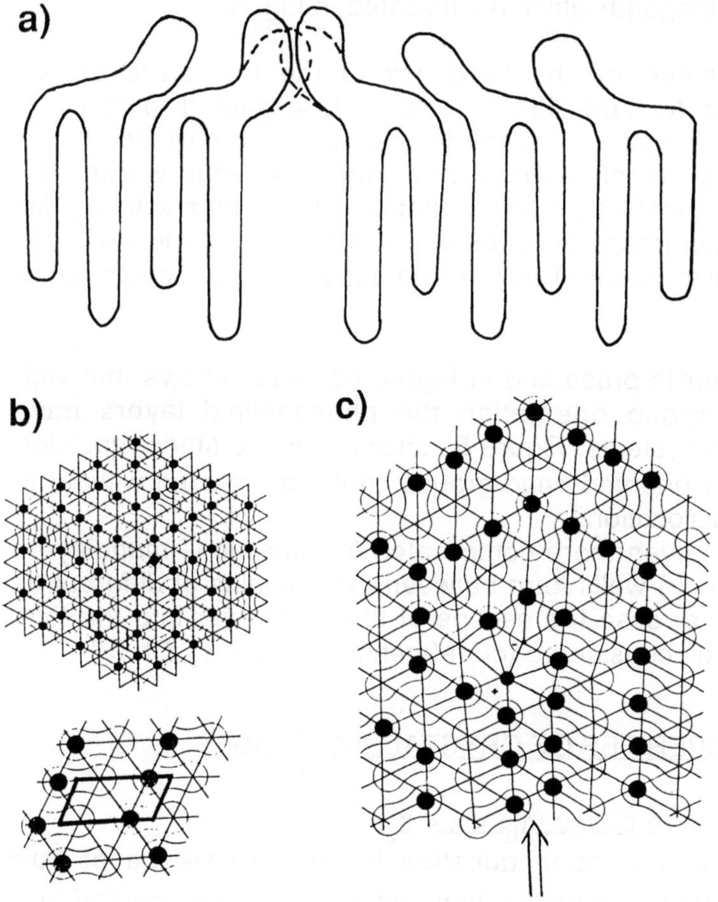

Fig. 5: **a**) Schematic view of phospholipid molecules stressing the non-equivalence of the two chains and the polar shape of the glycerol backbone with the phosphate group attached to it.
b)Discontinuity in orientation of head-groups of strength m = +1 (nomenclature of de Gennes) which is eqivalent to an isolated vortex in the xy-model. The black circle corresponds to the bulky phosphate group of the lipid head.
c) Edge dislocation within triangular lattice of chains. Note that the repulsion between the oppositely oriented phosphate groups is relaxed in the dilutation region of the dislocation (arrow).

(1) The two chains are not equivalent since the one distal to the phosphate group is bent at the C=O bond (cf. Fig. 5a) and may require a smaller area than the proximal chain.
(2) The polar head-group formed by the glycerol backbone and the phosphate group has a polar (that is wedge-like) shape and the head-group orientation may thus interfere with the chain packing.
(3) The hydrocarbon chains form a triangular lattice whereas the head-groups form an orthogonal lattice as indicated in Fig. 5.

The frustration introduced by the tendency of the two parts of the phospholipid molecules to form different lattices is expected to cause a high density of defects. In fact a hierarchy of defects can be envisaged:
(1) Edge-dislocations, disclinations, and point defects within the two-dimensional crystal lattice. (Note that the chains form a triangular lattice and the head-groups an oblique lattice (Fig. 5 b, thick line)).
(2) Discontinuities (disclinations) in the orientation of the dipolar head-groups.

An example of the latter is presented in Figure 5b which shows that with respect to the head-group orientation the phospholipid layers have similarities with xy-spin systems. Figure 5c shows how the steric repulsion between two differently oriented head-groups could be relaxed within the dilatation region of a dislocation.
The situation becomes even more complicated for smectic C monolayers where Neel (and/or Bloch) walls form between the domains of different tilt orientation. In the former case the escape of the monolayer into the third dimension has also to be considered (cf. reference [10] or [11], Ch.IV)

IV. ON ELASTIC PROPERTIES OF CRYSTALLINE PHASES

A) Low-pressure crystalline phase ($\pi_m < \pi < \pi_c$)

A heavily debated but still open question is the persistence of fluid domains at pressures up to about 5 $mN \cdot m^{-1}$ above π_m (corresponding to a temperature range of about 5°C [1]).
This behaviour has been explained in terms of an electrostatic repulsion of the platelets [7,14]. Another explanation suggested by the electron microscopic studies is that the persistence of the fluid domains is a consequence of the line tension associated with the grain boundaries betweeen the crystalline platelets of different crystal plane orientation. In the following, the shear elastic modulus and the Poisson ratio of the crystalline phase at $\pi < \pi_c$ is estimated by quantitative evaluation of this sintering process.

The low angle grain boundary between the crystal planes of the two adjacent monodomains (cf. Fig. 6a) can be described in terms of a quasi-linear array of edge dislocations of density $n_d = \Theta / b$, where b is the Burgers vector and Θ the angle between the crystal planes (cf. Fig. 6a). The elastic energy associated with this array of dislocations leads to a line tension per unit length of grain boundary of

$$\gamma = \frac{b\Delta\mu^{(3)}}{4\pi(1-\sigma)}\Theta \ln(\alpha/\Theta) \qquad (1)$$

where Δ is the membrane thickness ($\Delta \approx 2.5$ nm), σ the Poisson ratio, and $\mu^{(3)}$ is the three-dimensional shear elastic modulus, which is related to the two-dimensional constant $\mu^{(2)}$ of the monolayer by $\mu^{(2)} = \mu^{(3)} \cdot \Delta$ and α is a numerical factor of the order of one and is a measure for the core energy of the dislocation [11].

In order to crystallize the lipid between two domains of orientation Θ, additional mechanical work has to be supplied. Consider for example the elongated cleft of constant width h indicated in Fig. 6b by an arrow. This example is chosen since the angle Θ of the grain boundary is known to be approximately $\Theta \approx 30°$ from the electron diffraction (Fig. 6b). The fluid phase crystallizes at a lateral pressure π given by

$$\pi h - q/A \cdot h - \gamma = 0 \qquad (2)$$

where q is the heat of transition and A the area per lipid molecule.

For DMPA, the heat of transition is $q \approx 2 \cdot 10^4$ J/mole, $\pi_c \approx 15$ mN·m^{-1} and $A = 60$ Å^2 at 20 °C. By considering the narrow crack of Fig. 6 b, we arrive at a ratio

$$\mu^{(2)} / (1-\sigma) = 0.4 \text{ mN·m}^{-1} .$$

The Poisson-ratio may be estimated if the compression modulus K is known according to $\sigma = (3K - 2\mu) / (6K + 2\mu)$. In previous work [1], we measured slightly below π_c a value of $K \approx 100$ mN·m^{-1} which is by about a factor of 10 smaller than above π_c. With the above values we find a Poisson ratio of $\sigma = 0.5 - 0.001$.

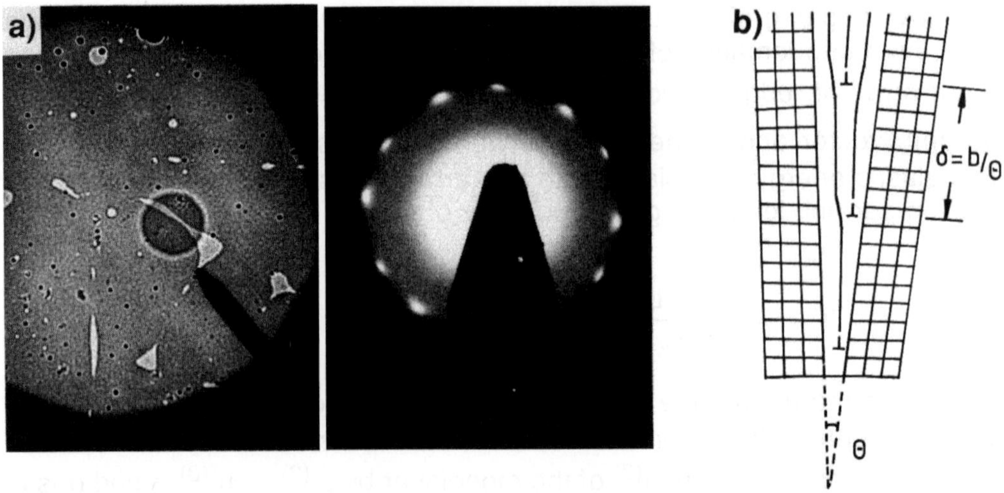

Fig. 6: **a**) Charge decoration electron micrograph of DMPA monolayer deposited from pressure π_c (at 33°C) and electron diffraction pattern taken from an area which covers both sides of a crack. Note that the crystal planes are mutually rotated by $\Theta = 30°$. **b**) Low angle grain boundary. Distance between dislocations $\delta = b/\Theta$.

B) High pressure crystalline phase $\pi > \pi_c$:

Owing to the compression in one direction (x), at second dimension (y) fixed, the crystalline monolayer suffers simultaneously a shear strain and a compression. If we assume that - due to hexagonal symmetry - the crystal is isotropic, the stress in the direction perpendicular to the moving boundary is $\sigma_{yy} = \pi \cdot \sigma$, where π is the applied lateral pressure. The strain in the x-direction is then :

$$\Delta A / A = u_{xx} = -(1-\sigma)\, \pi / \mu^{(2)}. \qquad (3)$$

For DMPE and DMPA, which exhibit non-tilted (smectic B) phases $\pi \cdot \Delta A/A \approx 2.5 \cdot 10^{-3}$ mN·m^{-1} and one obtains

$$\mu^{(2)}/(1-\sigma) \approx 200 \text{ mN·m}^{-1}.$$

For DMPA the lateral compression modulus K has now been separately measured at $\pi > \pi_c$ by X-ray scattering experiments as K = 1 400 mN·m^{-1}. Therefore $\sigma \approx 1/2$, that is $\mu \approx 100$ mN·m^{-1}.

V. ON THE NATURE OF THE CRYSTALLINE PHASES

The break of the π-A curves at π_c can be explained in two ways. First, as a true phase transition between two solid states (as is indeed the case for the fatty acid), and secondly, in terms of the merging of the domains of different orientations. The drastic decrease in the radial width of the X-ray deflections at $\pi \geq \pi_c$ [4] points to the first possibility whereas the electron microscopic studies favour the second explanation since it reveals the presence of fluid patches up to the pressure π_c. Provided that the radiation damage is small, the synchrotron radiation experiments provide evidence that the merging is indeed accompanied by a crystal-crystal phase transition at π_c. Further evidence for such an explanation comes from the increase in the shear elastic modulus by two orders of magnitude at the break point π_c. The question arises whether the low-pressure crystalline phases correspond to the hexatic phase introduced by Halperin and Nelson [15, 16]. This phase behaves liquid-like in the sense that the shear elastic modulus goes to zero for zero wave vector deformations ($\mu \propto q^2$). It behaves like a liquid crystal since the Frank elastic constant, characterizing the disclination elastic energy, is non-zero and of the order of k_BT.

The low pressure crystalline phase of phospholipids shows some characteristic features of a hexatic liquid crystal:

(1) It exhibits long-range orientational order but no translational order since the correlation length characterizing the latter is only some 10Å [2a,4].

(2) The orientational order decays rapidly with distance as follows from the large angular width (about 5 degrees) of the electron diffraction spots although a quantitative evaluation is not yet possible.

(3) The crystalline platelets at $\pi < \pi_c$ exhibit plastic behaviour, as follows from the growth kinetics of elongated crystallites in DPPC/cholesterol mixtures [8].

(4) The transition at π_c is continuous, as expected for a phase change from a two-dimensional crystal to a hexatic phase.

On the other hand, the above considerations show that the elastic modulus of the low pressure crystalline phase is small but finite. Moreover, the crystal-to-fluid transition is discontinuous at least at low temperatures. It is thus more likely that the condensed phase is a true two-dimensional crystal both below and above π_c. The fatty acids undergo a transition from a (tilted) smectic C to a smectic B phase. The situation for the phospholipids is not clear although both the X-ray diffraction studies [4] and the change in the shear elastic modulus point to a phase change also in this case.

There exist, however, similarities of the low pressure crystalline phase with a hexatic phase in the sense that the long-range order is strongly reduced by a high defect density. The very low shear rigidity is a consequence of this. It is interesting to note that the shear rigidity of the low pressure phase is about equal to that of the $L_{\beta'}$-phase of lecithins [17]. Certainly, besides the diffraction studies, more accurate measurements of the shear elastic properties would be very helpful in order to clarify the structural properties of the crystalline monolayer phases.

Finally, it should be pointed out that the defect structure may play at least an equally important role for the growth kinetics of the crystalline domains as the electrostatic effects [14]. Thus the dislocations are expected to accumulate at the outer rim of the platelets owing to image forces. In the case of tilted smectic C phases, such as the lecithines [2], wall defects owing to the transition from the tilted crystalline to the normal fluid phase must form at the interfaces. This could explain the spontaneous curvature of the elongated crystalline domains in DPPC/cholesterol mixtures [8].

Acknowledgement:

Financial support by the Fonds der Chemie and by the Freunde der TUM is gratefully acknowledged.

REFERENCES

1. O. Albrecht, H. Gruler and E. Sackmann, J. de Physique 39, 301 (1978)
2. A. Fischer and E. Sackmann
   a) J. de Physique 45, 517 (1984),
   b) J. Colloid Interf. Sci. 112, 1 (1986)
3. M. Lösche, E. Sackmann and H. Möhwald, Ber. Bunsenges. Phys. Chem. 87, 848 (1983)
4. K. Kjaer, J. Als-Nielsen, C.A. Helm, L.A. Laxhuber and H. Möhwald, Phys. Rev. Lett., in press
5. W. Frey, G. Schneider, H. Ringsdorf and E. Sackmann, Macromolecules, in press
6. G. Schneider, H. Joosten, W. Knoll and E. Sackmann, Europhysics Lett. 1, 449 (1986)
7. D.J. Keller, H.M. McConnel, V.T.Moy, J. Phys. Chem. 90, 2311 (1986)
8. H.E. Gaub, V.T. Moy and H.M. McConnell, J. Phys. Chem. 90, 1721 (1986)

9. Radiation damage cannot be excluded in the coexistence region. The exposure times are rather long and the electron diffraction experiments show that the damage occurs an order of magnitude faster below π_c than in the completely condensed state.
10. D. Rüppel and E. Sackmann, J. de Physique 44, 1025 (1983)
11. Ch. Kittel, Introduction to Solid State Physics, J. Wiley and Sons Inc. , New York 1971, 4th ed., Ch. 20
12. G. de Gennes, The Physics of Liquid Crystals, Oxford University Press, New York 1971, Ch. 4
13. L.D. Landau and E.M. Lifshitz Bd VIII, Addison Wesley, Reading Mass (1969)
14. a) A. Fischer, M. Lösche, H. Möhwald and E. Sackmann, J. Physique Lett. 45, 785 (1984)
15. J.M. Kosterlitz and D.J. Thouless, J. Phys. C. Solid State Phys 6, 1181 (1973)
16. D.R. Nelson and B.I. Halperin, Phys. Rev. B 19, 2457 (1979)
17. E. Evans and D. Needham, Faraday Disc. Chem. Soc. 81 (1986), in press

Surface-Density Transitions, Surface Elasticity and Rigidity, and Rupture Strength of Lipid Bilayer Membranes

E. Evans[1] *and D. Needham*[2]

[1]Pathology and Physics, University of British Columbia, Vancouver, B.C. V6T1W5, Canada
[2]Pathology, University of British Columbia, Vancouver, B.C. V6T1W5, Canada

Experimental advances have made possible direct measurements of cohesion, elasticity, and rigidity properties of surfactant double-layer membranes in aqueous media. Mechanical properties -- tension for rupture (expansion limit), elastic compressibility and surface rigidity -- are derived from pressurization of giant single-walled vesicles by micropipets. Area measurements as a function of temperature provide explicit definition of surface-density transitions at fixed state of stress. Special properties of crystalline-bilayer structures (e.g. ripple stiffness and energy of formation, shear rigidity and viscosity) are also measured. The effects of mixtures of phospholipids, cholesterol, and simple polypeptides on these properties have been established.

1. INTRODUCTION

Biological membrane structures are lamellar assemblies of amphiphilic molecules (e.g. lipids, cholesterol, proteins, etc.) as molecular double-layers. The strong preference for the lamellar configuration is evidenced by the negligible solubility - and extremely slow rate of exchange-of these molecules from a membrane capsule in aqueous media. Consequently, the membrane of a biological cell or artificial lipid vesicle behaves as a closed system for time periods on the order of hours or longer. In this condensed state, membranes exhibit solid-or liquid-like material behavior with the common feature of limited surface compressibility (i.e. great resistance to change in surface density). As a thin structure (3-4×10^{-7} cm), bilayer membranes can only support small forces without rupture. Hence, bilayer fragility and limited detectability by optical methods create significant obstacles to measurement of the physical-mechanical properties. Recently, micromechanical techniques have been developed to investigate the material structure of bilayer membranes *in situ*.[1,2] Combined with temperature changes, mechanical experiments can be used to directly quantitate thermal transitions, bilayer cohesion and elasticity, surface rigidity and heats of deformation.[1-3] Here, we outline prominent physical features of bilayer membrane materials as solid and liquid surface structures. We discuss how amphiphilic mixtures influence bilayer transitions, elasticity - rigidity - and cohesion. Lipids from two classes will be considered: phospholipids (phosphatidylcholine PC, and phosphatidylethanolamine PE) and a "sugar" lipid (digalactosyl diacylglycerol DGDG). The former are ubiquitous constituents of animal cell membranes whereas the latter exists predominantly in plant cell membranes.

2. MEMBRANE DEFORMATION, MECHANICS, AND CONSTITUTIVE RELATIONS[4-5,20]

Deformation and rate of deformation of thin materials are quantitated by variables that represent a sequence of simple shape changes for local regions (elements) of the surface: area dilation or condensation $\alpha = \Delta A/A_0$; in-plane extension $\lambda = L/L_0$ (i.e. surface shear) at constant surface density; and bending or curvature change $\Delta C = \Delta(1/R)$ without change in rectangular shape. Rates of deformation are time derivatives of area dilation, surface shear, and change in curvature. Modes of deformation are illustrated in Fig. 1. External forces applied to a membrane are distributed throughout the surface as forces and moments per unit length along membrane contours (specific actions are shown in Fig. 1). These include forces in the plane of the surface (i.e. tension τ_n normal to the contour and surface shear resultant τ_s tangent to the contour), a force normal to the surface (transverse shear Q_s), and moment resultants (membrane "couples" M). The external pressure normal to the membrane is balanced primarily by tension components multiplied by appropriate curvatures. External forces tangent to the membrane (e.g. fluid shear stresses) are opposed primarily by the surface gradients of tension. Membrane bending moments and transverse shear provide higher order contributions to these force balances.

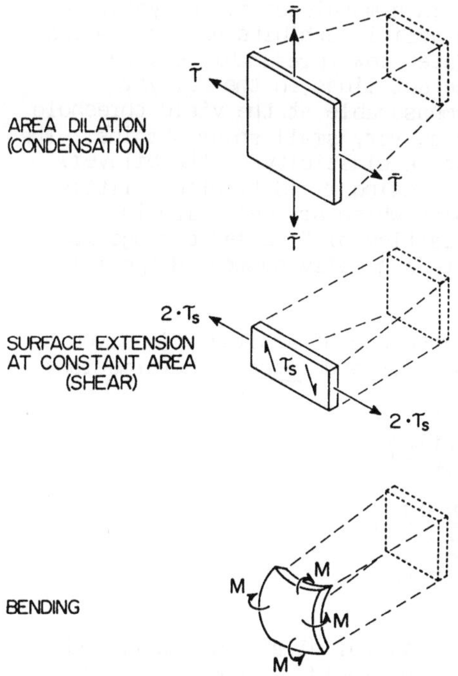

Fig. 1 Independent modes of deformation and force-moment resultants for an element of a thin-surface material.

Three first order elastic (conservative) relations characterize a surface-isotropic material: (1) mean tension $\bar{\tau}$ proportional to the fractional area dilation or condensation,

$$\bar{\tau} = K \cdot \alpha \quad ; \quad \bar{\tau} \equiv (\tau_1 + \tau_2)/2 ,$$

where τ_1 and τ_2 are principal values of tension for orthogonal directions tangent to the surface; (2) surface shear τ_s (i.e. the deviation between principal membrane tensions) proportional to in-plane extension or shear strain;

$$\tau_s = \mu\cdot(\tilde{\lambda}^2 - \tilde{\lambda}^{-2})/2 = 2\mu\cdot e_s \quad ; \quad \tau_s \equiv (\tau_1 - \tau_2)/2 \,;$$

(3) bending moments M proportional to changes in membrane curvatures,

$$M = B\cdot\Delta(1/R_1 + 1/R_2).$$

The first elastic relation represents the compressibility of the surface at constant temperature and does not differentiate solid and liquid structures except by relative value. Coupled with separate measurement of thermal area expansivity (fractional change in area with temperature $\partial\alpha/\partial T$), the surface compressibility modulus K yields the heat of expansion per unit surface area at constant temperature,[6]

$$\frac{1}{A}\cdot\left(\frac{\partial Q}{\partial \alpha}\right)_T = T\cdot K\cdot\left(\frac{\partial \alpha}{\partial T}\right),$$

which is related to the lateral pressure within the bilayer. The second elastic relation is unique to solid structures; liquids are defined by zero shear modulus ($\mu\equiv 0$). Thus, shear rigidity is peculiar to frozen (crystalline) bilayers. The frozen bilayer is a mosaic of microcrystalline domains; therefore, deformation of a frozen vesicle exhibits only a limited elastic response followed by yield and surface flow (presumably due to "melting" and "freezing" of crystal along defect lines in the bilayer surface). The surface rigidity $\hat{\tau}_s$ is only measurable at the yield threshold for the frozen bilayer since flow commences at very small shear strains. The third elastic relation represents the curvature elasticity of the bilayer. Due to the extreme thinness of the bilayer, bending rigidity offers little opposition to deformation for bilayer capsules which are cell size in dimension.[1,4] However, if there is a superlattice or "ripple" corrugation of the bilayer, the bending rigidity leads to a greatly augmented special-elastic stiffness.[7,14]

Ideal viscous-liquid (nonconservative) response of a bilayer is given by proportionalities between force-moment resultants and rates of deformation, i.e.

$$\bar{\tau} = \kappa\cdot\frac{\partial \ln(1+\alpha)}{\partial t},$$

$$\tau_s = 2\eta\cdot\frac{\partial \ln(\tilde{\lambda})}{\partial t},$$

$$M = \nu\cdot\frac{\partial(\Delta C)}{\partial t}.$$

The coefficients are surface viscosities (κ,η,ν) for dilation, shear, and bending respectively. When the acyl chains are liquid, viscous dissipation in the bilayer is difficult to measure because the deformation response of cell-size capsules is dominated by dissipation in the adjacent aqueous phases. The scale of regions dominated by membrane dissipation is given by the ratio of the surface viscosity to the external phase viscosity.[11] Surface diffusivity measurements of fluorescent membrane probes have been used to estimate the surface shear viscosity of lipids in the liquid state; the results are on the order of 10^{-6} - 10^{-5} dyn-sec/cm.[8-11] It is apparent that dissipation in the aqueous fluids will dominate the shear response of bilayer surfaces which are in excess of 10^{-4} cm in dimension. On the other hand, dynamic deformations of cell-size vesicles provide extremely useful methods for study of bilayer flow in the solid state where the shear viscosity is so large that measurements of probe diffusivity become questionable. Transition from solid- to liquid-surface behaviour is modeled by an ideal plastic relation where the material is solid below the yield threshold and a perfect liquid above the yield,

$$\tau_s - \hat{\tau}_s = 2\eta \cdot \frac{\partial \ell n(\tilde{\lambda})}{\partial t} \quad ; \quad \tau_s \geq \hat{\tau}_s \, .$$

Coefficients of viscosity for area dilation and curvature changes appear to be completely inaccessible to mechanical measurement because they are determined by relaxation times for acyl chain extension ($\lesssim 10^{-7}$ sec).[12] Since these deformations are usually elastic, viscosities for dilation and bending deformations may be estimated from the elastic moduli multiplied by the relaxation time t_e for chain extension,

$$\kappa \sim K \cdot t_e \, ,$$

$$\upsilon \sim B \cdot t_e \, .$$

3. VESICLE DEFORMATION TESTS AND METHODS OF ANALYSIS

Diacyl lipids exhibit the useful feature that they spontaneously form closed vesicular capules when hydrated from an anhydrous state. Although small in number, a few of these capsules are of sufficient size ($10\text{-}20 \times 10^{-4}$cm) that they can be subjected to direct micromechanical deformation and adhesion tests. Since the interior is a structureless liquid, tests of vesicle deformation can be used to isolate the bilayer mechanical properties. Production of big vesicles is most likely when lipids are hydrated by either water or aqueous solutions of nonelectrolytes (e.g. sugars) above the acyl chain crystallization temperature.[13-14] Electrostatic repulsion (even for zwitterionic species) acts to separate bilayers and reduce the likelihood of formation of multilamellar structures. Nonelectrolytes like sucrose also enhance optical images when vesicles are resuspended in salt solution and viewed with interference contrast optics (Fig. 2 shows vesicle images with/without refractile solutes). Single vesicles are selected from one chamber on a microscope stage and transferred into a solution (usually a buffered 0.1 M NaCl solution) which is made slightly more concentrated than the nonelectrolyte solution used for lipid hydration. Partial deflation of the vesicle follows which allows large projections to be aspirated into the selection pipet.

Vesicle deformation tests are based on simple pressurization by micropipet suction. After the vesicle is pressurized into a spherical shape, further suction acts to dilate the surface area and reduce internal volume. Because the pressures involved in these tests are several orders of magnitude smaller than the pressure required to filter water from the vesicle (against osmotic activity of the trapped solutes), vesicle volumes remain constant. Displacement of the vesicle inside the suction pipet is due to reduction in surface density of the amphiphiles. The area expansion is completely recovered after the suction is zeroed. Small changes in projection length L are proportional to the change in total area A of the vesicle scaled by pipet and vesicle dimensions (R_p, R_o),[1]

$$\Delta A \cong 2\pi \cdot R_p \cdot (1 - R_p/R_o) \cdot \Delta L \, .$$

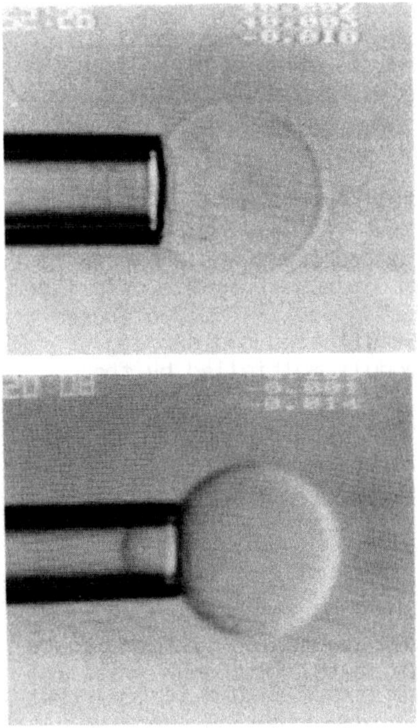

Fig. 2

Video micrographs of giant bilayer vesicle aspiration (vesicle diameter ~ 20 x 10^{-4}cm; pipet diameter ~6 x 10^{-4}cm). (a) Vesicle without refractile solutes in distilled water. (b) Vesicle with 0.1M sucrose inside and 0.1M NaCl outside.

Area changes are greatly amplified as changes in projection length which facilitates detection of the small elastic compressibility ($\Delta A/A \sim 10^{-2}$) of the surface. The mean tension $\bar{\tau}$ is uniform over the entire vesicle surface and is given by the suction pressure P and pipet/vesicle dimensions,[1]

$$\bar{\tau} = P\cdot R_p \,/(2 - 2\cdot R_p/R_o).$$

Figure 3 shows results for two different vesicle compositions; the slope of the tension versus area dilation is the elastic area compressibility modulus K. With selection first of the thinnest vesicles as imaged by interference contrast and then evaluation of the area elastic modulus, it is possible to discriminate between one -two -and more layers since the elastic modulus groups around discrete values where the lowest value is characteristic of a single bilayer.[1] The tension at rupture provides a measure of the cohesion limit or bilayer strength.

Another feature of vesicle pressurization is that the area can be observed as a function of temperature at constant bilayer tension. This allows direct quantitation of thermal transitions in the surface density at a controlled state of stress.[2,14] Large-initial projections of the vesicle into the pipet enable measurement of large condensations in surface density (~ 20%) without retraction from the pipet. Note, changes in length inside the pipet represent changes in the "projected" area of the bilayer. When submicroscopic structures like ripples are present (illustrated in Fig. 4), both changes in density of the molecules as well as ripple architecture contribute to the displacement of the length.[14] The ratio of the projected area A to the ideal area A_c for crystallized chains aligned normal to the surface is given by,

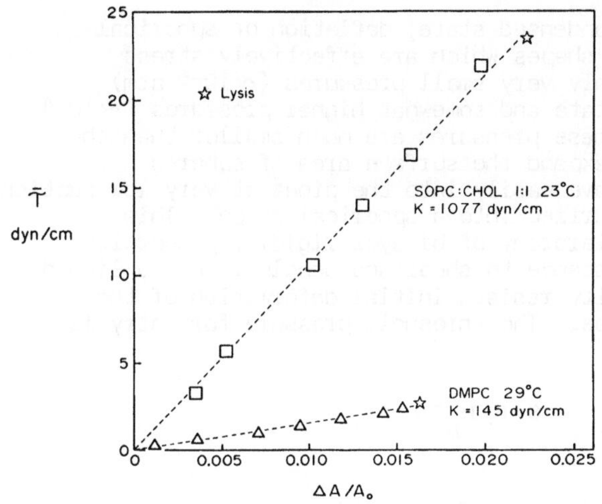

Fig. 3 Vesicle tension versus area dilation measured for single vesicles with different compositions. The vesicles were first pressurized to just below the level for subsequent rupture, then pressure was reduced to test elastic reversibility. Finally, the vesicles were stressed to the rupture point as shown.

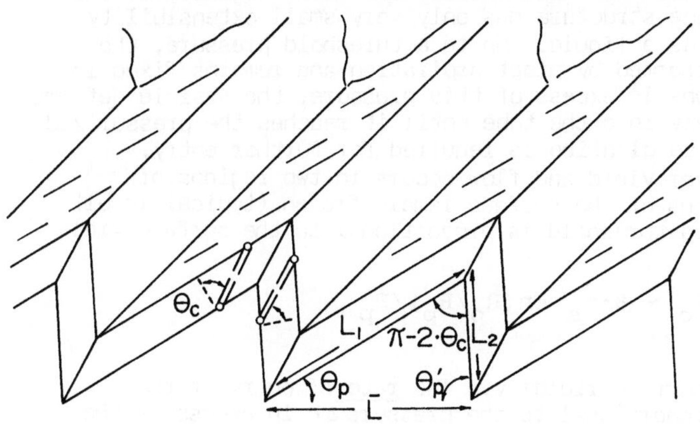

Fig. 4 Schematic illustration of superlattice architecture characteristic of a rippled-solid surface with tilted acyl chains. Measurements of P_{β}' structure indicate a more symmetric unit geometry (i.e. $x \sim 0.5\text{-}0.6$).

$$\frac{\bar{A}}{A_c} = x \cdot \frac{\cos\theta_p}{\cos\theta_c} + (1 - x) \cdot \frac{\cos\theta_{p'}}{\cos\theta_c} \,,$$

where x is the symmetry ratio $x \equiv \sin\theta_{p'} / (\sin\theta_p + \sin\theta_{p'})$.

Because the bilayer is a condensed state, deflation of spherical vesicles produces flaccid irregular shapes which are effectively stress free.[41] Initial aspiration takes only very small pressures ($\sim 10^{-6}$ atm) when the bilayer is in the liquid state and somewhat higher pressures ($\sim 10^{-4}$ atm) when the bilayer is frozen. These pressures are much smaller than the pressures ($\sim 10^{-2}$ atm) required to expand the surface area of sphered vesicles. Thus, flaccid vesicles move easily into the pipet at very low suction pressure until the vesicle is pressurized into a spherical shape. This deformation behavior reflects the hierarchy of bilayer rigidity properties: resistance to area dilation >> resistance to shear and bending. For a liquid bilayer surface, only bending rigidity resists initial deformation of the flaccid vesicle as it enters the tube. The threshold pressure for entry is given approximately by

$$P_o \sim 8 \cdot B / R_p^3 .$$

The suction required for initial aspiration of a liquid bilayer surface is 10^{-6} atm (1 dyn/cm^2) with a 5×10^{-4} cm pipet; thus the curvature elastic modulus B must be on the order of 10^{-12} dyn-cm (erg), consistent with values derived from analysis of thermal fluctuations of vesicle contours and theoretical predictions.[4,15-17]

When the acyl chains are crystallized, initial aspiration pressures for partially deflated vesicles are dominated by the shear rigidity of the surface.[18] The solid surface structure has only very small extensibility before it yields and flows as a liquid. Up to a threshold pressure, the vesicle is only slightly deformed by pipet aspiration and remains fixed in shape (Fig. 5). For suctions in excess of this pressure, the vesicle deforms continuously and flows slowly into the tube until it reaches the pressurized spherical geometry where area dilation is required for further entry. Analysis[18] shows that bilayer yield and flow occurs in two regions of the surface but other regions of the surface remain frozen (typical result shown in Fig. 6). The yield threshold is proportional to the surface yield shear $\hat{\tau}_s$,

$$P_o \sim 8 \cdot \hat{\tau}_s \cdot \ln(R_o/R_p)/R_p ,$$

which is a measure of the surface rigidity. The <u>rate</u> of entry of the solid-bilayer vesicle is proportional to the pressure ΔP in excess of the threshold with a coefficient determined by the surface shear viscosity,

$$\dot{L} \sim \Delta P \cdot R_p^2 / \left[4 \cdot \eta \cdot \ln(R_o/R_p) \right] \quad ; \quad \Delta P \equiv (P - P_o) .$$

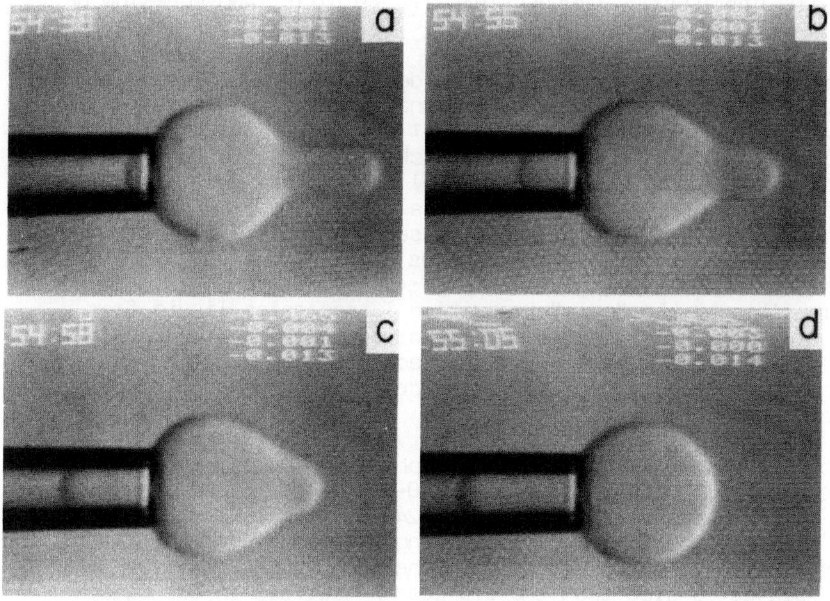

Fig. 5 Video micrographs of a frozen DMPC vesicle (13°C). (a) Vesicle as frozen replica of aspiration at high suction pressure, reversed for reaspiration. (b) Threshold suction sufficient to exceed surface shear rigidity (yield). (c) An image taken during continuous flow at elevated suction above yield threshold. (d) Final pressurization to a solid sphere.

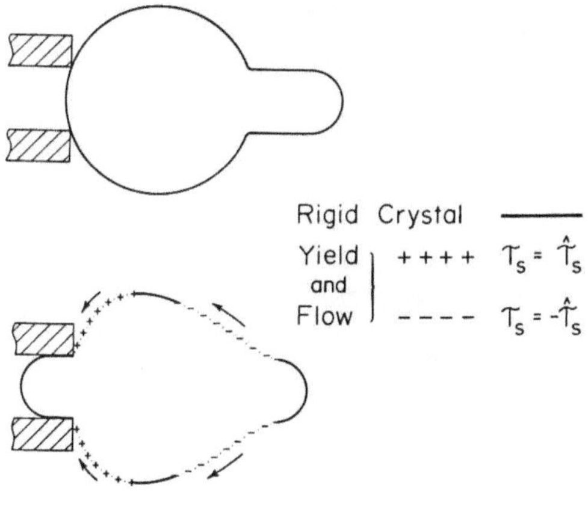

Fig. 6 Theoretical prediction for yield of a plastic-solid surface. The initial frozen geometry is in the upper panel. The lower panel shows the deformed geometry after yield with identification of the specific regions where yield and flow occur.

4. SURFACE-DENSITY TRANSITIONS, COMPRESSIBILITY, RIGIDITY, AND COHESION OF SINGLE COMPONENT BILAYERS

Lipid dispersions exhibit diverse phases that depend on temperature, water and ion content: i.e. lamellar and inverted hexagonal-and cubic-architectures.[19-20] Lamellar phases are either anisotropic liquids with fluid acyl chains or solids with crystallized chains. The liquid phase is labelled L_α whereas the solid phase has several descriptors related to chain tilt and surface topography, e.g. planar crystalline phases (L_β, L_β') and "rippled" or corrugated intermediate solid phases (P_β'). The prime superscript identifies layers where acyl chains are tilted with respect to the normal to the bilayer surface (e.g. as in Fig. 4). These phases are separated by enthalpic transitions: the "main" transition T_c separates liquid from solid phases whereas the "pre" transition T_p correlates with the separate existence of planar-solid (below T_p) and rippled-solid (above T_p) surfaces. Lower "sub" transitions have been identified and are currently under investigation.[21-22]

Saturated chain PC's exhibit L_β', P_β', L_α phases. For example, dimyristoyl phosphatidylcholine (DMPC) below 10-13°C is a L_β' solid at equilibrium, a P_β' solid above 13°C but below 24°C, and a L_α liquid surface above 24°C.[23] Figure 7 shows an individual single-walled vesicle at each temperature characteristic of these phases. The L_β' structure at the left shows a greater projected area than the rippled P_β' surface in the middle image. The intermediate P_β' structure (ripple) can be eliminated or prevented from formation by bilayer stress.[14] Also if vesicles are annealed at low temperatures (< 10°C) for a few days to ensure formation of L_β' structure and then heated to temperatures above T_p in a short time period, the P_β' ripple does not appear to form. Hence, a metastable (planar) phase can exist between T_p and T_c which is labelled as $L_\beta^{*}{}'$.[14] These observations are accumulated into a phase diagram (Fig. 8) that shows relative area changes for individual vesicles versus temperature as a function of bilayer tension. The work necessary to pull the ripple surface flat (given by the tension multiplied by the expansion of vesicle area) is only about 1/50 of the "pre" transition enthalpy: i.e. formation of the superlattice is not characterized by the pretransition heat exchange.

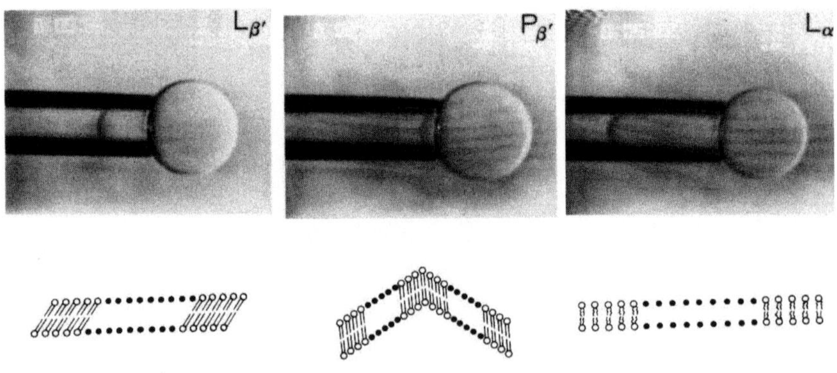

Fig. 7 Video micrographs of a DMPC bilayer vesicle in three structural states at different temperatures: left is L_β' state at 8°C; middle is P_β' state at 16°C; right is L_α state at 25°C.

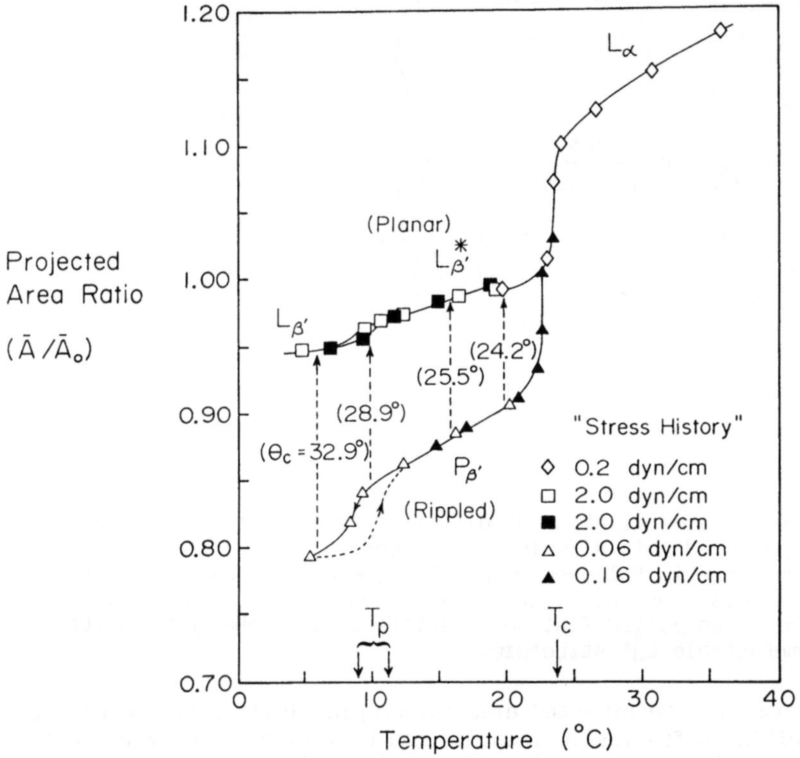

Fig. 8 Effect of applied stress on solid DMPC bilayer structure and phase formation. Relative vesicle area is plotted versus temperature for fixed bilayer tensions. Values of the acyl chain tilt to the bilayer normal are shown as angles in parentheses as derived from the ratio of projected areas for the rippled-solid and planar-solid surfaces at the same temperature.

Elimination of the superlattice is very useful. The ratio of areas between the rippled state and the metastable planar state at the same temperature provides a direct measure of the intrinsic tilt of the chains relative to the surface (as indicated in Fig. 4). This conclusion is based on x-ray diffraction studies which show that the acyl chains are aligned nearly normal to the projected area of the ripple in the P_{β}' phase ($x \sim 0.5$-0.6) but are tilted to the flat surface in the L_{β}' phase.[24] Hence, the ratio of upper to lower area curves in Fig. 8 below T_c is the cosine of the intrinsic crystal angle (angles are listed in Fig. 8). The results correlate very well with values derived from structural analysis by x-ray diffraction.[23] In addition, the fractional reduction in area from the L_α phase to the P_{β}' phase at the main transition is a direct measure of the condensation of the acyl chains (~20%) upon freezing; further condensation of the chains follows the low-stress curve.

These results clearly demonstrate the importance of stress history in formation of solid phase structure. In the domain of the P_{β}' phase, the tension versus projected area (as shown in Fig. 9) exhibits an initial soft-elastic response at low tensions, then plastic (permanent) expansion at high tensions which leads to elimination of the surface ripple, and finally a

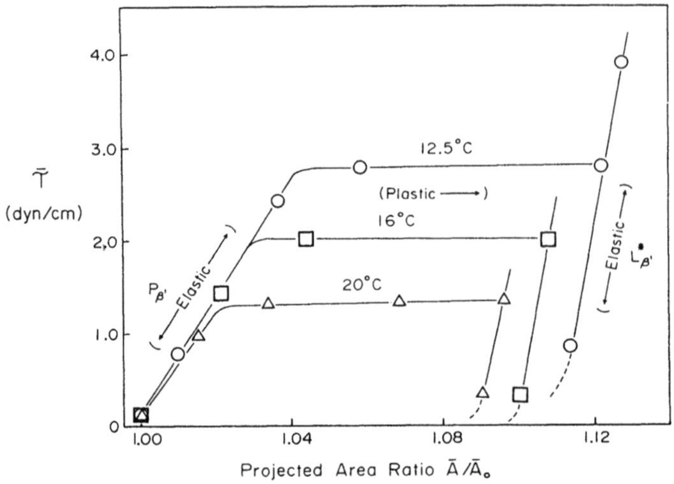

Fig. 9 Tension versus area dilation at different temperatures for single vesicles produced initially in the P_{β}' domain. The initial soft-elastic response is followed by plastic (permanent) expansion at constant stress. At the limit of expansion, the rippled-solid surface has been pulled flat and a stiff-elastic behavior results for the metastable L^{*}_{β}' structure.

stiff elastic response at the expanded area for compressibility of the planar surface.[14] The initial soft-elastic response can be rationalized by a model (illustrated in Fig. 10) where extension of the ripple is due to bending deflections of individual pleats in the corrugated surface. The soft-elastic modulus is proportional to the bending modulus divided by the square of the peak-to-peak amplitude "a" of the surface ripple,

$$K_{apparent} \sim 8 \cdot B/a^2 .$$

Measurements of the apparent compressibility modulus[14] for the initial elastic behavior (50-60 dyn/cm) plus values of ripple amplitude from x-ray diffraction[24] and SEM[25] studies (5-8 x 10^{-7} cm) indicate that the bending stiffness is about 3 x 10^{-12} dyn/cm at these temperatures.

Since the P_{β}' corrugation can be removed by stress, it is possible to compare the surface compressibility of the planar bilayer over all three phase domains. The area compressibility modulus increases only by about a factor of two from 1.2 x 10^2 dyn/cm to 2.8 x 10^2 dyn/cm when the surface is initially frozen from the L_{α} phase to the P_{β}' domain. Below the pretransition temperature, the compressibility is greatly reduced (K $\sim$ 10^3 dyn/cm). These results are consistent with spectroscopic observations that indicate a large number of residual gauche chain conformations exist in the intermediate range of temperatures ($T_p < T < T_c$).[26-28] Hence, chain packing "defects" in the surface greatly affect surface compressibility.

Partially unsaturated phospholipids (e.g. 1-stearoyl-2-oleoyl-phosphatidylcholine SOPC and 1-palmitoyl-2-oleoyl-phosphatidylethanolamine POPE) exhibit direct condensation from the L_{α} phase to the L_{β} phase without chain tilt or superlattice formation.[29-30] The area change upon freezing is about 20%

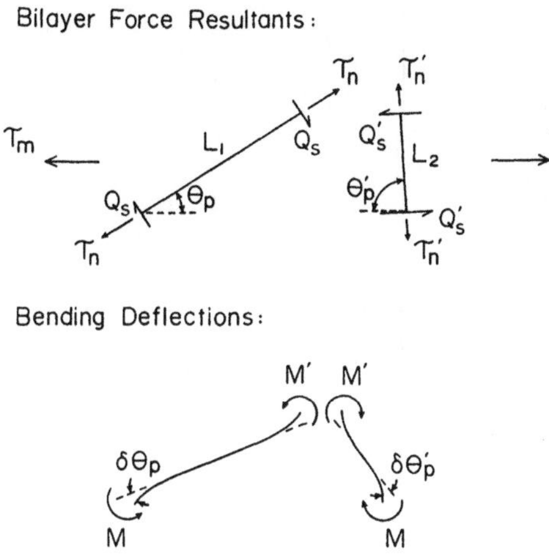

Fig. 10

Model of the soft-elastic response of a rippled-solid surface as bending deflections of individual pleats in the surface. The low stiffness is due to mechanical leverage as illustrated.

similar to DMPC. Area compressibility moduli in the L_α phase for these two lipids are comparable (2.0-2.3 x 10^2 dyn/cm) but higher than for DMPC. Values of the compressibility moduli in the L_β phase are similar to that for DMPC in the L_β' phase (~10^3 dyn/cm). The sugar-lipid DGDG (a natural product with three double bonds) has an area compressibility modulus which is slightly lower (1.9 x 10^2dyn/cm) than for SOPC or POPE. If surface moduli are divided by a representative value for the bilayer thickness (3 x 10^{-7} cm), a comparison can be made with values characteristic of volumetric compressibility for solids (10^{12} dyn/cm^2), liquids (10^{10} dyn/cm^2), and gases (10^6 dyn/cm^2). As a condensed state, the bilayer surface is 10-100 times more compressible than an ordinary liquid!

Like compressibility, the thermal area expansivity of bilayers is much greater than the volumetric expansivity (~10^{-4}/°C) of condensed liquids. In the L_α and P_β' domains, the fractional change in vesicle area with temperature describes the chain expansivity with temperature. In the L_β' domain, the change in vesicle area with temperature is related to the chain expansion and changes in chain tilt (e.g. Fig. 4). There is little difference in thermal area expansivity (5-6 x 10^{-3}/°C) for DMPC bilayers in the L_α and P_β' phases. Thermal area expansivity is significantly lower in the L_β' phase of DMPC but accurate values are difficult to obtain because of the limited range of accessible temperatures. Similarly, POPE and SOPC exhibit thermal area expansivities of about 3 x 10^{-3}/°C in the L_α phase and 1-2 x 10^{-3}/°C in the frozen L_β phase. The thermal area expansivity for DGDG is also lower (2 x 10^{-3}/°C) than for the L_α phase PC's. Hence in the absence of density transitions, bilayers expand a few percent in area when the temperature is increased by 10°C.

Surface rigidity is peculiar to solid bilayer phases. Figure 5 shows a frozen DMPC vesicle first pulled tight into the micropipet then ejected as a solid replica of the pipet cavity and reversed for reaspiration. Threshold pressures-and subsequent rates of vesicle entry into the pipet at higher pressures-are analyzed to determine the surface yield and coefficient of shear viscosity.[18] The yield threshold shows a slight dependence on domain structure (illustrated in Fig. 11) when vesicles are subjected to cyclic tests. Figure 12 shows the entrance-flow rate measured as a function of excess pressure which demonstrates the ideal-plastic behaviour. The surface shear rigidity and viscosity for solid DMPC surfaces increase significantly as

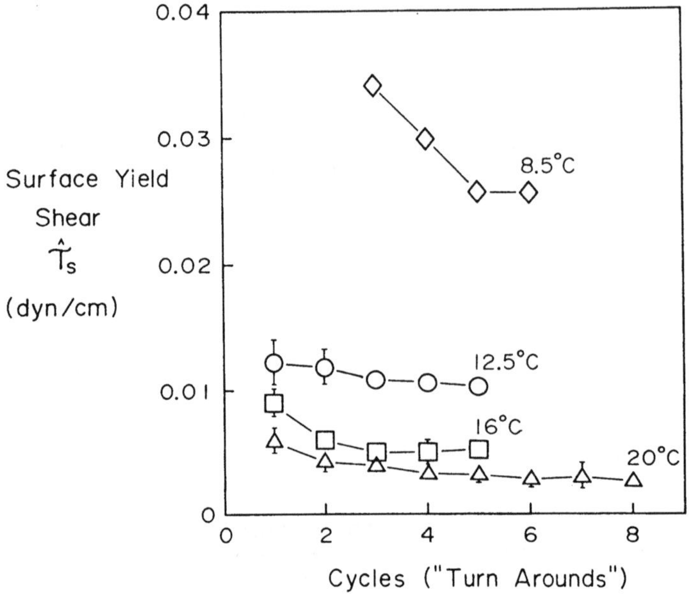

Fig. 11 Measurements of surface shear rigidity on single vesicles at different temperatures as a function of cyclic shear. There appears to be a slight "homogenization" of sub-microcrystalline domains. The surface is initially stressed to form a flat, ripple-free structure. Yield presumably occurs along defect lines that circumscribe the sub-microscopic domains.

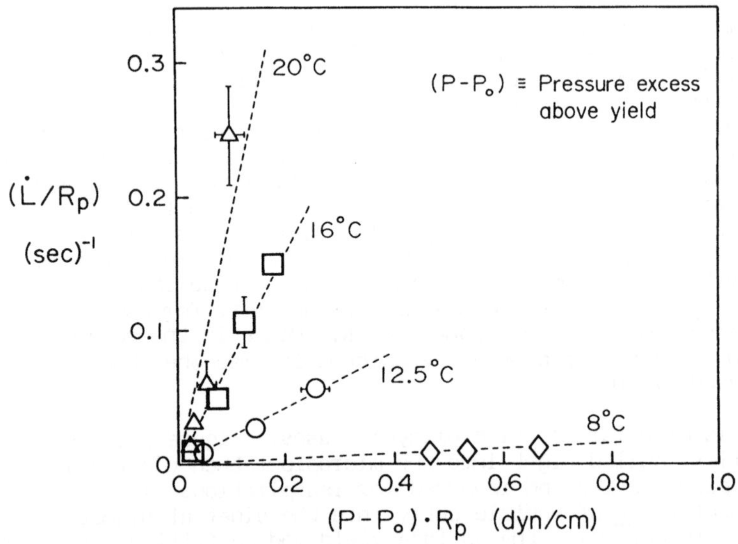

Fig. 12 Flow response of single vesicles to pressures in excess of the yield threshold for the frozen bilayer structure. The "Newtonian-like" response is consistent with ideal-plastic rheology of the surface. The strong effect of surface condensation is shown by the depression of the slope at lower temperatures.

the temperature is reduced below the pretransition T_p to the highly condensed low-temperature domain (see Fig. 13). Close to the main transition, the surface viscosity is four orders of magnitude larger than expected for the L_α phase based on measurement of lateral diffusivity of fluorescent probes.[11] Yield shear and viscosity primarily reflect the density of defect perimeters that circumscribe submicroscopic-crystalline domains in the surface.

Failure of solid and liquid bilayer materials by area expansion leads to vesicle rupture or lysis. Although lysis is expected to be a stochastic process which involves the time of exposure to the stress, experimental observations (e.g. Fig. 3) show precipitous failure when the tension reaches a critical value with reasonably long-term survival of vesicles below this tension. For DMPC vesicles, lysis occurs at 2-3 dyn/cm in the L_α phase but more than 15 dyn/cm is required in the L_{β}' phase. SOPC, POPE, and DGDG are more resistant to lysis in the L_α phase than DMPC (i.e. about 6 dyn/cm tension for rupture); similarly, SOPC and POPE bilayers become significantly more resistant to lysis in the frozen state.

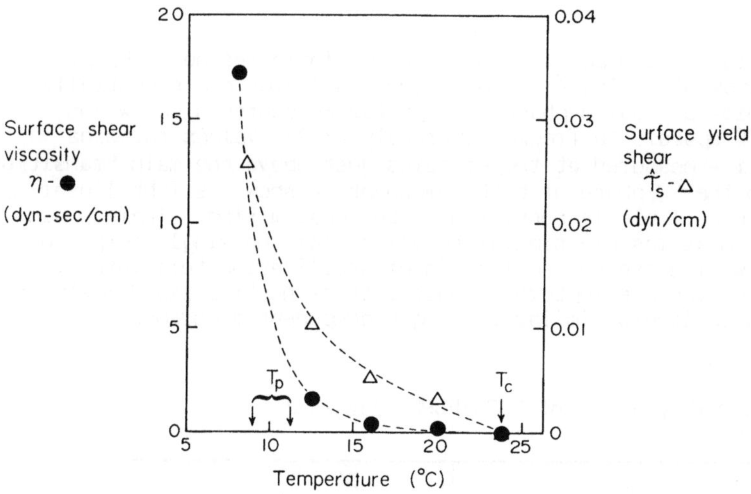

Fig. 13 Surface shear viscosity (left ordinate) and surface yield shear (right ordinate) are plotted as a function of temperature derived from measurements on frozen DMPC vesicles below the main acyl chain transition (T_c) and pretransition (T_p).

5. THERMOMECHANICAL PROPERTIES OF BILAYER MIXTURES

Mixtures of lipids, integral and peripheral proteins, cholesterol, etc. make up natural membranes. How do lipid mixture, cholesterol, and integral proteins affect membrane compliance, expansivity, and cohesion? The PC and PE classes of lipids, especially those with asymmetric chain composition and saturation, are major components of cell membranes. Mixtures of two of these lipids (SOPC and POPE) with similar chain length composition have been studied over a temperature range than spans both acyl chain crystallization transitions at pH 6.0 where both lipids were electrically neutral. Individual vesicle areas are plotted versus temperature and composition in Fig. 14 as determined from cooling experiments.[42] Based on limiting temperatures for each mixture

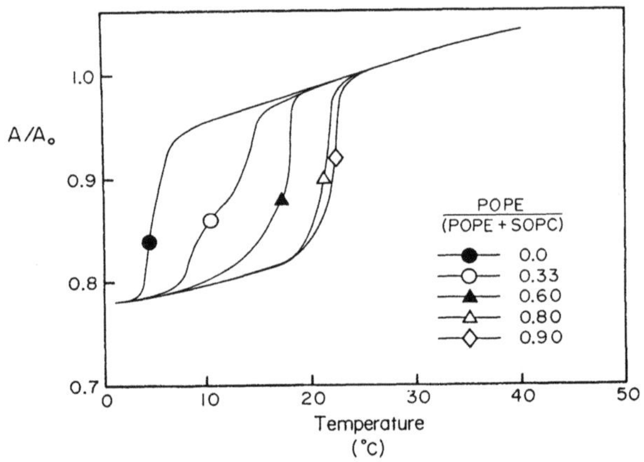

Fig. 14 Relative areas derived from single vesicle cooling experiments for mixed SOPC:POPE bilayers.

transition, a phase diagram can be constructed which indicates no enthalpy of mixing and ideal entropy of mixing.[31] Hence, the two lipids are essentially miscible in both liquid- and solid-phases except for PE contents below 33% where some of the SOPC appears to freeze separately.[32-33] Values for area compressibility moduli - measured at temperatures just above the main transition (T/T_c = 1.02-1.03) in the L_α phase of both components - show a slight linear compositional dependence (Table 1) consistent with ideal mixing. Thermal area expansivities for the mixtures are similar to values for the single components in the L_α phase when well above the acyl chain crystallization temperature. Based on observation of vesicle rupture strength, these neutral lipid mixtures exhibit bilayer cohesion levels similar to single component membranes.

Table 1 "Thermoelastic Properties of SOPC:POPE Bilayers"

POPE / (SOPC+POPE)	K Area Compressibility Modulus (dyn/cm)	$\dot{\alpha}$ Thermal Area Expansivity (x 10^3/°C)	T Temperature (°C)
0	199.6 ± 12.7	3.28 ± 0.68	15.0
0.33	209.9 ± 14.1	3.79 ± 0.24	22.0
0.60	221.4 ± 11.2	3.51 ± 0.57	25.0
0.80	233.7 ± 26.0	3.53 ± 0.49	26.0

Mixtures of DMPC and cholesterol (CHOL) have been used to examine the effects of cholesterol on the chain crystallization transitions.[34] Vesicle area versus temperature plots (Fig. 15) show that the main transition broadens and shifts to higher temperatures as the concentration of cholesterol is increased in agreement with regular solution theory.[31] Similar results have

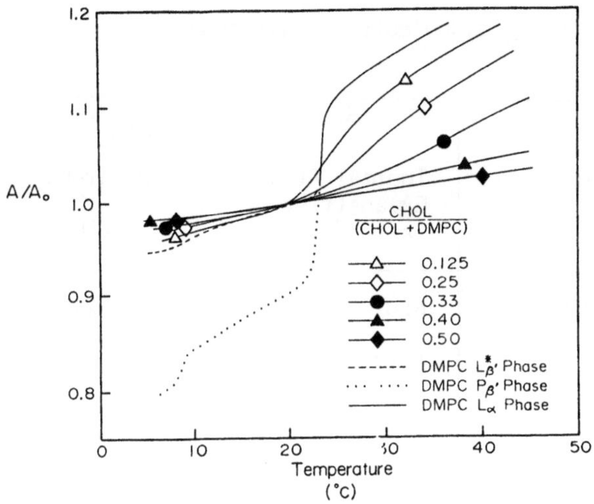

Fig. 15 Relative areas derived from single vesicle heating/cooling experiments for mixed DMPC:CHOL bilayers. Also shown is the area-temperature behaviour for pure DMPC vesicles from Fig. 5.

been obtained in scanning calorimetry and NMR studies.[35-36] Upon addition of cholesterol (even small amounts), the P_{β}' phase seems to disappear and a tilted geometry is indicated for isolated frozen lipid regions. Area changes over the broadened transitions are reduced by cholesterol and disappear with addition of 50 mol % to leave the thermal area expansivity at $1.3 \times 10^{-3}/°C$. The results in Fig. 15 are consistent with separate formation of a 1:1 (DMPC:CHOL) complex that does not condense - plus residual free lipid that condenses (freezes) normally.[37] Both above and below the broad transition, the elastic area compressibility modulus K is greatly increased with cholesterol addition (Table 2). The value for the 1:1 lipid:cholesterol complex is found to be 700-800 dyn/cm, comparable to that for DMPC in the L_{β}' crystalline phase. However for all concentrations above 12.5 mol % (which exhibits only very weak surface rigidity), cholesterol-rich bilayers behave as surface liquids with no surface rigidity even at temperatures well below the DMPC phase transition. Compared to bilayers in the liquid L_{α} state, introduction of cholesterol forms a tight complex with the lipid which greatly reduces bilayer compressibility (and permeability as evidenced by the extremely slow response of vesicles to osmotic dehydration) but maintains the bilayer in a liquid state well below the acyl chain crystallization temperature of the lipid. Similar results were obtained with the partially saturated lipid SOPC.

Addition of cholesterol to lipid bilayers not only reduces the area compressibility but also greatly enhances bilayer cohesion. This general feature has been seen in all of our vesicle studies, i.e. tension levels for rupture increase as the lipid area compressibility is reduced (e.g. Fig. 3). Figure 16 shows the correlation of rupture strength with compressibility moduli. The results indicate that critical fluctuations in surface density of the liquid phase may be the origin of failure.

In models of lipid-protein interaction, the mismatch between hydrophobic lengths of lipid chains and the hydrophobic peptide sequence plays a major role in the solution phase behaviour.[38-40] A few experiments on a simple

Table 2 "Thermoelastic Properties of Lipid:Cholesterol Mixtures"

Composition	K Area Compressibility Modulus (dyn/cm)	$\dot{\alpha}$ Thermal Area Expansivity (x 10^3/°C)	T Temperature (°C)
DMPC (L_α)	144.9 ± 10.5	6.81 ± 1.0	29
		4.17 ± 0.2	35
DMPC (L_{β}')	855.3 ± 140.1	1.0	8
$\frac{CHOL}{(DMPC+CHOL)}$			
0.125	396.9	2.83	15.5
0.33	646.8	1.97	15
	559.0	3.1	25
0.40	600	2.3	35
0.50	685	1.33	22
SOPC (L_α)	199.6 ± 12.7	3.28 ± 0.68	15
$\frac{CHOL}{(SOPC+CHOL)}$			
0.5	1077 ± 167	1.62 ± 0.16	23

lipid-peptide system have been carried out with two synthetic amphiphilic peptides: 16 and 24 hydrophobic leucine residues bounded by two lysines at the N and C terminals, i.e. long hydrophobic α- helices with hydrophilic ends.[43] The effects of peptide (at concentration of 1 mole%) on area compressibility modulus of both lipid and lipid:cholesterol bilayers are listed in Table 3. The introduction of $peptide_{16}$ into SOPC bilayers clearly increases the bilayer compressibility. Even though SOPC:CHOL (3:2) is much less compressible than pure lipid, the introduction of peptide produces an unexpectedly large increase in bilayer compressibility and reduced cohesion (Fig. 16). Here, it appears that the peptide contributes to chain-packing "defects" in the bilayer and interferes with formation of the lipid-cholesterol complex. Additional evidence that incorporation of the peptide alters the lipid-cholesterol interaction is obtained from measurements of area compressibility on DMPC:CHOL:$Peptide_{24}$ mixtures. At high temperatures (15-20°C above the main DMPC crystallization at 24°C), the lipid-cholesterol-protein vesicles commence a broad but significant area transition (not present for the lipid-cholesterol mixture); also, the area compressibility is much greater with the

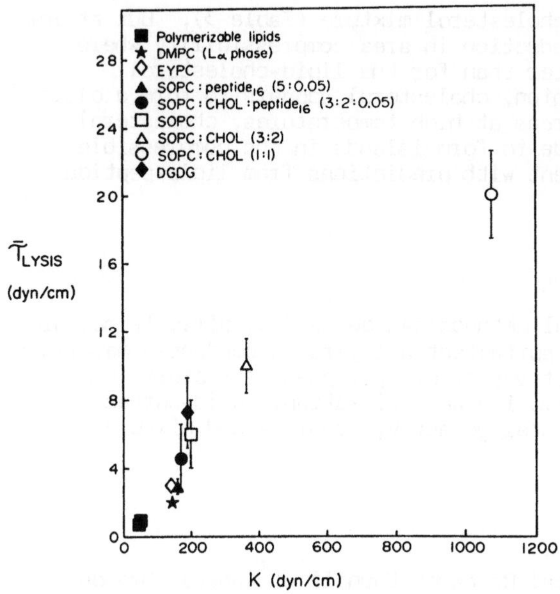

Fig. 16 Values for bilayer tension at vesicle rupture versus elastic area compressibility moduli for all lipid compositions studied to date.

Table 3 "Elastic Area Compressibility Modulus of Lipid, Lipid:Cholesterol, and Lipid:Cholesterol:Peptide Bilayer Mixtures"

Composition	K Area Compressibility Modulus (dyn/cm)	T Temperature (°C)
SOPC (L_α)	199.6 ± 12.7	15
SOPC:CHOL (3 : 2)	362.7 ± 20.6	15
SOPC:Peptide (5 : 0.05)	161.1 ± 23.6	15
SOPC:CHOL:Peptide$_{16}$ (3 : 2 : 0.05)	168.9 ± 24.6	15
DMPC (L_α)	144.9 ± 10.5	29
DMPC:CHOL (3 : 2)	600	35
DMPC:CHOL:Peptide$_{24}$ (3 : 2 : 0.05)	945 ± 218	12.5 ± 0.5
	245	29.5
	193 ± 31	34.4 ± 1.3

peptide present than for the lipid-cholesterol mixture (Table 3). But at low temperatures, there is a striking reduction in area compressibility where moduli increase to values even greater than for the lipid-cholesterol mixture. In the low-temperature region, cholesterol appears to form a close cohesive complex with the lipid whereas at high temperatures, cholesterol appears to associate with the peptide to form islands in the compressible liquid phase of the lipid - consistent with predictions from lipid-peptide wetting models.

6. SUMMARY

We have discussed how micromechanical methods can be used to directly measure surface-thermodynamic properties of surfactant bilayers as condensed materials. These methods provide unique perspectives of bilayer phase structure, e.g. stress-dependence of solid-phase formation and relaxation, solid-surface rheology, chain condensation and crystal geometry, impurity and mixture effects, strength and cohesion.

7. ACKNOWLEDGMENT

This work was supported by the Medical Research Council of Canada through Grant MT 7477. EAE is grateful to the Alexander von Humboldt Foundation, Federal Republic of Germany, for a Senior Scientist Award.

8. REFERENCES

1. R. Kwok and E. Evans: Biophys. J., 35, 637 (1981).
2. E. Evans and R. Kwok: Biochem., 21, 4874 (1982).
3. E. Evans and D. Needham: Faraday Discuss. Chem. Soc., No. 81 (1987, in-press).
4. E. Evans and R. Skalak: Mechanics and Thermodynamics of Biomembranes, (CRC Press, Boca Raton, Fla., 1980).
5. E. Evans: In NATO Advanced Study Institute Textbook on Biorheology, Hwang and Gross, Eds. (Sitjhoff and Noordhoff Int., 1981) p. 137.
6. E. Evans and R.E. Waugh: J. Colloid Interface Sci., 60, 286 (1977).
7. E. Evans: In Festkörperprobleme (Advances in Solid State Physics, Vol. 25), P. Grosse Ed. (Vieweg, Braunschweig, 1985) p. 735.
8. Z. Derzko and K. Jacobson: Biochem., 19, 6050 (1980).
9. P.G. Saffman and M. Delbruck: Proc. Natl. Acad. Sci. USA, 72, 3111 (1975).
10. H.J. Galla, W. Hartmann, U. Theilen and E. Sackmann: J. Memb. Biol., 48, 215 (1979).
11. E. Evans and R.M. Hochmuth: In Current Topics in Membranes and Transport, Vol. 10, F. Bronner and A. Kleinzeller Eds. (Academic Press, New York, 1978) p. 1.
12. M. Bloom and I.C.P. Smith: In Progress in Protein-Lipid Interactions, Watts, De Pont Eds. (Elsevier, 1985) p. 61.
13. J.P. Reeves and R.M. Dowben: J. Cell. Biol., 73, 49 (1969).
14. D. Needham and E. Evans: Biochem. (to be submitted).
15. R.M. Servuss, W. Harbich and W. Helfrich: Biochim. et Biophys. Acta., 436, 900 (1978).
16. H. Engelhardt, H.P. Durve and E. Sackmann: J. Phys. Lett. (Paris), 46, L395 (1985).
17. M.B. Schneider, J.T. Jenkins and W.W. Webb: J. Phys. (Paris), 5, 1457 (1984).
18. E. Evans and D. Needham: J. Colloid Interface Sci. (to be submitted).

19. V. Luzzati: In Biological Membranes, D. Chapman Ed. (Academic Press, New York, 1968) p. 71.
20. V. Luzzati and A. Tardieu: Ann. Rev. Phys. Chem., 25, 79 (1974).
21. S.C. Chen, J.M. Sturtevant and B.J. Gaffney: Proc. Natl. Acad. Sci. USA, 77, 5060 (1980).
22. M.J. Ruocco and G.G. Shipley: Biochim et Biophys. Acta, 691, 309 (1982).
23. M.J. Janiak, D.M. Small and G.G. Shipley: Biochem., 21, 4575 (1976).
24. J. Stamatoff, B. Fever, M.J. Guggenheim, G. Teller and T. Yamane: Biophys. J., 38, 217 (1982).
25. R. Krebecek, C. Gebhardt, M. Gruber and E. Sackmann: Biochim. et Biophys. Acta, 554, 1 (1979).
26. G.W. Brady and D.B. Fein: Biochim. et Biophys. Acta, 646, 249 (1977).
27. J.H. Davis: Biophys. J., 27, 339 (1979).
28. P.T. Wong: Ann. Rev. Biophys. Bioeng., 13, 1 (1984).
29. P.J. Davis, B.D. Fleming, K.P. Coolbear and K.M. Keough: Biochem., 20, 3633 (1981).
30. T.J. McIntosh: Biophys. J., 29, 237 (1980).
31. G.N. Lewis and M. Randall: Thermodynamics (McGraw Hill, New York, 1961).
32. E.J. Shimshick and H.M. McConnell: Biochem., 12, 2351 (1973).
33. D. Chapman, J. Urbina and K.M. Keough: J. Biol. Chem., 249, 2512 (1974).
34. D. Needham, T.J. McIntosh and E. Evans: Biochem. (to be submitted).
35. S. Mabrey, P.L. Mateo and J.M. Sturtevant: Biochem., 17, 2464 (1978).
36. M.R. Vist and J.H. Davis: (to be published).
37. F.T. Presti, R.J. Pace and S.I. Chan: Biochem., 21, 3831 (1982).
38. S. Marcelja: Biochim. et Biophys. Acta, 455, 1 (1976).
39. J.C. Owicki, M.W. Springgate and H.M. McConnell: Proc. Natl. Acad. Sci. USA, 75, 1616 (1978).
40. O.G. Mouritsen and M. Bloom: Biophys. J., 46, 141 (1984).
41. For chemically asymmetric bilayers, small induced bending moments act to wrinkle the membrane and produce tension in the surface.[4] Also, self-adherence of the membrane due to long-range attractive forces can cause the surface to wrinkle and to be stressed. In these situations, the unsupported vesicle maintains a weakly rigid spherical shape.
42. Some hysteresis in the area versus temperature curves is observed. At the (liquid)/(liquid-solid) boundary, T_{melt} is $< 1°C$ above T_{freeze} for heating and cooling rates of 0.5°C/min and increases to $< 2°C$ for higher rates of temperature change. At the (solid-liquid)/(solid) boundary, very little hysteresis is seen: $T_{melt}-T_{freeze} < 0.5°C$.

43. The peptides were gracious gifts from Dr. R. Hodges, University of Alberta, provided through Dr. M. Bloom, University of British Columbia.

Equilibrium Configurations of Fluid Membranes

W. Helfrich and W. Harbich

Fachbereich Physik, Freie Universität Berlin,
Arnimallee 14, D-1000 Berlin, Germany

It is shown that membranes in plenty of water, if they have certain properties, should assume one of three configurations in thermodynamic equilibrium: A lattice of passages (i.e. connections), a single big vesicle, or many small vesicles. The choice is controlled by the ratio of the elastic modulus of Gaussian curvature and the bending rigidity. The necessary membrane properties are specified and a few practical examples are given.

1. INTRODUCTION

Fluid membranes in a liquid environment, such as amphiphilic bilayers in water, can assume very different configurations. The lyotropic smectic phase, a dispersion of small vesicles, and cubic lattices formed by a single multiply connected membrane are among them. The same material (e.g. egg lecithin) may exhibit all those and other configurations, depending on sample preparation. It is of interest to know which of them, if any, represents equilibrium. Membrane bending elasticity furnishes a first answer to this question, provided the membrane has certain properties.

In the following we will show the equilibrium configuration of such membranes to be determined by the elastic modulus of Gaussian curvature, in particular by its sign and its size relative to the ordinary bending rigidity. Subsequently we will briefly discuss two examples. One is spontaneous vesiculation, a few years ago discovered by H. HAUSER [1] and F. EVANS [2] together with their coworkers. The other is the lattice of passages, i.e. membrane connections, formed by egg lecithin/water systems. New experiments, while not complete, conform with our earlier observation of such structures [3, 4]. Moreover, they strongly suggest that the modulus of Gaussian curvature of egg lecithin bilayers changes sign as a function of temperature.

2. BENDING ELASTICITY

The elastic energy of membrane bending per unit area may be written as [5]

$$g = \tfrac{1}{2}\kappa(c_1 + c_2 - c_0)^2 + \bar{\kappa}c_1c_2 \,. \qquad (1)$$

Here c_1 and c_2 are the principal curvatures ($c_i = 1/r_i$, r_i = principal radii of curvature), c_0 is the spontaneous curvature, κ the bending rigidity, and $\bar{\kappa}$ the elastic modulus of Gaussian curvature. The integral of Gaussian curvature over a closed surface is a topological invariant (Gauss-Bonnet theorem). It is 4π for a sphere and all shapes into which a sphere can be deformed without cutting the membrane. For a torus and derived shapes the integral vanishes. A general formula for it is

$$\oint c_1 c_2 dA = -4\pi(n-1),$$

where n is the number of "handles", i.e. of connections the surface makes with itself. We call these connections "passages", especially if they are between parallel bilayers. Obviously, the elastic energy of Gaussian curvature associated with a passage is

$$E_{\bar{\kappa}} = -4\pi\bar{\kappa}$$

and may be called the energy of membrane fusion. The definition will be seen to be exact for lattices of passages. It seems also good in the case of a passage connecting two parallel membranes. The elastic energy associated with the bending rigidity can then be made arbitrarily small by reducing the radius of the passage towards zero [3]. We have to keep in mind, however, that physically there is a lower limit to the size of a passage as to that of a vesicle (see below).

Only the case of positive rigidity, $\kappa > 0$, is considered here. (Negative rigidities are not known to us and would require a more complex theory.) There is reason to believe, however, that the modulus of Gaussian curvature can be negative as well as positive. This follows from [4]

$$\bar{\kappa} = \int x^2 s(x)\,dx \qquad (2)$$

and similar relationships [6] for membranes with a well-defined stress profile. The right-hand side is the second moment of the normal stress s(x) (dyn cm^{-2}) along a line perpendicular to the flat membrane. The formula applies if the two lower moments, i.e. lateral tension and spontaneous curvature of the bilayer, are zero. The stress profile includes the negative stress (i.e. pressure) of any electrostatic double layers apposed to the membrane. This will turn out to be relevant to spontaneous vesiculation.

3. EQUILIBRIUM MEMBRANE CONFIGURATIONS

Membranes have to break and close again in order to reach configurational equilibrium. These processes also ensure uniformity of the aqueous medium and zero spontaneous curvature. We still assume that there are practically no membrane edges. (If they are too frequent the membranes disintegrate into oblate micelles.) In addition, we consider only unilamellar structures. This seems reasonable if the interaction between membranes is repulsive at all spacings, e.g. because of long-range undulation forces [7]. Interaction energies, however, are supposed to be negligible relative to bending elastic energies. Finally, the membranes are taken to be fairly stiff ($\kappa > 10^{-13}$ erg, see below). Lecithin bilayers seem to satisfy all these criteria, apart, perhaps, from breaking too rarely.

Let us distinguish three possible equilibrium configurations of fluid membranes in plenty of liquid:

1. One big vesicle: The bending elastic energy of a vesicle is minimized for the spherial shape whenever $\kappa > 0$. The energy of the sphere is

$$E = 4\pi(2\kappa + \bar{\kappa})$$

as follows immediately from (1).

2. Many small spheres: This is the case of spontaneous vesiculation. The total elastic energy is given by

$$E = N_{ves} 4\pi(2\kappa + \bar{\kappa}),$$

where N_{ves} is the number of vesicles. For the present purpose, N_{ves} may be thought to be determined by the quantity of amphiphile and a minimum vesicle size given by a "hard" cut-off.

3. A lattice of passages: Here a multiple connected single membrane separates two equal, interwoven water networks. Very recently, J. CHARVOLIN [8] and H.T. DAVIS [9] have established by different means that there is a minimal surface ($c_1 + c_2 \equiv 0$) of this kind for every space group. Whenever the membrane forms a minimal surface its bending energy is simply given by

$$E = - N_{pass} 4\pi\bar{\kappa} \ .$$

We assume the number of passages, N_{pass}, to be controlled by a minimal size of the unit cell of the lattice which we treat again as a hard cut-off.

In thermodynamic equilibrium the membrane is in its configuration of lowest elastic energy. Therefore, we expect passages for $\bar{\kappa} \geq 0$, one big vesicle for $\bar{\kappa} < 0$, $2\kappa + \bar{\kappa} > 0$, and many small vesicles for $\bar{\kappa} < 0$, $2\kappa + \bar{\kappa} < 0$. The locations of the three regimes in the 2κ, $\bar{\kappa}$ plane are shown in Fig. 1. Clearly, at a given rigidity $\kappa > 0$ each of them can occur depending on sign and magnitude of the modulus of Gaussian curvature, $\bar{\kappa}$.

The planar multilayer system is not contained in our scheme. It is a poor alternative to the single big vesicle and has to be taken into consideration if the vesicle would be bigger than the total sample volume. The high energy of membrane edges enforces some kind of membrane closure at the border of the multilayer system. (In fact, a semicylindrical structure of concentric membranes has been observed with lecithin-water systems [10]). The bending elastic energy of a "closed" border is readily seen to surpass that of a single sphere, even if the system consists of only two parallel membranes. For certain sample geometries (e.g. for cubic cells) spherical or cylindrical liposomes, i.e. multilamellar vesicles, may be energetically more favorable than the planar multilayer system.

The translational entropy of vesicles and membrane fluctuations were entirely omitted in setting up the classification of Fig. 1. They should not matter much whenever the rigidity exceeds a few or several kT unless the excess of water is enormous. The next paragraphs serve to show this.

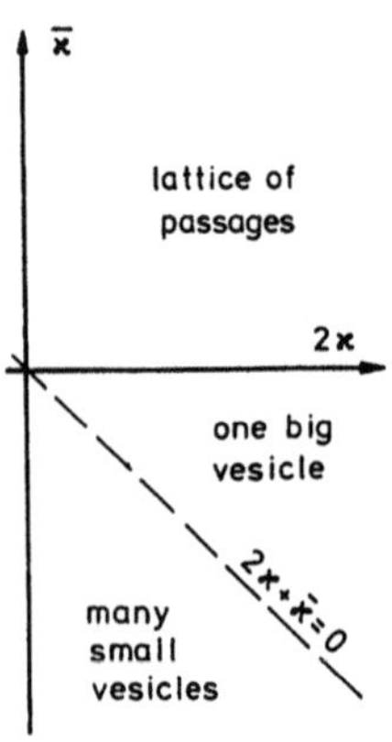

Figure 1

Locations of equilibrium configurations in 2κ, $\bar{\kappa}$ plane

Due to the translational entropy of vesicles and other entropies, the broken line $2\kappa + \bar{\kappa} = 0$ is not a real phase boundary. The transition from a single big vesicle (or very dilute multilayer system) to many vesicles of minimum size is gradual, the big vesicle breaking up into smaller and smaller ones. However, the concentration of vesicles should in general be very little for $2\kappa + \bar{\kappa}$ larger than a few kT. This limit may be verified by minimizing with respect to N_{ves} the following approximate formula for the total free energy G of a vesicle solution

$$G = N_{ves}[4\pi(2\kappa + \bar{\kappa}) - kT \ln \frac{N_{site}}{N_{ves}} - kT \ln \frac{N_{tot}}{N_{ves}}], \tag{3}$$

N_{ves} being the number of vesicles. The bracket of (3) refers to one of the vesicles and comprises its elastic energy and entropic terms due to the translational freedom and the size distribution [11] of the vesicle. kT is Boltzmann's constant times temperature and N_{tot} the total number of lipid molecules in the sample. N_{site}, the number of sites for the vesicles in a lattice model, may be put equal to the number of water molecules in the sample if the amphiphile is dilute enough to permit an ideal gas approximation.

It is also clear that a lattice of passages in excess water can, in principle, coexist with one or more vesicles. The vesicle dispersion should be very dilute for κ larger than a few or several kT, even as $\bar{\kappa}$ comes down to zero. The mean vesicle size will be near the minimum size as long as $\bar{\kappa}$ is sufficiently positive. This is because one has to destroy passages in order to create vesicles. We do not discuss here the disordered bicontinuous phase which may occur in a vicinity of a few kT around the origin of the 2κ, $\bar{\kappa}$ plane. Its monolayer counterpart in oil/water/surfactant systems is the well-known bicontinuous microemulsion. It can be understood in terms of monolayer bending elasticity and configurational fluctuations [12, 13].

A final remark concerns the effective rigidity, κ_{eff}, as opposed to the bare one, κ_o. The two are related by [11]

$$\kappa_{eff} = \kappa_o - \frac{kT}{8\pi} \ln \frac{N}{2}$$

in a first order approximation, where N is the number of molecules constituting the spherical vesicle. The distinction should not matter for $\kappa > kT$ unless the vesicle is very big. It seems important in the case of bicontinuous microemulsions [13]. In this context one may also wonder whether there is a size-dependent entropy of vesicle closure [14]. Polymer closure, e.g. when a loop fissions from an infinite line, becomes less likely, i.e. costs more entropy, as the length of the loop increases. There seems to be no size-dependent entropy effect as we create a new vesicle from a big membrane by subsequent deformation, constriction, and fission. This is because the constriction feels only the local bending fluctuations. Any size-independent effect could be lumped with $\bar{\kappa}$.

4. EXAMPLES

Let us briefly discuss examples of spontaneous vesiculation and bicontinuity. Spontaneous vesiculation has been discovered by HAUSER and GAINS [1] with egg phosphatidic acid when they increased the pH of the aqueous medium up to double dissociation of the acid. It was also found by TALMON, EVANS and NINHAM [2] with didodecylammonium hydroxide. In both systems the fluid membranes made of double-chain molecules are strongly charged, negatively in one case and positively in the other. The accompanying electrostatic double layers on the sides of the membrane tend to make the modulus of Gaussian curvature negative, apparently enough so to satisfy $2\kappa + \bar{\kappa} < 0$. In estimates of the electrostatic contribution to $\bar{\kappa}$ one may distinguish between an inner part of the double layer with a large potential drop and an outer part in which the Debye-Hückel approximation holds. The former contribution seems to be the more important (except perhaps for very large Debye lengths near 1 μm). It may reach a few times minus 10^{-12} erg, which seems necessary to outweigh 2κ if the rigidity is (1 to 2)$\cdot 10^{-12}$ erg as measured for lecithin membranes [15]. The estimates involve Maxwell stresses, ion gas pressures, and compensating dilatational stresses. There is also a change of κ, possibly to smaller values, which we cannot estimate.

The equilibrium configuration of electrically neutral egg lecithin membranes is probably a lattice of passages. Optically resolvable structures of this sort have been observed some time ago [3, 4]. They consisted of two interposed diamond lattices of water channels separated by a multiply connected membrane (space group no. 224, i.e. Pn3m, if the chirality of the molecules is disregarded). The lattices were seen only occasionally and had different lattice parameters. Nothing is known about the process of formation of the passages which, when visible, did not change in number or size. The lattices as seen are unlikely to represent thermodynamic equilibrium. They may have resulted from the expansion of a much denser lattice which, perhaps, was in equilibrium. In any event, the occurrence of huge numbers of passages, even if rarely observed, indicates that the membrane tends to fuse, which implies $\bar{\kappa} > 0$.

Very recently HARBICH and HELFRICH [16] found methods to reproducibly obtain, in swollen egg lecithin, structures which seem to be dense lattices of passages. These so-called "dark bodies" could not be resolved under the light microscope. A lattice order may be inferred from the fact that some of them were cornered. With one of the methods the "dark bodies" originated only along dislocations traversing the stack of membranes, thus making the defect lines visible. The dark bodies tended to vanish below ca. 15°C and above ca. 65°C. The observations appear to confirm that the equilibrium configuration of egg lecithin bilayers in excess of water is a lattice of passages, but only in the indicated temperature interval. In other words, they suggest that $\bar{\kappa}$ is positive at room temperature, but changes sign both at lower and higher temperatures.

Up to three regions have been distinguished in the stress profile of electrically neutral monolayers [6]: The hydrocarbon chains with a negative stress because of parallel chain ordering, the hydrocarbon/water interface with a δ-function like positive stress, and the polar heads with a negative stress due to a mutual repulsion of the heads. If the last region is neglected, as we have done previously [3, 4], eq. (2) predicts $\bar{\kappa}$ to be generally positive. If it is taken into account, both signs of $\bar{\kappa}$ become possible. However, a detailed theory will be necessary to understand the suspected twofold change of sign with temperature.

5. CONCLUSION

Our distinction of three equilibrium configuration of fluid membranes in plenty of water should be valid without major reservations if the membranes do not cohere, have negligible interaction energies, and are stiff enough. While there may be many such membranes it appears difficult, at least in the case of phospholipids, to go from one configuration to another so that thermodynamic equilibrium is not easily established. The main reason for this inertia seems to be the stability of the membranes which hardly break or fuse spontaneously. Membrane stability may also prevent the breakup of multilamellar structures such as liposomes into single lamellae, thus simulating membrane cohesion where actually none exists.

REFERENCES

1. H. Hauser, N. Gains, and M. Müller, Biochemistry 22, 4775 (1983)
2. Y. Talmon, D.F. Evans, and B.W. Ninham, Science 221, 1047 (1983)
3. W. Harbich, R.-M. Servuss, and W. Helfrich, Z.Naturforsch. 33a, 1013 (1978)
4. W. Helfrich, in Les Houches, Session XXXV, 1980 - Physics of Defects, R. Balian et al., eds., North-Holland (1981), p. 715
5. W. Helfrich, Z.Naturforsch. 28c, 693 (1973)
6. A.G. Petrov and I. Bivas, Progress in Surface Sci. 16, 389-512 (1984)
7. W. Helfrich, Z.Naturforsch. 33a, 305 (1978); R. Lipowsky and S. Leibler, this Workshop
8. J. Charvolin, this Workshop
9. H.T. Davis, this Workshop
10. W. Harbich and W. Helfrich, Chem.Phys. Lipids 36, 39 (1984)
11. W. Helfrich, J. Physique 47, 321 (1986); 48, 285 (1987)
12. P.G. de Gennes and Ch. Taupin, J.Phys.Chem. 86, 2294 (1982)
13. S.A. Safran, D. Roux, M.E. Cates and D. Andelman, Phys.Rev.Lett. 57, 491 (1986)
14. The question was raised by M.E. Fisher.
15. G. Beblik, R.-M. Servuss, and W. Helfrich, J. Physique 46, 1773 (1985) and references cited therein
16. W. Harbich and W. Helfrich, in preparation

Dynamics of Drying and Film-Thinning

P.G. de Gennes

Collège de France, F-75231 Paris Cedex 05, France

The spreading of non-volatile liquids on ideal, wettable, solid surfaces is relatively well understood[1]. Here we transpose the same theoretical ideas to two problems where a metastable liquid film tends to thin out.

a) A thick soap film may nucleate a patch of much thinner "black" film;

b) A liquid film lying on a non-wettable solid may nucleate a dry patch.

We consider here the dynamics of growth of the patch. They are basically the same for (a) and (b). The liquid which is expelled from the patch builds up a rim, which acts as a strong perturbation. The contact angle θ between patch and rim remains constantly much smaller than the equilibrium value θ_e ($\theta \to 0.7\ \theta_e$). The patch expands with a constant radial velocity, but the rim width b(t) (at time t) is expected to grow only like $t^{\frac{1}{2}}$.

I. PARTIAL / COMPLETE WETTING

Wetting and drying processes are important for many practical processes, involving paints, textiles, oil recovery, or the reinforcement of concrete by plastics... The simplest experiment amounts to putting a droplet on a flat solid, and watching its progressive spreading. But this requires a lot of care; impurities and surface irregularities can complicate matters enormously. Thus it is only recently that general laws have been clearly exhibited by experiments, and that the main dynamical classes have been identified.[1]

Let us start with a brief reminder of the equilibrium shape for a small liquid droplet on a horizontal solid plate (gravitational effects negligible).

In situations of partial wetting the optimal state is a spherical cap, with a certain contact angle θ_e (fig.1). This angle is related to the interfacial energies γ_{SV} (solid/vapor), γ_{SL} (solid liquid) and $\gamma_{LV} = \gamma$ (liquid vapor) by the Young condition[2]

$$\gamma_{SL} + \gamma\cos\theta_e = \gamma_{SV} \,. \tag{1}$$

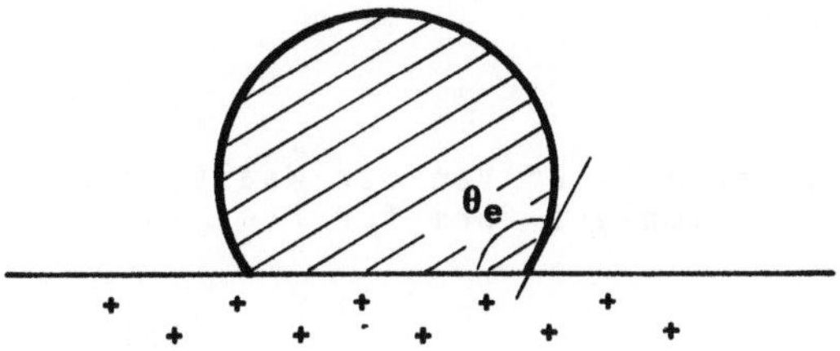

Fig.1 PARTIAL SPREADING

When $\gamma_{SV} - \gamma_{SL} > \gamma$ this gives a finite angle θ_e. On the other hand, when $\gamma_{SV} - \gamma_{SL} < \gamma$, no real angle θ_e exists, and we enter a completely different regime, called complete wetting. To understand the physics chemical factors underlying this distinction, see the review by Zisman[2]. Complete wetting may occur in two distinct regimes.

a) Moist spreading (when we have complete equilibrium with the vapor). Here, if the solid vacuum interfacial energy γ_{SO} is large, the real solid vapor interface will display an adsorbed liquid film, and the corresponding (renormalised) energy is

$$\gamma_{SV} = \gamma_{SL} + \gamma \cdot \qquad (2)$$

This implies that $\theta_e = 0$ exactly, and is known as the Antonov rule.

b) In dry spreading (no vapor) this renormalisation does not take place, eq.(2) does not hold, and the deviations from it are measured by the so-called spreading parameter

$$S = \gamma_{SO} - \gamma_{SL} - \gamma \cdot \qquad (3)$$

This S is positive for complete dry spreading.

What is the final state of a spreading droplet for complete wetting?

a) For moist spreading this is not a well-posed question: to maintain the liquid/vapor equilibrium in the experimental cell we need a reservoir of liquid, and an essential parameter is the difference H in altitude between the spreading plate and the reservoir. The final film thickness depends on H.[1]

b) For dry spreading, there is a well-defined answer, provided that S is known[3]. The final state is a thin film or "pancake" of area a, thickness e, (fig.2).

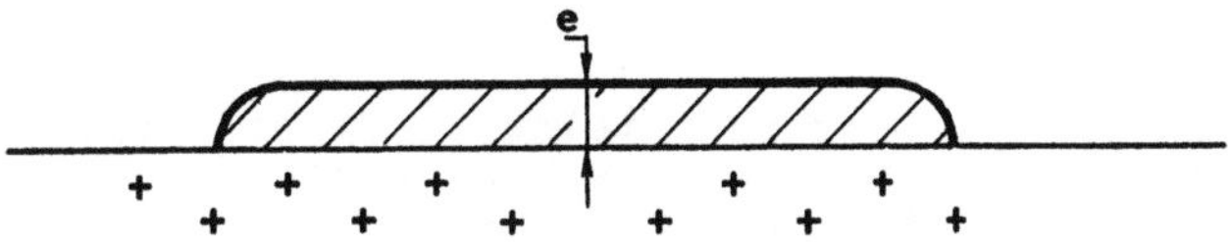

Fig. 2 "PANCAKE"

The free energy of the pancake is:

$$\mathcal{f} = \mathcal{f}_0 - S\mathfrak{a} + \mathfrak{a}P(e), \tag{4}$$

where P(e) represents the long-range effects of Van der Waals forces (which tend to thicken the film). The simplest form for P(e) is

$$P(e) = \frac{A}{12\pi e^2}, \tag{5}$$

where A is called the Hamaker constant[4]. Minimising (4) at constant volume (e) one finds

$$e = a\left(\frac{3\gamma}{2S}\right)^{\frac{1}{2}}, \tag{6}$$

where a is a molecular length defined by

$$a^2 = \frac{A}{6\pi\gamma}. \tag{7}$$

For small S/γ, this leads to sizeable thicknesses, which could be probed by ellipsometry: since He_4 spreads fast on solids, it may be a good candidate for these experiments. From general arguments[2] one should choose a solid of low polarisability (possibly H_2 solid) to reduce the S.

These considerations can be extended to films on vertical walls. The classical description of these films, constructed by the Russian school[5], and well verified on ^4He by thickness measurements[6] is based on a competition between Van der Waals and gravitational energies. However, this film should be truncated at some finite altitude, where the thickness reduces to the value of eq.(6): again an experimental check on ^4He should be extremely interesting, but the altitudes involved are large (meters).

II. DYNAMICS OF DRY SPREADING

Macroscopic observations on a spreading droplet indicate that the shape near the (nominal) contact line is not far from a simple wedge, with a certain "apparent contact angle θ_a (fig.2). Experiments on viscous liquids pushed into capillaries or on spreading droplets[7,8,9] show that, in many cases of dry spreading, the angle θ_a is related to the line velocity U by

$$\theta_a^3 \sim U\eta/\gamma, \qquad (U \to 0) \tag{8}$$

where η = viscosity. The law (8) is very surprising at first sight. In the language of irreversible thermodynamics we may write the entropy source $T\mathring{\Sigma}$ (per unit length of line) in terms of a flux (U), and of a conjugate force F, which is the unbalanced Young force:

$$T\mathring{\Sigma} = FU = [S + \gamma(1-\cos\theta_a)]U \cong [S + \tfrac{1}{2}\gamma\theta_a^2]U. \tag{9}$$

How is it that the major component in the force (S) does not show up in the result (8) ? The answer starts from an old observation by Hardy (1919) - summarised in Marmurs' review[9]: there is a precursor film moving ahead of the nominal contact line, with typical thicknesses 100 Å. The origin of this film is relatively obvious: it results from a balance between viscous forces and long-range Van der Waals forces. For the simplest situation (when eq.5 holds) the thickness $\zeta(x)$ of the film at a distance x from the contact line decreases like $\zeta \sim 1/x$.[1] For more complex structures of the VW forces, numerical calculations of the film have been performed[10].

Two essential features emerge:

a) the precursor film is truncated: it stops abruptly when the local thickness becomes comparable to e(S) as given in eq.(6);

b) the dissipation in the film is huge. In fact, all the available free energy described by S is burned out in the film region. It is only the weak force $\frac{1}{2}\gamma\theta_a^2$ which operates in the macroscopic wedge, and this explains why S does not show up in the macroscopic law (8).

III. THINNING OF FILMS AND DRYING OF WET SURFACES

1) Situations of metastability

In many practical situations we deal with fluid films which are metastable:

a) A thick soap film (fig.3) floating on or in a passive solvent tends to reach a much smaller equilibrium thickness ("black" film) corresponding to a minimum in it's long-range energy P(e)[11]. Here the Van der Waals contribution to the energy P(e) is negative (the film is in a symmetric environment) and the minimum (e_m) results from a competition: Van der Waals attraction against short-range repulsions between the two sides (fig.4). If a black film has nucleated at some point, it will expand later as shown in fig.3.

b) A liquid film deposited on a non-wettable solid (S < 0) is metastable: if a dry patch has nucleated (usually because of defects, or of mechanical noise) at some point on the solid, it will also expand (fig.5).

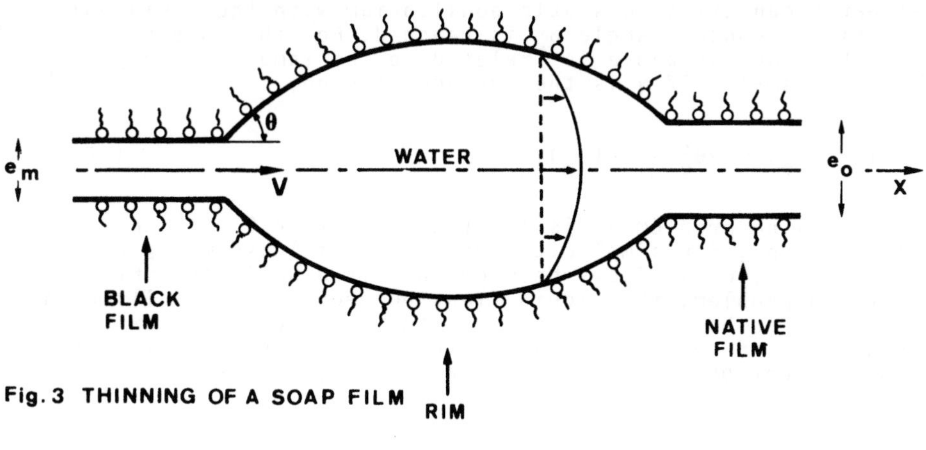

Fig. 3 THINNING OF A SOAP FILM

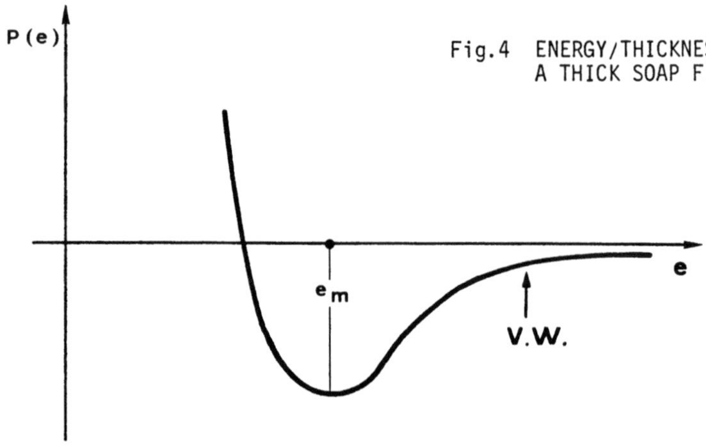

Fig.4 ENERGY/THICKNESS DIAGRAM FOR A THICK SOAP FILM

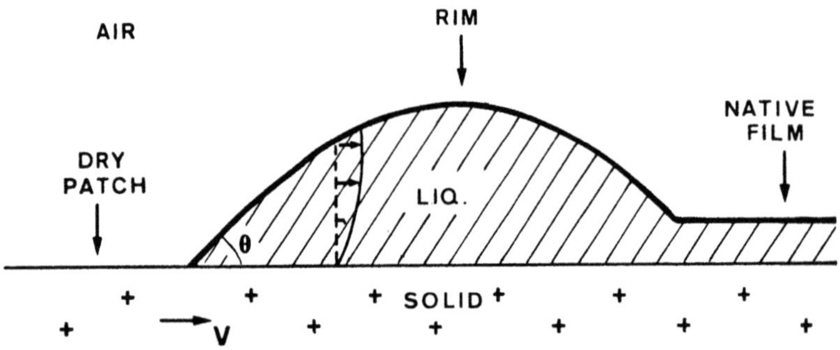

Fig.5 A SOAP FILM
DRYING OF A NON-WETTABLE SOLID

2) Contact angles and Poiseuille flows

In the present text, we are not concerned by the nucleation process, but only by the growth of the new region (black film or dry patch). F. Brochard was the first to notice that both systems are ruled by the same equations.[12] To understand this, we note first that the central patch can exist in static equilibrium with the fluid only when a definite contact angle θ_e is reached. For the soap film problem, this contact angle is related to the (small) energy $P(e_m)$ and is itself small. If γ is the surface tension,

$$2\gamma(1-\cos\theta_e) \sim \gamma\theta_e^2 = -P(e_m). \tag{10}$$

For the drying problem θ_e is related to S, $(S < 0)$, and need not be small: however, for simplicity, we shall still assume $\theta_e \ll 1$. In the soap film, water flows between two fixed walls of surfactant. In the drying problem, the liquid flows between a solid wall and a free surface. For both cases, we have simple Poiseuille flows, and the discussions of section II can be transposed. As before, we restrict our attention to one-dimensional flows (along x).

3) Dynamics of the rim

A rim must appear ahead of the patch, and store all the liquid which has been expelled from the patch. As before, we expect the pressure in the rim to uniformize fast: then the rim cross section is a portion of a circle[13], making an angle θ with the flat patch, and an angle ϕ with the unperturbed film (fig.6). For clarity we shall now discuss the film case which was analysed first[14]. The velocity V_F of the patch border has the form

$$V_F = V^*(\theta_e^2 - \theta^2)\theta . \quad (V^* = \text{const.}\gamma/\eta) \tag{11}$$

Eq.(11) is derived from eq.(9), applied here to a condition of partial wetting $2\gamma(\cos\theta - \cos\theta_e)$ is the uncompensated Young force. In situations of partial wetting, there is no precursor film, and all the dissipation takes place in the wedge of angle θ.

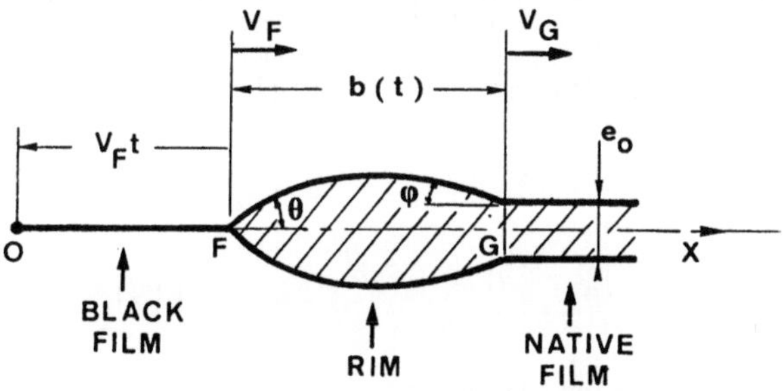

Fig.6 GEOMETRY OF PATCH AND RIM

A similar argument gives the velocity V_G of the outer end of the rim

$$V_G = V^*\phi^3 \tag{12}$$

because, at point G, the uncompensated Young force is simply $2\gamma(1 - \cos\phi)$.

We now assume, for simplicity, that the thickness of the patch (e_m) is very small ($e_m << e_o$), and set $e_m = 0$. The arc FG being a portion of a circle, we find

$$FG = b = \frac{e_o}{\theta - \phi} . \tag{13}$$

and a cross sectional area for the rim

$$A = \frac{e_o^2(2\theta - \phi)}{3(\theta - \phi)^2} . \tag{14}$$

The conservation of the volume inside the soap film imposes

$$A = (V_F t + b)e_o , \tag{15}$$

where t is the time counted from the instant where F was at $x = 0$.

4) Fully developed growth

Eq.(15) shows that, when the time t increases, the cross sectional area A of the rim becomes large. Thus, returning to eq.(14), we see that the angles θ and ϕ must become equal $\theta(\infty) = \phi(\infty) = \theta_\infty$.

We shall check later the fact that the rim width b grows more slowly than the size of the dry region (Vt). If we accept this assumption, we must have (at large t) $V_F = V_G = V$. Eqs.(11) and (12) then give

$$\theta_\infty = \theta_e/\sqrt{2} \,, \tag{16}$$

$$V = 2^{-3/2} V^* \theta_e^3 \,. \tag{17}$$

For a typical soap film, $V^* = 3.10^3$ cm/sec and $\theta_e = 10^{-2}$, giving velocities V of order 10 microns/sec. Knowing V, we can now return to a discussion of the rim width b(t). Eq.(15) is practically equivalent to $A = Vte_o$, and this fixes the difference in angles through eq.(14)

$$\varepsilon \equiv \theta - \phi \sim \left(\frac{\theta_\infty e_o}{3Vt}\right)^{\frac{1}{2}} , \tag{18}$$

$$b = e_o \varepsilon^{-1} = (3e_o Vt/\theta_\infty)^{\frac{1}{2}} \,. \tag{19}$$

All this holds when $\varepsilon \ll \theta_e$, or equivalently when

$$Vt \gg e_o/\theta_e \tag{20}$$

and we can then check that $b/Vt = \varepsilon/\theta_e$ is indeed small as announced. When (20) is satisfied, we say that we have a fully developed growth.

IV. CONCLUSIONS

The thinning and drying processes should be ruled by rather simple laws, with a constant growth velocity V and a rim growing more slowly ($b \sim \sqrt{t}$). The major surprise is that, even at large times, the contact angle θ does not reach the equilibrium value: the growing rim creates a strong perturbation.

The velocity V depends critically on θ_e: for the soap films, with very small θ_e values, V is a slow velocity. But for the drying problem, where θ_e may be large, the velocity V may be of order 1 meter/sec.

Clearly we need systematic studies on both systems, with a "patch" which is nucleated by some external means (e.g. a focused laser pulse).

a) free soap films suspended in air are difficult to manipulate, and are very sensitive to parasitic effects of gravity (even if the film is nominally horizontal). One should preferably use a water film embedded in an organic solvent (comparable to the films present in a polygonal emulsion O/W with high oil content): here, by a suitable choice of density, the gravity problem can be efficiently suppressed.

b) for the drying problem, the problem is to work with a clean solid surface - avoiding any pinning of the contact line[15]. But drying is of such importance in applications that a systematic program would really deserve to be launched. Another geometry of interest is obtained with thin fibers, for which a theoretical discussion has recently been produced[12].

We should emphasize that our discussion of drying was limited to a one-component liquid: more complex cases, with added surfactants or polymers, will require a separate analysis.

Acknowledgements: I have benefited from discussions with F. Brochard and H. Princen.

REFERENCES

1. P.G. de Gennes: In Rev. Mod. Phys. 57, 827, (1985)
2. W. Zisman: Advances in Chemistry 43, p.1, (ACS 1964)
3. J.-F. Joanny, P.G. de Gennes: CR Acad. Sci.(Paris) 299 II, 279, (1984)
4. See J. Israelashvili: Intermolecular and Surface Forces, Academic Press, NY, (1985)
5. L. Landau, I. Lifshitz: In Statistical Physics, Pergamon Press
6. D.F. Brewer: In "Physics of Solid and Liquid Helium", (Benneman Ketterson eds.), Wiley 1978
7. R. Hoffmann: In J. Colloid Interface Sci. 50, 228, (1975)
8. L. Tanner: In J. Phys. D 12, 1473, (1979)
9. A. Marmur: In Adv. Colloid Interface Sci. 19, 75, (1983)
10. G. Teletzke, phD Minneapolis 1983
11. K. Mysels, K. Shinoda, S. Frankel: In "Soap films", Pergamon, Lond., (1959)
12. F. Brochard: In C.R. Acad. Sci.(Paris) to be published
13. The pressure p in the rim is constant; outside we have the atmospheric pressure p_a. The difference $(p_a - p) = \gamma/R$ where R is the radius of curvature of the profile. Hence R is constant.
14. P.G. de Gennes: In C.R. Acad. Sci.(Paris), II 303, 1275, (1986)
15. J.-F. Joanny, P.G. de Gennes: In J. Chem. Phys. 81, 552, (1984)

Part II

Surface Thermal Fluctuations

Curvature and Fluctuations of Amphiphilic Membranes

S. Leibler[1*], *R. Lipowsky*[2], *and L. Peliti*[3]

[1]Baker Laboratory, Cornell University, Ithaca, NY 14853, USA
[2]IFF-KFA Jülich, D-5170 Jülich, Fed. Rep. of Germany
[3]Dipartimento di Fisica, Università "La Sapienza", I-00185 Roma, Italy

I. INTRODUCTION

Bilayer membranes and monolayer films are the simplest examples of various structures which are formed by amphiphilic molecules dissolved in water, or spread on an interface between water and another medium, such as oil or air. In order to describe these structures from a statistical point view, e.g. to obtain their phase diagram, or the dynamics of their fluctuations, it is necessary to introduce important simplifying assumptions.

The most fruitful approach consists in describing the membranes as two-dimensional, continuous objects and in constructing a statistical mechanics of such "surfaces". This approach would be particularly adequate for membranes in which internal degrees of freedom (e.g. configurations of hydrocarbon chains) do not play any important role. Such idealized, "thin" membranes will be called here *t-membranes*, in contrast to more realistic objects, for which one does not neglect completely their internal structure. In the statistical description of such "fat" membranes, or *f-membranes*, one should indeed add terms to the purely bidimensional free energy expression, such as the energy of stretching or tilting of the chains.

This paper reviews some recent developments in the theoretical modeling of behaviour of the membranes. We shall concentrate here mainly on the case of t-membranes, and put a special emphasis on the role of thermal fluctuations in the behaviour of such structures. After discussing in Section 2 the case of a single t-membrane, we shall consider two or more t-membranes in interaction (Section 3). Finally, to illustrate the concept of f-membranes, and the possibility of generalization of the continuum description to these systems, we shall consider in the last section a few other examples, such as the hexagonal phases in lyotropic liquid crystals.

II. SINGLE t-MEMBRANE

The formula which has been shown most often during this conference is due to Canham, Helfrich, and others[1] ; it gives an expression for the free energy density f of an elastic membrane :

$$f = \frac{1}{2}\kappa\,(H - H_0)^2 + \bar{\kappa}\,K, \tag{1}$$

where κ and $\bar{\kappa}$ are rigidity constants, H and K are local mean and Gaussian curvatures, and H_0 is the spontaneous curvature of the membrane. Equation (1) is the basis of the continuum description of idealized t-membranes. In fact, one can build the equilibrium theory starting from the statistical sum :

---

* On leave from: SPhT, CEN Saclay, 91191 Gif-sur-Yvette, France

$$Z = \int Du \, e^{-\beta \int f[u] \, dS} , \qquad (2)$$

where the integral is performed over all allowed configurations u of the membrane, and possibly subject to some additional constraints (e.g. fixed enclosed volume, self-avoidance condition etc). In Eq.(2) one neglects other possible terms in the free energy density functional; and in particular the term proportional to the area $A = \int dS$ of the membrane. One usually evokes the following argument in neglecting this "surface tension" term[2]: the membrane can adjust its total area in such a way that each amphiphilic molecule minimizes its energy ϕ with respect to the area Σ it occupies ; the surface tension term being equal to $\partial\phi/\partial\Sigma$ vanishes therefore at this point (provided the membrane does not exchange molecules with a reservoir, the membrane is not under a mechanical tension[2], etc.). This argument disregards several possible complications, such as: (i) the effect of thermal fluctuations, e.g. fluctuations in local density of amphiphiles; (ii) non-trivial boundary effects e.g. for closed vesicles or suspended films. The fact that in many experimental situations the area term seems not to play any important role may be connected to the fact that the minimum of $\phi(\Sigma)$ is very deep compared to thermal energies: the two-dimensional fluid membrane is highly incompressible[3]. It is important to notice that the absence of the area term in eq.(2), does *not* mean that the bulk phase built of membranes has a vanishing surface tension, if it coexists for instance with another phase (e.g. a bulk solvent).

The theories based on eqs.(1) and (2) were used to describe several phenomena observed for a single isolated membrane. The best known examples are:

(i) the equilibrium shape of closed vesicles [4,5]; shapes similar to that of red blood cells can be obtained by minimization of the curvature (bending) energy with appropriate constraints (e.g. constant enclosed volume).

(ii) the undulations of the enclosed vesicles[5,6], or microemulsion droplets[7]. The spectrum of such fluctuations (known in red blood cells as the "flicker phenomenon"[8]) can be obtained from simple hydrodynamic theories which include the energy given by eq.(1) as the dominant contribution[9].

In exploring any statistical theory based on eq.(1), a natural question arises: how do thermal fluctuations change the elastic properties of the membranes? In particular, what is their influence on the actual value of the rigidity constant κ?

To obtain an answer to this question one can consider the simplest case of a piece of membrane, and assume that H_0 vanishes. The difficulties in performing the sum in eq.(2) come in general from the following sources:

(i) the integrand is not a simple Gaussian, but contains also higher order nonlinearities;

(ii) taking into account all possible configurations with the right statistical weight is a non-trivial task since the membrane can be highly curved; one has also to impose on the membrane the self-avoiding condition;

(iii) additional physical constraints,such as the imposed conservation of the area or the incompressibility of the two-dimensional fluid, can modify the integral measures[9].

In the present state of the theory[10,11,12] one overcomes (ii) by simply considering quasi-flat membranes, i.e. by keeping only the lowest order term in gradients of the displacement vector in eq.(1). This approach also allows to treat the above-mentioned nonlinearities as a small perturbation around the Gaussian theory[11].

At the lowest order in the perturbation parameter $w = \frac{k_B T}{\kappa}$ one obtains :

$$\kappa = \kappa_0 \,[1 - (\alpha/4\pi) w \; ln(L/a)], \quad (d=3), \tag{3}$$

where κ_0 is the bare value of the rigidity constant; L is the linear size of the two-dimensional membrane; a is the microscopic cut-off.

The value of the numerical constant α is a controversial issue[9], mainly because of the difficulty (iii), which arises even in the perturbation scheme.

Equation (3) exhibits an interesting feature of fluctuating membranes first predicted by Helfrich[10], namely that the rigidity constant could depend on the length scale; it would decrease (slowly) with increasing L. It is tempting to search for such dependence in experimental systems: several recently developed techniques allow in principle to measure accurately the value of the rigidity κ.[13,14] One should however be cautious that the result (eq.(3)) contains severe theoretical shortcomings; for instance the decrease of κ with L implies that the perturbation parameter w becomes *large*: all perturbation schemes should thus break down. The same kind of caution should be applied in any theoretical application[15,16] of eq.(3), specially when treating films or membranes in interaction (which is completely neglected in the calculations leading to eq.(3)).

The perturbation calculation[11] leads however to two potentially useful results:

(i) it shows the existence of a *coherence length* ξ, first introduced by de Gennes and Taupin[17], which characterizes the decay of the correlations of the membrane orientation; such a length could play an important role in the behaviour of the assemblies of amphiphilic films[17] (e.g. microemulsions).

(ii) in higher dimensional spaces ($d > 3$), or, what is more important, in the presence of long-range interactions *within* the membrane, a *crumpling transition* can take place. It separates the state in which the membrane is flexible and crumpled on big scales, from another in which it is rigid and flat. Recent calculations, described during this Conference by one of us, show that this transition can in fact take place in membranes with crystalline or hexatic internal order[18].

III. INTERACTING t-MEMBRANES

Let us now consider an assembly of interacting t-membranes. We shall limit our discussion to the case of a stack of quasi-planar membranes, although in many experimental situations one can study the interactions between vesicles, tubes, etc.

In a stack of t-membranes two neighbours interact through various molecular forces[19]: short-range *hydration* repulsion, long-range *van der Waals* attraction, screened *electrostatic interactions*, etc. One can expect that the average distance between the membranes is determined by the balance of these molecular forces.

At the same time, however, the thermal fluctuations of the membranes induce a *steric repulsion* between them.[20] The interplay between the molecular and fluctuation-induced interactions is a quite subtle phenomenon and will be the subject of separate talks during this Conference[21]. Here, we shall only briefly summarize our recent results, which in particular include the prediction of *the unbinding transition* of the membranes.

In order to study the thermodynamic properties of a stack of t-membranes in equilibrium, and in particular to calculate the average spacing between them as a function of the external fields (e.g. the osmotic pressure P, the chemical potentials μ), i.e. its *equation of state*, one must again start with evaluating the statistical sum similar to eq.(2). This time however one has to integrate over the configurations of all considered membranes, and, more importantly, include the inter-membrane interactions in the energy density $f[u]$. This is in general quite difficult, therefore one is forced to use some approximation scheme. For instance, a well suited theoretical technique is here *the Functional Renormalization Group*, recently developed in the context of the interfacial phenomena[22].

The main results of such a calculation are the following.

(1) *In the absence of any external constraints, e.g. when the stack of membranes coexists with a bulk solvent*[23]:

a continuous transition can take place between the state in which the membranes are bound together, i.e. at a finite distance one from another, and the state in which they are unbound, i.e. they completely separated. This is the so-called *critical unbinding transition.* Of course, in realistic situations, on the unbound side of the transition, the distance between the neighbouring membranes is limited by some other factors: the boundary conditions, the appearence of another bulk phase different from a lamellar one, etc. In principle, however, one can enter the critical region before this happens, and verify the following critical behaviour of the mean separation $\bar{l}$ which is predicted to behave as : $\bar{l} \sim (\mathcal{W} - \mathcal{W}_{cr})^{-\psi}$, where $\mathcal{W}$ is a physical parameter which allows to modify the depth of the molecular potential $V(l)$ (e.g. the value of the Hamaker constant).

The most interesting feature of the critical unbinding transition is the fact that the membranes separate in the presence of a finite molecular attraction between them. One can think about this result as analogous to the quantum-mechanical behaviour of a particle in a well which is steep on one side, but becomes flat asymptotically on the other. In such a situation the quantum fluctuations can move the particle out of the well, in the same way as the thermal fluctuations "remove" one membrane from the "interaction well" of its neighbour. Thus the critical unbinding is a fluctuation-driven transition. This explains the fact why the mean field approximation which neglects the fluctuations fails in predicting it.

(2) *In the presence of an external constraint field P, e.g. the osmotic pressure*:

the non-zero external field P forces the membranes to be bound. The effect of thermal fluctuations, and thus the steric repulsion, is then to increase the equilibrium spacing $\bar{l}$, compared to its value given by the balance of the molecular forces[24]. However, an interesting situation occurs when one decreases the field P to zero in the case in which the molecular attraction is weak enough ($\mathcal{W} < \mathcal{W}_{cr}$). In such situation (which can be realized in practice e.g. in the process of *swelling of a lyotropic liquid crystal*[13]), one gets into the unbound state through a continous transition, called *complete unbinding.*[25]

It is important to stress the fact that the values of the "control parameters" , such as the rigidity κ and the Hamaker constant W, at which the unbinding transitions should take place (as predicted from the calculations[23]) lie in the physical region, which means that these transitions can in principle be observed for phospholipid bilayers, as well as in the (quasi)ternary systems oil/water/surfactant. It is of course tempting to speculate that Nature could adjust the values of the elastic constants and the interaction parameters of real membranes in order to put them in the vicinity of such transitions. Small changes of these values could then have drastic effects on the "interaction properties" of the membranes. This would be however quite naive, since as we know the interactions between the membranes in a cell are very specific.

IV. f-MEMBRANES

In this final section we would like to mention a more general case of f-membranes. The simplest example is here a lamellar structure with the mean spacing $\bar{l}$ comparable to the membrane thickness δ. In such systems there exists an interesting interplay between the phase transitions which happen within each membrane , and the mutual interactions between the membranes.[26] To illustrate this concept let us consider a more complicated case of an *inverted hexagonal phase* of phospholipids in water. In this phase water forms a hexagonal network of parallel tubes, surrounded by phospholipids. Of course, such structures hardly resemble a stack of membranes, and it would be erroneous to try to describe them as such.

However, instead of returning back to the molecular description, one can still build a phenomenological, continuum theory based on the curvature energy.[27] In fact, one can consider the surface of each tube as a two-dimensional amphiphilic film with elastic properties similar to those introduced before for membranes. Such films are again interacting through the solvent (e.g. via the hydration forces, etc), and fluctuate in their shape and position. In contrast to the case of t-membranes, however, one has to take into account the hydrocarbon chains which lie between such surfaces. Actually, the distance between the surfaces is not constant, and some chains must stretch more to fill up the space between them. This gives rise to an extra energy contribution: the *stretching energy* term. The existence of two terms : the curvature energy and the stretching energy, which cannot be "satisfied" simultaneously gives rise to the famous *frustration effects*, described in this Conference by J.Charvolin.[28] We can therefore think about f-membranes as of often *frustrated* ones.

In principle one should be able to develop theories, analogous to that build by S.Gruner et al.[29] for the hexagonal phases, describing other f-membranes. The most interesting seems the case of *the cubic phases* where the bilayers form a three-dimensional *crystal of surfaces*.[28] It needs to be shown that these kinds of phenomenological theories will be as successful as they are for the simple t-membranes.

Acknowledgements

It is a pleasure to acknowledge the excellent work of Dominique Langevin and Jacques Meunier,thanks to whom this Conference was as pleasant as it was interesting.

References

1. P.B. Canham, *J. Theor. Biol.* **26**, 61 (1970). W. Helfrich, *Z. Naturforsch.* **28c**, 693 (1973).
2. F. Brochard, P.-G. de Gennes, and P. Pfeuty, *J. de Physique* **37**, 1099 (1976).
3. R.E. Goldstein and S. Leibler (unpublished).
4. See e.g. H.J. Deuling, and W. Helfrich, *Biophys. J.* **16** , 861 (1976).
5. See e.g. E. Sackmann , this Conference.
6. M. Schneider, J.T. Jenkins and W.W. Webb, *J. de Physique* **45**, 1457 (1984).
7. S. Milner and S. Safran (unpublished).
8. F. Brochard and J.-F. Lennon, *J. de Physique* **36**, 1035 (1975).
9. W. Helfrich *J. de Physique* **48**, 285 (1987), D. Foerster (unpublished).
10. W. Helfrich (unpublished),and *J. de Physique* **46**, 1263 (1985).
11. L. Peliti and S. Leibler *Phys. Rev. Lett.* **54**, 1690 (1985).
12. See e.g. F. David *Europhys. Lett.* **2**, 577 (1986) and references therein.
13. D. Roux , this Conference.
14. F. Nallet, this Conference. J.-M. diMeglio, this Conference.

15. S. Safran, this Conference.
16. W. Helfrich, *J. de Physique* **47**, 321 (1986). See also : H. Kleinert, *Phys. Lett.* **A116**, 57 (1986).
17. P.-G. de Gennes and C. Taupin, *J. Phys. Chem.* **86**, 2294 (1982).
18. D. Nelson and L. Peliti (unpublished) ; L. Peliti , this Conference.
19. J.N. Isrelachvili, *Intermolecular and Surface Forces*, (Academic Press, Orlando, 1985).
20. W. Helfrich. *Z. Naturfosch.* **33a**, 305 (1978).
21. See e.g. Ref (13), and R. Lipowsky, this Conference.
22. R. Lipowsky and M.E. Fisher , *Phys. Rev. Lett* **57**, 2411 (1986), and (to be published).
23. R. Lipowsky and S. Leibler, *Phys. Rev. Lett.* **56**, 2541 (1986).
24. See e.g. E. Evans , this Conference.
25. S. Leibler and R. Lipowsky, (*Phys.Rev. B*, in print).
26. R. Goldstein, S. Leibler and R. Lipowsky (unpublished).
27. G.L. Kirk, S.M. Gruner and D.L. Stein *Biochemistry* **23**, 1093 (1984).
28. J.-F. Sadoc and J. Charvolin, *J. de Physique* **47**, 683 (1986)
29. D. Anderson, S. Gruner and S. Leibler, (unpublished).

Fluctuations and Interactions Between Membranes

D. Sornette

Laboratoire de Physique de la Matière Condensée (CNRS UA 190),
Université des Sciences, Parc Valrose, F-06034 Nice Cedex, France

Abstract: The influence of fluctuations due to thermal excitations on the interactions between membranes is reviewed. The complex interplay between the bare interactions (for example Van der Waals, hydration or electrostatic forces) and the membranes' thermal fluctuations is analyzed within a linear functional renormalization group (FRG), for membranes with a significant surface tension and for membranes with a vanishing surface tension and non-zero curvature rigidity.

In all cases, the fluctuations erase the details of the interactions which characterize each particular system and produce universal behaviour.

1 - INTRODUCTION

Interface fluctuations provide important contributions to the thermodynamic behaviour of systems in such different problems as the structure of lamellar phases and membranes[1], critical wetting transitions[2], commensurate-incommensurate transitions[3] as well as the paramagnetic-ferromagnetic transition in the random field Ising model[4] (which constitutes a paradigm for , say , binary fluid phase separation in gels or porous media).

The purpose of this paper is to review some of the concepts and results developed recently to analyze the interactions between membranes, interfaces and walls. The presentation will be organized as follows.

First, I briefly recall the harmonic approximation (§2.1) and introduce the notion of nonlinear coupling between the bare interactions and the membrane undulations in §2.2. To go beyond the harmonic approximation, as is necessary to tackle these nonlinear effects, I then introduce the renormalization group approach (§3). Several theoretical treatment have been proposed recently and we recall the main results in §3.1. The linear functional renormalization group , which will be used throughout the paper, is then derived in §3.2. In §4, the FRG is applied to the problem of the steric interaction between interfaces having a non-zero surface tension in 3D. The case of membranes with zero surface tension ($r=0$) and non-zero curvature rigidity ($K\neq0$) is then considered (§5). In §6, I present a simple picture for the thermal softening of the curvature rigidity which allows the description of membranes at scales much larger than the persistence length. Its consequence on the expression of the steric potential is briefly discussed.

In the problems which we will face, it is enough to work with an effective interface model consisting of a single interface or membrane placed at an average distance $\underline{z}$ from a rigid wall and compressed by a pressure p on the other side. This models a multilamellar lyotropic system, a stack of membranes, a liquid-liquid or liquid-gas interface in a wetting problem or an array of uniaxial solitons in a commensurate-incommensurate transition. The pressure can be osmotic (due to other membranes or particles) or may stem from a difference in chemical potential describing departure from coexistence.

One has to compute the partition function

$$Z=\int d\{z(\rho)\} .\exp\{-F/kT\}, \qquad (1)$$

where the functional integration is performed over all functions $z(\boldsymbol{\rho})$ describing the interface shape. In eq.(1), D is the bulk space dimension and $\boldsymbol{\rho}$ denotes the (D-1)-dimensional in-plane position vector. The relevant free energy F of the system takes the general form

$$F\{z(\boldsymbol{\rho})\} = \int d^{D-1}\rho \,[1/2\, r(\nabla z)^2 + 1/2\, K(\Delta z)^2 + V(z) + pz]; \qquad (2)$$

r is the interface surface tension, K is the curvature rigidity, $z(\boldsymbol{\rho})$ is the local membrane displacement along the z-axis and V(z) is the total "microscopic" potential, which is the sum of the different repulsive and attractive (when they exist) potentials.

One can expect intuitively that the presence of the repulsive wall/interface forces prevents the approach of the interface near the wall and therefore hinders the interface fluctuations. On the contrary, attractive forces enhance the fluctuations. In both cases, this results in a coupling between the bare microscopic interactions and the thermal fluctuations. In the next section, I review briefly the consequences of these couplings within the harmonic approximation.

2-GENERAL IDEAS

Our goal is to determine the expression for the steric interaction F_{st}[5], which is the name for the potential obtained from the microscopic repulsive interaction by renormalization by the thermal fluctuations.

Let us introduce two relevant correlation lengths $\xi_\perp$ and $\xi_{//}$[11] ,describing respectively the correlation along the z-direction for $\xi_\perp$ and along the directions parallel to the interface for $\xi_{//}$ of the displacements of two points on the interface, and defined from the following correlation function $g(\boldsymbol{\rho})$:

$$g(\boldsymbol{\rho})=<\{z(\boldsymbol{\rho})-z(\mathbf{0})\}^2>$$

In the mean field approximation which amounts to replacing V(z) in eq.(2) by $V(\underline{z})+(1/2)(z-\underline{z})^2\delta^2V/\delta z^2|_{\underline{z}}$, $g(\boldsymbol{\rho})$ has the following expression:

$$g(\boldsymbol{\rho})=(kT/K(2\pi)^{D-1}) \int d^{D-1}q \,\{1-e^{i\mathbf{q}\boldsymbol{\rho}}\}/\{q^2\xi_r^{-2} + q^4 + \xi_{//}^{-4}\} \qquad (3)$$

where $\xi_{//}=(K^{-1}\,\delta^2V/\delta z^2|_{\underline{z}})^{-1/4}$ and $\xi_r=(K/r)^{1/2}$. $\xi_{//}$ is the in-plane correlation when r=0. If $r\neq0$, the in-plane correlation length is $\xi_{//}^{\,2}/\xi_r$. ξ_r separates two regimes: for scales $L<\xi_r$, the surface tension contribution is negligible. For $L>\xi_r$, the surface tension dominates over the rigidity K. Therefore, computations performed for r=0 also apply to $r\neq0$ as long as the in-plane correlation length $\xi_{//}$ is found less than ξ_r, that is, for sufficiently small surface tension $r<r^*$, where r^* is such that $\xi_{r^*}=\xi_{//}$.

The correlation length $\xi_\perp$ is defined by $\xi_\perp^{\,2}=g(+\infty)$.

In order to distinguish clearly between the two cases ($r\neq0$) and ($r=0;K\neq0$), it is interesting to introduce an orientation correlation function $g_{or}(\boldsymbol{\rho})$:

$g_{or}(\rho)=<\mathbf{n}(\rho).\mathbf{n}(0)>$.

Within the harmonic approximation and in the limit $\xi_{//}\rightarrow+\infty$ corresponding to isolated membranes, one has

$$g(\rho) \sim (kT/r)\mathrm{Log}\{|\rho|\Lambda\}, \qquad (r\neq 0)$$

$$g(\rho) \sim (kT/K)\,.\,|\rho|^2, \qquad (r=0;K\neq 0) \qquad (4)$$

$$g_{or}(\rho) \sim 1-(kT/K), \qquad (r\neq 0)$$

$$g_{or}(\rho) \sim 1-(kT/K)\,\mathrm{Log}\{|\rho|\Lambda\}\simeq(|\rho|\Lambda)^{-kT/K}, \qquad (r=0;K\neq 0)$$

$l=\Lambda^{-1}$ is a microscopic scale acting as the ultraviolet cutoff. These limiting behaviours show clearly that $(r\neq 0)$ membranes keep a finite orientation correlation at infinite distance $\rho\rightarrow+\infty$, even if their normal fluctuations diverge (logarithmically). On the contrary, both translational and orientational correlations are lost in $(r=0;K\neq 0)$-membranes at sufficiently large distances. These features control the expression of the steric potential as will become clear in the following.

For a membrane placed at a finite distance $\underline{z}$ from other membranes or from a wall, $\xi_{//}$ and $\xi_{\perp}$ are not independent. They describe the coupling between fluctuations and short range repulsive interactions in the large separation limit $\underline{z}>>\lambda$ where λ is the range of the microscopic potential. For a mean membrane separation $\underline{z}\rightarrow+\infty$, each isolated interface freely undulates due to thermal excitations. Very long wavelength excitations (capillary modes) propagate along the interface resulting in infinite correlation lengths $\xi_{\perp}$ and $\xi_{//}$. For $\underline{z}$ i.e. $\xi_{\perp}$ finite, the total potential resulting from the repulsive interactions (structural, electrostatic, hard or soft core...) hinders the interface undulations whenever the wall-interface distance becomes locally of the order of the potential range.
The global effect of the short range interactions thus gives birth to an effective steric interaction $F_{st}(\underline{z})$ leading to a decrease of the in-plane correlation length $\xi_{//}$.
The evaluation of $F_{st}(\underline{z})$ relies on the determination of $\xi_{//}(\underline{z})$.

2.1 HARMONIC APPROXIMATION

The harmonic theory essentially amounts to identifying $\xi_{//}$ as being the largest wavelength (infrared cut-off) compatible with the condition

$$\sigma^2 = \mu\,\underline{z}^2, \qquad (5)$$

where $\sigma=<(\Delta z)^2>^{1/2}$ is the root mean square displacement of the interface and μ is a numerical factor of order unity. In other words, $\xi_{//}$ is the size of the plaquettes which are decoupled by membrane collisions.
The harmonic theory replaces, as usual, the many body problem by a one-interface system plunged into an harmonic well of "spring force" $m^2(\xi_{//})$. The hamiltonian thus reads

$$H = \int dxdy\,\{\,1/2\,r(\nabla z)^2 + 1/2\,K\,(\nabla^2 z)^2 + 1/2\,m^2(\xi_{//})\,z^2\,\}. \qquad (6)$$

From eq.(6), with the equipartition theorem, we can deduce σ^2:

$$\sigma^2 \simeq (kT/4\pi r)\,\mathrm{Log}\,(r\pi^2/l^2m^2), \qquad (r\neq 0; K=0) \quad (7)$$

$$\simeq kT\,/(8\,K^{1/2}\,m). \qquad (r=0; K\neq 0) \quad (8)$$

Condition (5) then yields

$$\xi_{//}=(r/m^2)^{1/2} \sim l\exp\{2\pi r\mu \underline{z}^2/kT\}, \qquad r\neq 0, K=0 \quad (9)$$

$$\xi_{//}=(K/m^2)^{1/4} \sim (K/kT)^{1/2}\,\underline{z}. \qquad r=0, K\neq 0 \quad (10)$$

$\xi_{//}$ becomes finite due to the suppression of $q_{//}\to 0$ modes caused by the constraint (5). The steric energy is given by $F_{st} = -kT\,\mathrm{Log}\,Z_m/Z_0 - 1/2\ m^2\ \sigma^2$, where

$$Z_m/Z_0 = \exp\left[-1/2\ \Sigma_{q_{//}}\ \mathrm{Log}\{(m^2+rq_{//}^2+Kq_{//}^4)/(rq_{//}^2+Kq_{//}^4)\}\right] \quad (11)$$

is the ratio of the partition functions of the interface with and without the harmonic well. The substraction of the mean spring potential energy $(1/2)m^2\sigma^2$ from the total free energy gives access to the steric part F_{st} of the free energy.
A straightforward integration in fourier space yields

$$F_{st}/A = (8\pi)^{-1}\ kT/\xi_{//}^2 \quad (12)$$

with $\xi_{//}$ given by eq. (9) $(r\neq 0, K=0)$ and (10) $(r=0, K\neq 0)$.
This gives the results obtained previously by other authors[5-8]:

$$F_{st}/A = (\pi kT/8l^2)\ \exp\{-4\pi r\ \mu\ \underline{z}^2/kT\}, \qquad r\neq 0, K=0 \quad (13)$$

$$F_{st}/A = (kT)^2/128\mu K\underline{z}^2. \qquad r=0, K\neq 0 \quad (14)$$

2.2 HYPERSCALING AND GENERAL EXPRESSION OF THE STERIC INTERACTION

The hyperscaling relation (12) can be derived more generally as follows. It is enough to estimate the surface tension energy or the bending energy for typical configurations of the membranes[11]. These configurations consist of bumps of longitudinal and transverse dimensions $\xi_{//}$ and $\xi_{\perp}$.
For $r\neq 0$, the surface tension energy is

$$F_{st} \sim r\{\xi_{\perp}/\xi_{//}\}^2 \quad (15)$$

via the gradient term, with $\xi_{//} \sim l\exp\{(2\pi r\mu/kT)\ \xi_{\perp}^2\}$.
For $r=0, K\neq 0$, the bending free energy per unit area is

$$F_{st} \sim r\{\xi_{\perp}/\xi_{//}^2\}^2 \quad (16)$$

via the laplacian term, with $\xi_{//} \sim (K/kT)^{1/2}\,\xi_{\perp}$. Using $\xi_{\perp} \sim \underline{z}$, we reobtain the previous results (13) and (14). The steric energy can then be interpreted as a surface tension (eq. 15) (or curvature (eq. 16)) contribution for the typical membrane configurations created by thermal excitations in presence of the steric constraints[11].
The hyperscaling relation (12) holds both within the harmonic theory and the FRG. However, it breaks down for fluctuations induced by quenched disorder as occurs in the random field ising model[4]. This problem will be discussed elsewhere.

2.3 ENHANCEMENT OF THE SHORT RANGE HYDRATION FORCE BY THE FLUCTUATIONS

We now discuss the "non-universal" case where the average distance z between membranes is not large but on the contrary is on the order of the range λ of the bare interactions (hydration, electrostatic... interaction). This case $\underline{z} \sim \lambda$ is relevant experimentally since lamellar phases are often characterized by an interlamellar spacing of the order of a few nanometers. In these cases, the fluctuations also play a very important role but do not completely erase the form of the microscopic interactions.

Consider two parallel membranes separated by the average distance $\underline{z}$ and whose thermal undulations are described by the displacement fields $u_1(x,y)$ and $u_2(x,y)$. The probability density for finding any of the two variables $u_1(\boldsymbol{\rho})$ or $u_2(\boldsymbol{\rho})$ in the interval $[u,u+du]$ is noted $w_\sigma(u)$, where σ characterizes the r.m.s. of u as defined in the preceding section.

Due to thermal fluctuations, two undulating membranes will have some part of their surfaces almost in contact, and other portions much further apart, the r.m.s. of the mutual distance fluctuations being $\sigma\sqrt{2}$.

Now, if we assume that the external field (the hydration interaction for the present) is essentially local, i.e. involves only adjacent portions of two opposite membranes, the "renormalized" hydration free energy accounting for the undulations is given by

$$F_H(\underline{z}) = \int_0^{+\infty} w_{\sigma\sqrt{2}}(u)\ F_H^{\,b}(\underline{z}+u)\, du\ , \qquad (17)$$

where $F_H^{\,b}(\underline{z})$ is the bare hydration free energy per unit surface corresponding to rigid planar membranes, separated by the distance $\underline{z}$.

Since in general $\sigma \leqslant 0.3$ d for all separations of interest[9], the integral (17) may be extended from $-\infty$ to $+\infty$. An analytic expression of $F_H(\underline{z})$ may be obtained for a particular choice of $w_\sigma(u)$ corresponding to the mean field approximation of the previous section :

$$w_\sigma(u) = (2\pi\sigma^2)^{-1/2}\ e^{-u^2/2\sigma^2} \qquad (18)$$

Performing integral (17) with $F_H(z) = Ae^{-z/\lambda}$, we find for the renormalized hydration free energy between two membranes of size L

$$F_H(z) = A_H\ e^{-z/\lambda}\ , \qquad (19)$$

where $A_H = Ae^{\sigma^2/\lambda^2}$. For typical values of the parameters ($z \sim 10$Å , $\lambda \sim 2$Å), this result implies a tenfold increase of the hydration interaction between membranes when going from the solid state ($\sigma \simeq 0$) to the fluid state ($\sigma \simeq 3$Å).

Note however that the precise numerical value of this multiplicative factor A_H/A depends crucially on the r.m.s. σ of the undulation displacement. σ must be estimated by taking self-consistently into account the presence of the hydration force. Indeed, not only is the hydration interaction renormalized by the fluctuations but also it has the effect of diminishing the fluctuations. These competing aspects can be studied within a smectic liquid crystal theory (see ref.[9] for further details).

We only sketch the main results: for membranes at low temperatures below the melting point T_m, the membrane rigidity K is very large (typically $K \sim 10^{-10}$ erg for a solid membrane) and it suppresses almost all undulations. Note that it is not clear whether the form (2) for the membrane deformation energy is the correct one for a "solid" membrane. However, it is clear that the fluctuations are much hindered in the solid phase and the experimental measurement of the hydration force should therefore give access to the bare values of A and λ.

At higher temperatures above T_m, the membranes are in a fluid state and the rigidity is much smaller ($K \sim 10^{-12}$erg). The undulations are much stronger and renormalize the hydration interaction according to eq.(17) with the correct self-consistent value of σ. These ideas and the quantitative effects which can be estimated from the smectic analogy are found to be in fair agreement with the experimental observations of ref.[10]. These ideas have been reconsidered recently and proved to correlate well with experimental studies of osmotic (and mechanical) compression of multilamellar lipid arrays, mechanical compression of bilayers immobilized on mica substrates, and controlled adhesion of giant bilayer membrane vesicles(see ref.[9b]).

This elementary "renormalization" approach breaks down for large undulation amplitudes as occurs for large membrane separations $z >> \lambda$ and in presence of attractive forces. A more general formalism is thus needed, which is discussed in the next section.

3-THE FUNCTIONAL RENORMALIZATION GROUP.

3.1 SUMMARY OF THE MAIN RESULTS

The full problem involves taking into account both the repulsive and attractive forces which control the equilibrium structure of multilayers systems. A naive expectation would be to add the steric interaction going as $\underline{z}^{-2}$ (for r=0) at a large distance to the other bare interactions described in §2.1. Since this repulsion dominates at large $\underline{z}$, this would result in the prediction of the existence of a bound membrane state distinct from an isolated membranes state where the membranes are infinitely far from each other, the two states being separated by a first order transition.

This assumption is very questionable, especially in the light of the discussion of the previous section, underlying the special importance of the nonlinear coupling between the membranes. The careful description of these nonlinear effects has been done within various renormalization groups[2,8,11,12]. We summarize here the main results.

The case $r \neq 0$ has been mostly studied in the context of the critical wetting and commensurate-incommensurate transitions, and has been clarified several years ago[2]. In order to understand the effects of the fluctuations in the presence of both attractive and repulsive forces, one has to compute thermodynamic variables which involves summations over the wave vector degrees of freedom and estimate integrals (or propagators) of the form

$$G=\int_{q_{min}} d^{D-1}q.\{rq^2 + Kq^4\}^{-1},$$

where $q_{min} \sim \pi/L$ is the infrared cut-off inverse of the membrane size L. For $r \neq 0$,

$$G \sim q_{min}^{D-3} \qquad \text{as } q_{min} \rightarrow 0.$$

G is therefore finite (non-infrared divergent) for $D>D_c=3$ and diverging for $D \leqslant D_c$. This defines the upper critical dimension $D_c=3$ (see ref.[2] for a careful discussion of this point). Three different scaling regimes have been found in the presence of attracting forces:

1)For $D>D_c$, the fluctuations are negligible and the phase structure involves only the competition of the bare interactions. This is the mean field regime for which the steric interaction is vanishing.

2)For $D_2<D<D_c$, the phase boundaries are as in mean field but the critical exponents have non-classical values. This is the "weak-fluctuating" regime.

3) For $D<D_2$, both the phase boundaries and the critical exponents are shifted compared to their classical values. This is the "strong fluctuating" regime.

The marginal or "weak fluctuating" case $D=D_c$ is very peculiar since, in contrast to usual critical transitions where only logarithmic corrections are found, the fluctuations change completely the mean field predictions and produce a line of continuous critical transitions with non-universal critical exponents varying continuously with the temperature[2,8]. We come back to this case in the analysis of the RG computation of the steric interaction in the case ($r\neq 0$;$D=D_c=3$). The peculiarity of this case is again found in the important corrections[8] to the harmonic approximation of §2.1.

For the case $r=0$,

$G\sim q_{min}^{D-5}$ as $q_{min}\to 0$.

G is therefore finite (non-infrared divergent) for $D>D_c=5$ and diverging for $D\leqslant D_c$. This defines the upper critical dimension $D_c=5$ for membranes without surface tension. The discussion follows that of the case $r\neq 0$.

1)For $D>D_c=5$, mean field theory is valid and the transition is first order.

2)For $D_2=11/3<D<D_c$[11,13], one is in a "weak fluctuating" regime and a continuous transition takes place when the amplitude of the attractive potential (Hamaker's constant) vanishes.

3)For $D<D_2$ (hence in $D=3$), the transition is still continuous but takes place with a non zero value of the Hamaker constant (the fluctuations and therefore the steric repulsion are so strong that a finite attraction is necessary to hold the membranes together). In addition, a line of first order phase transition may exist in some cases. We will discuss these results in more details when applying the technique of the linear FRG which we now treat.

3.2 THE LINEAR FUNCTIONAL RENORMALIZATION GROUP

Let us consider, as above, a single membrane plunged into an interaction potential $V(z)$ describing the influence of neighbouring membranes.

I now recall the functional renormalization group approach of Fisher and Huse [10], for the case of non-vanishing surface tension membranes ($r\neq 0$) and adapt it to the case of zero surface tension membranes ($r=0$;$K\neq 0$). The approach [14,12] can be viewed as a development in powers of the "small" parameter V/r for ($r\neq 0$) and $V/K\Lambda^{D_c-D}$ for ($r=0$;$K\neq 0$), where Λ^{-1} is a microscopic length. In the "non-perturbed" case ($V=0$), the hamiltonian is gaussian and constitutes a fixed point in the renormalization group which

i) integrates out the degrees of freedom z_q with $\Lambda/b<|q|<\Lambda$

ii) rescales the system to new coordinates

$$\rho' = \rho/b,\ q' = bq,\ z'_{q'} = b^{(D-D_c)/2} z_q \qquad (20)$$

The choice (20) for $z'_{q'}$ insures the invariance of r (respectively K) in the transformation i)- ii).

The linear FRG of Fisher and Huse is a perturbation expansion about this line of Gaussian fixed points to lowest order in $V_0(z)$. Following ref. [14], this is done by dividing $z(\rho)$ into a fast part $z_f(\rho)$ and a slow part $z_s(\rho)$:

$$z_f(\rho) = \int_{\Lambda/b<q<\Lambda} d^{D-1}q \,.\, e^{-i\mathbf{q}\boldsymbol{\rho}}\, z_q\,, \qquad (21)$$

$$z_s(\rho) = z(\rho) - z_f(\rho)\,. \qquad (22)$$

The partition function reads

$$Z_\Lambda = \Pi_{0<q<\Lambda} \int dz_q \exp\{-\beta \int d^{D-1}\rho\ [\Delta(z_s+z_f) + V(z_s) + z_f\,.\,(\delta V/\delta z)_{z_s}$$

$$+ 1/2.\, z_f^2\ (\delta^2 V/\delta z^2)_{z_s}]\}, \qquad (23)$$

where

$$\Delta(z)=(r/2).(\nabla z)^2 \qquad \text{for } r\neq 0,$$

$$=(K/2).(\nabla^2 z)^2 \qquad \text{for } r=0;K\neq 0.$$

We have developed $V(z)$ around $V(z_S)$ and stopped the expansion to second order. With the condition $V/r<1$ (respectively $V/K\Lambda^2 < 1$), eq.(23) transforms into

$$Z_\Lambda = \Pi_{0<q<\Lambda/b} \int dz_q \exp\{-\beta\int d^{D-1}\rho\, H_S\}.\ \Pi_{\Lambda/b<q<\Lambda} \int dz_q[1+z_f(\delta V/\delta z)_{zS} +$$

$$z_f^2/2\ .(\delta^2 V/\delta z^2)_{zS}]\,.\exp\{-\beta\int d^{D-1}\rho\, \Delta(z_f)\}, \qquad (24)$$

where $H_S=\int d^{D-1}\rho\ \{\Delta(z_S)+V(z_S)\}$.

The momentum shell integration yields

$$<z_f^2>_{\Lambda/b<q<\Lambda} \sim (b-1).\mathbb{C}^{-1},$$

where

$$\mathbb{C}\sim kT/4\pi r \qquad \text{for } r\neq 0, D=3,$$

$$\sim kT/K\Lambda^{D_c-D} \qquad \text{for } r=0;K\neq 0. \qquad (25)$$

The renormalization group equation for $V(z)$ is obtained by considering continuous rescaling $b=1+l$ with l infinitesimal :

$$V_l(z_S) = b^{D-1}\ V_0(b^{(D_c-D)/2}\, z_S) + (b-1)\mathbb{C}\ (\delta^2 V_0/\delta z^2)_{zS}. \qquad (26)$$

The first term in eq.(26) is just the dimensional rescaling (20) and the second term comes from the momentum-shell integration.
For l infinitesimal, eq.(26) yields :

$$\delta V/\delta l\ = (D-1)\ V(z) + (D_c-D)/2.\ z\, \delta V/\delta z + \mathbb{C}\ \delta^2 V/\delta z^2. \qquad (27)$$

The general solution of eq. (27) is

$$V_l(z) = (2l\mathbb{C})^{1/2}\, e^{2l} \int_{-\infty}^{+\infty} dz'\ V_0(z')\exp\{-(z'-z)^2/4l\mathbb{C}\} \qquad (28a)$$

for $r\neq 0, D=3$;

$$V_l(z)=[e^{(D-1)l}/\sqrt{2\pi}\,\delta(l)].\int_{-\infty}^{+\infty} dz'\ V_0(z')\exp\{-(z\zeta(l)-z')^2/2\delta(l)^2\} \qquad (28b)$$

for $r=0;K\neq 0$, with

$$\delta^2(l)\ = 2\mathbb{C}\ [e^{(D_c-D)l}-1]/(D_c-D), \qquad (29)$$

$$\zeta(l) = \exp\ \{l(D_c-D)/2\}.$$

Note that a different FRG has been used by several authors, based on Wilson's approximate recursion relation. It can be shown[12] that the linear FRG derived above is obtained as the continuous rescaling limit of the nonlinear FRG used in ref.[11].

The interest of this derivation is to exhibit the condition of its validity. The practical advantage of expression (26) is that its solution can be given analytically in complete form by eq.(28). The existence of this solution simplifies considerably the analysis of the phase diagram.

We are now in position to discuss the corrections to the harmonic approximation.

4-RENORMALIZED STERIC INTERACTION FOR $r\neq 0$ IN 3D

This case has been studied in details in ref.[8], by applying a different renormalization scheme developed in the context of the critical wetting transition[2].
The "harmonic" result

$$F_{st}/L^2 \sim kT.\exp\{-\underline{z}^2/(kTr^{-1})\}$$

given by eq.(13) is changed into

$$F_{st}/L^2 \sim kT.\exp\{-\underline{z}/(kTr^{-1})^{1/2}\}.$$

This gives a much longer range for the steric interaction than predicted from the harmonic treatment. Within the linear FRG, this result can be recovered in a few lines. We first give a simplified heuristic argument followed by the FRG treatment.

4.1 HEURISTIC ARGUMENT

We develop a simple physical picture for understanding why the "harmonic" result, that the correlation length $\xi_{//}$ defined in §2.2 scales as $\xi_{//} \sim \exp(z^2)$, is wrong and must be replaced by $\xi_{//} \sim e^{\underline{z}}$[16].

We start from the interface model (2) with a microscopic bare potential $V_0(z)=Ae^{-z/\lambda}+pz$. Around the equilibrium distance given by $\delta V_0/\delta z=0$, the harmonic part of $V_0^\circ(z)$ is

$$V_0(z) \simeq (1/2)m^2z^2\ , \qquad \text{where } m^2 \simeq 2A\lambda^{-2}e^{-z/\lambda}$$

One has to perform integrals of the form $\Sigma_q kT/\{m^2+rq^2\}$.
In the summation over the wave vectors q, all modes such that $m^2<rq^2$, do not feel the wall and therefore are not constrained. This defines a wavevector cutoff

$$q_c \simeq (m^2/r)^{1/2} \simeq \{2A(r\lambda^2)^{-1}e^{-z/\lambda}\}^{1/2}.$$

For $q<q_c$, the modes feel the steric constraint. A crude assumption is to fix the amplitude of the fluctuations of these modes with $q<q_c$ to zero and let the modes with $q>q_c$ free. This amounts to replacing $\xi_{//}$ by q_c^{-1} and from the hyperscaling relation this leads to the announced result.

In sum, the steric constraint is not $\sigma^2 \sim \underline{z}^2$ as posed within the harmonic treatment, but rather $rq^2 \sim V_0(\underline{z})$ where V_0 is a suitable short range potential.

4.2 FRG TREATMENT OF THE STERIC INTERACTION

We start from the renormalization flow equation (28a) for the potential V(z). In the following, we do not distinguish in notation between z and $\underline{z}$.

Supposing a microscopic bare potential of the form $V_0(z)=Ae^{-z/\lambda}$, the integral in eq.(28a) calculated within the saddle point approximation yields

$$V_l(z)=A\,e^{(2+\omega)l-z/\lambda}, \qquad \text{for } z>2\omega l$$
$$=A\{2\omega l\lambda/z(4\pi\omega l)^{1/2}\}\,e^{2l-z^2/4l\omega\lambda^2}, \qquad \text{for } z<2\omega l \qquad (30)$$

where $\omega=kT/4\pi r\lambda^2$. In order to maintain the interface at a fixed distance z from the wall, we add a pressure term pz which is renormalized to $(4\pi\omega l)^{-1/2}\,e^{2l}\,pz$. The condition that the interface is in equilibrium at the minimum of the renormalized total potential $V_T(z)$, sum of the exponential interaction and the pressure term, is $\delta V_T/\delta z=0$, which determines p. The second condition that the rescaling must be performed self-consistently up to the scale of the correlation length $\xi_{//}\sim e^l$ reads $\delta^2V_T/\delta z^2=1$. This fixes l and therefore $\xi_{//}$:

$$\xi_{//}\sim e^{z/(2+\omega)\lambda}, \qquad \text{for } \omega<2,$$
$$\sim e^{z/\lambda(8\omega)^{1/2}}, \qquad \text{for } \omega>2. \qquad (31)$$

From the hyperscaling relation $F_{st}\sim kT\xi_{//}^{-2}$, we recover the result

$$F_{st}\sim kT\,e^{-z/\lambda_R} \qquad (32)$$

with

$$\lambda_R=\{(\omega+2)/2\}\lambda=\lambda\{(kT/4\pi r\lambda^2)+2\}/2, \qquad \text{for } \omega<2,$$
$$\lambda_R=(2\omega)^{1/2}\lambda=(kT/2\pi r)^{1/2}, \qquad \text{for } \omega>2. \qquad (33)$$

Note that when $\omega>2$, thermal fluctuations are strong and the renormalized range of the steric interaction becomes independent of the microscopic bare interaction. Any other choice for $V_0(z)$ could have been made as long as it were short ranged (shorter than the steric interaction range $\lambda_R=(kT/2\pi r)^{1/2}$).
In short, the change from the $\exp\{-z^2/\delta^2\}$ to $\exp\{-(r/kT)^{1/2}z\}$ comes about from the fact that $\delta^2\sim\log\xi_{//}$ with $\xi_{//}\sim\exp\{-(r/kT)^{1/2}z\}$, in the FRG.
Note finally that the exponential form of the steric interaction has been proved rigorously[8b] by determining upper and lower bounds for the expression of the mass m.

5-STERIC INTERACTION AND TRICRITICAL BEHAVIOUR IN MEMBRANES WITH ZERO SURFACE TENSION

5.1 STERIC INTERACTION IN ABSENCE OF ATTRACTIVE FORCES

We have seen that the harmonic treatment of the steric constraint results in an entropic repulsive interaction decreasing as z^{-2} at large membrane separations. The z^{-2} powerlaw dependence only reflects the scale invariance of the free energy for the undulation of a membrane with vanishing surface tension. Consider a closed membrane (vesicle) constrained between two concentric rigid spheres of radius r_1 and r_2. We may keep the inner shell of fixed size and imagine the outer one inflatable to measure the pressure. Performing the scaling $x\rightarrow bx$ keeps the free energy invariant except that the size of the spheres gets inflated and the total free energy $F_s(r_1,r_2)$ is changed according to

$$F_s(r_1,r_2)=F_s(br_1,br_2) \qquad (34)$$

for $r_1-r_2<< r_1,r_2$. The free energy density obeys the corresponding scaling

$$4\pi r_1^2 f_s(r_1-r_2) = 4\pi(br_1)^2 f_s(br_1-br_2) \tag{35}$$

Therefore, $f_s(bz)=(1/b^2)f_s(z)$. One thus concludes that $f_s(z)\sim z^{-2}$ in agreement with the harmonic result. It is possible that violation of this scale invariance may exist, similarly to what happens in critical phenomena (see §6).

5.2 FRG IN THE PRESENCE OF ATTRACTIVE FORCES

Two different approaches have recently studied this case. The first treatment[11] uses an extension of the non-linear renormalization group developed initially by Wilson. The corresponding non-linear functional renormalization group (FRG) yields an equation for the total renormalized potential in terms of the bare potential in a way similar to eq.(28b). The procedure then consists in determining numerically the "fixed points" in the space of potentials of the FRG flow equation. From the minimum of the "fixed point" potential, one finds the equilibrium spacing z. This procedure shows the existence of a continuous unbinding transition separating a phase of bound membranes for $W_C<W$ and a phase of isolated membranes ($z\rightarrow\infty$) for $W_C>W$, where W_C is a critical value of the Hamaker coefficient which depends on the temperature and the microscopic potential parameters. For typical values of these parameters, the critical point occurs for $W_C\sim kT$, and in the scaling region, the spacing scales as $z\sim a(W-W_C)^{-\psi}$, where ψ is determined numerically[11]: $\psi=1.00\pm0.03$. The influence of the strong fluctuations and the corresponding long ranged steric repulsion reveals itself in the non-zero value of the Hamaker constant at the critical unbinding transition: the fluctuations and therefore the steric repulsion are so strong that a finite attraction is necessary to hold the membranes together.

The second approach[12] starts from the renormalization flow equation (28b) for the potential V(z). Since the FRG is linear, it is not possible to determine directly from it the fixed points in the space of the potentials. However, I proposed to construct an "effective" renormalized potential whose minimization yields the equilibrium spacing z [12]. Briefly, the arguments run as follows.

First, the linear FRG allows one to infer that short ranged potentials (i.e. decreasing faster than z^{-2}) are renormalized into short ranged potential(s), whose precise form cannot be obtained within the linear FRG.

The second step is to consider the influence of large wavelength capillary waves. From the previous discussions, we know that , due to the effect of the large wavelength capillary waves, a repulsive short-ranged potential results in steric entropic repulsion (14). Therefore, a bare "short ranged" potential leads to a <u>short-ranged renormalized</u> potential but to a <u>long ranged effective</u> "steric" potential.

In real world, the potential can also have an attractive part. In this case, the discussion follows similar lines. The renormalized potential is short ranged and should have both an attractive and a repulsive part whose precise forms are not accessible within the linear FRG (eq.(20)). However, its precise form is not so important since the "capillary waves" rub out the details of its shape. Now, the effect of the long wavelength capillary waves is more tricky to evaluate, since "collisions" between the membrane and the wall lead to a loss in free energy of $\simeq$kT for the part of the membrane in the repulsive part of the potential but also to a gain of $\simeq$-W for the part of the membrane in the attractive part of the potential. From this simple reasoning, one expects that the hard core plus hydration interactions in presence of the attractive Van der Waals interaction leads to an effective potential of the form

$$V_E(z) \sim (kT/K).\ (kT-\alpha W)/z^2, \qquad (36)$$

where α is a number of the order of 1.

Note that these ideas can be checked by an exact computation carried out for a "similar" problem of a one-dimensional interface in 2D bulk space near a wall[12]. The term "similar" has the following meaning: for an interface with non-zero surface tension, it is easy to show that the effect of the capillary waves in the presence of a short-ranged potential leads also to an effective repulsive steric interaction which decays as z^{-2}. This comes from the larger low-momentum phonon density in two dimensions. This can be computed exactly by mapping the configuration z(x) of the interface to a problem of a quantum path trajectory, in which z becomes the 1D position of the particle and x is the time. The z^{-2} law is checked by adding a pressure term pz to the potential. The computation of the average position <z> of the quantum particle is obtained once the wave function (with proper norm) is determined. By this method, it is possible to check the influence of an attractive part in the potential, which indeed results in a correction given by eq.(36).

5.3 COMPUTATION PROCEDURE

To be able to predict the multilamellar phase diagram and the equilibrium membrane spacing z, we have to estimate the leading corrections to expressions (36). The result is

$$V_R(z) = kT/K\,.\ \{ (kT-\alpha W)/z^2 - 2\lambda\, kTe^{-l}/z^3 + (kT)^2/K\Lambda^2.\ 1/z^4 \}, \qquad (37)$$

valid for D=3. Eq.(37) was the main result of ref.[12].

The second term in z^{-3} stems from the possibility of a partial overlap within a scale λ between adjacent membranes. λ is typically of the order of a few angstroms. $\lambda>0$ models a soft membrane allowing partial penetration. This can occur for lamellar phases swollen by water having small hydrophilic groups[17]. Also in systems swollen by oil, the short range repulsion is modulated by the affinity of the aliphatic part with the oil. By changing the temperature, the oil or the chains, the aliphatic-oil interaction may vary from good solvent to bad solvent behaviour. Crossing the "Θ" point corresponds to changing the sign of λ. $\lambda>0$ is close to a bad solvent and $\lambda<0$ corresponds to good solvent behaviour. $\lambda<0$ means that matter is not allowed within a distance λ of the conventionally defined z=0 plane of each membrane. In the water-swollen case, this correspond to a very strong hydration force created by large hydrophilic groups[18].

The third term in z^{-4} of eq.(37) is the leading correction to the z^{-2} powerlaw given by eq.(36).

We now describe the self-consistent treatment for the derivation of the equilibrium membrane spacing. Note the scale dependence (exp(-l}) of the coefficients of the effective potential. This is similar to the phenomenological expression of the leading correction to the Landau theory of critical phenomena where the coefficients of the Landau expansion become dependent on the scale of the coarse graining procedure. The e^{-l} is non-trivial and can only be obtained from a renormalization group.

The equilibrium condition is $\delta V/\delta z = 0$ and the consistent FRG condition is $\delta^2 V/\delta z^2 = \xi_{//}^{-4}$ where, as usual, $\xi_{//}$ has the meaning of an in-plane correlation length. Using the scaling (20), one has

$$z = e^l z_R;\ V = \xi_{//}^{-2} V_R\ ;\ \xi = e^l \xi_{//R}, \qquad (38)$$

which yield the condition on the renormalized variables :

$$\delta^2 V_R/\delta z_R{}^2=\xi_{//R}{}^{-4}. \quad (39)$$

The usual procedure [14] would be to renormalize up to the scale $\xi_{//}$ which amounts to imposing $\xi_{//R}=1$. However, in a strong fluctuating regime, this may not be allowed. Indeed, as we will see below, the renormalization must be stopped to some scale $e^{-l}=\xi_{//}/\xi_{//R}$ such that the two equations $\delta V/\delta z=0$ and $\delta^2 V_R/\delta z^2=\xi_{//R}{}^{-4}$ find solutions.

5.4 TRICRITICAL BEHAVIOUR

The potential (37) has the structure describing a tricritical point at ($\varepsilon=kT-\alpha W=0$; $\lambda=0$). This is most easily seen by posing $M=(\Lambda z)^{-1}$. The tricritical point occurs at ($\alpha W-kT=0,\lambda=0$). Let us examine briefly the different regimes[12].

5.4.1 The unbinding transition($\lambda<0$)

We first consider the case $\lambda<0$ ($b<0$) corresponding to a strong "exclusion". Taking V_R given by eq.(37) yields

$$z_R \sim e^l \sim |\varepsilon|^{-1/2}$$
and
$$\xi_{//R} \sim |\varepsilon|^{-3/4}. \quad (40)$$

Using the scaling (20), it gives

$$z \sim \lambda\, |\varepsilon|^{-1} = \lambda\, kT/|kT-\alpha W|\,, \quad (41)$$

$$\xi_{//} \sim \lambda\, |\varepsilon|^{-5/4}.$$

Eq.(41) shows the existence of a continuous phase transition from a state where the lamella are bound together , to a state where they separate ($z\rightarrow\infty$). This transition is characterized by a divergence of the mean lamellar spacing $z\sim\varepsilon^{-\psi}$ with a critical exponent $\psi=1$ according to our theory. This compares well with the numerical determination of ref.[11]. The driving parameter is $kT-\alpha W$. The $kT\sim W$ critical point can be reached either by varying the temperature or by changing the polarizability of the aqueous medium (i.e. changing W). However, reaching the unbinding transition by varying the temperature should be exceptional since kT varies by no more than 20°/ₒ ($\sim$4-5 10^{-21} J) for attainable temperatures (20-100°C). Typical values for W are : W$\sim$2-6 10^{-21} J [19,20] well within the range of kT. This unbinding transition does not seem to have been observed experimentally. However, two experimental groups[19,20] have sustained a controversy on the problem of the equilibrium spacing of multibilayers systems, one[20] claiming that membranes naturally separate in excess water ($z_{eq}\rightarrow\infty$) , whereas the other group[19] found a small finite separation ($\sim$20-30Å).
The Hamaker constant estimated by each group is different and seems in qualitative agreement with the expected trend :
ref[20] : W$\sim$2-4 10^{-21}J $\leqslant$ kT, ($z\rightarrow\infty$),
ref[19] : W$\sim$4-6 10^{-21}J $\geqslant$ kT (z finite).
More experiments are needed to ascertain the existence and features of the unbinding transition in multibilayers.

In microemulsions, most lamellar phases have been shown to be characterized by finite small lamellar spacings (~20-50A). However, some of them have been shown to swell upon addition of brine (water+salt) or mixtures of oil and alcool, up to extremely large spacings (~1μm)[21] indicating unbound states in excess diluant. It would be also extremely interesting to carry on systematic experiments in these systems in order to ascertain the existence of these phase transitions.

5.4.2 PHASE SEPARATION BETWEEN DIFFERENT LAMELLAR PHASES ($\lambda>0$)

A first order phase separation between two different lamellar systems can occur when λ is positive (soft membranes). The reader is referred to ref.[12] for details in this regime.

The first order phase transition separates an unbound membrane phase from a finite spacing lamellar phase. However, two coexisting finite spacing lamellar phases can be observed if an additional external pressure p is applied. Then, the ($z=\infty$) phase gives place to a swollen lamellar phase whose spacing scales as

$$z \sim p^{-1/3}. \qquad (42)$$

The stabilizing external pressure may be an osmotic pressure exerted by small particles, unable to penetrate the space between the lamellae because of steric hindrance, or electrostatic repulsion [22]. Therefore one can expect from the previous results that in the presence of micellar and submicellar particles, a phase separation within a lamellar phase is possible : one with a large spacing admitting these small particles and another with a smaller spacing expelling them. This indeed has been observed in several ternary and quaternary systems (see ref [1] and [23]). For example, in the decanol-water-potassium caprylate system, the main lamellar phase D (small spacing) is found to be in coexistence , in a certain concentration range, with a swollen lamellar phase B which itself is in coexistence with a solution of micelles. Some doubts remain however on the exact nature of the B phase. In view of our results, other studies should be carried on to confirm these features.

Another interesting experimental fact is that this mesophase B has only been found in ternary and not in binary systems [1]. This is in agreement with our ideas on the role of the membrane-overlap parameter λ. The fourth component (the alcohol serving as a cosurfactant) , which is known to decreases strongly the short range repulsion[18] and to lower the membrane curvature rigidity, could have the role of allowing partial overlaps between adjacent membranes. Experimentally, one could explore the λ-axis by varying the amount of cosurfactant.

6-FRACTAL MEMBRANES DUE TO THERMAL SOFTENING OF THE BENDING RIGIDITY

The influence of thermal fluctuations has recently been recognized as leading to a softening of the bending elastic modulus at large scales[26]. This result was obtained by taking into account the natural non-linear couplings between capillary modes present even in absence of steric constraints.

To first order in kT/K_0 (where K_0 is the bare curvature rigidity), one has[26]

$$K(L) = K_0\{ 1- \alpha kT/4\pi K_0 \,\mathrm{Log}\,(L\Lambda) \} ,$$

where α= 1 or 3 depending on the authors[26]. This can be rewritten as

$$K(L) \sim K_0 (\Lambda L)^{-\chi} \quad \text{for } L < \xi_p = \Lambda^{-1} . e^{+4\pi K_0/\alpha kT} . \qquad (43)$$

The exponent χ is

$$\chi = \alpha kT/4\pi K_0 \text{ for } L < \xi_p , \qquad (44)$$

where ξ_p is a persistence length for a single membrane.

For $L > \xi_p$, the normals to the membranes are decorrelated and the bending rigidity must involve entropy reductions similar to the problem of polymer elasticity. This can be estimated as follows. On scales $L >> \xi_p$, the membrane is crumpled and can be characterized by its radius of gyration $R_g(=L)$:

$$R_g \sim L_t^{\nu} , \qquad (45)$$

where L_t^2 is the true area of the membrane and $\nu \leqslant 1$. The typical curvature is $\underline{C} \simeq R_g^{-1}$. The probability distribution for the curvature should be gaussian

$$P(C) \sim \exp\{-C^2/2\underline{C}^2\} = \exp\{-H(C)/kT\} ,$$

where

$$H(C) = (kT/2\underline{C}^2 L_t^2) . L_t^2 C^2 .$$

Comparing with eq.(2), this leads to the following value for the effective curvature rigidity at the scale L_t:

$$K \simeq kT . (R_g/L_t)^2 \sim kT\, L^{2(\nu-1)/\nu} , \qquad (46)$$

where we have used $L \sim R_g$. This defines

$$\chi = 2(1-\nu)/\nu , \qquad \text{for } L > \xi_p . \qquad (47)$$

If $\nu = 1$, $K \simeq kT$. However, this is not in general true. Indeed, for self-avoiding surfaces, $\nu \simeq 6/(4+D) = 6/7$ in $D=3$[27], whereas for tethered surfaces, $\nu \simeq 4/(2+D) = 4/5$ in $D=3$[28]. This leads to

$$K \sim kT\, L^{-1/3} \quad \text{for self-avoiding surfaces in 3D[27]},$$
$$\sim kT\, L^{-1/2} \quad \text{for tethered surfaces in 3D[28]}. \qquad (48)$$

This softening of the membrane rigidity by thermal fluctuations implies a fractal structure, with Hausdorf dimension $D_f = 2/\nu$. For $L < \xi_p$, $D_f \simeq 2 + \chi$ with $\chi = 3kT/4\pi K_0$, in agreement with the result of ref.[26].
Starting from eq.(10) reading $L = (K/kT)^{1/2} \xi_\perp$ valid for $L < \xi_p$, using eq.(43), we obtain

$$\xi_\perp \sim (kT/K_0)^{1/2} \Lambda^{\chi/2} L^{1+\chi/2} , \quad \text{for } L < \xi_p \qquad (49)$$

For $L > \xi_p$, one expects $\xi_\perp \sim L$.

This allows one to derive the correction to the steric interaction due to bending rigidity softening:

$$F_{st} \sim \{(kT)^2/K_0\}\, z^{-2+2\chi} \quad \text{for } L<\xi_p ,$$
$$\sim kT\, z^{-2} \quad \text{for } L>\xi_p . \tag{50}$$

These corrections should have measurable consequences for the stability of hyperswollen lamellar phases [21].

Also, it is not clear whether the scaling form (48) for K , valid for $L>\xi_p$ and which corrects expression (43) valid for $L<\xi_p$, will induce important changes in the theory of middle phase microemulsions based on the effect of the thermal softening of K[29].

7-CONCLUSION

This review was aimed at showing the importance of fluctuations , due to thermal excitations, in the determination of interface interactions. The important fact to recognize is that, due to the low dimensionality of the systems (2D-interfaces in 3D-bulk), the fluctuations are very important and therefore change the interactions drastically. The complex interplay between fluctuations and the bare microscopic interactions must be addressed fully to be able to understand the physics of systems made of interfaces. The renormalization group is therefore a natural tool to develop in this context. Note that these systems offer an intuitive illustration of the meaning and functioning of the RG.

The main result is that in many cases the fluctuations rub out the details of the microscopic interactions which characterize each particular system and they lead to universal behaviour.

Acknowledgments: Some of the ideas which have been presented are shared with N.Ostrowsky. I am grateful to D.Forster, W.Helfrich, D.Roux and J.Villain for stimulating discussions and to P.G.de Gennes for his hospitality at Collège de France in Paris where part of this work was completed.

REFERENCES :

[1] P. EKWALL "Composition, Properties and Structures of Liquid Crystalline Phases in Systems of Amphiphilic Compounds". in Advances in Liquid Crystals. Vol. 1, p. 1, ed. G.H. BROWN (1975).
D. ROUX, PhD Thesis (1984), Bordeaux.
See for a review, " Physics of Amphiphiles: Micelles, Vesicles and Microemulsions". North-Holland, ed. DEGIORGIO V. and CORTI M. (1985)
[2] E.BREZIN, B.I.HALPERIN and S.LEIBLER, Phys.Rev.Lett.50,1387(1983); J.Physique,44,775(1984)
[3] V.L.POKROVSKY and L.TAPALOV, "Theory of incommensurate crystals", Soviet Scientific Review, Physics, Vol.1(1984).
[4] J.VILLAIN, in "Scaling Phenomena in Disordered Systems", ed. by R.Pynn and A. Skjeltorp, Plenum Publishing Corporation (1985)
[5] W. HELFRICH, Z.Naturforsh,33a,305(1978).
[6] M.E. FISHER and D.S. FISHER, Phys. Rev. B25,3192(1982).
[7] D. SORNETTE and N. OSTROWSKY, J. Physique, 45,265 (1984)
[8] a)D. SORNETTE, Euro. Phys. Lett.2,715(1986)
b)J.BRICMONT, A.EL MELLOUKI and J.FROHLICH, J.Stat.Phys.42,743(1986)

[9] a)D.SORNETTE and N. OSTROWSKY,Chem.Scr.,25,108(1985) and J.Chem.Phys. 84,4062(1986)
b)E.A.EVANS and V.A.PARSEGIAN, Proc.Natl.Acad.Sci.USA,83,7132,(1986)
[10] L.J.LIS,M.McALLISTER,N.FULLER,R.P.RAND and V.A.PARSEGIAN, Biophys.J., 37,657(1982).
[11] R.LIPOWSKY and S.LEIBLER, Phys.Rev.Lett.56,2541(1986)
[12] D. SORNETTE, "Tricritical behaviour in lamellar phases", to appear
[13] D.SORNETTE and N. OSTROWSKY, "Renormalized interactions between membranes", to be published in the Proceedings of "The international symposium on surfactants in solution", New Dehli, August 1986, ed.K.L.MITTAL
[14] D.S. FISHER and D.A. HUSE, Phys. Rev. B32, 247 (1985)
[15] R.LIPOWSKY and M.E.FISHER, Phys.Rev.Lett.57,2411(1986)
[16] J.VILLAIN, in Proceedings of the summer school Beg-Rohu,June 1985, Edition de Physique, ed. C.GODRECHE
[17] D. ROUX, PhD Thesis (1984), Bordeaux.
[18] J.ISRAELACHVILI and D.SORNETTE, J.Physique,46,2125(1985)
[19] S.NIR, Progress.Surf.Sci.8,1(1976).
[20] W.HARBICH and W. HELFRICH, Chem.Phys.Lipids,36,39(1984).
[21] F.C.LARCHE,J.APPEL and G.PORTE, Phys.Rev.Lett.56,1700(1986).
[22] R.KJELLANDER and S. MARCELJA, Chemica Scripta,25,1(1985) and Chem.Phys.Lett. in press.
[23] W. HELFRICH, "Amphiphilic mesophases made of defects", Les Houches,XXXV, ed. R.BALIAN et al (1980).
[24] D.S. FISHER, Phys. Rev.Lett. 56,1964(1986).
[25] R.LIPOWSKY and M.E.FISHER, Phys.Rev.Lett.56,472(1986)
[26] W.HELFRICH, J. Physique, 46,1263 (1985) and
L.PELITI and S.LEIBLER, Phys. Rev.Lett. 54,1960(1985).
D.SORNETTE, Thèse d'état, Nice(1985)
[27] A.MARITAN and A.STELLA, Phys. Rev.Lett. 53,123(1984).
[28] Y.KANTOR, M.KARDAR and D.R.NELSON, Phys. Rev.Lett. 57,791(1986).
[29] .A.SAFRAN, D.ROUX, M.CATES and D.ANDELMAN, Phys. Rev.Lett. 57,491(1986)

Tangential Flows in Fluid Membranes and Their Effect on the Softening of Curvature Rigidity with Scale

D. Foerster

Institut für Physik, Freie Universität Berlin,
Arnimallee 14, D-1000 Berlin, Germany

ABSTRACT

During perpendicular displacements of fluid membranes, lipids must flow along the membrane to conserve lipid density. The kinetic energy of this flow must be taken into account when setting up the canonical phase space needed in the thermal statistics of fluid membranes. We confirm an earlier calculation of Helfrich on the precise amount of softening of curvature rigidity of fluid membranes as a function of distance scale.

(To be published in Europhysics Letters).

Unbinding of Membranes

R. Lipowsky[1] *and S. Leibler*[2]

[1]Institut für Festkörperforschung, KFA Jülich, D-5170 Jülich, Fed. Rep. of Germany

[2]Baker Laboratory, Cornell University, Ithaca, NY 14853, USA

I. INTRODUCTION

Ordered lamellar phases are formed in many-component systems containing amphiphilic molecules. In these phases, roughly parallel lamellae or membranes are separated by a fluid medium. Examples are (i) phospholipid bilayers separated by water, and (ii) swollen bilayers separated by oil (or water). In the latter case, the membranes consist of thin layers of water (or oil) sandwiched between two monolayers of surfactant. Recent swelling and dilution experiments have shown /1-4/ that the membrane spacing can become large compared to the membrane thickness. During such swelling or dilution experiments, the membranes undergo an unbinding process from a state where they are bound together to a state where they are completely separated.

The membrane separation is controlled by the intermembrane interactions. It is now generally accepted that one may distinguish two different types of interactions: (i) *direct* interactions between planar membranes which reflect the microscopic forces between the molecules /5,6/, and (ii) *fluctuation-induced* interactions arising from thermally excited undulations of the membranes as pointed out by Helfrich /7/. From a theoretical point of view, it is clearly desirable to obtain a *systematic* description for the *interplay* between these two types of interactions. This paper reviews some recent work in this direction /8,9/.

The next two sections contain a brief description of direct and fluctuation-induced interactions. Then, an effective Hamiltonian (or free energy functional) for the membrane configurations is introduced which serves as a starting point for a systematic theory, see Sec. IV. The statistical properties of this model can be investigated by a variety of theoretical methods. So far, the most powerful method is a functional renormalization group approach /8,10/ which is described in Sec. V. Then, two different types of unbinding transitions are discussed in Sec. VI and VII.

II. DIRECT INTERACTION BETWEEN PLANAR MEMBRANES

First, consider two *planar* membrane segments which are roughly parallel and, thus, have a constant separation, $\ell > 0$. The internal structure of the membranes will be ignored, i.e., they will be regarded as featureless sheets or drumheads of thickness δ. The direct interaction between these membranes has the generic form

$$V(\ell) = P\ell + V_m(\ell) \tag{1}$$

for $\ell > 0$. The parameter P is (i) an external pressure /5/, or (ii) a Lagrange multiplier resulting from a constraint on the system /9/. The direct interaction, $V_m(\ell)$, for $P = 0$ arises from microscopic forces between molecules. In addition to these interactions which operate for $\ell > 0$, $V(\ell)$ also contains a *hard wall repulsion* given by

$$V_{HW}(\ell) = \infty \quad \text{for } \ell < 0$$
$$= 0 \quad \text{for } \ell > 0 \tag{2}$$

resulting from the steric hindrance of the membranes.

For two *neutral* or *uncharged* membranes, the interaction $V_m(\ell)$ is usually governed, for large ℓ, by van der Waals forces. As long as retardation effects can be ignored, one has $V_m(\ell) \simeq -W\delta^2/\ell^4$ for large ℓ where W is the Hamaker constant. Retarded van der Waals forces, on the other hand, will lead to $V_m(\ell) \simeq -1/\ell^5$ which is expected to apply for $\ell \gtrsim 50$ nm /8/. Note that these effective van der Waals interactions between two identical membranes are always *attractive*. For small values of ℓ, two membranes separated by a water layer interact by repulsive short-ranged hydration forces /5,6/. For swollen bilayers separated by oil, one expects an additional attractive interaction as soon as the hydrocarbon chains of the surfactant molecules start to overlap. If the membranes are *charged*, they repel each other by electrostatic interactions. For ℓ large compared to the Debye length, λ_E, of the ionic solution, this electrostatic repulsion decays exponentially $\simeq \exp(-\ell/\lambda_E)$. If there are no ions between the membranes apart from counterions, λ_E is infinite (in practise, $\lambda_E \simeq 10^2$-10^3nm in water) and the electrostatic interactions lead to the Langmuir repulsion which behaves as $\simeq 1/\ell$ for large ℓ /6/.

The direct interactions described so far have been known for a long time. More recently, it has been argued that, for $\ell \simeq 1$-10 nm, the intermembrane interactions are more complex and not completely understood /11/. However, the process of unbinding studied here does not depend on the detailed form of the interaction $V(\ell)$ but only on its gross features. Furthermmore, the theoretical approach reviewed in this paper is quite general and can be applied to any form of $V(\ell)$. In fact, one important result of this approach is a *classification scheme* by which one can distinguish different classes of direct interactions, $V(\ell)$.

In principle, the shape of $V_m(\ell)$ could be determined by experiments since, for planar membranes, the pressure P is given by

$$P = P(\ell) = -\partial V_m/\partial\ell. \tag{3}$$

Thus, one could calculate $V_m(\ell)$ from the measured values of $P(\ell)$. Indeed, such a procedure has often been used for bilayers in water /5/. One must realize, however, that real membranes are, in general, not planar but deformed as a result of thermal fluctuations. In such a situation, the measurement of $P(\ell)$ will determine an *effective* interaction which is already renormalized by membrane fluctuations. On the other hand, force measurements in the absence of thermal fluctuations are possible when the membranes are attached to mica surfaces /6,11/.

III. FLUCTUATION-INDUCED INTERACTIONS

Some years ago, Helfrich pointed out /7/ that thermal fluctuations lead to an effective *repulsive* interaction between membranes. For spatial dimensionality d=3, he predicted that these undulations together with a hard-wall interaction as in (2) lead to the fluctuation-induced interaction/7/

$$V_{FL}(\ell) \simeq (k_B T)^2/\kappa \ell^2, \tag{4}$$

where κ is the rigidity constant. Recent experiments indicate that this parameter is roughly given by $\kappa \simeq 10^{-12}$erg for lipid bilayers /1,12/ and by $\kappa \simeq 10^{-14}$erg for swollen bilayers in quasi-ternary systems /2-4/.

It is instructive to rederive the effective repulsion as given by (4) from scaling arguments /8/. First, assume that the membranes are completely separated. Then, each membrane is governed by the effective Hamiltonian (or free energy functional) /13,14/

$$H_o\{\ell\} = \int d^{d-1}x \, \frac{1}{2} \, \kappa \, (\nabla^2 \ell)^2, \tag{5}$$

where $\ell = \ell(\underset{\sim}{x})$ measures the distance of the membrane from a reference plane with coordinate $\underset{\sim}{x}=(x_1,...,x_{d-1})$. Here it is assumed that the spontaneous curvature of the membranes is negligible. Likewise, higher order terms arising from the mean curvature are omitted which is justified as long as $(\nabla \ell)^2 << 1$. Now, it follows from (5) that a membrane segment of longitudinal dimension, $L_{\parallel}$, will typically make a transverse excursion, $L_{\perp} \sim L_{\parallel}^{\zeta}$, with the spatial anisotropy exponent /8/

$$\zeta = (5-d)/2 \qquad \text{for } d < d_o \equiv 5. \tag{6}$$

On the other hand, if the membranes are bound together, the transverse excursions are limited and the largest humps have a typical size, $\xi_{\perp}$. Then, scaling arguments show that these largest humps or fluctuations have a longitudinal extension, $\xi_{\parallel}$,with /8/

$$\xi_{\perp} \approx (\mathscr{C}_{\infty} \, k_B T/\kappa)^{1/2} \, \xi_{\parallel}^{\zeta}, \tag{7}$$

where $\mathscr{C}_{\infty} \sim O(1)$ is related to the difference correlation function. For the Gaussian model in d = 3, one has $\mathscr{C}_{\infty} \simeq 0.125$. Similar ideas have been presented by Sornette /15/.

The largest humps or fluctuations of the confined membrane lead to an increase of the free energy per unit area as given by /8/

$$V_{FL}(\xi_{\perp}) \approx c_V \, k_B T \, (k_B T/\kappa)^{\tau/2} \, \xi_{\perp}^{-\tau} \tag{8}$$

for d<5 with $c_V \sim O(1)$ and decay exponent

$$\tau = 2(d-1)/(5-d) \tag{9}$$

For d=3, the Helfrich interaction (4) is recovered provided one assumes that $\ell \sim \xi_{\perp}$. This relation is indeed valid for a hard wall repulsion, i.e., for $V(\ell)=V_{HW}(\ell)$ as given by (2) but does not hold in general: sufficiently long-ranged interactions $V_m(\ell)$ lead to $\xi_{\perp} << \ell$, /8,9/.

IV. EFFECTIVE HAMILTONIAN FOR INTERACTING MEMBRANES

So far, two types of interactions have been considered: (i) direct interactions between planar membranes, i.e., in the absence of thermal fluctuations, and (ii) fluctuation-induced interactions where the explicit form of the direct interaction $V(\ell)$ has been ignored. Now, an effective Hamiltonian, $\mathscr{H}(\ell)$, will be introduced in order to study the interplay between these two interactions in a systematic way.

First, the local distance of two membranes from a reference plane is denoted by $\ell_1(\underset{\sim}{x})$ and $\ell_2(\underset{\sim}{x})$ with $\underset{\sim}{x}=(x_1,\ldots,x_{d-1})$ as before. Now, deformations of these membranes are governed by $\mathscr{H}_o\{\ell_1\}+\mathscr{H}_o\{\ell_2\}$ where the effective Hamiltonian $\mathscr{H}_o$ has been defined in (5). Furthermore, the direct interaction beween the deformed membranes is taken to be $V[\ell_1(\underset{\sim}{x})-\ell_2(\underset{\sim}{x})]$. It is then convenient to introduce new variables, $\ell=\ell_1-\ell_2$ and $\ell_o=(\ell_1+\ell_2)/2$. Then, the effective Hamiltonian separates into two terms, $\mathscr{H}_o\{\ell_o\} + \mathscr{H}\{\ell\}$, with $\mathscr{H}_o$ as given by (5) and /8,9/

$$\mathscr{H}\{\ell\} = \int d^{d-1}x \; \{ \; \tfrac{1}{2} \kappa \; (\nabla^2\ell)^2 + V[\ell(\underset{\sim}{x})] \; \} \, . \tag{10}$$

This expression implicitly contains a small-distance cutoff, $1/\Lambda \simeq \delta$, where δ is the membrane thickness. The Hamiltonian as given by (10) governs the unbinding process of the membranes. Obviously, it contains both an elastic free energy as in (5) associated with membrane deformations and an interaction free energy which is given in terms of $V(\ell)$. Now, one can use the general rules of statistical mechanics in order to calculate thermodynamic quantities and expectation values of ℓ via the Boltzmann factor, $\exp[-\mathscr{H}/k_BT]$.

V. FUNCTIONAL RENORMALIZATION

The effective Hamiltonian (10) resembles the interface models studied in the context of wetting phenomena, see, e.g., Refs.16-18. Such models have been studied by a variety of field-theoretic methods. Not surprisingly, the same methods can be used in the present context. Here, I will describe a functional renormalization group (RG) approach /10,8/ which has been found to be most widely applicable.

This functional RG which represents an extension of Wilson's approximate recursion relation /19/ preserves the form of the effective Hamiltonian (10): it acts as a nonlinear map in the function space of direct interactions, $V(\ell)$, while the elastic term, $(\nabla^2\ell)^2$, remains unchanged. This implies that the rigidity, κ, stays finite at the unbinding transition or, in field-theoretic language, that the variable ℓ does not acquire an anomalous dimension. This is indeed valid for the effective Hamiltonian as given by (10) as one can show by scaling arguments /8/ and within the loop expansion /9/.

In order to write the recursion relation for the RG in a transparent way, let us introduce the free energy density scale

$$v \equiv k_B T \int_{\Lambda/b}^{\Lambda} d^{d-1}p \;/\; (2\pi)^{d-1} \tag{11}$$

and the length scale, $a_\perp$, defined by

$$a_\perp^2 \equiv (k_B T/\kappa) \int_{\Lambda/b}^{\Lambda} d^{d-1}p \;/\; (2\pi)^{d-1}\, p^4, \tag{12}$$

where $\Lambda \simeq 1/\delta$ is the high-momentum cutoff and $b>1$ is the usual spatial rescaling factor. The length scale $a_\perp$ can be regarded as the roughness of the membranes arising from small-scale excitations with wavenumber $\Lambda/b \lesssim p \lesssim \Lambda$. Then, the initial interaction $V^{(0)}(\ell) \equiv V(\ell)$ is renormalized by successive applications of

$$V^{(N+1)}(\ell) = \mathfrak{R}[V^{(N)}(\ell)] \tag{13}$$

where /8,20/

$$\mathfrak{R}[V(\ell)] \equiv -vb^{d-1} \ln\{ \int_{-\infty}^{+\infty} dz \exp[-\tfrac{1}{2}(z/a_\perp)^2 - G(\ell,z)]/[2\pi a_\perp^2]^{1/2} \} \tag{14}$$

with

$$G(\ell,z) \equiv [V(b^\zeta \ell - z) + V(b^\zeta \ell + z)] \;/\; 2v \tag{15}$$

and $\zeta=(5-d)/2$ as in (6). In the infinitesimal rescaling limit $b=\exp(\delta t)$ with $\delta t \to 0$, one obtains the nonlinear flow equation given by

$$dV/dt = (d-1)V + \zeta \ell \partial V/\partial \ell + \tfrac{1}{2}B \ln[1+(A^2/B)\partial^2 V/\partial \ell^2] \tag{16}$$

with cutoff-dependent scale parameters A and B.

These recursion relations are very similar to the nonlinear recursion relation used in the context of wetting /10/. Indeed, these two RG's differ only (i) in the spatial anisotropy exponent, ζ, which determines the rescaling factor of ℓ; and (ii) in the choice for the length scale $a_\perp$ which ensures that this RG is exact to linear order in V for all values of b and d. Therefore, most features of the functional RG found for wetting/10/ have immediate analogies for the RG used here for membranes.

By definition, a RG fixed point, V^*, satisfies $\mathfrak{R}[V^*(\ell)]=V^*(\ell)$. This relation is trivially fulfilled for the Gaussian fixed point $V_G^*=0$. Furthermore, numerical iterations of the recursion relation as given by (13)-(15) reveal two nontrivial fixed points, $V_o^*(\ell)$ and $V_c^*(\ell)$, for $d<d_o=5$. In $d=d_o=5$, these two fixed points do not bifurcate from the Gaussian fixed point, $V_G^*=0$, but rather from an unusual line of drifting fixed points as will be described elsewhere /21/, see also /10/.

Within the subspace of direct interactions, $V_m(\ell)$, which satisfy

$$V_m(\ell) \ll 1/\ell^\tau \quad \text{for large } \ell \quad , \tag{17}$$

with $\tau = 2(d-1)/(5-d)$, there is no relevant perturbation at the fixed point $V_o^*(\ell)$. Therefore, this fixed point has a large domain of attraction. In particular, a hard wall repulsion, $V_{HW}(\ell)$, as given by (2) is mapped onto V_o^*. Therefore V_o^* may be called the "hard wall fixed point". In contrast, there is one relevant perturbation at the other fixed point, $V_c^*(\ell)$, which governs the critical effects at a critical unbinding transition /8/. These features apply to general d < 5. In the remaining sections, I will discuss the physical case of 3-dimensional systems.

VI. COMPLETE VERSUS INCOMPLETE UNBINDING

Consider two membranes which are in a bound state as the result of an external pressure or constraint corresponding to $P > 0$. As $P \to 0$, the mean separation, $\overline{\ell}$, of the membranes can (i) attain a finite limit, or (ii) become arbitrarily large. These two cases correspond to *incomplete* and *complete* unbinding, respectively. It is intuitively clear that the behavior of the membranes for $P \to 0$ will, in general, depend on their direct interaction, $V_m(\ell)$. If this interaction is *repulsive*, i.e., $V_m(\ell) \approx V_R(\ell) \geq 0$ for large ℓ, the membranes will completely unbind as $P \to 0$. Such a situation occurs, for example, for the Langmuir repulsion $V_m(\ell) \sim 1/\ell$ arising from the pressure of counterions. On the other hand, the direct interaction can also be *attractive* for large ℓ. Then, one may write

$$V_m(\ell) \approx V_A(\ell) + V_R(\ell) \tag{18}$$

with $|V_A(\ell)| \gg V_R(\ell)$ for large ℓ where $V_A(\ell)$ and $V_R(\ell)$ represent the attractive and the repulsive part of the interaction. In this case, the membranes can *not* unbind completely if $V_A(\ell)$ is sufficiently long-ranged and satisfies $|V_A(\ell)| \gg 1/\ell^2$ for large ℓ (in d=3). For real systems, the attraction comes from van der Waals forces which lead to $V_A(\ell) \sim 1/\ell^4$ or $\sim 1/\ell^5$ for large ℓ as mentioned. Thus, one has

$$|V_A(\ell)| \ll 1/\ell^2 \text{ for large } \ell. \tag{19}$$

In the latter case, the membranes can unbind completely or incompletely depending on the *strength* of the attractive interaction, $V_A(\ell)$ /8/. What happens can be determined within the functional RG approach described in Sec. V. Indeed, this RG provides a *simple criterion for complete unbinding*: the membranes unbind completely whenever they interact by a direct interaction, $V_m(\ell)$, which is mapped onto the hard wall fixed point , V_o^*.

Now, assume that $V_m(\ell)$ leads to complete unbinding. Then, the mean separation, $\overline{\ell}$, behaves as

$$\overline{\ell} \sim 1/P^\psi \quad \text{as} \quad P \to 0. \tag{20}$$

The critical exponent, ψ, is correctly given by mean-field (MF) theory as long as $V_m(\ell) = V_R(\ell) >> 1/\ell^2$ (in d=3). This applies, for instance, to the Langmuir repulsion which leads to /9/

$$\psi = 1/2 \qquad \text{for } d = 3 \quad . \tag{21}$$

Within MF theory, the exponent ψ reflects the nature of the underlying microscopic forces. On the other hand, for sufficiently short-ranged interactions with $|V_m(\ell)| << 1/\ell^2$ for large ℓ, one finds the *universal* value /8,9/

$$\psi = 1/3 \qquad \text{for } d = 3 \ . \tag{22}$$

Thus, there are two scaling regimes for complete unbinding /9,18/: (i) a MF regime, and (ii) a weak-fluctuation (WFL) regime which is characterized by nonclassical critical exponents while the phase boundary is still given by P=0 as in MF theory.

These scaling regimes can be distinguished experimentally via the behavior of the scattering intensity, I, of X-rays which exhibit Laudau-Peierls singularities:

$$I(q_z) \sim (q_z - q_m)^{-(2-X_m)} \quad ,$$

where $q_m = 2\pi m/\bar{\ell}$ and q_z is the momentum transfer perpendicular to the membranes /3,4,9/. In the WFL-regime, the exponent X_m is independent of $\bar{\ell}$ for large $\bar{\ell}$. In contrast, it depends explicitly on $\bar{\ell}$ within the MF regime: e.g., one finds $X_m \sim 1/\bar{\ell}^{1/2}$ in the presence of the Langmuir repulsion /9/. This qualitatively different behavior has indeed been observed in recent high-resolution X-ray experiments /22/.

VII. CRITICAL UNBINDING TRANSITION

Now, assume that P=0 and consider again a direct interaction $V_m(\ell) = V_A(\ell) + V_R(\ell)$ with $|V_A(\ell)| << 1/\ell^2$ for large ℓ as in (19). The parameters of such an interaction span a low-dimensional space. Within this parameter space, there is a region where the membranes are completely separated and another region where they are bound together /8/. Obviously, these two regions must be separated by a phase boundary. In fact, it turns out /8/ that the transition along this phase boundary is typically a second order transition at which the mean separation, $\bar{\ell}$, of the membranes diverges in a continuous manner. Thus, the parameter space contains a manifold of critical unbinding points. The critical behavior along this manifold which depends on two different scaling fields has been described in Ref. 8 and will be discussed in more detail in Ref. 21. Recent experiments on a quasi-ternary mixture indicate that such a critical unbinding transition can be studied experimentally by changing the membrane thickness, δ /23/.

VIII. SUMMARY AND OUTLOOK

In summary, the interplay between direct and fluctuation-induced interactions leads to critical phenomena associated with the unbinding of amphiphilic membranes. We have theoretically studied these critical effects both for complete unbinding, see Sec.VI, and for unbinding critical points, see Sec.VII. Our theoretical results are confirmed by recent experiments.

The theory for unbinding of membranes is far from complete. Some interesting problems which we currently investigate are: (i) the marginal case with $V_m(\ell) \sim -1/\ell^2$ which should lead to an unbinding transition of infinite order, (ii) tricritical unbinding transitions which should be governed by yet another fixed point, compare /10/, and (iii) the influence of higher order gradient terms arising from the mean curvature.

ACKNOWLEDGEMENTS

We would like to thank the organizers of this conference, Dominique Langevin and Jacques Meunier, for the possibility to participate in this exciting event. We are grateful to Michael E. Fisher, Ray Goldstein, Geoffrey Grinstein, Frederic Nallet, Jacques Prost, Didier Roux, Cyrus R. Safinya, Marilyn Schneider, Daniel Wack, and Ben Widom for helpful and stimulating discussions. One of us (S.L.) acknowledges the support of the National Science Foundation through the Materials Science Center at Cornell University.

REFERENCES

1. W. Harbich and W. Helfrich, Chem.Phys.Lipids **36**, 39 (1984)
2. J.M. di Meglio, M. Dvolaitzky, L. Leger, and C. Taupin, Phys. Rev. Lett. 54, 1686 (1985)
3. F. Larche, J. Appell, G. Porte, P. Bassereau, and J. Marignan, Phys.Rev.Lett. 56, 1700 (1986)
4. C.R. Safinya, D. Roux, G.S. Smith, S.K. Sinha, P. Dimon, N.A. Clark, and A.M. Bellocq, Phys.Rev.Lett. 57, 2718 (1986).
5. R.P. Rand, Ann. Rev. Biophys. Bioeng. **10**, 277 (1981)
6. J.N. Israelachvili, *Intermolecular and Surface Forces* (Academic Press, Orlando, Fla., 1985)
7. W. Helfrich, Z. Naturforsch. **33a**, 305 (1978)
8. R. Lipowsky and S. Leibler, Phys. Rev. Lett. **56**, 2541 (1986)
9. S. Leibler and R. Lipowsky, Phys. Rev. B. (in Press)
10. R. Lipowsky and M.E. Fisher, Phys. Rev. Lett. **57**, 2411 (1986), and (to be published)
11. D.F. Evans and B.W. Ninham, J. Phys. Chem. **90**, 226 (1986)
12. M.B. Schneider, J.T. Jenkins, and W.W. Webb, J.Physique **45**, 1475 (1984)
13. W. Helfrich, Z. Naturforsch. **28c**, 693 (1973)
14. P.G. de Gennes and C. Taupin, J. Phys. Chem. **88**, 2294 (1982)
15. D. Sornette, presentation at this conference
16. R. Lipowsky, D.M. Kroll, and R.K.P. Zia, Phys. Rev. B27, 4499 (1983)
17. E. Brezin, B. Halperin, and S. Leibler, Phys. Rev. Lett. **50**, 1387, (1983)
18. R. Lipowsky, Phys. Rev. Lett. **52**, 1429 (1984)
19. K.G. Wilson, Phys. Rev. B4, 3184 (1971)
20. Eq. (20) and (21) of Ref. 8 contain several misprints, and the scale factors v and $a_\perp$ have been absorbed in V and ℓ.
21. R. Lipowsky and S. Leibler (to be published)
22. D. Roux, presentation at this conference
23. D. Roux and C. Safinya, private communication

Elasticity of Crystalline and Hexatic Membranes

L. Peliti and D.R. Nelson*

Lyman Laboratory of Physics, Harvard University,
Cambridge, MA 02138, USA

1. Introduction

We report the results of recent investigations on the statistical mechanics of membranes with in-plane crystalline or hexatic order[1]. Our motivation was to understand the interplay between thermally excited shape fluctuations (undulations) and degrees of freedom associated with the internal ordering of the membrane. The more usual membranes, in particular those described by the Helfrich[2] free energy, are fluid in the sense that the molecules forming them freely and quickly rearrange to comply with shape fluctuations. As a consequence, it becomes possible to describe them as geometrical surfaces, whose free energy depends only on their shape.

The situation is very different if the molecules are not free to move about, as e.g. in a rubber sheet or in polymerized membranes. It then costs a certain amount of elastic energy to modify the mutual distances of the molecules, even if the overall membrane shape remains invariant. The elastic energy due to these deformations is called *stretching energy*, to distinguish it from the *bending energy* associated with undulations. Stretching and bending energies are sensitive to different properties of the deformation of the surface. Let us consider a membrane whose equilibrium shape is a plane. One can give it a cylindrical shape without affecting the mutual distances of the molecules forming it, by bending it in the way one bends a sheet of paper. The *intrinsic geometry* of the surface is not affected by this transformation. This deformation does require bending energy, because the final state has a nonzero mean curvature H. On the other hand, if we warp an initially plane membrane to form a minimal surface, there is no bending energy cost (H remains equal to zero), but there is a stretching energy . It is in fact impossible to perform this deformation without modifying the mutual distances between molecules in the membrane, as is witnessed by the fact that the *Gaussian* curvature K does not vanish in a minimal surface. Stretching elasticity is sensitive to fluctuations in the *intrinsic geometry* of the surface, whereas bending elasticity is sensitive to its *embedding* in ambient space.

The behavior of fluid membranes, where only bending energy is present, is the subject of several contributions to this Workshop. They are crumpled at any nonzero temperature[3,4] and lose orientational order in their normals over distances of the order of the *persistence length* ξ_b introduced by de Gennes and Taupin[5]. Surfaces with stretching energy, but without bending energy, have been considered by several authors[6,7]. The result is that these membranes are also crumpled at all temperatures. It came therefore as a surprise that if *both* bending and stretching energy effects are present the surface is rigid

* Permanent address: Dipartimento di Fisica and Unità GNSM-CNR, Università "La Sapienza", Piazzale Aldo Moro 2, I-00185 Roma (Italy).

and flat at low temperatures[1]. A *crumpling transition*[4] takes place at finite temperature, separating a flat phase from a crumpled phase. This effect has been numerically confirmed by Kantor and Nelson[8].

We have also found arguments, however, to believe that a nonpolymerized membrane cannot sustain crystalline order, hence that the simplest form of stretching energy is not applicable in general. Dislocations, which destroy positional order, are rather energy-costly in flat crystalline arrays, yet may be introduced with no stretching if the array is allowed to buckle, assuming a warped shape. We argue that the energy cost of an isolated disclination remains finite even in presence of bending energy. One would expect free dislocations to be present at all temperatures in nonpolymerized membranes, destroying the long-range positional order of the molecules forming it.

Even if the membrane is fluid, a residual order in the orientations of the bonds connecting neighboring atoms might be present[9]. In this *hexatic phase*, there is a different kind of elasticity, which is also sensitive, like the stretching energy, to fluctuations of the intrinsic geometry of the surface. This *orientational stiffness* tends to counteract the softening of the bending rigidity due to undulations. It turns out that if this stiffness is sufficiently large, the membrane is warped at large scales in a self-similar way[10], with exponents which depend continuously on the stiffness constant. The actual shape will be dictated in practice by self-avoidance effects. Much work remains to be done to understand this intriguing system.

2. Crystalline membranes

We take Helfrich's expression[2] for the bending energy F_b of crystalline membranes:

$$F_b = \int dS \; (\frac{1}{2}\kappa H^2 + \bar{\kappa} K), \tag{1}$$

where dS is the area element of the membrane, H its mean curvature, and K is Gaussian curvature. The coefficients κ and $\bar{\kappa}$ are known as the (bare) rigidity and Gaussian rigidity respectively. The stretching energy is a functional of the *strain tensor* u_{ij} which is in turn related to the variation in the mutual distances of the molecules upon deformation[11].

We consider a membrane only slightly deformed from its equilibrium state, where it lies on the (x_1, x_2) plane, and we identify the molecules by their coordinates at equilibrium. The position of the molecule in the deformed state will be given by $(x_1 + u_1, x_2 + u_2, x_3 = f(x_1, x_2))$, which defines the *deformation vector* (u_1, u_2, f). The strain tensor u_{ij} is given by

$$u_{ij} = \frac{1}{2}[\partial_i u_j + \partial_j u_i + (\partial_i f)(\partial_j f)], \quad i,j = 1,2. \tag{2}$$

The third term of eq.(2) is crucial. Its neglect has recently led some authors[12] to the incorrect conclusion that crystalline order is irrelevant to the behavior of membranes with bending rigidity. The stretching energy F_s is given by[11]

$$F_s = \frac{1}{2}\int d^2x \; (\lambda u_{ii}^2 + 2\mu u_{ij}^2). \tag{3}$$

If the crystalline membrane were thought as a thin plate of an isotropic, elastic solid, the rigidities κ, $\bar{\kappa}$ would not be independent of λ and μ. One would have in fact[11]

$$\kappa = \frac{h^2}{12}(\lambda + 2\mu); \quad \bar{\kappa} = -\frac{h^2}{12} \cdot 2\mu, \tag{4}$$

where h is the membrane thickness. This relation is not valid in general for amphiphilic membranes. We can consider for the present purposes the membrane as a thin plate of an *anisotropic* solid, possessing in-plane hexagonal symmetry. The bulk elastic energy density per unit volume then has the expression

$$\mathcal{F} = \frac{\lambda_{\|}}{2} u_{ii}^2 + \mu_{\|} u_{ij}^2 + \frac{1}{2} B u_{33}^2 + \lambda_{\perp} u_{33} u_{ii} + \mu_{\perp} u_{3i} u_{3i}, \quad i,j = 1,2. \tag{5}$$

By going through the derivation[11] of the bending and the stretching energy for a thin plate, one obtains

$$\begin{aligned} \kappa &= \frac{h^3}{12}(\lambda_{\|} + 2\mu_{\|} - \frac{\lambda_{\perp}^2}{B}); \quad \bar{\kappa} = -\frac{h^3}{12} \cdot 2\mu_{\|}; \\ \lambda &= h\,(\lambda_{\|} - \frac{\lambda_{\perp}^2}{B}); \quad \mu = h\mu_{\|}. \end{aligned} \tag{6}$$

We can thus consider κ, λ, μ as essentially independent quantities. Note that eq.(6) implies that the Gaussian rigidity $\bar{\kappa}$ is *negative* for a thin solid plate. The negative sign of $\bar{\kappa}$ can be understood as follows: if a planar plate is slightly deformed into a minimal surface with a small negative Gaussian curvature, its free energy should *increase*, even though only the second term of eq.(1) will contribute to the energy. Although $\bar{\kappa}$ vanishes with the in-plane shear modulus in this model, this is not necessarily the case for actual amphiphilic layers, where splay elasticity effects can contribute. We shall neglect from now on the effects of Gaussian rigidity.

The stretching energy given by eq.(6) is quadratic in the tangential displacements (u_1, u_2), which may be thus easily integrated out. One thus obtains an effective elastic energy F_{eff} which depends only on the *transverse* deformation f, and has the following expression:

$$F_{eff} = F_b + \frac{1}{2} K_0 \int d^2x \, [\frac{1}{2} P_{ij}^T (\partial_i f)(\partial_j f)]^2, \tag{7}$$

where

$$K_0 = \frac{4\mu(\mu + \lambda)}{2\mu + \lambda}, \tag{8}$$

and P_{ij}^T is the two-dimensional transverse projector operator. It is useful to remark that

$$-\nabla^2 [\frac{1}{2} P_{ij}^T (\partial_i f)(\partial_j f)] = \det(\partial_i f \partial_j f) = K, \tag{9}$$

where K is the Gaussian curvature of the surface. The second term of F_{eff} expresses a long-range interaction between Gaussian curvature at different points.

This interaction contributes very strongly to a *stiffening* of the membrane. To lowest order in K_0 the effective rigidity at wavenumber $\mathbf{q}$ acquires a term which *diverges* like q^{-2} at small $|\mathbf{q}|$. A simple, self-consistent expression for the effective rigidity κ_{eff} is $\kappa_{eff} \propto \sqrt{K_0 k_B T} |\mathbf{q}|^{-1}$. This expression for κ_{eff} implies long-range order in the normals $\mathbf{n}$ to the membrane. Let θ be in fact the angle between $\mathbf{n}$ and the x_3 axis. Then $< \cos\theta > \simeq 1 - \frac{1}{2} < (\nabla f)^2 >$, if the membrane is only slightly warped. Now one has

$$< (\nabla f)^2 > = k_B T \int \frac{d^2 q}{(2\pi)^2} \frac{1}{\kappa_{eff}(\mathbf{q}) q^2}. \tag{10}$$

The integral appearing in this equation is *finite* if κ_{eff} is given by the expression above. We are thus led to the conclusion that $< \cos\theta >$ is different from zero at low enough temperatures, which makes the above calculation self-consistent.

Kantor and Nelson have recently produced[8] numerical evidence for a finite temperature crumpling transition in tethered surfaces with bending elasticity. The system is defined as a two-dimensional array of molecules, embedded in three dimensions, such that distances between neighboring atoms are constrained between two bounds (no excluded volume constraint is imposed on *non-neighboring* molecules). Bending elasticity is introduced by adding to the energy a term given by

$$F_b = -\kappa \sum_{<\alpha,\beta>} (\mathbf{n}_\alpha \cdot \mathbf{n}_\beta - 1). \tag{11}$$

The sum runs over all pairs of elementary triangles in the array sharing a common side, and $\mathbf{n}_\alpha$ is the normal to triangle α in the fluctuating surface. The simulations show that the membrane is highly crumpled at low values of κ, and its gyration radius R_G increases like $\sqrt{\ln L}$, where L is the linear size of the membrane, in agreement with the behavior of surfaces with vanishing bending rigidity. If κ increases, however, one observes a crossover to a behavior in which R_G is proportional to L, characteristic of a flat surface. The transition is marked by a pronounced peak in the specific heat. Although the samples are quite small, since simulating these surfaces is extremely time-consuming, this is a good evidence that a crumpling transition actually takes place in crystalline membranes.

3. Defects in crystalline membranes

In order to understand the relevance of the above results for actual amphiphilic membranes, it is important to investigate the stability of crystalline order against thermally excited defects. The basic defect of a two-dimensional array is the *disclination*, in which a given molecule is surrounded by an exceptional number of neighbors (five or seven, instead of the normal six, for the triangular lattice). Inserting an isolated disclination in a flat array costs a stretching energy proportional to $L^2 \ln L$. It is easy to realize, however, by cutting and glueing a 60° sector in a sheet of paper on which a triangular lattice is drawn, that it is possible to produce a disclination at *no* stretching energy cost, if the surface is allowed to assume a nonplanar shape. Paper does not in fact allow stretching to occur. The bending energy of the resulting shape increases only like $\ln L$, because the mean curvature H at a distance r from the disclination is inversely proportional to r.

A *dislocation* is a weaker defect, and may be thought of as a bound pair of disclinations. It is again possible to build up a paper model of a buckled membrane with an isolated dislocation, and to check that it would cost no stretching energy. The corresponding *bending energy* is more difficult to estimate, though, but we may expect it not to increase faster than $\ln L$. By taking a close look at the paper model, one may convince oneself that the deformation f at distance r from the disclination, averaged over all directions, increases roughly like $r^{\frac{1}{2}}$. The average mean curvature H should thus decrease like $r^{-\frac{3}{2}}$, yielding a finite bending energy. In actual amphiphilic membranes it will become energetically favorable for the membrane to buckle (giving away some stretching energy in exchange for a smaller amount of bending energy) only if it is sufficiently large. Adapting the result of a calculation by Mitchell and Head[13], valid for thin metal plates, one obtains that L must be larger than a critical value R_c, which is itself estimated to be bounded by

$$R_c \leq 120 \frac{\kappa}{K_0 b}, \tag{12}$$

where b is the Burgers vector of the dislocation. We expect therefore free dislocations to be present in all sufficiently large membranes, causing them to melt.

4. Hexatic membranes

Although it may be impossible for membranes to sustain crystalline (positional) order, they may still exhibit the residual *orientational* order characteristic of the hexatic phase[9]. Free disclinations cost in fact a bending energy of order $\ln L$ and cannot be present at low enough temperatures. The *directions* of the bonds connecting neighboring atoms would then possess quasi long range order. In a planar hexatic, this order appears in the correlation functions of the angle θ between the bonds and, say, the x_1 coordinate axis. The correlation function $< e^{6i\theta(0)} e^{-6i\theta(x)} >$ vanishes at large distances like a power of distance. It goes to a constant in the crystalline phase, and vanishes exponentially in the fluid phase. The factor 6 expresses the fact that there are *six* equivalent preferred directions in a perfect array. Planar hexatic exhibit orientational elasticity, since the energy required to produce a nonuniform value of θ is given by

$$F_h = \frac{1}{2} K_A \int d^2x \, (\nabla\theta)^2, \tag{13}$$

where the *hexatic stiffness constant* K_A is proportional to the square of the average distance ξ_T between dislocations.

The continuum theory of hexatic elasticity can be generalized to undulating membranes by observing that if the *Gaussian* curvature does not vanish at a certain point, it will be impossible to arrange the bonds in its vicinity in such a way that their directions are mapped on one another by parallel transport. This frustration effect is taken into account by the following expression for the hexatic free energy:

$$F_h = \frac{1}{2} K_A \int d^2x \, (\nabla\theta + \mathbf{A})^2, \tag{14}$$

where θ is the angle between the bonds and the first axis of a local coordinate frame, and the "vector potential" $\mathbf{A}$ is related to the Gaussian curvature K by $\nabla \times \mathbf{A} = -K$. Since this energy is again only quadratic in θ, one can easily integrate over it, and one is led to the following effective elastic energy:

$$F_{eff} = F_b + \frac{1}{2} K_A \int d^2x \, [\frac{1}{2} \nabla P_{ij}^T (\partial_i f)(\partial_j f)]^2. \tag{15}$$

The second term in F_{eff} expresses a long-range interaction between Gaussian curvatures, which increases logarithmically with separation.

The most convenient way to represent the effect of this interaction on the effective bending rigidity is by means of a renormalization group equation, which expresses the variation of the inverse rigidity $\alpha = \frac{k_B T}{\kappa}$ as a function of the renormalization wavenumber μ. One may take μ to be proportional to the inverse of the linear size L of the membrane. It turns out[10] that the hexatic stiffness constant K_A is not renormalized by thermal fluctuations, and may be kept as a free parameter. To lowest order in a loop expansion (i.e. to first order in the temperature T) one obtains[10]

$$\mu \frac{d\alpha}{d\mu} = -\frac{3}{4\pi}\alpha^2 + \frac{3}{16\pi}K_A \alpha^3 + \ldots \tag{16}$$

The first term expresses the thermal softening of κ due to undulations[3,4], and the second one its stiffening due to the hexatic elasticity. (A similar result was obtained in ref.1 as the first order in an expansion in powers of K_A). If we take eq.(16) at face value (which is warranted in the limit of very large K_A), it implies that hexatic membranes are governed at large size by a fixed point for which $\kappa^* \sim K_A k_B T$. At this fixed point the gyration radius R_G increases like L^ν where $\nu = 1 - \frac{k_B T}{8\pi\kappa^*} + \ldots$, and the correlation functions of the normals decay with exponents which also depend on K_A. All these results do not take self-avoidance effects into account. It would be interesting to know how these effects would modify the behavior of hexatic membranes.

5. Discussion

The coexistence of bending rigidity and of stretching or hexatic elasticity appears to have striking consequences and to lead to novel behaviors. The nature of the crumpling transition in crystalline membranes — and of the new phase of hexatic ones — should be more thoroughly investigated. Self-avoidance effects should be better understood.

It would be interesting to identify systems which are accurately described by the models discussed above. One might think to polymerize amphiphilic solutions in the lamellar phase, and then to separate the lamellae by dilution, to produce crystalline membranes. One should also check the prediction that unpolymerized, but crystalline, lamellae should go over to a fluid (possibly hexatic) phase, when they are forced apart. The observation by Sackmann[14] of hexatic order in Langmuir monolayers is an encouragement to the search for hexatic membranes, which should exhibit the striking behavior that we have discussed.

This research has been supported by the National Science Foundation through the Harvard Materials Research Laboratory and through grant DMR-85-14638. Fruitful discussions during the Les Houches workshop with E. Guitter, S. Leibler, D.Roux, and E. Sackmann are gratefully acknowledged. We are also indebted to F. David and E. Guitter for having communicated the results of ref.10 prior to publication. LP warmly thanks D. Langevin and J. Meunier for having organized such a stimulating and pleasant meeting.

References

1. D. R. Nelson and L. Peliti, *J. Physique* (in press).
2. W. Helfrich, *Z. Naturforsch.* **28c**, 693 (1973).
3. W. Helfrich, *J. Physique* **46**, 1263 (1985).
4. L. Peliti and S. Leibler, *Phys. Rev. Lett.* **56**, 1690 (1985).
5. P.-G. de Gennes and C. Taupin, *J. Phys. Chem.* **86**, 2294 (1982).
6. See e.g. J. Fröhlich, in *Applications of Field Theory to Statistical Mechanics,* edited by L. Garrido, *Lecture Notes in Physics,* **216** (Berlin: Springer, 1985).
7. Y. Kantor, M. Kardar and D. R. Nelson, *Phys. Rev. Lett.* **57**, 791 (1986) and *Phys. Rev. B* (in press).

8. Y. Kantor and D. R. Nelson, unpublished.
9. D. R. Nelson and B. I. Halperin, *Phys. Rev.* **B19**, 2457 (1979); see also D. R. Nelson, *Phys. Rev.* **B27**, 2902 (1983).
10. F. David and E. Guitter, unpublished.
11. L. Landau and E. M. Lifshitz, *Theory of Elasticity* (New York: Pergamon, 1970), secs. 11-15.
12. S. Ami and H. Kleinert, *Phys. Lett.* **A120**, 207 (1987).
13. L. H. Mitchell and A.K. Head, *J. Mech. Stat. Solids* **9**, 131 (1961).
14. E. Sackmann, this workshop.

Fluctuations on Crystal Surfaces

S. Balibar

Groupe de Physique des Solides de l'Ecole Normale Supérieure,
24 Rue Lhomond, F-75231 Paris Cedex 05, France

At first sight, there is no connection between the physics of amphiphilic layers and that of crystal surfaces. However, both are two-dimensional systems in which fluctuations play an important role. Surface quantities like surface tensions, stiffnesses or rigidity constants are affected by the existence of these fluctuations. In both systems similar renormalization group methods have been successfully used to calculate their effect. I would like to describe the case of crystal surfaces, in which such calculations lead to the description of the "roughening transition", with a very short reference to our experiments on helium crystals. Then I will examine interactions between crystalline steps which may be very similar to the steric interaction between membranes.

1-FLUCTUATIONS of a CRYSTAL SURFACE and the ROUGHENING TRANSITION

A crystal is periodic in space. This means that the height z of its surface may have preferential values which are multiples of some lattice spacing a. In a continuous version of the "solid-on-solid" model /1/ /2/, one thus writes a hamiltonian for a surface deformation as

$$H = (\gamma/2)(\nabla z)^2 + V\cos(2\pi z/a).$$

In the first term, γ is the surface stiffness which governs the surface curvature ($\gamma = \alpha + \partial^2\alpha/\partial\phi^2$ where α is the surface energy and ϕ its crystalline orientation) and V is a potential energy due to the lattice. The problem is to know whether V is relevant or not, i.e. whether the surface localized or not, or more precisely whether the height correlation function $G(r) = \langle (z(r) - z(0))^2 \rangle$ is a bound or diverging function of r.The principle of the renormalization calculation is the following.
One supposes that the partition function $Z = \exp(-H/k_BT)$ does not depend on the way the summation over all configurations z_k is done. Setting $z = \Sigma_k z_k e^{ikr}$ one thus writes

$$Z = \Sigma_{k<2\pi/L} \exp(-H(L)/k_BT) = \Sigma_{k<2\pi/L'} \exp(-H(L')/k_BT),$$

where the hamiltonian H(L) now has size-dependent coefficients $\gamma(L)$ and V(L). After choosing L'=L+dL, tedious algebra leads to the determination of γ and V at any scale L, through recurrence or "renormalization" equations. Without the cosine term, H would be linear in z and no renormalization of γ would occur. All the interesting physics comes from the fact that the surface modes z_k are coupled together by the nonlinear term $V\cos(2\pi z/a)$.The physics is simple: either the fluctuations are small (low

T) and the surface is localized in one of the minima of V which is relevant, or they are large enough to delocalize it, so that the cosine term is averaged to zero at large scale and the surface is as free as a liquid surface. A roughening temperature exists at which the height correlation on the surface disappears.

The results are often illustrated /2/ by renormalization trajectories which show how, starting from microscopic values, the stiffness $\gamma(L)$ and the lattice potential V(L) change with scale L. Three different situations may be encountered, depending on the value of T. If $T>T_R$, increasing the scale brings the potential V to zero and the surface stiffness to a finite value; the surface is rough or free. If $T<T_R$, V and γ tend to infinity; this means that the surface is localized under the influence of the lattice which kills the fluctuations of size larger than the one for which V(L) becomes larger than k_BT. Also γ being very large at large scale, the surface is a smooth and flat facet. The marginal case at $T=T_R$ is such that γ tends to a universal value $\gamma_R=\pi k_B T_R/2a^2$. Said another way, a surface is rough at any temperature T higher than $2a^2\gamma(T)/\pi k_B$, which is the modern roughening criterion. Other important quantities may be obtained, namely the height correlation function G(r) which is found to diverge as $(k_BT/\pi\gamma)\ln(r/L_0)$ for $T>T_R$. For $T<T_R$, G(r) saturates near $(2a^2/\pi^2)\ln(L^*/L_0)$ when r becomes larger than the correlation length L^*. This latter quantity has an interesting and simple interpretation, since the intrinsic thickness ξ of steps on smooth surfaces is about $L^*/2$. Indeed, the shape of steps is obtained by minimizing the surface energy with boundary conditions like $z(-\infty)=a$ and $z(+\infty)=0$, and one finds a surface profile $z(x)=(2a/\pi)\mathrm{arctg}\{\exp(x/\xi)\}$. One also finds a step energy ß of order $1.6k_BT/\xi$ which has the asymptotic behaviour

$$ß \cong \exp(-\pi/2(tt_c)^{1/2})$$

for $T\leq T_R$ (here $t = (T_R-T)/T_R$ and t_c is a parameter proportional to V). Such an exponential behavior indicates that the roughening transition is an infinite order transition: the surface free energy and all its temperature derivatives are continuous at T_R.

I would like to make two other remarks. First, the above theory describes the statics of a crystal surface, but our experiments /3/ /5/ show that the dynamics is very sensitive to the roughness: submitted to the same driving force, a rough surface and a smooth one have growth velocities which may differ by many orders of magnitude. As an example, with an applied difference in chemical potential between liquid and solid helium of the order of 50 cm^2/s^2, and around 0.5K, we observed facets growing at 5µ/s, when arbitrarily oriented (rough) surfaces may grow at 5 m/s /5/. In order to describe such effects, one needs a dynamic theory which is obtained by applying the same renormalization methods to a Langevin equation /2/:

$$(\rho_S/K)\partial z/\partial t = \gamma\Delta z - (2\pi/a)V\sin(2\pi z/a) + R(z,t) + \rho_S\Delta\mu ,$$

where R(z,t) is a random force with gaussian distribution, ρ_s is the solid density, K a surface mobility or growth rate and $\Delta\mu$ the difference in chemical potential between liquid and solid. One supposes here that this equation is renormalizable,which means that it may keep the same form whatever the scale as soon as its coefficients γ, V and K are allowed to vary with the scale L. The same results are obtained for the static quantities γ and V but a third equation is found for K. A remarkable result is

$$d\gamma/\gamma = - C(t)\, dK/K ,$$

showing that the curvature $1/\gamma$ and the growth rate K have similar critical behaviours (C(T) is a coefficient close to 2 for $T>T_R$).

The second remark concerns finite size effects. Let us indeed consider where the renormalization has to start and stop. It starts at a microscopic scale L_0 which I understand as a zero temperature thickness of the surface, or minimum wavelength of the fluctuations. This is a parameter which we adjusted when fitting our experimental results. For an infinite size system, renormalization stops either when V=0 ($T>T_R$), which means that the modes become uncoupled, or when $V \cong k_B T$ ($T<T_R$) in which case the fluctuations are killed by the increasing influence of the lattice. But the system may have a finite size or possess a proper length scale at which the renormalization also stops. It obviously happens if the crystal has a finite size itself. If we had included a gravity term $(1/2)\rho g z^2$, the capillary length $\lambda=(\gamma/\rho g)$ appeared. A more subtle length appears in the dynamic situation: when the growth proceeds at a velocity v, the surface goes up one lattice spacing a in a time $\tau=a/v$. As written above, the Langevin equation supposes that fluctuations diffuse along the surface with a diffusion constant $D=K\gamma/\rho_s$. All inertia effects have been ignored since it was applied to helium crystals above 1K, a temperature range where surface deformations do not propagate as crystallization waves (such a propagation would need to be considered for helium at lower temperatures). During the time $\tau=a/v$, the surface deformation diffuses on a length $l=(D\tau)^{1/2}=(K\gamma a/\rho_s v)^{1/2}$, which is a new length scale in the problem, where renormalization stops.

I do not want to describe here the experiments which we performed on the roughening transitions of helium crystals. All details can be found in the two successive articles of Wolf et al. /3/ and Gallet et al. /5/. My purpose here was mainly to show how fluctuations can be taken into account for the calculation of the surface properties of a crystal, how renormalization proceeds and what the main theoretical results are. Still, I have to mention that all these predictions are confirmed by our experiments: we measured the surface stiffness γ of (0001) surfaces on ^4He crystals, near their roughening temperature T_R=1.28K; it agrees with the universal relation $k_B T_R = 2\gamma a^2/\pi$ if one carefully takes into account finite size effects on surfaces close to the (0001) orientation. The step energy is found to vanish exponentially when T goes up to T_R. Finally the growth rate around T_R and above is found to increase with temperature in a way which is well described by the above theory. The roughening transition is broadened at finite growth velocity and this broadening is well interpreted as the subtle finite size effect we also mentioned above.

2-STEP-STEP INTERACTIONS

Consider now a cubic crystal and a vicinal surface on it, that is a surface which is slightly tilted, for example an angle ϕ with respect to a [100] four fold symetry axis. It may have Miller indices of type (10n) and it is then made out of a certain number of steps, all with the same sign, separating successive terraces. Each of these terraces has a mean width of n lattice spacings a (ϕ=1/n). At low enough temperature, the steps are anchored on the lattice and have negligible fluctuations. At the roughening transition T_{Rn} of the vicinal surface, the steps become free to fluctuate with respect to the lattice. Close to T_{R0}, the roughening temperature of the (100) surface, the steps have a diverging width and they all merge into each other: the vicinal surface is no longer modulated by the array of steps. Let us focus on the intermediate temperature range, $T_{Rn}<T<T_{R0}$.

There are various interactions existing between neighbouring steps, mainly two for simple crystals like helium crystals: an elastic one and a statistical or steric one. The elastic one arises from the fact that all steps create a deformation of the underlying lattice. When two steps come close to each other, their two strain fields interact. Marchenko and Parshin /6/ find a repulsion between steps whose typical energy per unit step length is

$$U(d) = 2(1-\sigma^2)A^2/\pi E d^2$$

where d is the distance between steps, σ is the Poisson ratio, E is the Young modulus and A is an energy which does not seem easy to calculate.

The statistical repulsion between steps is also inversely proportional to the square of their separation d, and, as we will see, it is very similar to the steric interaction between fluctuating membranes. Indeed steps of the same sign are undistinguishable objects so that a configuration with two steps crossing each other is the same as another one with redrawn steps only touching each other. When calculating the step free energy of a step with neighbours at a mean distance d, one may thus limit the mean amplitude Δ of fluctuations to d. This entropy reduction leads to an increase in the step free energy ß which is (to first order)

$$\delta ß = k_B{}^2T^2 / 2\pi^2 ß d^2 .$$

The way to derive this result is not so difficult. Indeed the amplitude of the fluctuations is

$$\Delta^2 = \int_{kmin}^{kmax} \Delta_k{}^2 \,(L/2\pi)\, dk$$

with $(1/2)\Delta^2 k^2 ßL = k_B T$ from the energy equipartition theorem (L being the step length), so that if $\Delta^2 = d^2$, $k_{min} = k_B T/ß\pi d^2 L$, instead of π/L for an isolated step, and $\delta ß$ is obtained by integrating $k_B T$ per mode k from π/L to k_{min}.

In principle the statistical interaction can be distinguished from the elastic one by measuring a temperature dependence. But how do you measure it ? The surface energy or stiffness of ordinary crystals is not easy to measure, the step energy even more difficult and the step-step interaction would seem impossible to obtain. However, I think that helium 4 crystals offer an opportunity to go that far in the general study of crystal surfaces. Indeed the $1/d^2$ behaviour of the interaction has a remarkable effect on the two components γ_1 and γ_2 of the surface stiffness. In the plane perpendicular to the steps, γ_1 is proportional to ϕ; in the plane parallel to them γ_2 is inversely proportional to ϕ. The measurement of these two quantities would give all the desired information. One way is to measure the two curvatures of equilibrium crystal shapes, near the edge of a facet, with great accuracy /7/. Another method would be to study the propagation of crystallization waves on vicinal surfaces as a function of ϕ (and T). These waves are the exact analog of capillary waves on liquid surfaces. They only propagate on helium crystals below about 0.6K, thanks to the superfluid nature of their melt. This experiment is in progress in our laboratory.

3- CONCLUSIONS and REFERENCES

I have been, myself, surprised to realize that some aspects of the physics of crystal surfaces are similar to that of fluctuating membranes. My purpose here was just to present them briefly . More details can be found in the following references.

1- J.D.Weeks and G.H.Gilmer Adv. in Chem. Phys.40,157(1979)
2- P.Nozières and F.Gallet J.de Physique 48,353(1987)
3- P.E.Wolf, F.Gallet, S.Balibar, P.Nozières and E.Rolley J.de Physique 46,1987(1985)
4- M.J.Graf and H.J.Maris to be published in Phys.Rev. B
5- F.Gallet, S.Balibar and E.Rolley J.de Physique 48, 369 (1987)
6- V.Marchenko and A.Y.Parshin Sov.Phys. JETP 52,129(1980)
7- C.Rottman, M.Wortis, J.C.Heyraud and J.J.Métois Phys.Rev.Lett. 52,1009(1984)

Surface Tension and Rigidity: Role of the Fluctuations and Optical Measurements

J. Meunier

Laboratoire de Spectroscopie Hertzienne de l'Ecole Normale Supérieure, 24 Rue Lhomond, F-75231 Paris Cedex 05, France

Thermal fluctuations play a fundamental role in the optical properties of liquid interfaces. We show that the coupling between thermal modes must be taken into accound to avoid any difficulty in the interpretation of optical measurements. First, this leads us to propose a new point of view for critical interfaces : the Van der Waals diffuse profile is depicted as a thin but rough interface with capillary coupled modes of small wavelength. Secondly, we deduce the rigidity constant and persistence length of surfactant monolayers from ellipticity measurements at oil/brine interfaces which are in good agreement with the phase structure observed.

1. INTRODUCTION

A liquid interface is commonly depicted as a diffuse layer of thickness L_p between two phases, where the density changes from ρ_1 to ρ_2, the densities of the two phases /1-3/. This layer is constantly distorted by thermal motion and should therefore present a certain roughness /4/. At a given instant, the vertical displacement of the layer from the equilibrium plane is $\zeta(r,t)$ and can be expressed as a sum of Fourier components $\zeta_{\vec{q}}$.

Two optical techniques give information about the thickness and the roughness of liquid interfaces :

- Reflectivity measurements which only give information about very thick interfaces as critical ones. In this case, the quantity $L_p^2 + \langle\zeta^2\rangle$ is measured /5-6/.

- Ellipsometry allows us to measure the ellipticity $\overline{\rho}$ of the light reflected at the Brewster angle. The ellipticity is the sum of two terms : $\overline{\rho^L}$ which is the contribution of the thickness of the interface and whose value is given by the Drude formula (7) ; $\overline{\rho^R}$ which is the contribution of the roughness of the interface /8-10/

$$\overline{\rho^R} \sim \sum_q q \langle\zeta_q^2\rangle \ .$$

It is interesting to note that the roughness contribution is very different for the two measurements. Ellipsometry is more sensitive to the roughness of short wavelength than reflectivity measurements. It can be a means to distinguish roughness and thickness of a liquid interface, as we shall see later for critical interfaces /11/.

The roughness of liquid interfaces is connected to interfacial parameters : surface tension γ and rigidity coefficient K, so its study by optical methods allows us to measure K /10//12/. The aim of this paper is how to reanalyse the connection between roughness and macroscopic liquid parameters of interfaces to overcome the difficulties that appeared in the interpretation of the measurements. We shall avoid details of calculations that will soon appear in a more complete paper /13/.

2. THERMAL FLUCTUATIONS OF LIQUID INTERFACES

The calculation of the two quantities $<\zeta^2>$ and $\Sigma q<\zeta_q^2>$ needs that of $<\zeta_q^2>$ and is a statistical mechanical problem. The probability of a configuration ζ of the interface is proportional to :

$$\exp - E(\zeta)/k_BT ,$$

where $E(\zeta)$ is the energy of a fluctuation ζ. It is the sum of the gravity energy $E_G(\zeta)$, the capillary energy $E_C(\zeta)$ and the curvature energy $E_K(\zeta)$:

$$E_C(\zeta) = \int \gamma \, dS \quad \text{and} \quad E_K(\zeta) = \int \frac{K}{2} C^2 \, dS ,$$

where C is the mean curvature of the interface. The calculation of $<\zeta_q^2>$ needs approximations. We first examine the approximation of independent modes and show why it is insufficient to analyse the optical measurements. Then, we examine a coupled modes theory and compare it with the experimental results.

2.1. Approximation of independent modes

This approximation is commonly used and only takes into account the ζ^2 terms. (For instance $dS = \sqrt{1 + (\nabla\zeta)^2}\, dx\, dy \approx [1 + 1/2\,(\nabla\zeta)^2]\, dx\, dy$). With this approximation $\zeta_{\vec{q}}$ are independent modes :

$$<\zeta_q^2> = \frac{k_BT}{\Delta\rho.g + \gamma q^2 + Kq^4} . \quad (1)$$

where $\Delta\rho$ is the density difference between the two phases, $g = 981\ cm/s^2$. With this approximation the parameters γ and K are independent of the observation scale. This formula was used to deduce K from ellipsometric measurements /10//12/.

At this point two remarks lead us to take into account the coupling between modes :

-In the case of critical interfaces we wait for K = 0 /14/. A difficulty appears in this case when we have to calculate $<\zeta^2>$ and $\Sigma q<\zeta_q^2>$: the summations lead to divergences of theses quantities at high q because of the lack of the q^4 terms in the expression of $<\zeta_q^2>$ and two high cut-offs q_R and q_C must be introduced :

$$<\zeta^2> = \frac{k_BT}{4\Pi\gamma} \mathrm{Ln}\,(\ell_C q_R) \quad \text{and} \quad \sum_q q <\zeta^2> = \frac{k_BT}{2\Pi\gamma} q_e , \quad (2)$$

where ℓ_C is the capillary length $\ell_C^2 = \gamma/\Delta\rho.g$. We wait for q_R and $q_e \sim 1/L_p$, but these two parameters are not well defined and this is a difficulty for the analysis of experimental results. Of course, the origin of the lack of the q^4 terms in the denominator of $<\zeta_q^2>$ is the absence of coupling between modes.

- K in the formula (1) is a constant independent of q. It is well known that this is not true if we take into account the coupling between modes /15//16/.

2.2. A coupled mode theory for interfacial fluctuations

The next approximation takes into account the ζ^4 terms (for instance, dS is written $dS \approx [1 + (\nabla\zeta)^2/2 - (\nabla\zeta)^4/8]\, dy\, dy$). We still assume that each mode can be considered as independent and its energy can be written with the same function of q as before. The coupling between modes (or ζ^4 terms) is taken into account through its average over all the configurations. The difference with the case before is that now γ and K are decreasing functions of q (that is to say that γ and K are renormalized) /13/

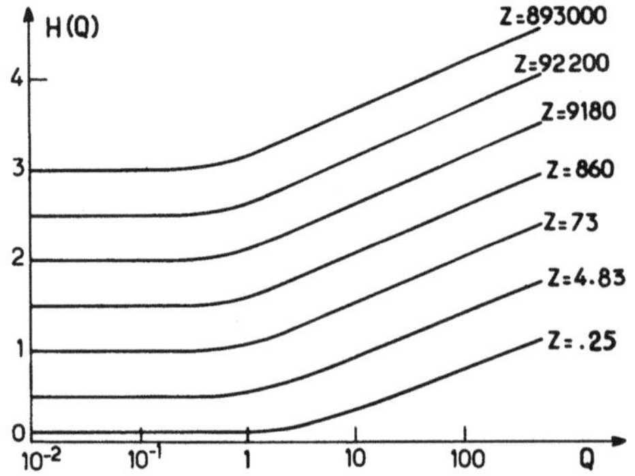

Figure 1 : The reduced rigidity coefficient H versus the reduced wavevector Q. Z is a parameter to identify the solutions.

$$\gamma = \gamma_\infty + a\ k_B T\ q^2, \tag{3}$$

where $a = 3/8\Pi$ and γ_∞ is the macroscopic surface tension.

At large scales, when the capillary energy dominates, we found $K = K_\infty$, a constant value. At small scale, when the curvature energy dominates

$$K = K_0 + a\ k_B T\ \mathrm{Ln}\ q^2/K\ . \tag{4}$$

This decrease of K with the scale is represented in fig. 1 in reduced parameters :

$$H = K/k_B T \quad \text{and} \quad Q = q\ \sqrt{\gamma_\infty / k_B T} \tag{5}$$

and agrees with the Peliti and Leibler result /16/. We finally find :

$$<\zeta_q^2> = \frac{k_B T}{\Delta\rho . g + \gamma_\infty\ q^2 + [\,a + K(a)]\ q^4}\ . \tag{6}$$

3. OPTICAL PROPERTIES of LIQUID INTERFACES : THEORY and EXPERIMENTS

Formula (6) allows us to calculate the reflectivity and the ellipticity of reflected light on liquids interfaces. We will examine successively two cases : 1) $K \equiv 0$ (critical interfaces). This case is very important to improve the validity of the theory, because all the parameters needed to calculated optical properties are macroscopic ones (macroscopic surface tension γ_∞, $\Delta\rho$ the density difference between phases, the refractives indices of the phases n_1 and n_2) and these parameters are separately measured. Unfortunately few measurements have been performed on critical interfaces : only two reflectivity measurements and two ellipsometric ones (and one of the two is not very near the critical point). These measurements are reported in Tables 1 and 2. A question appears when we want to compare these measurements to the calculation : the summations over q take into account very large q ($q >> 1/L_p$) and it would be stupid to add a thickness L_p. So we compare the experimental values with the calculated ones for an interface without thickness ($L_p = 0$) and we find a good agreement. The discrepancy can be explained by experimental incertitudes. So, we have now a new point of view : a critical interface can be considered as thin but rough /23/. Is it possible to reconciliate the two points of view ? For a thin but rough interface with coupled modes, the reflectivity measurements give $<\zeta^2>$, which can be written :

$$<\zeta^2> = \sum_0^{q_{max}} <\zeta_q{}^2> + \sum_{q_{max}}^{\infty} <\zeta_q{}^2>\ . \tag{7}$$

The first summation takes into account the capillary waves with a cut-off q_{max}. The second one can be seen as a diffuse layer with a density profile ERF /5-6/. This profile is very close to the Fisk and Widom one /17-18/ for a critical interface and has the same thickness if

$$q_{max} \simeq \frac{1}{2.55\xi}, \qquad (8)$$

where ξ is the correlation length over T_c.

Consequently the two points of view are equivalent for the reflectivity measurement if q_{max} is given by (8).

We must now examine the case of ellipsometry. The ellipticity of a thin liquid interface with coupled modes is given by :

$$\overline{\rho^R} \sim \sum_0^\infty q \langle \zeta_q^2 \rangle .$$

Now, taking into account the other point of view, the high q fluctuations ($q > q_{max}$) are considered as a density profile ERF as before. The ellipticity is :

$$\overline{\rho} = \alpha \sum_0^{q_{max}} q \langle \zeta_q^2 \rangle + \overline{\rho^L},$$

where ρ^L is deduced from the density profile by the Drude formula. We found /11/ :

$$\overline{\rho} \approx .65\ \overline{\rho^R} .$$

This value does not agree with the experimental values of the tables 1 and 2. In conclusion, the two points of view are equivalent for reflectivity measurements but the available ellipsometric ones are in better agreement with the second point of view : a critical liquid interface is a thin but rough interface with capillary coupled modes.

2) Generally $K \neq 0$ at a liquid interface and ellipsometry is a good technique to have the rigidity constant at the scale $q = q_e$ ($q_e \approx \Pi/2 \sqrt{\gamma_\infty / k_B T}$). We used this technique to measure the rigidity of surfactant monolayers at oil/brine interfaces /10//12/. This rigidity allows us to calculate an important parameter of monolayers : the persistence length /22/ ξ_K, which is the length over which the monolayer loses its orientational memory when its macroscopic surface tension γ_∞ vanishes.

In a brine, oil, surfactant mixture as a bicontinuous phase, the monolayer parting the oil and brine domains is such that $\gamma_\infty = 0$ and when the spontaneous curvature of the monolayer vanishes, the dominant parameter for the shape of the monolayer must be ξ_K. For $\xi_K \sim 100$ Å, the monolayer is very flexible and we wait for small brine or oil domains (a bicontinuous phase). For large ξ_K, the monolayer is planar over a large scale and we wait for a birefringent phase (as lamellar phase). To test this point of view,we have deduced ξ_K for three differents monolayers from ellipsometric measurements through our coupled mode theory of thermal fluctuations,and we compare it to ξ_K', the size of the oil and water domains in the bulk, deduced from X-ray scattering (table 3). The two last results are preliminary ones. A good agreement between ξ_K and ξ_K' is obtained.

Table 1 : Ellipticity of critical interfaces

Critical cyclohexane-aniline mixture			
measured $\overline{\rho}$ /19/	calculated $\overline{\rho^R}$	temperature range	discrepancy
$(4.28\times10^{-3})\Delta T^{-0.375}$	$(4.39\times10^{-3})\ \Delta T^{-0.4}$	$0.35 < \Delta T < 3.5°$	5%

Liquid-vapor-argon interface. The critical temperature is T_c = 150.9°K			
T	measured $\overline{\rho}$ /20/	calculated $\overline{\rho^R}$	maximum difference
85	5.4 ± 0.4	4.87	15%
90	5.7 ± 0.4	5.43	
100	6.5 ± 0.4	6.06	
110	7.5 ± 0.4	6.5	
120	3.8 ± 0.6	7.8	

Table 2 : Reflectivity of critical interfaces. $<\zeta^2>^{1/2}$ is calculated for a collection angle $\theta = 5\times10^{-3}$ /13/

Liquid vapor SF_6 interface		
measured L (Å) /17/	calculated $<\zeta^2>^{1/2}$	temperature range
$(148 \pm 9)\ \Delta T^{-0.62}$	$(132 \pm 7)\ \Delta T^{-0.587}$	$0.03 < \Delta T < 20°$
Cyclohexane methanol critical mixture		
$186 \pm 6\ \Delta T^{-0.665}$/21/	$237 \pm 40\ \Delta T^{-0.580}$	$0.1 < \Delta T < 3°$

Table 3 : The persistence length ξ_K for three monolayers and the size ξ_K' of the oil or water domains in bulk.

Monolayer	Oil	Salinity	ξ_K	ξ_K'
SDS - Butanol	Toluene	6.8%	350 Å	250 Å
AOT	Heptane	0.225%	∿ 5000 Å	Birefringent phase
DTAB-Butanol	Toluene		∿ 75 Å	∿ 80 Å

CONCLUSION

We have pointed out the origin of the difficulties of analysis of optical measurements on liquid interfaces and we have taken into account the coupling between thermal modes to solve them.

This analysis leads us to propose a new point of view to describe the critical interfaces, in better agreement with the couple of measurements : reflectivity and ellipsometry. The Van der Waals diffuse profile is replaced by a rough but thin interface with coupled capillary waves of small wavelength.

The ellipsometric study of surfactant monolayers at oil-brine interfaces allows us to measure a rigidity constant in good agreement with the one expected to explain the bulk structures of brine, oil, surfactant mixtures.

REFERENCES

1. J. Van der Waals: Z. Phys. Chem. 13, 657 (1894)
2. J. W. Cahn, J.E. Hilliard: J. Chem. Phys. 28, 258 (1958)
3. S. Fisk, B. Widom: J. Chem. Phys. 50, 3219 (1969)
4. L. Mandelstam: Ann. Physik 41, 609 (1913)
5. J. Meunier: C.R. Acad. Sci. 292 II, 1469 (1981)
6. J. Meunier, D. Langevin: J. Phys. Lett. 43, L-185 (1982)
7. P. Drude: Ann. Phys. 43, 126 (1981)
8. D. Beaglehole: Physica B100, 163 (1980)
9. B.J.A. Zielinska, D. Bedeaux, J. Vieger: Physica A107, 91 (1981)
10. J. Meunier: J. Phys. Lett. 46, L-1005 (1985)
11. J. Meunier: to be published
12. J. Meunier, B. Jerome: In Surfactant in Solution, ed. by K.L. Mittal and B. Lindman, Plenum Press, to be published
13. J. Meunier: presented to the Journal de Physique
14. D. Sornette: In These d'Etat, Nice 1985
15. W. Helfrich: J. Phys. 46, 1263 (1985)
16. L. Peliti and S. Leibler: Phys. Rev. Lett. 54, 1690 (1985)
17. J.S. Juang and W.W. Webb: J. Chem. Phys. 50, 3677 (1969)
18. D. Beyssens, M. Robert: to be published
19. D. Beaglehole: Physica 112B, 320 (1982)
20. D. Beaglehole: Physica 100B, 173 (1980)
21. E.S. Wu and W.W. Webb: Phys. Rev. A8, 2065 (1973)
22. P.G. de Gennes and C. Taupin: J. Phys. Chem. 86, 2294 (1982)
23. This point of view was introduced by : F.P. Buff, R.A. Lovett, J.F.M. Stillinger: Phys. Rev. Lett. 15, 621 (1965)

Part III

Ordered Phases

Geometrical Basis of Cubic Structures

J. Charvolin and J.F. Sadoc

Laboratoire de Physique des Solides, Bâtiment 510,
Université Paris-Sud, F-91405 Orsay, France

We consider cubic liquid crystalline phases formed by amphiphilic molecules in the presence of water and which can be found in some phase diagrams in the immediate vicinity of lamellar phases [1,2]. These cubic phases have symmetries [2] and topologies [3] strikingly different from those of lamellar phases. However we argue here that the basis of these structures can be understood in the same geometrical terms. They are periodic systems of fluid films separated by interfaces: lamellar phases are periodical stackings of flat interfaces at constant distances and cubic phases can be seen as periodical stackings of symmetrically curved interfaces at optimized distances.

1 Physical Forces and Geometrical Frustration

We do not detail the forces acting in amphiphile/water systems [4]. We just consider that they have components normal to the interfaces, which impose constant distances between them when moving along them, and components parallel to the interfaces, whose values in the aqueous and paraffinic media control the interfacial curvature. The structures of these fluid films must therefore conciliate constant interfacial distances and curvatures and they can be understood with a knowledge of the geometrical configurations which satisfy these constraints.

When the thermodynamical conditions are such that the interfaces are flat the obvious configuration is that of the periodical stacking along one dimension of the lamellar phase shown in Fig.1a. When these conditions move away from the above the interfaces become curved and a typical case of geometrical frustration is met, as shown in Fig.1b. This frustration has no direct solution in flat Euclidean space R_3 but has one in curved space S_3, or the hypersphere. However, real systems exist in R_3 and the possible configurations are to be found in this space. They will be obtained by introducing defects of rotation, or disclinations, to decurve S_3 [5]. Following this process the structures of amphiphile/water liquid crystals can be looked at as structures of disclinations. We shall study here the particular process leading to configurations having the topology of cubic phases [6].

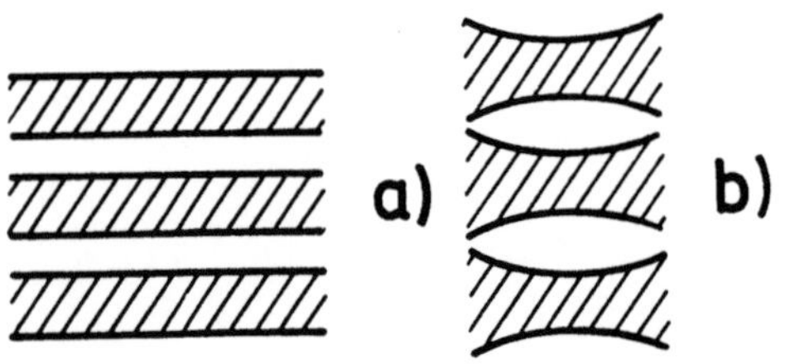

Figure 1. (a) Schematic representation of a periodic system of fluid films with flat interfaces: constant interfacial distances and zero curvatures are compatible. (b) A periodic system of fluid film with curved interfaces: constant interfacial distances and curvatures are no longer compatible (frustration)

2 Relaxation of the Frustration in S_3

The whole periodic system of frustrated fluid films can be represented without frustration in S_3 by transforming the middle surface of the film in one cell onto the spherical torus T_2. This is a particular surface of S_3 separating it into two identical subspaces. It can be obtained by identification in S_3 of the opposite sides of a square sheet as shown in Fig.2 and, therefore, admits a {4,4} regular

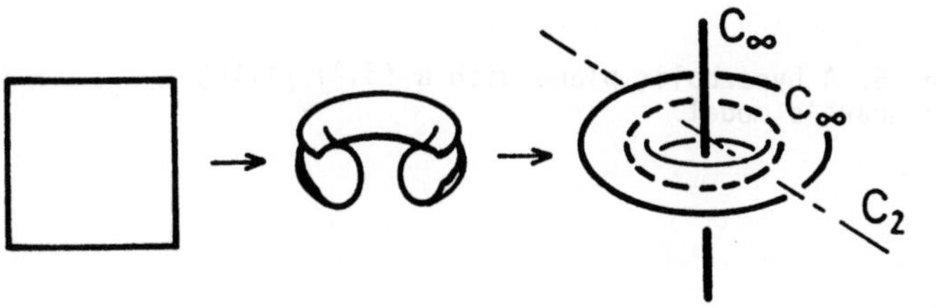

Figure 2.The spherical torus can be built by identification of a square sheet in S_3, as this is done in a curved space the sheet suffers no distortion; the torus represented in this figure is a stereographic projection in R_3.

tiling [7].The symmetry of the relaxed structure in S_3 imposes the nature and the mode of introduction of the disclinations needed to décurve S_3.

3 Disclinations

Disclinations are to be introduced around the symmetry axes of the relaxed structure in a manner respecting the symmetries [8]. As shown in Fig.3 half a torus has to be introduced in between the lips of a cut limited by a C_2 axis normal to the surface of the torus. This is a $-\pi$ disclination which does not affect the property of the surface which separates the space into two identical sub-spaces.

This process leads to a configuration which has the bicontinuous topology of cubic phases. In the course of this process the squares of the {4,4} tiling of the spherical torus are transformed into hexagons without any modification of the number of polygons per vertex or the $\pi/2$ values of the angles, as shown in Fig.4.

A new surface is therefore generated by the disclination process which admits a {6,4}tiling. Obviously such a tiling is not Euclidean, or planar, and the new surface has a constant negative Gaussian curvature, it is a hyperbolic plane.

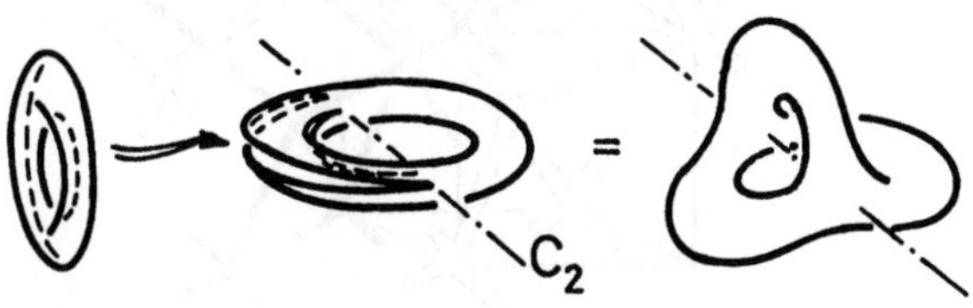

Figure 3.A-π disclination around a C_2 axis of a torus, half a torus is inserted between the lips of the cut.

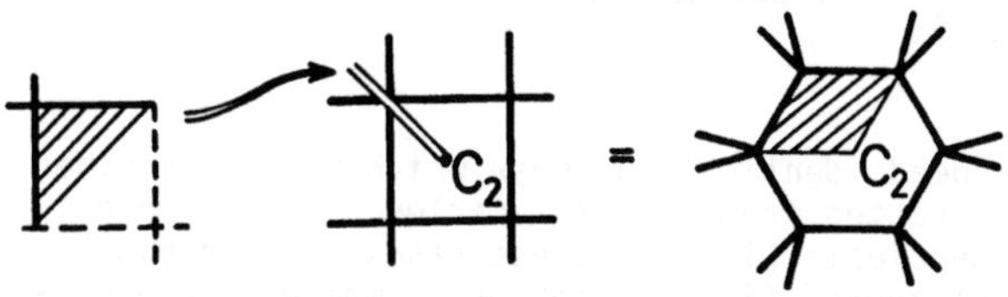

Figure 4.A $-\pi$ disclination in a square transforms it into a hexagon, the 4-connectivity is preserved at every vertex.

4 Hyperbolic Plane

A bi-dimensional space with constant negative Gaussian curvature cannot be embedded in R_3 without metric distortions. However, the properties of such hyperbolic planes can be studied using Poincaré's model in the Euclidean plane [9]. A hyperbolic plane with a {6,4} tiling is represented in Fig.5, together with its orthoscheme triangles, or asymmetric units.

Figure 5. A hyperbolic plane with a {6,4} tiling in Poincaré's model.

This triangle has angles of $\pi/2$, $\pi/4$, $\pi/6$ and is non-Euclidean. The analysis of its possible configurations, straight or curved sides, informs about the possible surfaces in R_3. We consider here one case only, where the two sides of the right angle are straight; other cases are discussed in [6]. These straight sides build a lattice of intersecting straight lines whose element is a non-planar quadrangle (a, b, c, d) with three angles of $\pi/2$ and one of $\pi/3$. These elements can be assembled four by four in a more symmetrical skew quadrangle (d,e,f,g) with four angles of $\pi/3$ and four equal sides.

5 Schwarz's F Surface in R_3

These quadrangles are regularly organized on the hyperbolic plane and it was shown by Schoenflies [10] and Schwarz [11] that, when embedded in R_3, they build a cubic lattice whose translation cell is shown in Fig.6. Schwarz has shown that this cubic lattice is the support of an infinite periodic minimal surface (IPMS), called F or D, separating R_3 into two identical sub-spaces whose labyrinths are the two interwoven but not connected lattices of rods shown in Fig.7.

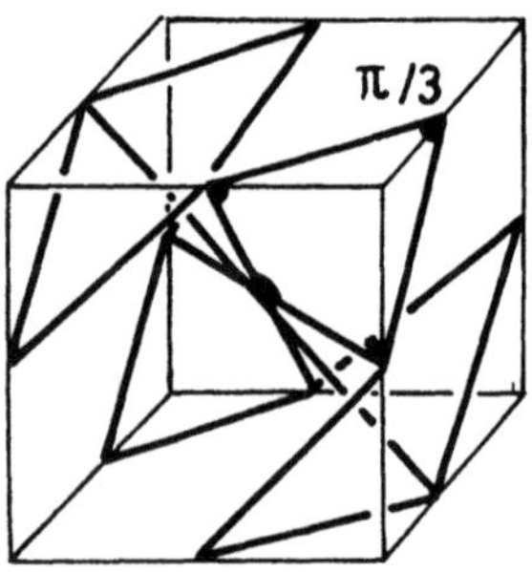

Figure 6. The translation cell of the cubic structure built in R_3 with quadrangle (d,e,f,g), it spans surface F.

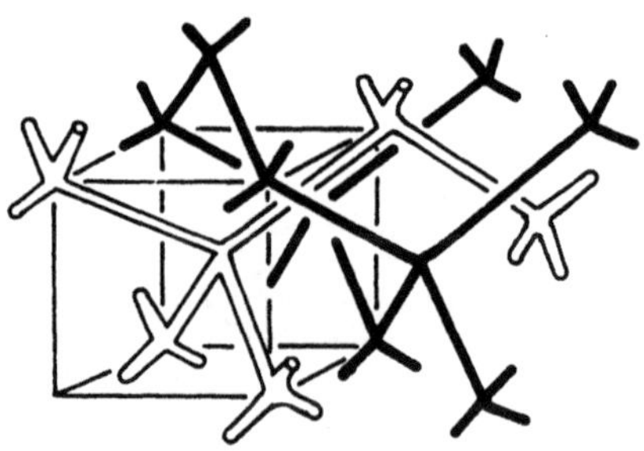

Figure 7. The two labyrinths separated by surface F.

6 Relations with Cubic Phase Pn_3

These two interwoven labyrinths are indeed identical to those which were determined for the cubic phase built by lipids extracted from insect cuticles [12] and more recently for that built by glycerol monooleate [13]. In these cases the labyrinths are channels of water separated by a bilayered film of amphiphile whose middle surface should follow the F surface very nearly.

7 Final Remark

We looked for geometrical configuration of periodic systems of fluid films with bicontinuous topologies and which conciliate constant interfacial distances and curvatures. Such configurations can be found in curved space S_3 only. The decurving of that space to find the possible configurations in R_3 implies the introduction of defects, which modulate distances and curvature so that their values are no longer homogeneous along the interfaces.

REFERENCES

1. V.Luzzati: this conference
2. V.Luzzati: In Biological Membranes 1 71 (1968), edited by D.Chapman, Academic Press
3. J.Charvolin: In Summer School on Microemulsions, Erice (1985), to be published
4. Several lectures in this conference and J.N.Israelachvili: In Intermolecular and Surface Forces Academic Press (1985)
5. J.F.Sadoc and J.Charvolin: J.de Physique 47, 683 (1986)
6. J.Charvolin and J.F.Sadoc: J.de Physique, to be published
7. The Schlafli notation {p,q} means that the tiling is made of regular polygons having p edges that meet q by q at vertices
8. J.Friedel: In Proceedings of the 6th General Conferences of the E.P.S., Prague (1984)
9. H.S.M.Coxeter: In Introduction to Geometry, J. Wiley, New York (1961)
10. A.Schoenflies: Comptes Rendus 112 478 (1882)
11. H.A.Schwarz: In gesammelte Mathematische Abhandlungen Band 1, Springer Verlag, Berlin (1890)
12. A.Tardieu: Thesis, Orsay (1972)
13. W.Longley and J.MacIntosh: Nature 303 612 (1983)

Periodic Surfaces of Prescribed Mean Curvature

D.M. Anderson, H.T. Davis, J.C.C. Nitsche, and L.E. Scriven

Departments of Chemical Engineering and Materials Science
and of Mathematics, University of Minnesota,
421 Washington Avenue S.E., Minneapolis, MN 55455, USA

Summary of Talk Presented at Conference

While there are eighteen triply periodic minimal surfaces that reportedly are free of self-intersections, to date there is no known example of a triply periodic surface of constant, *nonzero* mean curvature that is embedded in $\mathbf{R}^3$ (three dimensional Euclidean space). We have computed and displayed [1,2] five families of such surfaces, where every surface in a given family has the same space group, the same Euler characteristic per lattice-fundamental region, and the same dual pair of triply periodic graphs that define the connectivity of the two labyrinthine subvolumes created by the infinitely connected surface. Each family comprises two branches, corresponding to the two possible signs of the mean curvature, and a minimal surface. The branches have been tracked in mean curvature, and the surface areas and volume fractions recorded, with the region $dA = 2HdV$ carefully checked to hold. The three families that contain the minimal surfaces P and D of Schwarz and the I-WP minimal surface of Schoen terminate in configurations that are close-packed spheres. However, one branch of the family that includes the Neovius surface C(P) contains self-intersecting solutions and terminates at self-intersecting spheres. On approaching the sphere limit, whether self-intersecting or close-packed, the gradual disappearance of small 'neck' or 'connector' regions between neighboring 'sphere-like' regions is in close analogy with the rotationally symmetric unduloids of Delauney. We give what we suspect are analytical values for the areas of the I-WP and F-RD minimal surfaces, and a possible limit on the magnitude of the mean curvature in such families is proposed and discussed. We also report that the I-WP and F-RD minimal surfaces each divide $\mathbf{R}^3$ into two subspaces of *unequal* volume fractions.

The numerical method is based on a new approach to the formulation of the Galerkin, or weak form of the problem of prescribed — not necessarily constant — mean curvature. The Surface Divergence Theorem is applied directly to a vector-valued function that is the product of a scalar weighting function and a vector field chosen to enforce the boundary conditions. This formulation applied in the context of the finite element provides a robust algorithm for the computation of a surface with: 1) mean curvature a prescribed function of position, and 2) contact angle against an arbitrary bounding body a prescribed function of position or of arc length. A parametrisation scheme for triply periodic surfaces is described that calls only for knowledge of the two 'skeletal' graphs; this is demonstrated by the computation of the triply periodic minimal surface $S' - S''$ hypothesized by Schoen, who described only the skeletal graphs associated with the surface. The parametrisation allows for easy calculation of the scattering function for various density profiles based on the solutions, as well as of areas and volume fractions. For the three minimal surfaces – P, D, and C(P) – whose areas and volume fractions are known analytically, the numerical results are in agreement with these values. Furthermore, we review the history of such surfaces, and clear up some inconsistencies over the D minimal surface.

References Cited

1. D. M. Anderson, Ph.D. Thesis (University of Minnesota 1986).
2. D. M. Anderson, H. T. Davis, J. C. C. Nitsche and L. E. Scriven, Phil. Mag. (to be published).

The Cubic Phases of Liquid-Containing Systems: Physical Structure and Biological Implications

V. Luzzati, P. Mariani, and T. Gulik-Krzywicki

Centre de Génétique Moléculaire, CNRS,
F-91190 Gif-sur-Yvette, France

1 Introduction: the Cubic Phases

The first observation of a cubic phase in lipid-water systems was reported in the early days of lipid polymorphism [1], at a time when the very existence of non-lamellar phases still met widespread skepticism. That cubic phase (Q^{230}, see below) was the object of controversial reports: its structure was firmly established in 1967 [2,3]. Several other cubic phases were discovered since. In 1972, TARDIEU [4] listed five different phases of that class: more recently some of those phases, and others as well, have been observed in a variety of systems (see [5] and below).

At the present date six different cubic phases have been identified on the basis of X-ray scattering experiments, and the list may still be incomplete. According to the spacing ratios of the reflections observed in the powder spectra (confirmed in a few cases by single crystal experiments) it is possible to ascribe one of a few space groups to each of the phases. We use here a notation (introduced in [6]) based upon the letter Q (for cubic) and a superscript specifying the number (according to the International Tables [7]) of the space group of highest symmetry among the possible ones. Since, moreover, the volume of the cubic phases can generally be subdivided into two topologically distinct regions, the structures may be either of type I (oil-in-water) or type II (water-in-oil) [8]. Freeze-fracture electron microscopy observations have also been used to identify some of the phases ([9] and also unpublished results by H. DELACROIX).

It is worth noting that in all the cubic phases the conformation of the hydrocarbon chains is liquid-like [8]. These phases, moreover, provide a vivid illustration of the extraordinary polymorphism of lipids - a characteristic property of this class of chemical compounds - and of the unusual occurrence of both excellent long-range order and extreme short-range disorder (as a rule long-range order arises from propagation of short-range order) [5].

For each of the cubic phases, we report below the systems in which the phase has been observed, its identification, and the structure - when it is known.

Q^{230} (Ia3d)

This space group is determined unambiguously by the X-ray powder spectra. The spacing ratios are $\sqrt{6}:\sqrt{8}:\sqrt{14}:\sqrt{16}:\sqrt{20}:\sqrt{22}:\sqrt{24}:...$. This phase was observed in a variety of systems: anhydrous soaps of divalent cations [10] (type II); soap (and detergent)-water, in the intermediate region between the lamellar and the hexagonal phases [3] (type I); diacylated lipids-water, at low hydration and high temperature [3] (type II); monoglycerides-water [11] and tetraether lipids-water [6], at fairly high water content (type II) (see Fig. 1).

The structure of this phase [2,3] is described in Fig.2. Freeze-fracture electron micrographs have recently provided support for the structure (unpublished). This phase has been the first example of a bicontinuous lipid-water structure [3], and also an illustration of the effects of structural frustrations [2,8]. It is also worth stressing that this structure corresponds to the G-type three-dimensional periodic minimal surface [12].

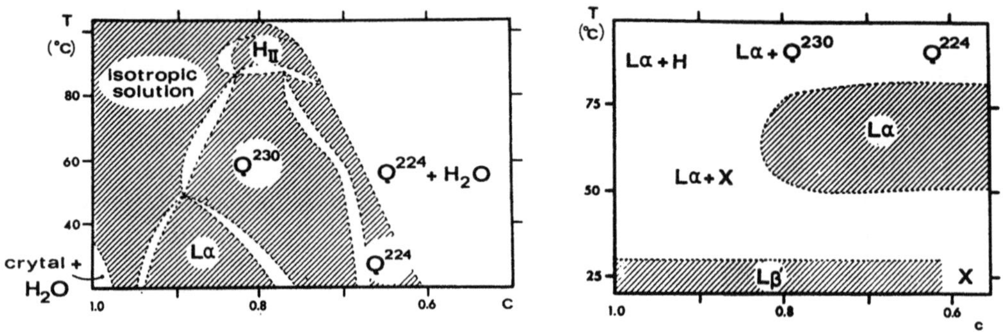

Figure 1 - The phase diagrams of two lipid-water systems: monoolein (left, redrawn from [11]) and polar lipid extract (PLE) from *S. solfataricus* (right, from [6]). c is the (weight) concentration lipid/(lipid+water). The one-phase regions are hatched; L means lamellar, H hexagonal. X shows the presence of unidentified sharp reflections.The conformation of the chains is disordered (α) in all the phases, with the exception of Lβ'.

Q^{224} (Pn3m)

Other possible space group N°201 (Pn3). The spacing ratios are $\sqrt{2}:\sqrt{3}:\sqrt{4}:\sqrt{6}:\sqrt{8}:\sqrt{9}:\sqrt{10}:...$. This phase was observed in several lipid-water systems - monoglycerides [11,13], tetraether lipids [6] - on the hydrated side of the phase Q^{230} (see Fig.1). The structure, tentatively proposed by TARDIEU [4], was confirmed by LONGLEY and McINTOSH [13]. The structure is represented in Fig.2. This structure, like that of the phase Q^{230}, is bicontinuous; it corresponds to the D-type three-dimensional periodic minimal surface [12].

Q^{223}(Pm3n)

Other possible space group N°218 ($P\bar{4}3n$). Spacing ratios $\sqrt{2}:\sqrt{4}:\sqrt{5}:\sqrt{6}:\sqrt{8}:\sqrt{10}:\sqrt{12}:...$ Observed in several lipid-water (and lipid-water-oil) systems, in the concentration range intermediate between the hexagonal phase and the

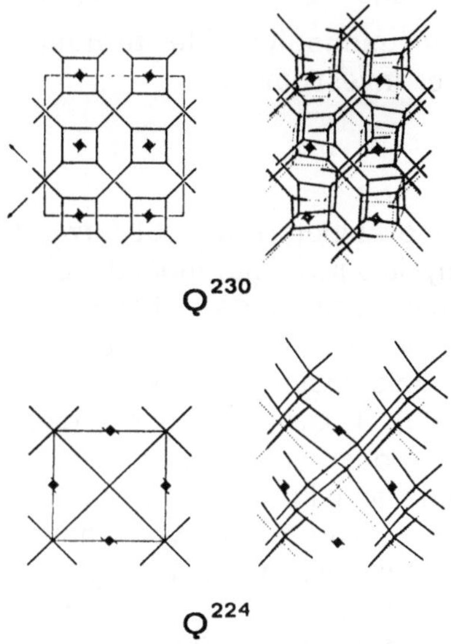

Figure 2 - The structure of the phases Q^{230} and Q^{224}. The two structures consist of a pair of continuous 3-dimensional networks, mutually intertwined and unconnected. The rods may contain the water and be embedded in a continuous hydrocarbon medium (type II), or *vice versa* (type I, but only in the case of Q^{230}); the surface of the rods is coated by the polar groups of the lipid molecules. The thick lines represent the axes of the rods. *Left frames* : representaion of the unit cells with the positions of the axes of the rods and of some of the symmetry elements. *Right frames* : perspective view of the structures. *Upper frames* : phase Q^{230} ; note that the rods are linked coplanarly three by three. *Lower frames* : phase Q^{224}; note that the rods are linked tetrahedrally four by four (from [6])

micellar solution [14,15,16]. The structure is under investigation: an early proposal [15] has indeed been shown to be incompatible with NMR experiments [16].

Q^{229} (Im3m)

Other possible space groups N°s 217 ($I\bar{4}3m$), 211 (I432), 204 (Im3), 199 ($I2_13$), 197 (I23). Spacing ratios $\sqrt{2}:\sqrt{4}:\sqrt{6}:\sqrt{8}:\sqrt{10}:\sqrt{12}:\sqrt{14}:...$Observed in several monoglyceride-(protein-salt)-water systems (unpublished). The structure is under investigation.

Q^{227} (Fd3m)

Other possible space group N° 203 (Fd3). Spacing ratios $\sqrt{3}:\sqrt{8}:\sqrt{11}:\sqrt{12}:\sqrt{16}:\sqrt{19}:\sqrt{24}:...$ Observed in one glycolipid-water system [4] and in monoglyceride-(fatty acids)-water systems (unpublished). The structure is under investigation.

Q^{212} ($P4_332$)

Other possible space groups N°s 208 ($P4_232$) and 198 ($P2_13$). Spacing ratios $\sqrt{2}:\sqrt{3}:\sqrt{5}:\sqrt{6}:\sqrt{8}:\sqrt{9}:\sqrt{10}:.....$ Observed in monoglyceride-(protein)-water systems (unpublished). The structure is under investigation.

2 Biological Implications

Since the very discovery of lipid polymorphism, some authors have put forward the hypothesis that this phenomenon may play a physiological role. Indeed, lipids constitute one of the major classes of biological compounds - lipids are conspicuous components of membranes - and the possibility of structural transitions, occurring *in vivo* and involving the lipid moiety may well have interesting physiological implications (reviewed in [5]). This view, though, does not seem to have made much of an impact on the widespread opinion that the essential role of lipids is to provide passive barriers to diffusion, and that the only structure biologically relevant is the lamellar one. The fact is worth noting [17] that the phases Q^{230} and Q^{224} (and probably other cubic phases), like an isolated lipid bilayer, display two disjointed and continuous water volumes, separated from each other by a continuous hydrocarbon septum (see Fig. 2). This property is not shared with other lipid phases (for example lamellar and hexagonal) in which the water or the hydrocarbon media (or both) are subdivided into an infinite number of disjointed regions. In other words, the phases Q^{230} and Q^{224} can be visualized as topological generalizations of the isolated lipid bilayer. This property may well be relevant to the physiological role of lipids: the two examples below illustrate this possibility.

The digestion of fats

As a rule, any particular lipid-water system may display one cubic phase, more rarely two; besides, most of these phases are stable at water contents and temperatures remote from physiological. Two exceptions (see Fig.1) are the tetraether lipids from *Sulfolobus solfataricus* (see below) and monoolein; note, moreover, that five of the six cubic phases discussed above are observed in a variety of monoglyceride-containing systems, most of them at high water content and room temperature.

It is of interest to speculate why monoglycerides, of all lipids, have such phase behaviour. Monoglycerides play a physiological role in fat digestion (reviewed in [18]): enzymatic lypolysis transforms each triglyceride molecule into one monoglyceride and two fatty acids. A peculiar aspect of this enzymatic reaction is that neither the substrate nor the final products are soluble in the reaction medium: how, in these conditions are the access of the enzyme(s) to the substrate and the disposal of the reaction products preserved? As originally suggested by PATTON [18], the answer may well reside in the structure of the cubic phases. As the lipolytic reaction proceeds, the triglycerides give way to the monoglycerides which in the presence of excess water take the form of the phase

Q^{224} (Fig.1). This phase is insoluble in water and its 3-dimensional networks of water channels (Fig.2) leave access of the lipase(s) to their substrate. The observation of a cubic phase in the course of triglyceride digestion by pancreatic lipases *in vitro* [19], and also the structure of other cubic phases observed in systems containing monoglycerides, fatty acids, proteins and water (work in progress) strongly support this suggestion.

Life at high temperature

Over the last few years, a variety of organisms growing in unusual habitats - characterized by high or low pH, high temperature, high ionic strength, etc. - have become the object of intensive studies. Not unexpectedly, some of the most striking chemical alterations are observed in the membranes and in the membrane lipids (reviewed in [20]). Thermophilic prokaryotic organisms are either eubacteria (the classical prokaryotes), or archaebacteria (a novel class of microorganisms growing in exceptional ecological niches [21]). Among the different phylogenetic criteria used to distinguish archaebacteria from eubacteria, the chemical composition of the lipids - schematically represented in Fig.3 - is the most direct one [20]. We are concerned here with the lipids extracted from a thermophilic archaebacterium, *S. solfataricus*, growing at 85°C and pH3: these lipids are entirely tetraethers (type *III* in Fig.3). We summarize below the main conclusions of a recent X-ray scattering analysis of those lipids [6,19,21].

In spite of the conspicuous chemical difference, the phase diagram of the tetraether lipids (see Fig.1) is similar to that of other biological lipid extracts, in the sense that it displays an extensive polymorphism, and that the hydrocarbon chains undergo a transition from a partly ordered conformation (β') at low temperature to a disordered (α) conformation at "physiological" temperature. These analogies, though, should not overshadow other remarkable differences.

Firstly, the phases which are predominant at "physiological" conditions are not lamellar, as generally observed with other lipid extracts, but cubic (Q^{230} and Q^{224}). Note, in this respect, that the phase diagram of the tetraether lipids, shifted to a lower temperature, is similar to that of monoolein (see Fig.1).

Secondly, in these tetraether lipids a high proportion of the headgroups (approximately 1 out of 5) consists of an unsubstituted glycerol: this fact is noteworthy since diglycerides are virtually absent among the membrane lipids of types *I* and *II* (Fig.3). These unsubstituted glycerol headgroups are found to partition preferentially in the middle of the hydrocarbon region, rather than at the lipid-water interface. This partition has the effect of imparting novel properties, some of potential physiological interest, to the lipid phases.

Thirdly, the cubic phases of these lipids are strikingly metastable, to a degree never observed before in any lipid phase transition involving two phases in both of which the hydrocarbon chains are in the α (disordered) conformation. This

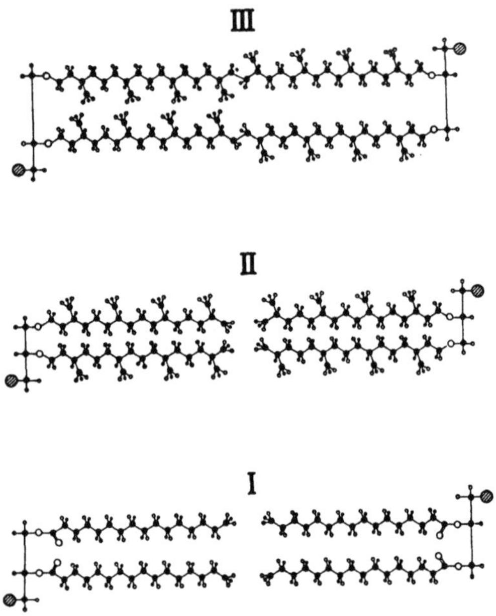

Figure 3 - Three examples of lipid molecules, apposed as in lipid bilayers. Filled, open, small and hatched circles represent respectively the carbon, oxygen and hydrogen atoms, and the polar residues. *I* : Lipids of eukaryotes and eubacteria: the hydrocarbon chains usually are linear (with exceptions), ester linked to glycerol (also with exceptions). *II* and *III* : Lipids of archaebacteria; the hydrocarbon chains are branched (isopranyl) and ether linked to glycerol. The molecules of type *II* (diethers) contain two C_{20} chains linked to one glycerol. The molecules of type *III* (teraethers) are dimers of the diethers, and consist of two C_{40} chains linked at both ends to one glycerol group (from [6]).

phenomenon has been explained by the interplay of the chemical structure of the lipid molecules and the structure of the cubic phases. Moreover, this unusual metastability may play a physiological role, related to the extensive temperature shifts that the archaebacterial colonies are likely to undergo in their natural habitat.

Fourthly, in the presence of excess water these lipids are likely to form lumps of the phase Q^{224}. These lumps can be visualized as sponge-like objects highly penetrable to water, with an extended lipid-water interface (approximately 500 m^2/g), with potentially interesting catalytic properties.

Fifthly, the fact that the highly organized protein envelope of *S. solfataricus* cells (which has been studied by electron microscopy [23]) displays a remarkable epitaxial coincidence with the phase Q^{224} has been interpreted to suggest that the structure of the plasma membrane may be reminiscent of that of the lipid phase Q^{224} and be different from the canonical "lipid bilayer". This hypothesis has led to a variety of speculations.

References

1. V. Luzzati, H. Mustacchi, A. E. Skoulios, F. Husson: Acta Cryst. 14, 660 (1960)
2. V. Luzzati, P. A. Spegt: Nature (London) 215, 701 (1967)

3. V. Luzzati, A. Tardieu, T. Gulik-Krzywicki, E. Rivas, F. Reiss-Husson: Nature (London) 220, 485 (1968)
4. A. Tardieu: Thesis, Université Paris-Sud (1972)
5. V. Luzzati, A. Gulik, T. Gulik-Krzywicki, A. Tardieu: In Lipids and Membranes: Past, Present and Future, ed. by J. A. F. Op den Kamp, B. Roelofsen, K. W. A. Wirtz (Elsevier Science Publishers B. V., Amsterdam, 1986) p.137
6. A. Gulik, V. Luzzati, M. DeRosa, A. Gambacorta: J. Molec. Biol. 182, 131 (1985)
7. International Tables for X-Ray Crystallography, Vol. 1 (Kynoch Press, Birmingham, U. K., 1952)
8. V. Luzzati, A. Tardieu: Ann. Rev. Phys. Chem. 25, 79 (1974)
9. T. Gulik-Krzywicki, L. P. Aggerbeck and K. Larsson: in Surfactants in Solution, ed. by K. L. Mittal and B. Lindman, Vol. 1 (Plenum Press, New York, 1984) p.237
10. V. Luzzati, A. Tardieu, T. Gulik-Krzywicki: Nature (London). 217, 1028 (1968)
11. S. T. Hyde, S. Andersson, B. Ericsson, K. Larsson: Z. Krist. 168, 213 (1984)
12. S. Andersson, S. Hyde, H. G. von Schnering: Z. Krist. 168, 1 (1984)
13. W. Longley, T. J. McIntosh: Nature (London). 303, 612 (1983)
14. R. R. Balmbra, J. S. Clunie, J. F. Goodman: Nature (London). 222, 1159 (1969)
15. A. Tardieu, V. Luzzati: Biochim. Biophys. Acta. 219, 12 (1970)
16. P.O. Eriksson, G. Lindblom, G. Arvidson: J. Phys. Chem. 89, 1050 (1985)
17. V. Luzzati, A. Gulik, M. DeRosa, A. Gambacorta: Nobel Symposium"Membrane Proteins:Structure, Function, Assemby, Chimica Scripta (1987), in press
18. J. S. Patton: In Physiology of the Gastrointestinal Tract, ed. by L. R. Johnson (Raven Press, New York, 1981) p.1123
19. J. S. Patton, M. C. Carey: Science, 204, 145 (1979)
20. V. Luzzati, A. Gambacorta, M. DeRosa, A. Gulik : Ann. Rev. Biophys. and Biophys. Chem. 17 (1987), in press
21. C. R. Woese, R. S. Wolfe: The Bacteria, Vol. VIII (Academic Press, New York, 1985)
22. V. Luzzati, A. Gulik: System. Appl. Microbiol. 7, 262 (1986)
23. J. F. Deatherage, K. A. Taylor, L. A. Amos: J. Molec. Biol. 167,, 823 (1983)

Acknowledgements

This work was supported in part by grants from the Commission of the European Community (Stimulation Plan) and the Ministère de l'Industrie et de la Recherche. We are grateful to Hervé DELACROIX, Annette GULIK and Annette TARDIEU for discussions and suggestions, and also for the communication of unpublished results.

Interactions in Lyotropic Lamellar Phases: A High Resolution X-Ray Study

D. Roux[1*] *and C.R. Safinya*[2]

[1]Chemistry Department, UCLA, Los Angeles, CA 90024, USA
[2]Exxon Research and Engineering Co., Annandale, NJ 08801, USA

* Permanent address, CRPP and GRECO Microémulsions
Domaine Universitaire, 33405 Talence, France

The understanding of interactions between membranes has attracted a lot of attention in the last decade, the reason being the relevance of these interactions in biological processes such as cell-cell interactions. The basic forces between membranes are the van der Waals attraction and the electrostatic repulsion [1]. Besides these two well known forces PARSEGIAN, RAND and coworkers [2] have shown in the last decade that there is another short range interaction between phospholipid bilayers embedded in water. This repulsive interaction is attributed to the hydration of the head groups and decays exponentially with a typical length of 3 Å. More recently, HELFRICH [3] has proposed that thermally induced out of plane fluctuations of membranes in a multilayer system can lead to a long range repulsive interaction. The crucial elastic modulus which controls the strength of this interaction is k_c associated with the bending energy of a single membrane. For this steric interaction to compete with the microscopic forces (that is, van der Waals and electrostatics), k_c has to be of order k_BT simply because the interaction is induced by thermal fluctuations. Steric repulsion is also known to be the dominant interaction associated with wandering walls of incommensurate phases [4]. Despite much theoretical [5,6] and experimental [7,8] effort the experimental relevance of this interaction to biological membranes remains controversial.

In this paper, we present results of a comprehensive x-ray scattering study of a multi-lamellar system (water, SDS and pentanol). This system exhibits a lyotropic lamellar phase (smectic A) consisting of bilayers of surfactant and alcohol separated with water. We have studied three different dilutions: an "oil" dilution where the bilayers are separated with a mixture of oil (dodecane) and alcohol (pentanol), a "water" dilution where the spacing between bilayers is increased by adding water to the initial mixture and a "brine" dilution where salt (NaCl) is added to the water.

Due to the very special property that the marginal dimension of a smectic A phase (or lyotropic lamellar) is equal to three [9], the structure factor of a lyotropic lamellar phase is a strong function of the bulk elastic moduli. Very importantly, we are able to measure the bulk modulus of layer compression B(d) as a function of the intermembrane distance d; this modulus is **directly** related to the interactions between neighboring membranes. We find [14] that our experimental results are accurately described by the Helfrich theory along the "oil" dilution. On the contrary, unscreened electrostatic interactions dominate along the "water" dilution. Finally, the addition of a monovalent salt to the water has the effect of screening the electrostatic interactions, in this last case the undulation interaction is shown to dominate in the same way as in the "oil" dilution.

1 X-ray Technique

X-ray measurements usually give information on the structure of the sample studied. In certain cases (such as for the smectic A phase [11,12,13]) however, thermodynamic properties can be obtained from the analysis. It has been known for a long time that the lower marginal dimension for the smectic A phase is three [9]. The situation is named marginal since in this case the thermal fluctuations diverge very weakly with the size of the sample (as the logarithm of the size). This divergence of the fluctuations has a dramatic effect on the structure factor. The normally observed Bragg peaks of the structure factor are replaced by power-law singularities which signal the lack of true long range order. In two dimensions this property has been clearly measured [10]. In smectic A phases a very similar phenomenon also exists. Caillé [11] has calculated the expected structure factor for smectic A phase. The singularities replacing the Bragg peaks appear to be strong functions of the elastic constants B and K [12] of the smectic A phase. The asymptotic behavior of the structure factor in the directions parallel (//) and perpendicular (⊥) to the layers is given by simple power laws

$$S(0,0,q_{//}) \propto \frac{1}{|q_{//} - q_m|^{2-\eta_m}} , \tag{1a}$$

$$S(q_\perp,0,q_m) \propto \frac{1}{q_\perp^{\ 4-2\eta_m}} , \tag{1b}$$

where q_m is the position of the mth harmonic of the structure factor (q_m=mq_o, m=1, 2, ...), and η_m is an exponent related to the elastic constants B and K

$$\eta_m = m^2\, q_o^{\ 2} \frac{k_B T}{8\pi \sqrt{BK}} . \tag{2}$$

This behavior has been recently [13] confirmed experimentally for the first harmonic of a smectic A phase. The expressions given by (1) are simple but the asymptotic behavior of the structure factor does not completely give the whole physical imformation that can be extracted from the experimental data. Indeed, the structure factor is given by the fourier transform of the correlation function [11]

$$G(\vec{R}) = G(\rho,z) = \left[\frac{1}{\rho}\right]^{2\eta} e^{-\eta\,[2\gamma + E_1(\rho^2/4\lambda z)]} . \tag{3}$$

Here, γ is Euler's constant, $E_1(x)$ is the exponential integral function, $R^2 = z^2 + \rho^2$, η has been defined previously (formula (2)) and λ is a length related to the elastic constants ($\lambda=\sqrt{K/B}$). It is the limiting behavior of this correlation function for $\rho<<\sqrt{\lambda z}$ and $\rho>>\sqrt{\lambda z}$ which gives the simple expressions of equation (1). When these two limiting regimes are considered the function G(R) is independent of λ, but in general G(R) does depend on both parameters η and λ. In fact, we will show that experimentally the fit of the experimental data is sensitive to λ when $\lambda<2\pi/q_o$.

We have seen how we can, in principle, extract the elastic constants of a smectic A liquid crystal from the x-ray structure factor. To unambiguously study power-law behavior one must, in general, use a special spectrometer. The experimental set-up must have a very high resolving power capable of determining length scales of the order of a micron to probe the interesting long wavelength behavior of the system. In addition, the resolution function must decrease as a function of the wave vector q faster than $1/q^2$ which is the fall off expected for the structure factor of a "Landau-Peierls" system with a small value of η. These two requirements are fulfilled when using the high resolution set-up we used. The experiments were carried out at the Stanford Synchrotron Radiation Laboratory on beam line VI-2 and at the National Synchrotron Light Source on the Exxon beam line X-10A. The monochromator consists of a double bounce Si(1,1,1) crystal. The analyser is a triple bounce Si(1,1,1) channel cut crystal. We have used crystals of silicon which enables us to resolve longitudinal length scales as large as 1.5 µm. The channel cuts were used to cut down the tails of the resolution function.

2 Oil Dilution

We show in fig. 1 a series of scattering profiles for longitudinal scans through the first harmonic of samples corresponding to the "oil" dilution [14] for the weight fraction of oil (x) increasing from 0 to 0.54. The first three samples exhibit in addition to the first harmonic, a second harmonic whose intensity decreases rapidly with dilution (the second harmonic was not observed for x>0.15). From the variation in the peak position we are able to get the repeat distance d_u (d_u=$2\pi/q_o$) in the lamellar phase. A very important and striking feature of the profiles of the peaks shown in Fig. 1 is the tail scattering which becomes dramatically more pronounced as d_u increases. This effect is due to the thermal fluctuations as previously explained and will be elucidated in detail. Beside the "Landau-Peierls" effect, and for the most diluted samples studied, a small angle scattering (SAXS) appears whose intensity increases as the dilution increases. In fact, for x greater than 0.65, although the samples are in the lamellar phase (the texture between crossed polarisers exhibits characteristics of a smectic A phase: oily streaks and focal conics), the x-ray scattering is dominated by the SAXS which overwhelms the first harmonic. We believe that this SAXS is associated with an isotropic phase L_1 [14]. However, for the purpose of this paper we will only present results far from the phase transition, when a peak is evident.

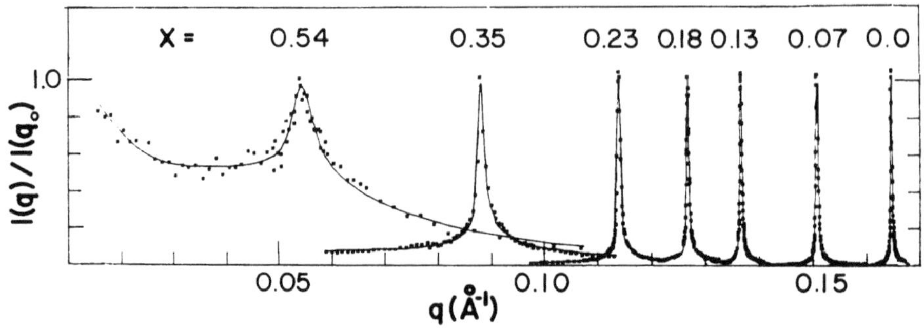

Figure 1: Longitudinal profiles through the first harmonic of seven different samples along the dodecane dilution path, the percent dodecane by weight of the mixture (x) is indicated above each profile. All peak intensities are normalized to 1. The solid lines are fits by the Caillé expression of the structure factor (equation (4))

We now discuss the evolution of the shape of the peak with dilution. We show in Fig. 2c on a semi-log scale the scattering profile for three samples (x=0.07, 0.23 and 0.35). The difference in the profiles over the entire dilution range is now immediately clear. It is qualitatively clear that the exponent η, which characterizes the scattering profile and which is a direct measure of the ratio of the tail to peak intensity, increases with $d_\perp$. It is possible to analyse the profile quantitatively, using the Caillé calculation [11]. Since we are dealing with an unoriented sample we cannot use the simple asymptotic expressions given by equations (1). We fit our data with the fourier transform of the correlation function given by equation (3). We now present details of the fitting procedure used.

Firstly, because we have extremely good resolution (resolving length scales up to 1.5 µm), we see the rounding of the scattering data due to the finite size of the domains. The dashed lines in Fig. 2 show the longitudinal resolution function. The widths of the profiles are normally two or three times larger than resolution. In addition, different preparations of the same mixture may yield widths differing by as much as 50%. This broadening is just the consequence of the finite size of a lamellar domain and is associated with defects. However, we stress that the typical domain sizes are unusually large between 2500 Å and 12 500 Å, which allowed for meaningful analysis of the profiles. We incorporate the finite size effect in our analysis in a straightforward manner outlined by Dutta and Sinha [15]. This then modifies the structure factor

$$S_{FS}(\vec{q}) \sim \int d\vec{R}\; G(\vec{R})\; e^{-\pi \frac{R^2}{L^2}}\; e^{i(\vec{q} - q_m \hat{z}) \cdot \vec{R}} ,$$

where L^3 is the domain volume and $q_m = m\, q_o$. $G(\vec{R})$ is the correlation function and is given by expression (3). Finally, because the samples consist of randomly oriented domains, we perform a powder average over all solid angles in reciprocal space. The finite size and powder averaged structure factor is given by:

$$S(q) = \langle S_{FS}(\vec{q}) \rangle \sim \int_{-\infty}^{+\infty} dz \int_0^{+\infty} d\rho\; G(z,\rho)\; e^{-\pi \frac{R^2}{L^2}}\; \frac{\sin(qR)}{q R}\; e^{-i q_m z} \tag{4}$$

with $G(z,\rho)$ given by expression (3).

The analysis consists of a least squares fit of equation (4) convoluted with the resolution function (represented as gaussian functions) to the experimental profiles. The solid lines in figs. 1 and 2 are results of the best fits to the data with four fitting parameters (η, λ, L and q_m). Figure 2a details the shape of the experimental and theoretical profile of a typical sample (x=0.23). The results of a typical fit plotted on a log-log scale elucidates the main features of the profile (only one side of the peak is plotted). At large q the profile is linear indicating a power law decay of the scattering intensity, as expected from Caillé's expression

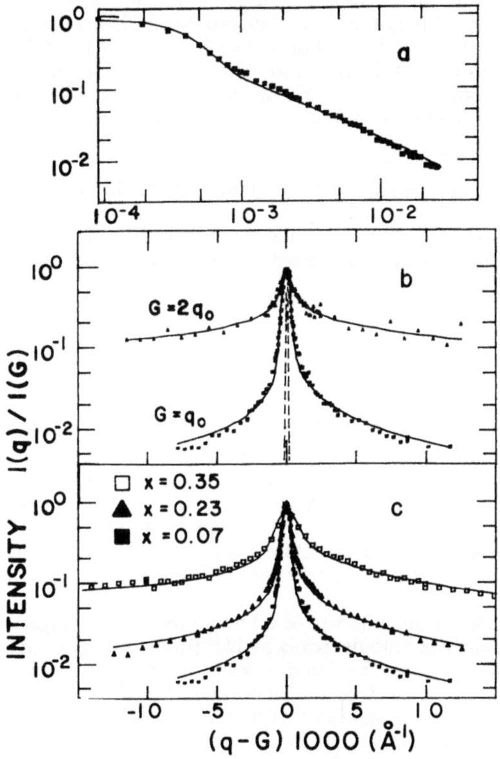

Figure 2: (a) Profile of the first harmonic for the mixture x=.23 of the oil dilution on a log-log scale which shows finite size rounding at small q followed by power law behavior at larger q. (b) Profile of the first and second harmonic for the mixture x=0.07 on a logarithmic intensity scale. The solid lines are fits to the Caillê expression (eq. 4).

$$S(q) \sim |q - q_0|^{-p}, \quad \text{for } |q - q_0| > 2\pi/L \quad ;$$

the exponent p is approximately equal to 1-η due to the powder averaging. For larger distances ($|q - q_0| < 2\pi/L$) the finite size effects dominate and round off the observed profile with a characteristic width equal to 1/L.

A further more subtle aspect of the data is shown in Fig. 2c where the profiles are slightly asymmetric with the high q region more intense than the low q. This effect which is also seen in the theoretical fit is due to the parameter λ ($\sqrt{K/B}$) which enters in the real space correlation function (G(ρ,z) formula (3)). This parameter is a measure of the anisotropy present in S(q). This effect is only observable while $\lambda < 2\pi/q_0$; for larger values of λ the asymmetry is negligible and S(q) is not sensitive to the value of this parameter. In our case, λ remains small enough to be measured (except for the last sample), however, the determination of λ is less accurate than the determination of the other parameters (η for example).

The three first samples (for x < 0.15) exhibit a second harmonic peak in addition to the first harmonic. This feature, which is different from the regular thermotropic smectic A systems [13], allows us to self-consistently verify the validity of Caillé's harmonic model. Indeed, as shown in expression (2), η scales with m^2 where m is the harmonic number. The value of η should be four times larger for the second harmonic than for the first harmonic. We show in Fig. 2b the profiles and the fits for the first and second harmonic for a typical sample (x=0.07). The fit gives the following values for the parameters:

- first harmonic: $\eta = 0.14 \pm 0.02$, $\lambda = 8.6 \pm 1$ Å, L = 10,000 ± 1000 Å
- second harmonic: $\eta = 0.58 \pm 0.02$, $\lambda = 8.1 \pm 1$ Å, L = 9,100 ± 1500 Å.

This result indicates clearly that η is effectively four times larger for the second harmonic while λ and L remain comparable. This result also rules out the hypothesis that the broadening of the peaks with dilution would be due to intrinsic disorder, since in this case L should scale with m^2 [16]. Similar results are also obtained for the samples with x=0 and x=0.13.

For the most dilute samples (x > 0.45), the scattering is complicated due to the appearance of the small-angle scattering (SAXS) at small q. As explained previously, we believe this SAXS is due to fluctuations of the isotropic phase. The very dilute samples (x > 0.62) do not exhibit a peak but only small angle scattering; we are able to fit satisfactorily the SAXS for these samples with a model of polydispersed spheres [16]. We have used this same model to fit the SAXS of the samples x=0.47, 0.54 and 0.62 in order to be able to fit the complete structure factor. This leads to larger errors on the values of η and λ for these samples.

To summarize the essential points in the fits, from the experimental data we are able to extract four parameters: q_o, η, L and λ corresponding respectively to the peak position, the power law behavior away from the central peak, the central peak width and the asymmetry in the profile around the peak. Using the definition of η and λ we can calculate B and K, the elastic constant of the lamellar phase:

$$B = \frac{kT}{8\pi\,\lambda\,\eta}\, q_o^2 , \qquad K = \frac{\lambda}{\eta}\,\frac{kT}{8\pi}\cdot q_o^2 \tag{5}$$

For all the samples the value of k_c (k_c=Kd) ranges around k_BT (from 2 k_BT to 0.5 k_BT). The fact that k_c remains of the order k_BT confirms previous measurements [17] and differs strongly from measurements of k_c in double layers of phospholipids [18] (~20 k_BT). The value of K is then one order of magnitude less than the value measured in thermotropic liquid crystals (~10^{-6} dyn) [13]. Even more striking is the value of B which varies by more than two orders of magnitude along the dilution (probably even more since we were unable to get values for the most dilute samples). The values of B range from typical smectic A values (10^9 - 10^8 erg/cc) to extremely weak values (10^5 erg/cc).

We will now interpret quantitatively the x-ray results in terms of interactions between layers. The van der Waals and undulation energies per unit surface both decrease as $1/d^2$ (in the range we are interested in) and usually the coefficients of these forces favor the van der Waals interactions. This is not the case in this system. Comparing both interactions, we find that the undulation forces range between 0.12 k_BT/d^2 and 0.46 k_BT/d^2, whereas the van der Waals interaction remains of the order of -0.03 k_BT/d^2 (or even less for large separations). The interactions are then completely dominated by the undulation forces. Using the Helfrich model it is then possible to calculate B(d), since B is the second derivative of the free energy per unit volume with respect to d_u. Using the Helfrich theory [3], we calculate η:

$$\eta = \frac{4}{3}\,\frac{d^2}{d_u^2} = \frac{4}{3}\left(1 - \frac{\delta}{d_u}\right)^2 . \tag{6}$$

The simplicity of this expression comes from the fact that the interactions are inversely proportional to k_c and consequently the product KB is not a function of k_c. The Helfrich theory predicts that the exponent η should be a universal function of δ/d_u which increases continuously to saturate at η=1.33 for large separation between the lamellae. Since we measure η accurately from the power law behavior of S(q), we can directly compare the experimental values of η to the prediction of the Helfrich theory for the "oil" dilution. Figure 3 is a plot of the experimental data and the curve given by the expression (6) assuming δ = 29 Å. The remarkable agreement between the experiments and the theoretical prediction is a direct proof of the importance of the undulation forces and the precise functional dependence on d. It is clear that the interactions in this multimembrane system are dominated by the phenomenon predicted by Helfrich. The value of δ slightly larger than the thickness of the water (d_w = 18 Å) takes into account the finite thickness of the film (the polar head and a part of the hydrophobic tail corresponding roughly to the alcohol length). We stress that the comparison between theory and experiment is based on the value of η alone, a parameter whose value is determined accurately from the algebraic decay of the structure factor. The length λ and its experimental value has no consequence on our basic conclusions.

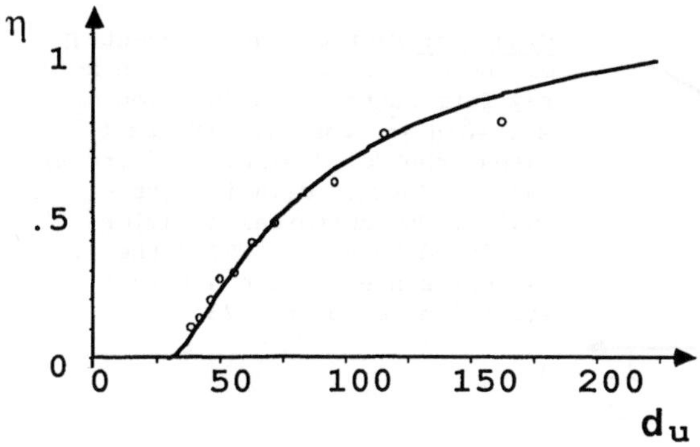

Figure 3: Power law exponent η plotted as a function of the intermembrane distance for mixtures along the dilution path of the "oil" dilution. The solid line is the prediction of the model of HELFRICH of entropically driven steric interactions ($\eta = 1.33\ (1-\delta/d_u)^2$).

3 "Water" and "Brine" Dilution

In order to investigate the competition between undulation and electrostatic interactions, we have diluted the same multimembrane system with water or brine [19] (in this case no oil is added and the system remains in the lamellar phase). Using the same technique we have measured the exponent η as a function of the separating distance along the two dilutions. Figure 4 shows the variation of η as a function of d_u for the "water" and "brine" dilutions. In both cases η increases as a function of d_u but in a different way. For the water dilution η saturates rapidly at a value around 0.3, while for the "brine" dilution η continues to increase and saturates at a value larger than 1. A comparison of these experimental results with the prediction given by the theory of the undulation forces (solid line) very clearly shows that the interactions in the "brine" dilution (open squares) can be accurately described as undulations, whereas in the "water" dilution (open circles) such an interpretation is not satisfactory. When diluting with pure water, it is possible to show [20] that the Debye length scales with the distance between layers and increases with it. Taking into account that the electrostatic interactions are unscreened when no additional ions and counterions (salt) are present, it seems reasonable to use the electrostatic theory to interpret the data [20]. We show in Fig. 4 η predicted by the solution of the Poisson-Boltzmann equation [19]. The result indicates that in the "water" dilution the interactions are dominated by the electrostatic forces.

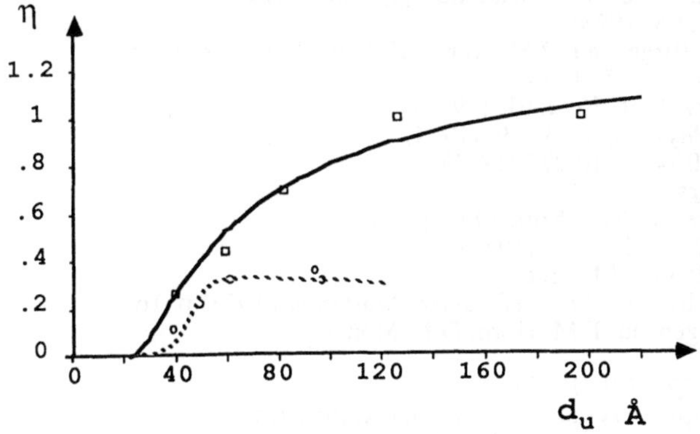

Figure 4: Variation of the exponent η as a function of the repeating distance d_u for the "water" dilution (circles) and for for the "brine" dilution (squares). The solid line corresponds to the value of η when undulation forces dominate ($\eta = 1.33\ (1-\delta/d_u)^2$), the dashed line corresponds to the numerical solution of the Poisson-Boltzmann equation

4 Conclusion

When undulation interactions dominate we have shown that a universal behavior is expected. An elegant way of seeing this behavior is to plot the experimental values of η as a function of $(1-\delta/d_u)^2$, where δ and d_u are determined experimentally. Indeed, using the HELFRICH theory, we can predict that $\eta = 4/3\ (1-\delta/d_u)^2$. Figure 5 represents such a plot for the oil dilution (δ=29Å) and the "brine" dilution (δ=20Å). The solid line corresponds to a straight line of slope 4/3. The agreement between our data and the theory is remarkably good, indicating that the d dependence of the undulation interaction is correct ($\Delta F/S \propto 1/d^2$) as is the absolute prefactor multiplying this d-dependence, which was determined from the free energy of a smectic A system.

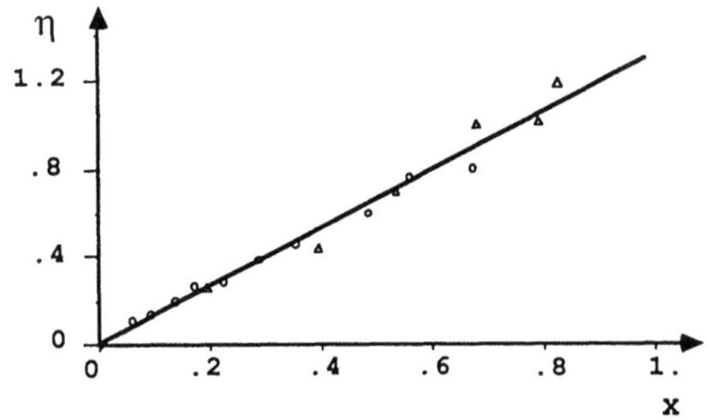

Figure 5: Plot of the exponent η of the algebraic decay of the x-ray form factor as a function of $x=(1-\delta/d_u)^2$ for two different cases: the "oil" dilution (circle) and the "brine" dilution (triangle). The universal behavior predicted by the HELFRICH theory is represented by the straight solid line of slope 4/3.

We have shown that, in a multimembrane system made of a mixed film of surfactant and alcohol (pentanol), the undulation forces predicted by HELFRICH are dominant if no other long-range interactions (such as unscreened electrostatic) are present. This behavior is directly related to the very weak value of the elastic constant of the film ($k_c \sim k_B T$). For comparison, we have also studied, using the technique of freely suspended films, a model membrane system made of DMPC (lipid) plus water [21]. In this study, we find that the dominant interaction is probably due to the hydration force (in the range of d-spacing accessible to the experiment d < 20Å). In this system, the undulation forces remain weak probably due to the large value of k_c ($\sim 20\ k_B T$). This difference in k_c explains the difference of behavior between a pure lipidic membrane and a mixed film. The alcohol seems to be essential in tuning the value of k_c from large values to small ones. Recent theoretical [22] and experimental results [23] confirm this idea.

The authors would like to acknowledge useful discussions with D. Andelman, S. Leibler, R. Lipowsky, S. Milner and S. Safran. The National Synchrotron Light Source, Brookhaven National Laboratory, and the Stanford Synchrotron Radiation Laboratory are supported by the U.S. Department of Energy.

References

1. J.N. Israelachvili: Intermolecular and Surface Forces, (Academic, Orlando, Fla 1985)
2. A. Parsegian, N. Fuller, R.P. Rand: Proc. Natl. Acad. Sci. 76 2750 (1979)
3. W. Helfrich: Z. Naturforsch. 33a, 305 (1978)
4. S.G.J. Mochrie, A.R. Kortan, R.J. Birgeneau, P.M. Horn: Z. Phys. B 62, 79 (1985)
5. D. Sornette, N. Ostrovsky: J. Phys. 45 265 (1984)
6. R. Lipowsky, S. Leibler: Phys. Rev. Lett. 56, 2561, (1986)
7. W. Harbich, W. Helfrich: Chem. Phys. Lipids 36, 39 (1984)
8. R.P. Rand: Annu. Rev. Biophys. Bioeng. 10 277 (1981)
9. a) L.D. Landau: In Collected Papers, ed.by D. Ter Haar (Gordon and Breach, New York,1965) p. 209
   b) R.E. Peierls: Helv. Phys. Acta 7 suppl. 11,81 (1974)
10. P. Dutta, S.K. Sinha, P. Vora, L. Passel, M. Bretz: In Ordering in two dimensions,ed. by S.K. Sinha (Elsevier Amsterdam 1978) p. 169
    P.Heiney, P.N. Stephens, R.J. Birgeneau, P.M. Horn, D.E. Moncton: Phys. Rev. B 28, 6416 (1983)
11. A. Caillé: C. R. Acad. Sci., Ser. B 274 891 (1972)
12. P.G. de Gennes: The Physics of Liquid Crystals (Clarendon, Oxford 1974)
13. J. Als-Nielsen, J.D. Litster, R.J.Birgeneau, M. Kaplan,C.R. Safinya Lindegaard-Andersen, S. Mathiesen: Phys. Rev. B 22, 312 (1980)
14. C.R. Safinya, D. Roux, G.S. Smith, S.K. Sinha, P. Dimon, N.A. Clark, A.M. Bellocq: Phys. Rev. Lett. 57, 2718 (1986)
15. P. Dutta, S.K. Sinha: Phys. Rev. Lett. 47, 50 (1981)
16. A. Guinier: X-Ray Diffraction (Freeman, New York 1963)
17. J.M. Dimeglio, M. Dvolaitsky, C. Taupin: J. Phys. Chem. 89 871 (1985)
18. M.B. Schneider, J.T. Jenkins, W.W. Webb: J. Phys. 45, 1457 (1984)
19. D. Roux, C.R. Safinya: J. de Phys., submitted
20. A.C. Cowley, N.L. Fuller, N.P. Rand, V.A. Parsegian: Biochem. 17, 3163 (1978)
21. G.S. Smith, C.R. Safinya, D. Roux, N. Clark: Mol. Crystal. Liquid. Crystal. 144, 235 (1987)
22. A. Ben-Shaul, I. Szleifer, W.M Gelbart, D. Roux: to be published
23. C.R. Safinya, D. Roux, E. Sirota, G. Smith: to be published

Stability of Brine Swollen Lamellar Phases

G. Porte[1], *P. Bassereau*[1], *J. Marignan*[1], *and R. May*[2]

[1]Groupe de Dynamique des Phases Condensées –
GRECO Microémulsions U.S.T.L., F-34060 Montpellier Cedex, France
[2]Institut Laue Langevin, 156X Centre de Tri, F-38042 Grenoble Cedex, France

1. INTRODUCTION

Swollen lamellar phases show long-range smectic order and consist of thin lamellae with a wide lateral extension separated by very large average distances $\bar{d}$. Correlatively their stability involves two successive requirements. First the individual membrane must be the optimum morphology for the aggregation of the surfactant in the given experimental conditions. And secondly, some minimum interaction potential ($\Delta F/A$) between membranes must be present in order to induce the long-range smectic order. This second more collective aspect of the stability of the L_α swollen phase is presently the subject of intense theoretical and experimental investigations : in particular several contributions to this conference (W.Helfrich; E.Evans; D.Roux; D.Sornette; S.Leibler; R.Lipowsky) are devoted to the relevance of the Helfrich steric interaction in the effective intermembrane forces. Nevertheless, the first individual aspect is also important. It is related to the morphological transformations of the individual membrane when going into the phases that are adjacent to L_α in the diagrams. In general, the membrane morphology and the intermembrane interaction are both affected by the variations of any composition variable of the system. And it is usually difficult to define the adequate procedure to investigate separately these two intricate features of the stability.

The situation is different with the brine-swollen L_α phase of some well-chosen ternary surfactant/alcohol/brine systems. The phase diagram of these systems (such as in figure 1) often shows a domain of stability for the regular L_α phase extending very far towards the brine corner : here (CPCl system) the L_α phase remains monophasic and birefringent up to at least $\phi_w = 0.995$ where ϕ_w is the volume fraction of the brine. Otherwise, the radial geometry of the diagram in the dilute range indicates that, irrespective of the degree of dilution ϕ_w, the stability of the phases and of the primary surfactant structures is entirely determined by the alcohol to surfactant ratio (ϕ_A/ϕ_S).

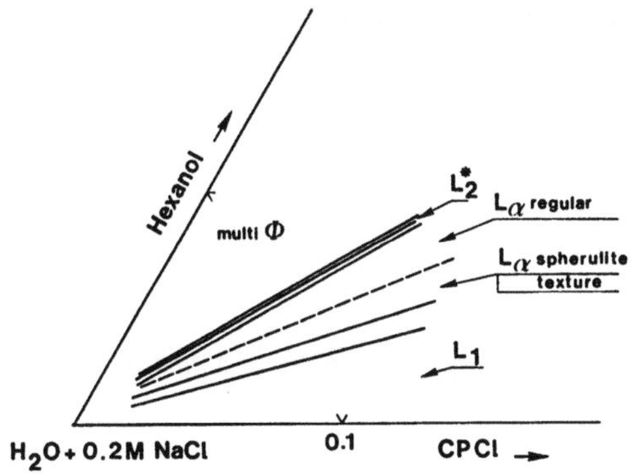

Fig. 1. Brine-rich corner of the phase diagram of the ternary system CPCl/hexanol/brine (0.2M NaCl)

Here we take advantage of this favorable situation in order to investigate separately the individual and the collective levels of the stability of the swollen lamellar phases. In a first step, ϕ_W is varied at constant adequate ϕ_A/ϕ_S value : doing so we essentially modulate the mean distance $\bar{d}$ between membranes which remains constant in thickness and chemical composition. The ultimate purpose of the X-ray scattering study is to check the theoretical predictions about the steric interaction /1,2/ between fluctuating membranes.

In a second step, ϕ_A/ϕ_S is varied at any convenient dilution ϕ_W. We so investigate the structure of the "anomalous isotropic phase" (L_2^* on the diagram) and of the "spherulites mixture" which are contiguous to L_α in the diagram. The purpose here is to determine what different morphologies replace the regular membrane at higher and lower alcohol content.

2. SWOLLEN L_α PHASE : STABILITY AGAINST DILUTION

The X-rays scattering experiments are performed with samples where the wall of the capillary cell induces homeotropic alignment. Due to the cylindrical shape of the cell the obtained orientation is cylindrically symmetrical with the layers everywhere parallel to the capillary axis.

The swelling behavior of the L_α phase of three systems was studied :

- Cetylpyridinium chloride (CPCl)/hexanol/brine (0.2M NaCl)
- Na-octylbenzene sulfonate (OBS)/pentanol/brine (0.5M NaCl)
- n-dodecyl betain/pentanol/water (uncharged system).

The L_α phase of all systems show focal conics texture up to $\phi_W = 0.96$ at least.

2.1 Characterisation of L_α

A typical 2D scattering pattern of swollen L_α phase is given in figure 2. It shows two well-defined Bragg patches and a diffuse pattern which extends far from the Bragg position in both q_z and $q_\perp$ directions. The diffuse pattern is not fully understood and is probably related to the positional fluctuations of the stacking. The small extention of the Bragg spots in the direction parallel to the layers ($q_\perp$ direction) indicates that the membranes are smooth and defect free over in-plane distances of 10^3 Å at least. This regularity is further confirmed by ESR experiments /3/ which could not detect any appreciable density of highly curved defects in the present lamellar samples.

For the three systems, the evolution of the Bragg position q_B with the dilution Φ_W shows that the stacking swells with constant membrane thickness d_o : CPCl system d_o = 26.5 Å; OBS system d_o = 20.5 Å; betain system d_o = 25 Å.

Both the morphology of the membrane and the dilution procedure are thus adequate for the study of the interactions between fluctuating membranes.

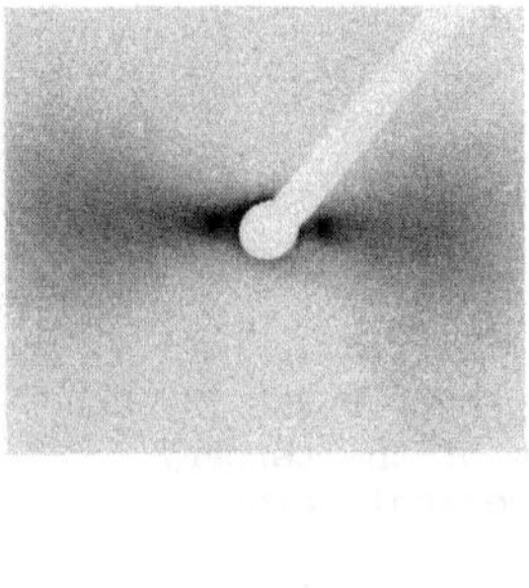

Fig. 2. 2D scattering pattern of an oriented OBS L_α sample with $\bar{d}$ = 112 Å

2.2 Steric interaction and smectic order

Figures 3 and 4 show the evolution with the dilution of the scattering pattern in the q_z direction for the CPCl and betain systems. In case of the CPCl system, the apex of the first Bragg maximum remains very sharp at all dilutions while its foot broadens smoothly with the dilution. For the betain system the picture is different: the foot broadens much faster with ϕ_w and for the most swollen samples a strong small-angle scattering almost overwhelms the first Bragg maximum. The OBS system (data not represented here) corresponds to a situation which is intermediate between the two others.

As shown in /4,5,6/, the X ray intensity scattered by smectic media should show a power law divergence at the Bragg position. More precisely, for a sample with the present cylindrical alignment one expects the divergence to scale as

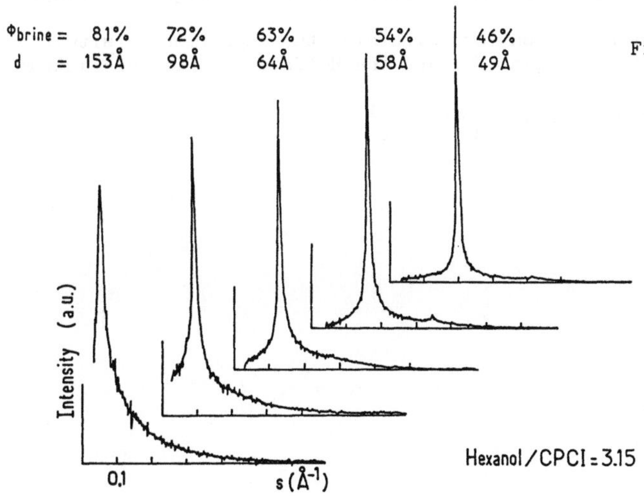

Fig. 3. Scattering pattern in the q_z direction for the CPCl L_α phase at various dilutions

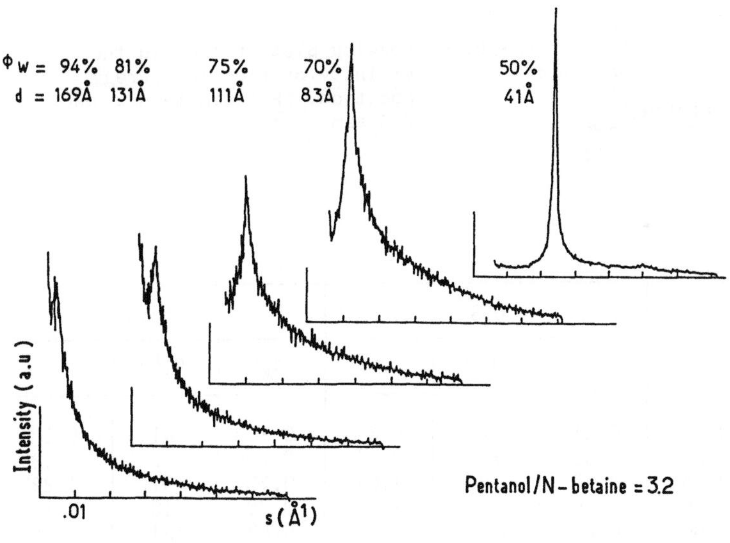

Fig. 4. Scattering pattern in the q_z direction for the betain L_α phase at various dilutions

$$I(q) = (q - q_B)^{-(1.5 - \eta)},$$

where η is simply related to the mechanical properties of the stacking (the bending rigidity of the membrane k_c and the effective intermembrane potential $\Delta F/A$). Accordingly, the X-rays scattering data obtained for the three systems at various dilutions (50 Å < d < 200 Å) have been plotted on a log-log scale (figure 5) for an estimation of η as function of $\bar{d}$ (table 1). The obtained values provide a quantitative representation of strong differences in behavior between systems.

The evolution of η ($\bar{d}$) is indeed controlled by the d dependence of the direct molecular forces (Van der Waals, electrostatic, hydration). In the present situations where the electrostatic contribution is either screened (brine systems) or absent (zwitterionic system) the two following conditions are fulfilled :

- direct molecular potentials decreasing faster than d^{-2}
- complete unbinding situation (since the phases swell up to very high dilution).

According to the analysis of R.Lipowski and S.Leibler the experimental situations correspond to the weak fluctuation regime /7/ and the Helfrich steric interaction /1/:

$$(\Delta F/A)_{St} = a_o (k_B T)^2 \cdot k_c^{-1} \cdot d^{-2}$$

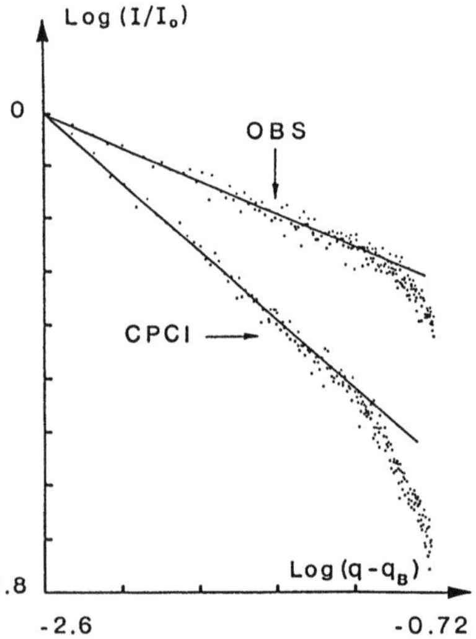

Fig. 5. Log-log plot of I(q) in the vicinity of the first Bragg position. CPCl sample $\bar{d}$ = 98Å OBS sample $\bar{d}$ = 84 Å

Table 1

CPCl			OBS			Betain		
ϕ_W	d(Å)	η	ϕ_W	d(Å)	η	ϕ_W	d(Å)	η
0.49	49	< 0.6	0.51	40	< 0.6	0.50	41	< 0.6
0.62	64	< 0.6	0.63	52	0.89	0.70	83	1.18
0.66	73	0.75	0.70	71	1.04	0.75	97	1.29
0.75	98	0.84	0.75	84	1.17	0.78	110	-
0.83	153	0.86	0.80	112	1.17	0.82	131	-
						0.84	169	-

should dominate in the effective potential. One then expects that $\eta(\bar{d})$ should scale for all systems according to :

$$\eta(\bar{d}) = c_o(1 - d_o/\bar{d})^2 , \tag{1}$$

where c_o is a universal constant of the order of 1.

In contrast with the experimental results of the Exxon group /6/, our data in table 1 do not fit to expression (1) unless we assume that c_o is strongly system dependent. The discrepancy is truly puzzling since it contradicts a very strong prediction of the theory. However, an explanation may be proposed when noting that prediction (1) holds for the asymptotic condition of small osmotic pressure of the stacking. The here investigated dilution range (50 Å < d < 200 Å) could well correspond to intermembrane distances $\bar{d}$ which are not large enough for this asymptotic condition to hold. And clearly, data at much larger dilution must be obtained before claiming any definite statement about the validity of the theoretical prediction.

2.3 Persistence of the Bragg maximum at very high dilution

To investigate much larger spacings, neutron scattering is used (ILL Grenoble). However, neutron scattering is a low-resolution technique ($\Delta\lambda/\lambda$ = 10%) and quantitative treatments are out of reach.

Figure 6 represents the scattering profile for a very swollen CPCl sample (d = 970 Å) using again the cylindrical geometry. In spite of the very high dilution the first Bragg maximum is still well defined. Additionally we observe a strong central scattering which spreads mainly in the q_z direction. This anisotropy indicates that the central scattering is also relevant to positional fluctuations rather than to some density of defects in the sample /6/.

The evolution of the Bragg position q_B as function of ϕ_W (figure 7) reveals that the stacking swells with a constant membrane thickness on the entire range : 50 Å < d < 1100 Å).

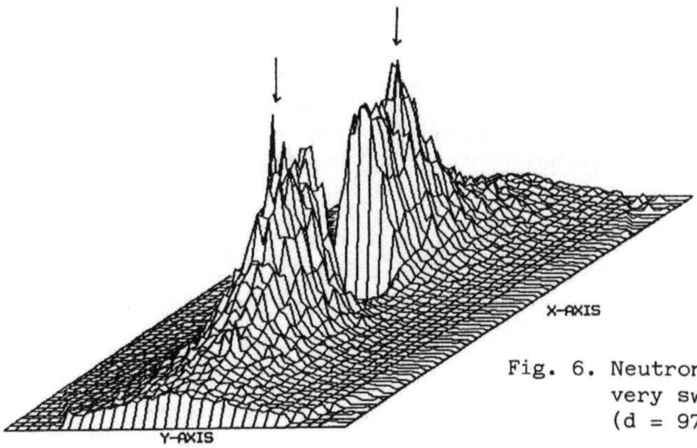

Fig. 6. Neutron scattering patter of a CPCl very swollen oriented sample (d = 970 Å)

3. ALTERNATIVE MORPHOLOGIES AT HIGHER AND LOWER ϕ_A/ϕ_S

ϕ_W is now fixed at some convenient value and the effect of variations of ϕ_A/ϕ_S on the membrane morphology is investigated. The structure of the alcohol-poor "spherulite mixture" and of the alcohol-rich "anomalous isotropic phase" are successively studied for the CPCl system.

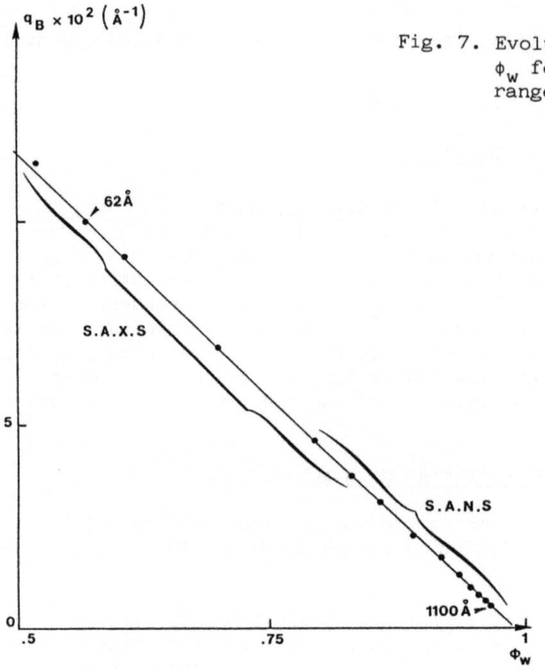

Fig. 7. Evolution of q_B as function of Φ_w for the CPCl L_α phase in the range 50 Å < $\bar{d}$ < 1100 Å

3.1 Spherulite texture /8/

As observed in polarized light microscopy this "phase" consists of "birefringent droplets" (maltese cross) embedded into a continuous birefringent background. When sealed in the scattering cell the background orients cylindrically like the regular L_α phase while the droplets remain undeformed. The X-ray scattering pattern (figure 8) then shows two small Bragg spots related to the oriented background and an isotropic ring related to the droplets. The important point is that both the spots and the ring correspond to the same Bragg wave vector q_B at all dilutions (50 Å < d < 170 Å). We thus conclude that both droplets and the background are lamellar stackings with the same interlayer distance. And the spontaneous formation of the droplets reveals the tendency for the membrane to bend with positive both mean and Gaussian curvature when the alcohol content is too low.

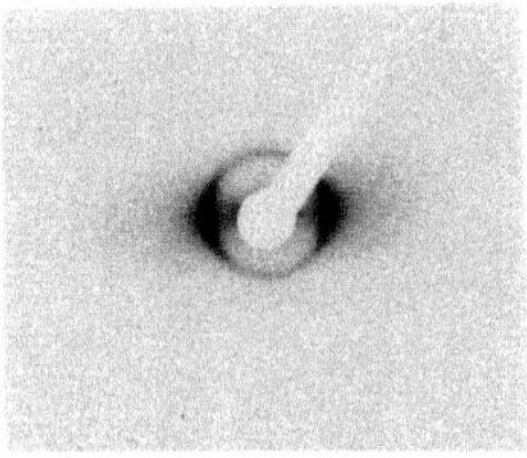

Fig. 8. 2D scattering pattern of the CPCl spherulite texture

3.2 The anomalous isotropic phase : L_2^* (= L_3).

It is stable for higher relative alcohol contents than the L_α phase. It shows very special physical properties which suggest a structure different from that of the classical L_1 isotropic phase. It expells excess brine and concentrates spontaneously upon further addition of alcohol /9/. Its electrical conductivity is constant over the whole range $0.80 < \phi_w < 0.98$ with a value thirty per cent smaller than this of the pure brine. When submitted to a shear stress or to a magnetic field, it exhibits a strong transient birefringence which increases with the dilution ϕ_w.

Its structure is investigated using neutron scattering (ILL Grenoble). The scattering profiles show a broad maximum at low q and a long decreasing tail at higher q. The intensity in the higher q range ($0.02\,Å^{-1} < q < 0.15\,Å^{-1}$) for the most diluted samples (ϕ_w = 0.96 and 0.98) scales as (figure 9) :

$$I(q) \propto q^{-2} \exp\left(\frac{-q^2 d_o'^2}{12}\right) ,$$

which together with the conductivity properties /10/ indicates that at a local scale (∿ 100 Å) it consists of flat bilayers of thickness d_o' = 25 Å (similar to d_o in L_α) with random orientations.

This local structure can accommodate several different arrangements at larger scales. We have considered four different possibilities : large disconnected disks /10/, foam-like structure /9/, lamellar packing with no long-range order /11/, and disordered cubic arrangement of Schwartz minimal surfaces /12/. To discriminate between these models, the position of the maximum has been plotted as function of ϕ_w (figure 10). It scales as :

$$q_M = \frac{2\pi}{\alpha d_o'} (1 - \phi_w) \text{ with } \alpha = 1.5 \text{ and } d_o = 25 \text{ Å} .$$

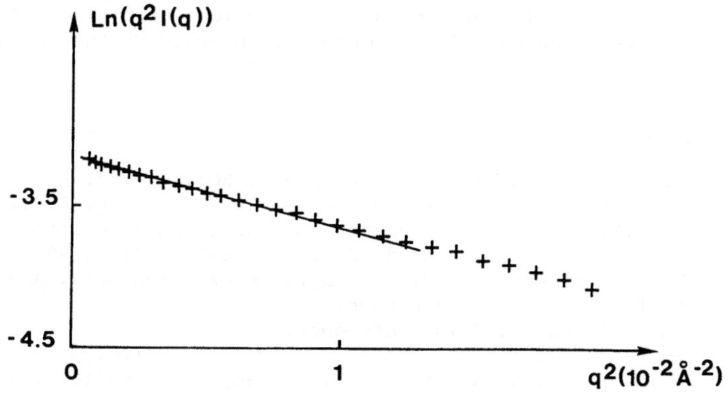

Fig. 9. Ln (q^2 I(q)) as function of q^2 for the L_2^* phase in the range $0.02\,Å^{-1} < q < 0.15\,Å^{-1}$ (ϕ_w = 0.98)

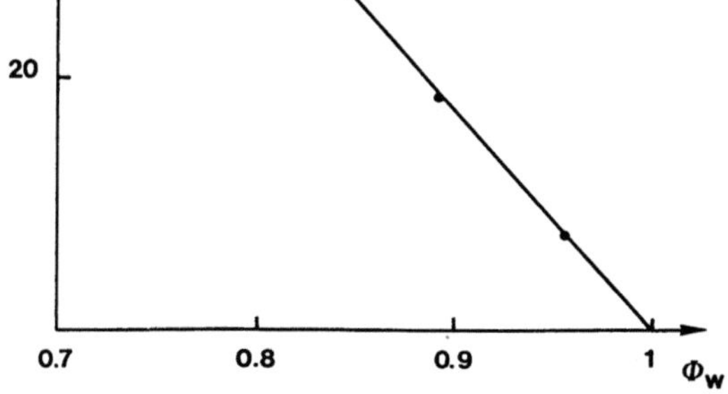

Fig. 10. Position of the I(q) maximum as function of ϕ_w in the L_2^* phase

A detailed discussion shows that, amongst the four models, only the last one (disordered cubic arrangement) is quantitatively consistent with this particular swelling behavior and also with the other physical properties of the phase. Indeed the series of considered models is not exhaustive and we cannot claim a clear determination of the L_2^* structure. Nevertheless the consistency of the disordered cubic structure with both physical and scattering data is encouraging. It would indicate that at high enough alcohol content the regular membrane spontaneously bends with negative Gaussian curvature everywhere /12/.

4. CONCLUSION

These brine-swollen ternary systems provide adequate situations to investigate the stability of lamellar stacking from both points of view of the membrane morphology and of the long-range regularity of the smectic stacking. The L_α phase consists of regular membranes which keep constant thickness and chemical composition upon dilution in a very large $\bar{d}$ range (50 Å < d < 1100 Å). The persistence of the first Bragg maximum up to 10^3 Å at least demonstrates the long-range character of the Helfrich steric interaction between fluctuating membranes. On the other hand, the strong deviations observed at moderate dilutions ($\bar{d}$ < 200 Å) from the predicted universal Bragg exponent indicate that this prediction should be experimentally checked at higher dilution using a high-resolution technique (very small angle X-rays scattering).

The spontaneous formation of birefringent droplets in the "spherulite mixture" reveals the spontaneous tendency for the membrane to bend with positive Gaussian curvature when the alcohol content is too low. While the plausible disordered cubic structure of the anomalous isotropic phase probably indicates the opposite tendency at too high alcohol content. The physical mechanism underlying this behavior is not yet well understood and certainly deserves further experimental investigations.

REFERENCES

1. W. Helfrich, Z. Naturforsch 33a, 305, (1976).
2. R. Lipowski, S. Leibler, Phys. Rev. Lett. 56, 2541, (1986).
3. J.M. Di Meglio, P. Bassereau, unpublished results.
4. A. Caille, C. R. Acad. Sci. Ser. B274, 891, (1972).
5. L. Gunther, Y. Imry, J. Lajzerowicz, Phys. Rev. A22, 1733, (1980).
6. D. Roux, This Conference.
7. R. Lipowski, This Conference.
8. C.A. Miller, O.G. Ghosh, W.J. Benton, Colloids and Surfaces 19, 197, (1986).
9. G. Porte, R. Gomati, O. El Haitamy, J. Appell, J. Marignan, J. Phys. Chem. 90, 5746, (1986).
10. B. Lindman, This Conference.
11. O. Parodi, in Colloïdes et Interfaces, ed. A.M. Cazabat and D. Langevin, p. 355 (Editions de Physique, Paris, 1983).
12. J. Charvolin, Journal de Physique C3, 46, 173, (1985).

Phase Diagram of Lamellar Phases: Rigidity and Curvature

J.-M. di Meglio

Laboratoire de Physique de la Matière Condensée, UA 792 and GRECO Microémulsions du CNRS, Collège de France, 11 Place Marcelin Berthelot, F-75231 Paris Cedex 05, France

We propose to relate in this article some patterns of phase diagrams of lyotropic systems (birefringent microemulsions and nonionic binary aqueous systems) to the rigidity and the curvature of the interfacial film.

1. BIREFRINGENT MICROEMULSIONS

Microemulsions are thermodynamically stable dispersions of oil (O) and water (W) stabilized by a surfactant (S) often associated to a cosurfactant (CS) (a short chain alcohol), with a characteristic size much smaller than the wave length of visible light (typically of order 100 A). Much attention has been first paid to the study of droplet structure (O/W or W/O) : the surface over volume ratio is extraordinarily large (=3/R, R the droplet radius), so theinterfacial tension has to be very low [1]. One model even assumes [2] that the surface pressure of the film is equal to the bare interfacial tension between oil and water so that the resulting actual interfacial tension is zero. Now more interest is paid to liquid crystalline phases. Figure 1 represents in a schematic way the phase diagram of a typical ionic microemulsion (W, O=cyclohexane, S=sodiumdodecylsulfate (SDS), CS=1-pentanol).

The role of the cosurfactant is a challenging problem : why such a small amount of cosurfactant induces such a drastic topological change in the structure ? An accurate description of the phase diagram would in principle require the knowledge of many different energies :

- entropy of the film
- electrostatic energies between surfactant polar heads
- interactions between interfacial films
- curvature energies.

All these interactions are small, dependent on each other and of the same order of magnitude. DE GENNES and TAUPIN [3] have nevertheless pointed out the role of curvature energy and we will see that this will allow us to describe some special paths in the phase diagram.

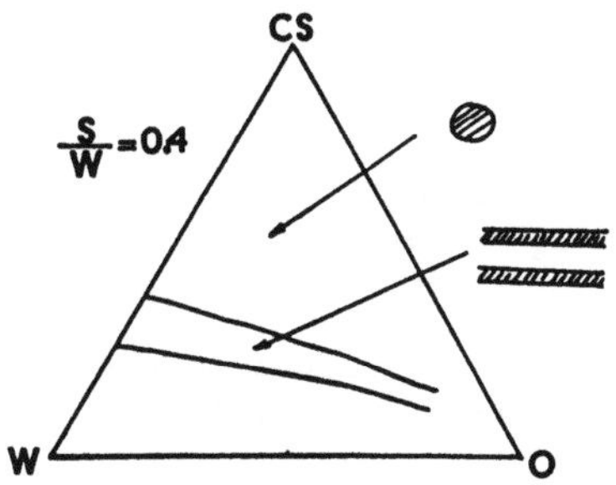

Figure 1
Schematic phase diagram

The curvature energy per unit area reads as [3] :

$$E = \frac{1}{2} K \left[\frac{1}{R} - \frac{1}{R_o} \right]^2$$

with K the rigidity constant, R the radius of curvature of the interfacial and R_o the spontaneous natural radius due to packing constraints of surfactant molecules inside the film. One can derive from the above expression a "persistence length" of the film :

$$\xi_K = a \exp \left[\frac{2\pi K}{kT} \right] ;$$

a is a molecular size. For r taken along the film less than ξ_K the film is flat while it is undulated for r larger then ξ_K. ξ_K depends on K through an exponential so a slight decrease of K (adding of cosurfactant?) would induce a drastic decrease on ξ_K and explains some of the phase transition (Fig. 2).

How to measure K ? : as microemulsions are very curved objects, the rigidity constant likely is very small and the interface of a lamellar phase close to the droplet phase in the phase diagram may be undulated. This was the starting point of our study. We have used the spin technique method. Its basic concept is to study the electronic resonance of a nitroxide probe grafted on the alkyl chain of a surfactant (one surfactant labelled for one thousand unlabelled). This method enables to determine the orientation of the interfacial film with respect to a permanent magnetic field [4]. By comparison of two different resonance geometries (isotropic in a 4 mm-diameter spherical container and oriented in a 100 μm-pathlength rectangular cross section container) we have developed a method [5] which gives the amplitude of the undulations (Fig. 3-a).

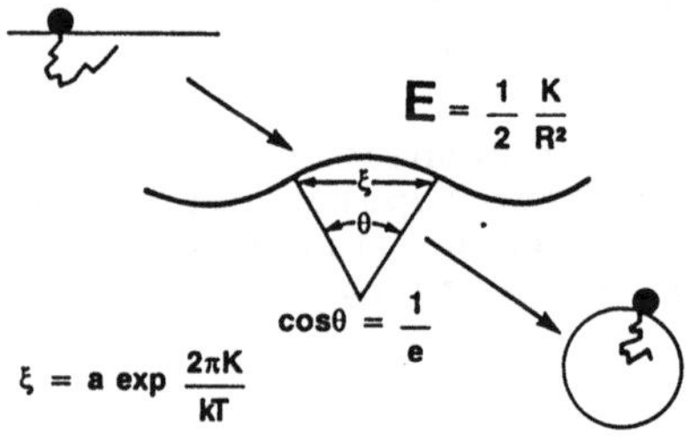

Figure 2 : From lamellar to micellar by decreasing K

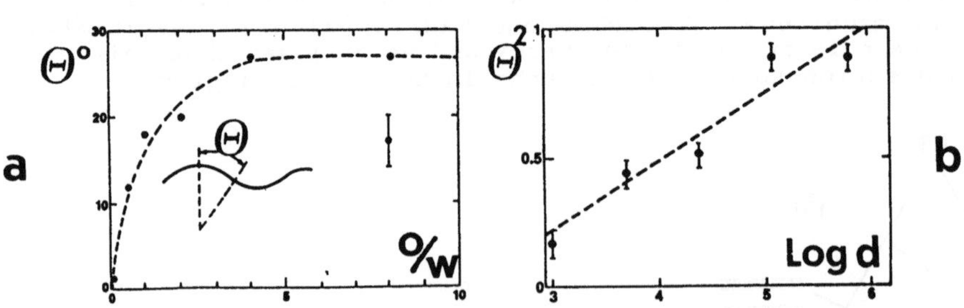

Figure 3 : a) Amplitude of undulations versus swelling ratio b) Quadratic mean of θ versus the logarithm of the distance between lamellae

The problem is now to relate this angle of disorientation with the rigidity constant K : this addresses the multilayer problem. DE GENNES and TAUPIN [3] wrote the energy as :

$$E = \frac{1}{2}\,\frac{K}{R^2} + \frac{1}{2}\,U''z^2 \quad \text{with } U'' = \left[\frac{\partial^2 V}{\partial z^2}\right]_{z=0} \;;$$

V is the interaction potential and z is the out-of-reference z-component of the interfacial film. The quadratic mean of the undulation is now :

$$\theta^2 = \frac{kT}{\pi K}\,\mathrm{Log}\left[\frac{K}{U''}\right]^{1/4} a \;.$$

We have considered two types of interactions:
- attractive van der Waals interactions
- repulsive entropic interactions first invoked by HELFRICH [6] reflecting the fact that the lamellae cannot cross each other. For our system, the latter are predominant and are equal to :

$$V_H = 0.42\,\frac{(kT)^2}{Kd^2}\,, \quad d \text{ the distance between sheets .}$$

We did check that θ^2 is linear with temperature, but a more elegant test of the theory is shown on Fig. 3-b : θ^2 is linear with Log d and we can compute the rigidity constant from the slope of the curve, we found K = kT which is very low compared to the rigidity of lecithins (10^{-12} ergs, the classical lyotropic system). We use this spin labelling method to obtain the variation of K with cosurfactant (Figure 4) : the effect of cosurfactant is indeed to lower the rigidity constant and can explain the transition from lamellar to isotropic. This transition also is favored by the occurrence of very curved defects which were described elsewhere [7]. Our results are in good qualitative agreement with the very recent theoretical approach of BEN SHAUL [8].

Spin labelling (of characteristic time 10^{-8}s) sees the undulations as static so we have performed a quasielastic light scattering (QELS) study on oriented samples which is sensitive to low frequencies ($< 10^4$ s^{-1}) system. The theoretical study of the hydrodynamical mode frequencies for which the scattering vector q lies inside the lamellae planes was achieved by BROCHARD and DE GENNES [9] for swollen lecithin systems and we have extrapolated it to our four-component system. The relevant mode is the undulation mode :

$$\tau^{-1} = \frac{K}{\eta(d+e)}\,q^2$$

with d+e the reticular distance and η the viscosity of the interlayer solvent. The modulation of the scattering is due to the modulation of

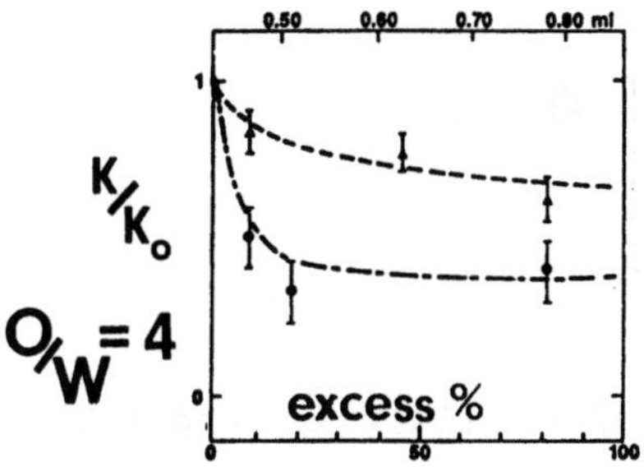

Figure 4 : Relative rigidity coefficient versus cosurfactant excess, ● = spin labelling, ▲ = QELS

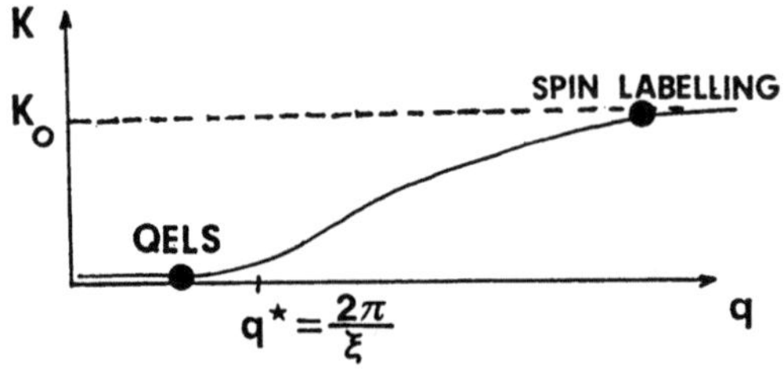

Figure 5 : Renormalization of K versus observation q-vector

the birefringence. Results [10] indicate a rigidity for O/W = 2 and 4 equal to 2.10^{-15} ergs which is one order of magnitude lower than the value we got by spin labelling. This may be due to many different reasons (QELS only sees collective motions, the viscosity of the interlayer solvent is taken equal to the viscosity of bulk cyclohexane but not measured, possible presence of defects, existence of other modes in a 4-component system with respect to the lecithin + water system) but might also be the first proof (Fig. 5) that the rigidity constant should be renormalized according to the observation scale :

$$K = K_o - [\alpha]\ kT \ \mathrm{Log}\left[\frac{q_{max}}{q}\right];$$

α is a numerical coefficient whose value can be found in this conference. For spin labelling, the measurement is very local and $q = q_{max}$ while for QELS, $q = 10^4 cm^{-1}$. At least, the effect of the cosurfactant is shown to be the same, as revealed by spin labelling (Fig. 4).

2. NONIONIC SYSTEMS

$C_{12}E_5$ nonionic surfactant exhibits a very fascinating phase diagram (Fig 6-a) when mixed with water [11] : lower consolute point and reentrant lamellar phase in the water-rich region, and, in general , nonionic systems form microemulsions without the help of cosurfactant.

The spin labelling method revealed a very striking feature of the lamellar phase (73 % w/w surfactant, reticular distance = 43 Å) : the lamellae are undulated at low temperature and become flat when temperature is increased (Fig 6-b) [12], the transition between these two states occurring at the same temperature where the system enters the lamellar phase in the water-rich region. This transition might be due to a change of the rigidity so we have undertaken a QELS study analogous to the one described in the previous chapter.

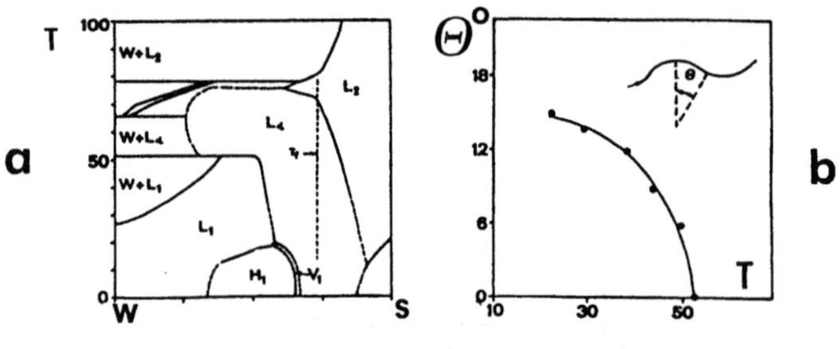

Figure 6 : a) Phase diagram of $C_{12}E_5$/water (after [11]) b) Undulation amplitude (after [12])

2.1 Room temperature

Results [13] (Fig. 7-a) show a feature which was not present with birefringent microemulsions : the undulation mode frequency does not decrease to zero for q = 0. This is due to the quenching of the wall [14], the compressibility coefficient being larger than in the birefringent microemulsion (the reticular distance is smaller). The theoretical expression for the mode frequency is now :

$$\tau^{-1} = \frac{K}{\eta(d+e)} \left[q^2 + \frac{q_c^2}{q^4} \right] \quad \text{with } q_c^2 = \frac{\pi}{\lambda D} ;$$

D is the distance between the glass walls and λ the penetration length of DE GENNES :

$$\lambda = \sqrt{\frac{K}{B(d+e)}} , \quad \text{B the compressibility coefficient ;}$$

we find that λ is equal to 6.3 A, in very good agreement with mechanical measurements [15]. B was measured by OSWALD and ALLAIN [15] and is equal to 2.5 10^7 ergs/cm^3 : this involves K = kT and η = 1 Poise. We should take care that the rigidity value is obtained by QELS and thus indicates that this nonionic system is rigider than the birefringent micremulsions. The viscosity is the same as the macroscopic viscosity.

2.2 Effect of temperature

As the temperature is increased, we have not found a sharp increase of the frequency of the undulation mode as Fig. 4 would suggest. The analysis of the dynamical signal indicates an increase of the rigidity by a factor of about 2. Meanwhile, the intensity of the scattered light decreases strongly with temperature (Fig. 7-b). An elegant way for matching our set of three experiments has been suggested to us by M. ALLAIN : because of the hydration of the hydrophilic part of the surfactant at low temperature, the interfacial film is bent towards the oil and the consequence is that the spontaneous radius of curvature is no more infinite and would induce large-scale undulations. When temperature is increased, water is expelled from the surfactant and the interfacial film becomes flat.

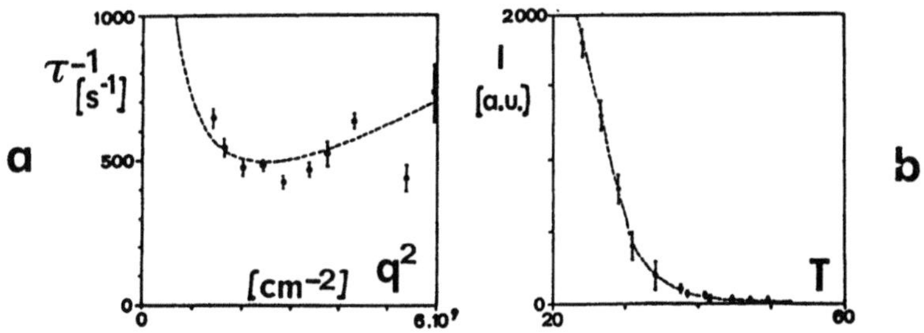

Figure 7 : a) Frequency of the undulation mode versus the square of the scattering vector b) Scattered intensity versus temperature

3. CONCLUSION

We have shown for the first time that the rigidity constant of microemulsions forming systems is of order kT and proved that the role of the cosurfactant is to lower the rigidity as assumed generally. The description of nonionic systems is more complex and requires the knowledge of the detailed mocular interactions.

ACKNOWLEDGMENTS

This work was carried out with M. Allain, A. Brun, M. Dvolaitzky, L. Léger, R. Ober, L. Paz and C. Taupin. It is a great pleasure to thank all of them.

REFERENCES

1. A. Pouchelon, D. Chatenay, J. Meunier and D. Langevin : J. Coll. Int. Sci., 82, 418 (1981)
2. J.H. Schulman and J.B. Montagne : Ann. N.Y. Acad. Sci., 92, 366 (1961)
3. P.G. de Gennes and C. Taupin : J. Phys. Chem., 86 2294 (1982)
4. L.J. Berliner : Spin Labeling (Academic Press, New York 1976)
5. J.M. di Meglio, M. Dvolaitzky, R. Ober and C. Taupin : J. Phys. Lettres (Paris), 44, L-229 (1983)
6. W. Helfrich : Z. Naturforsch., 33a, 305 (1978)
7. L. Paz, J.M. di Meglio, M. Dvolaitzky, R. Ober and C. Taupin : J. Phys. Chem., 88, 3415 (1984)
8. A. Ben Shaul : this conference
9. F. Brochard and P.G. de Gennes : Pramana, 1, 1 (1975)
10. J.M. di Meglio, M. Dvolaitzky, L. Léger and C. Taupin : Phys. Rev. Lett., 54, 1686 (1985)
11. D.J. Mitchell, G.J.T. Tiddy, L. Warring, T. Bostock and M.P. McDonald : J. Chem. Soc., Faraday Trans. I, 79, 975 (1983)
12. J.M. di Meglio, L. Paz, M. Dvolaitzky and C. Taupin : J. Phys. Chem., 88, 6036 (1984)
13. A. Brun and J.M. di Meglio : submitted to Mol. Cryst. Liq. Cryst.
14. G. Durand : Pramana, 1, 23 (1975)
15. P. Oswald and M. Allain : J. Phys. (Paris), 46, 831 (1985)

Edge Dislocations and Elasticity in Swollen Lamellar Phases

J. Prost and F. Nallet

Centre de Recherche Paul Pascal, Domaine Universitaire,
F-33405 Talence Cedex, France

1. INTRODUCTION

A lamellar phase, i.e. a smectic A liquid crystal, is often present in the phase diagram of multicomponent systems which contain at least one amphiphilic species. It may be viewed as a stack of surfactant films separating the lamellae of hydrophilic components from those of lipophilic ones. In some cases swollen lamellar phases exist : one kind of lamella is much thicker than the other one and the reticular distances of some of these smectic phases have been reported to be close to one micrometer /1/.

As any other smectic A, swollen lamellar phases are a priori characterized, as regards their static elastic properties, by two elastic constants : the layer compression modulus B and the layer curvature modulus K /2/. The behaviour of these elastic constants as swelling proceeds is theoretically predicted : for a swollen lamellar phase, described as interacting elastic membranes, both B and K are function of a single elastic constant κ, the membrane bending constant, and of the mean spacing between membranes /3,4/.

A new method for measuring the smectic penetration length $\lambda = (K/B)^{1/2}$ has been recently proposed /5/.

Based upon the direct observation of edge dislocations in the sample, it is not subject to the usual drawback of other techniques, namely that defect motions might affect elastic properties /6/. The aim of this contribution is threefold : we outline the method ; we give experimental results, obtained with quaternary mixtures of water, dodecane, sodium dodecylsulphate and pentanol ; we comment upon our results.

2. THE CANO WEDGE METHOD

Cano wedges, first described by GRANDJEAN /7/ and later revived by CANO /8/, are cells of variable thickness holding a smectic or a cholesteric liquid crystal sample. In such a cell, the boundary conditions (homeotropic alignment, for smectics ; strong, non degenerate, planar anchoring, for cholesterics) are not compatible with the requirement of a constant layer spacing (smectics) or helical pitch (cholesterics). As a result, defects arise. In the case of a smectic sample, the simplest defect is the edge dislocation : the abrupt termination of a number b of layers ; b is then the Burgers vector of the dislocation.

Due to the repulsion by the boundaries each dislocation sits at an equal distance from the upper and lower walls of the cell /9/. But the spatial distribution of them along a direction where the thickness of the cell varies is not fixed by geometry alone, as would be the case for the light intensity in a pattern of "equal thickness" interference fringes. Here, to the array of edge dislocations is associated an elastic energy, which arises mainly from the gradually decreasing compression strains, or increasing dilation ones, between neighbouring dislocations and also from the curvature strains located near the core of each dislocation /10/.

The equilibrium distribution is obtained in Ref. /6/ by minimizing this elastic energy. We quote the result, relevant for a Cano wedge made of a spherical lens (radius of curvature : R) sitting on a plane plate : dislocations located near the centre, i.e. at a distance r less than some critical value r^*, have a constant Burgers vector b_{min} ; the distance between adjacent dislocations then decreases as r increases. On the other hand the Burgers vector of the dislocations far from the centre is an increasing function of the distance r :

$$b = \frac{r}{a}\left(\frac{3\pi\ \lambda^2}{R^2}\right)^{1/3} \mathcal{L}\left(\frac{r}{r_0}\right), \qquad (1)$$

where a is the smectic reticular distance, $\mathcal{L}(x)$ the one-third power of the largest root of the equation $X = \ell n\ x + \frac{1}{3}\ \ell n\ X$, and $r_0/(3\pi\ \lambda^2\ R)^{1/3}$ is related to the microscopic core energy of a dislocation. Now the distance between adjacent dislocations increases as r increases.

By means of an optical polarizing microscope, the locations $\{r_k\}$ of consecutive dislocations are measured as a function of their rank k. A fit of the theoretical distribution, based upon $b = b_{min}$ or (1), to the experimental data $\{r_k\}$ results in the output of the three fitting parameters : λ (or $(K/B)^{1/2}$), from the mean slope of the location vs rank curve ; r_0 from its upward curvature, for r_k greater than r^* ; $a \cdot b_{min}$, from its downward curvature, for r_k less than r^*.

In the experimental results that follow we shall focus on the behaviour of λ.

3. EXPERIMENTAL RESULTS

The Cano wedge method was used to measure the smectic penetration length λ of six samples of the quaternary mixture of water (W), dodecane ("oil", O) sodium dodecylsulphate ("surfactant", S) and pentanol ("alcohol", A). Denoting x_α the weight fraction of the α species, the samples are prepared by dilution of the ternary system $x_W/x_S = 1.55$, $x_A = 0.22$ by a binary oil/alcohol system with $x_A/x_O = 0.087$; the final oil content of the mixtures varies from $x_O = 0.54$ to $x_O = 0.75$. According to ROUX et al. /11/, all these samples are located in a one-phase, L_α (i.e. lamellar, or smectic) region of the phase diagram of this quaternary system at room temperature. The last one has a composition near that of the boundary of the L_α region : from $x_O^* \simeq 0.755$, the swelling of the lamellar phase by the oil/alcohol mixture stops and there is a two-phase separation between the lamellar phase and an isotropic one /12/.

Figure 1 shows two typical experimental data, and the corresponding theoretical fits, for the less and the most dilute samples. The radius of curvature R of the lens used in the Cano wedge was, respectively, R = 1063mm and R = 528mm. Note that the positions of more than two hundred dislocations have been recorded.

Displayed on Fig. 2 is the smectic penetration length λ as a function of the oil content x_0. The error bars take into account only the statistical uncertainty due to the fit. The strong increase of λ as one goes close to the lamellar /isotropic first order phase transition is noticeable.

4. DISCUSSION

The theoretical description of a swollen lamellar phase as a pile of interacting, elastic membranes implies that the compression and curvature moduli are function of a single elastic constant, the membrane bending constant κ, and of the reticular distance of the pile, a, according to /3,4/ :

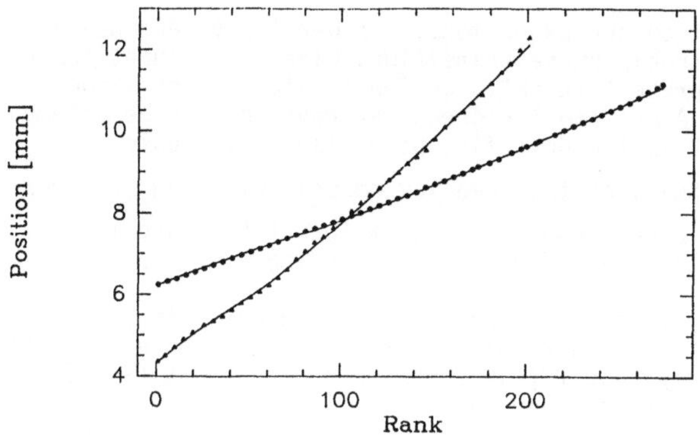

Figure 1. Position vs rank of dislocations in two samples ; triangles : $x_0 = 0.54$; circles : $x_0 = 0.75$

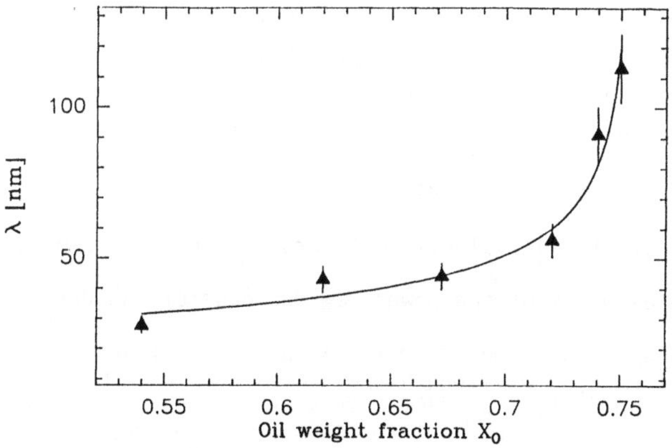

Figure 2. Smectic penetration length vs oil weight fraction ; triangles : experimental data ; continuous curve : see paragraph 4

$$B = C_0{}^4 \frac{(k_BT)^2}{\kappa} \frac{a}{(a-\delta)^4} , \tag{2}$$

$$K = \frac{\kappa}{a} , \tag{3}$$

where δ is the membrane thickness and C_0 is an unspecified numerical constant /4/ or $C_0 = (3\pi/8)^{1/2}$ /3/. Combining (2) and (3), the smectic penetration length is then given by :

$$\lambda = \frac{\kappa}{C_0{}^2\, k_BT}\, a\, (1 - \frac{\delta}{a})^2 . \tag{4}$$

A direct comparison of (4) with experiment is not possible, because a is not known everywhere in the phase diagram but only for the samples studied by SAFINYA et al. /13/, with a maximum oil content $x_0 = 0.62$. An indirect use of (4) is to notice that λ becomes proportional to a for highly swollen phases. A prediction of the theory of interacting membranes is the possible occurrence of a phenomenon called *complete unbinding* /4/, when a goes continuously to infinity as the swelling proceeds. If the parameter controlling the swelling is, e.g., the oil weight fraction x_0, complete unbinding is then described by the scaling law /4/ :

$$a \;\alpha\; (x_0 - x_0{}^c)^{-1/3} , \tag{5}$$

where $x_0{}^c$ is here the oil fraction for infinite separation between membranes.

If complete unbinding is relevant to our experiment, a power-law divergence of λ is expected near the lamellar/isotropic phase transition. Clearly, a quantitative check of (5) via (4) is beyond our present reach : too few λ values, over a too restricted dynamic range are known. A *qualitative* check may nevertheless be attempted : the continuous curve on Fig. 2 is the best fit, power-law curve, using $x_0^c \equiv x_0^* = 0.755$ (experimental boundary of the lamellar phase), yielding an exponent $\psi \simeq -0.4$. It is therefore not unreasonable to expect that future and more complete experiments might confirm the idea of complete unbinding.

Other techniques can provide values of λ : in particular, the analysis of the asymmetry of the quasi-Bragg peaks in a high-resolution X-ray scattering experiment /13/. The difference between the X-ray result (λ = 2.5nm) and our result (λ = 28nm), for the same system, with an oil weight fraction $x_0 = 0.54$, is striking. The nature of this discrepancy is not understood. The specific length scales in the X-ray experiment are different from those in the Cano wedge method : X-rays probe typical wavevectors of the order of 10^{-2} nm^{-1} /13/ ; Cano wedge method probes essentially cut-off wavevectors for K ($q_c \simeq 10^{-1}$ nm^{-1}) but much smaller ones (of the order of the inverse thickness of the sample, i.e. about 10^{-5} nm^{-1}) for B. Under such circumstances, a renormalization of the elastic constants /14/ may come into play. It is controlled by a dimensionless coupling constant, $w = \frac{k_B T}{B_0 \lambda_0^3}$, where B_0 and λ_0 are the "bare" values, i.e. the short wavelength parameters.

An estimate of the coupling constant (equivalently but more conveniently expressed as $w = \frac{2 a^2 \eta}{\pi \lambda^2}$, where η is the exponent of the power law decrease of the density-density correlation function) may be obtained from the X-ray data, if we assume that they are provided at a "short enough" wavelength. For $x_0 = 0.54$, a = 11.5nm, $\eta = 0.76$ and λ = 2.5nm /13/ ; one then gets : w = 10.2. This is a large value. The renormalization of the elastic constants is relevant for wavevectors q (component parallel to the optical axis q_z ; perpendicular to it $q_\perp$) such that /14/ :

$$(q_z^2 + \lambda^2 q_\perp^4)^{1/2} \leqslant q^* \tag{6}$$

with

$$q^* = \lambda q_c^2 \, e^{-\frac{128\pi}{5w}} \, . \tag{7}$$

From (7) we get : $q^* \simeq 10^{-5}$ nm^{-1}. Since typical X-ray wavevectors are much larger than q^*, our estimate of w is sensible provided that one can trust the X-ray data. On the other hand, the characteristic wavevector q of the Cano wedge measurement of B is of the order of q^* : thus B is decreased with respect to the short scale X-ray value by a factor /14/ :

$$\left[1 + \frac{5w}{64\pi} \ln \frac{\lambda^{1/2} q_c}{(q_z^2 + \lambda^2 q_\perp^4)^{1/4}} \right]^{4/5} . \tag{8}$$

This leads to a small increase (about 30%) of the expected Cano wedge λ value as compared with the X-ray one.

However, the coefficient $\frac{5w}{64\pi}$ in (8) (but *neither* the logarithmic dependence *nor* the 4/5 power-law exponent /14/)results from a first order expansion : this requires $\frac{5w}{64\pi} \ll 1$, which is not satisfied here. There are some indications that the actual factor in front of the logarithm might be much larger than expected from (8), perhaps as much as twenty times larger /15/. This would have two consequences : to bring q^* close to the X-ray regime ; to raise the expected ratio $\lambda_{Cano\ wedge}/\lambda_{X\text{-}ray}$ to about 3. This is still not enough to explain the discrepancy between the two results, but at least points in the right direction. If a single membrane regime is relevant to the description of the water layers near the dislocation core, K might be larger than expected : in this regime, K *decreases* with increasing wavelength /16/, before the smectic, many-membrane regime, characterized by the opposite wavelength dependence /14/, is reached. This points again in the right direction.

These considerations are very conjectural and very approximate : they require a closer attention. They suggest however that swollen lamellar phases might be good candidates for exhibiting signatures of the breakdown of hydrodynamics in smectics /17/ : according to (2) and (3) for instance, $w = C_0^2 \left(\frac{k_B T}{\kappa}\right)^2$ in the limit of large swelling, i.e. may be quite large for soft membranes. Different methods of measurement of λ, and still more of B and K at different wavelengths are obviously called for.

REFERENCES

1. F.C. Larché, S. El Qebbaj, G. Porte, P. Bassereau, J. Marignan: Phys. Rev. Lett. 56, 1700 (1986)
2. P.-G. de Gennes: The Physics of Liquid Crystals (Clarendon, Oxford 1974)
3. W. Helfrich: Z. Naturforschung 33a, 305 (1978)
4. S. Leibler, R. Lipowsky: to be published
5. F. Nallet, J. Prost: submitted to Europhys. Lett.
6. P.S. Pershan, J. Prost: J. Appl. Phys. 46, 2343 (1975)
7. F. Grandjean: C.R. Hebd. Séan. Acad. Sci. 172, 71 (1921)
8. R. Cano: Bull. Soc. Franç. Miner. Cristall. 91, 20 (1968)
9. P.S. Pershan: J. Appl. Phys. 45, 1590 (1974)
10. C.E. Williams, M. Kléman: J. Physique Colloq. 36, C1-315 (1975)
11. D. Roux, A.-M. Bellocq: Physics of Amphiphiles, V. de Giorgio, M. Corti editors (North-Holland, Amsterdam, 1985)
12. A.-M. Bellocq: private communication
13. C.R. Safinya, D. Roux, G.S. Smith, S.K. Sinha, P. Dimon, N.A. Clark, A.-M. Bellocq: Phys. Rev. Lett. 57, 2718 (1986)
14. G. Grinstein, R.A. Pelcovits: Phys. Rev. A26, 915 (1982)
15. J.-C. Rouillon, J.-P. Marcerou, J. Prost: unpublished results
16. W. Helfrich: J. Physique 46, 1263 (1985) ; L. Peliti, S. Leibler: Phys. Rev. Lett. 56, 1690 (1985)
17. G.F. Mazenko, S. Ramaswamy, J. Toner: Phys. Rev. A 28, 1618 (1983)

Lamellar Lyotropic Phases: Rheology, Defects

M. Kléman

Laboratoire de Physique des Solides, Université Paris-Sud, Bâtiment 510, F-91405 Orsay, France

1. INTRODUCTION

Rheological properties of anisotropic fluids are determined by two factors : a) properties of symmetries : in that respect SmA thermotropic phases and Lα lyotropic phases should have common rheological features, b) specific material properties : we can find larger differences in flow properties between phospholipid-based Lα phases (1), non-ionic surfactants (2), and possibly (but the experiments have not yet been done) ternary systems where the layers are made of some kinds of micellar disruptions (3) (by contrast, we speak of smooth layers for the other lamellar phases) than between a typical SmA phase like 8 OCB (4) (4-n-octyl-4'-cyanobiphenyl) and a lyotropic phase like lecithin-water.

The same properties of symmetry impose that the same types of topological defects play a similar role in flow properties, where they act by their friction with the crystal when they move – their interactions, their multiplication. Topological defects are : a) screw dislocations which are perpendicular to the layers and connect them (Fig. 1) ; their presence is also essential to understand diffusivity properties of chemical constituants through the layers ; b) edge dislocations (Fig. 2), which have long straight segments or can form loops (pores) in the layers ; c) loops of mixed character with segments of dislocation perpendicular to the layers ; d) disclinations (of the focal conic-type essentially).

The differences between various flow properties have to be related to different material constants (splay K_1, compressibility modulus B, permeation λ_p, viscosity

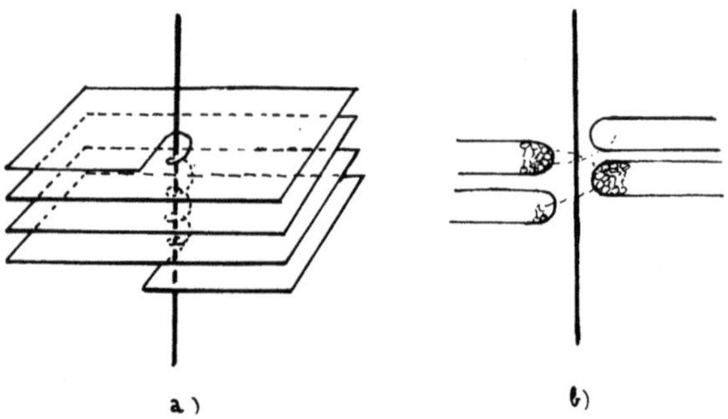

Fig. 1 Screw dislocation : a) how layers are connected, b) a core model in a lyotropic L_α phase.

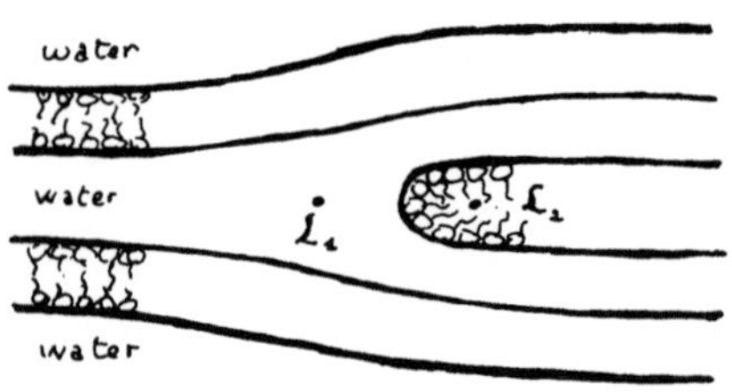

Fig. 2 Edge dislocation line ; the core topology, which is made of two disclinations of opposite signs, is imposed by the necessity of separating water and hydrophobic chains.

tensor...) and possibly different transport mechanisms at microscopic or/and submacroscopic levels (diffusivities $D_{//}$ and $D_{\perp}$, activation energies,...). They are revealed in the process of preparation of good samples. As-grown samples are always full of imperfections ; most of them can disappear in the course of time, and only those required by the boundary conditions (if they are incompatible with perfect single crystallization) might remain. But this annealing process, which is driven by the interactions between imperfections, takes weeks in lecithin-water samples (1) or can even be unsuccessful, while it takes only a few minutes in $C_{12}E_5$ or $C_{12}E_6$-water non-ionic surfactants (2). It is also rapid in swollen $L\alpha$ phases (5). Annealing processes reveal therefore elementary flow processes at vanishing frequencies.

K_1 has a usual value ($\sim 10^{-6}$ dyne/cm) in lecithin and is twenty times smaller in $C_{12}E_6$, at least (6,7). We shall see that the small $\lambda/d = (K_1/B)^{1/2}/d$ ratio leads to a particular mode of edge dislocation core in non-ionic surfactants. This is certainly at the origin of different annealing behaviours, but going further would require also understanding the molecular origins of the activation energies of transport properties, which exhibit large differences from one lamellar phase to another.

To understand flow properties at large stresses, large velocity gradients and/or large frequencies, necessitates including interactions between defects, their multiplication during flow, the interplay between dislocations and disclinations. Interesting observations and related theories have been reported on the nucleation and growth of focal conics in homeotropic geometries under large dilations (8,9) and shear (9,10,11) and in planar geometries (12,13). Simple experiments are however still lacking ; they would have to explain how dislocations are emitted or absorbed when a disclination moves (Fig. 3), since this is the topological basic mechanism which is in play here. Anomalous unexplained behaviours have been reported for very small (few Å) displacements (14) imposed on the layers. The present article will remain at the level of rheological properties at vanishing frequencies and large wavelengths : we consider the behaviour of individual non-interacting defects. This is a prerequisite for further studies ; this is also the domain of rheology which is closer to plasticity and where therefore the metallurgical concepts developed for the study of the plasticity behaviour of common crystals can be used and have to be extended. The first experiments to be done in this area considered the nucleation of dislocations and their relaxation under very small dilations (15). Sect. 2 and 3 will analyze simple mechanisms of motion of defects and review some related measurements of mobility in various materials. In the last section we shall go to a specific example and shall try to relate some features of the variation of viscosity of $C_{12}E_5$ with temperature (16) with the models established in the former section ; as required, these measurements were made on samples which do not contain textural defects and in a domain of shears where such defects do not nucleate. We shall also see that structural dislocations are not enough to explain all the measurements, and that the fluctuations of the layers discovered by di Meglio (7,17) play a role.

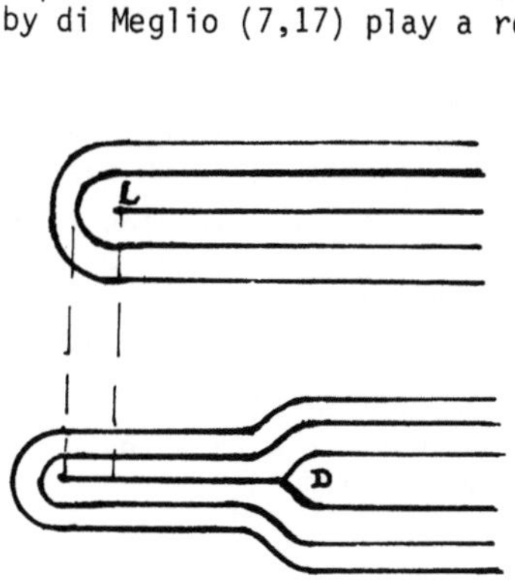

Fig. 3 The motion of a disclination L (or of a focal conic) requires the emission or absorption of a dislocation D.

2. CLIMB AND GLIDE OF EDGE DISLOCATIONS.

The main mechanism of motion of edge dislocations is by climb: i.e. the line moves in a layer, perpendicularly to its Burgers' vector. This is in strong contrast to the main mechanism of motion of dislocations in solids, where glide is dominant. Climb is non-conservative, i.e. requires the displacement of matter by diffusion or flow process of molecules through the layers, while glide is conservative. It is topology (separation of water and hydrophobic chains) which favours easy climb. However, glide might be easy in low K_1 materials like non-ionic surfactants, as we shall see.

Let us define the mobility of a dislocation line as the ratio $m = v/\sigma$, where v is the velocity and σ the stress which is at the origin of the displacement. Stress may be either of mechanical origin or of hydrodynamical origin. The only stress components acting on an edge dislocation along the Oy direction are σ_{33}, σ_{13}, through configurational (Peach and Kohler) forces (18) :

$$f_x = b\,\sigma_{13}\,, \qquad f_y = b\,\sigma_{33}\,. \tag{1}$$

Consider first mechanical stresses. Near static equilibrium, we have in the elastic approximation :

$$\sigma_{13} = -K_1 \frac{\partial^3 u}{\partial x^3}\,, \qquad \sigma_{33} = B\frac{\partial u}{\partial z} \tag{2}$$

with the equilibrium condition :

$$\sigma_{13,1} + \sigma_{33,3} = 0\,. \tag{3}$$

This relationship implies that σ_{13} is much smaller than σ_{33}. We have indeed :

$$\sigma_{13} \sim \sigma_{33}\frac{\delta x}{\delta z} \sim \sigma_{33}\frac{\lambda}{\delta x} \tag{4}$$

since any perturbation δx extends over $\delta z \sim (\delta x)^2/\lambda$ along the z-direction. Therefore, we expect that mechanical stresses act in the same way as topology and favour only climb: a compressive stress σ_{33} acting on the dislocation of Fig. 2 would move it to the right.

Hydrodynamical stresses of the type $\sigma_{13} = \mu_2(\partial v_x/\partial z + \partial v_z/\partial z)$, $\sigma_{33} = \mu_3\,\partial v_z/\partial z$ would play a role comparable to the mechanical stresses only for very large velocity gradients, which cannot be established easily on large distances. This remark justifies a 'metallurgical' approach of the mobility of edged dislocations in climb (19), which leads to the relationship :

$$m = \frac{D_{/\!/} V_m}{k_B T}\,\frac{1}{\ell}\,, \tag{5}$$

where V_m is a molecular volume, $D_{/\!/}$ a diffusivity perpendicular to the layers, and ℓ a molecular length. The quantity $(D_{/\!/} V_m)/(k_B T)$ is akin to λ_p, the Helfrich coefficient of permeation (20). In this approach, it is assumed that the molecules jump from one layer to the next near the tip of the supplementary layers, thought of as having an infinite density of jogs of one molecule each. There is a constant probability of adding (removing) molecules along this "ragged" tip if the driving force is constant and the line moves parallel to itself. According to the case, this diffusion consists in a flip-flop from one monolayer to the next or a shift from one bilayer to the next (Fig. 2), followed in each case by a rapid diffusion $D_\perp$ along the layer ($D_\perp > D_{/\!/}$) ; see ref. 1 for more details.

Eq. 5 gives a mobility which is independant of the Burgers' vector. Models which consider that the permeation process takes place all along the layer, not only at the tip, lead to a mobility proportional to the Burgers' vector (21). Conclusive experiments are lacking.

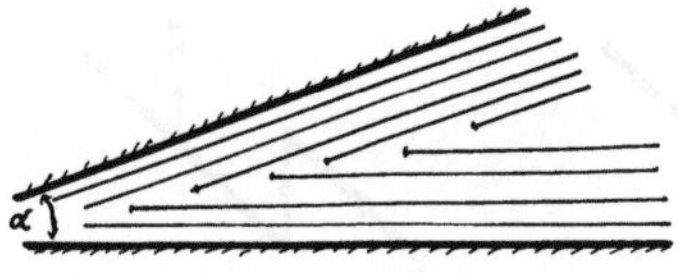

Fig. 4 The lubrication wedge. Edge dislocations gather in the mid-plane, where they form a grain boundary.

A measurement of the mobility of edge dislocations in climb leads to a direct measurement of the order of magnitude of λ_p. This is achieved in the "lubrication wedge" method (22), which is described in Fig. 4 : when a liquid is sheared between two planar surfaces making a small angle α, a large pressure is set up, which is at the origin of the lubrication process ; by replacing the liquid by a SmA, one expects an even more spectacular effect since, apart from the easy flow of the layers past one another, there is genuine solid-like strength perpendicular to the layers. The lubrication process with shear $\gamma = v/d$ is also equivalent to a compression process with a vertical velocity $v_z = \alpha v$ imposed on one of the plates. One observes that the dislocations move in the plane of the subgrain boundary where they sit, after a transitory regime of duration :

$$\tau_1 = d/\alpha m B \, . \tag{6}$$

It is this time which is measured (22). Typical values obtained this way are given in Table I.

Table I: Climb of edge dislocations.

Compound	m(cm^2 sec g^{-1})	λ_p(cm^2/poise)	U(eV)
8OCB	$3 \cdot 10^{-7}$ at 25°C	10^{-15}	1.8
$C_{12}E_5$	$6 \cdot 10^{-7}$ at 50°C	$4 \cdot 10^{-14}$	0.4

The essential difference is with the activation energy U which is very low in $C_{12}E_5$. This system anneals very easily indeed when the temperature increases.

The values of λ_p in Table I differ markedly with the results of Chan and Webb (1) for the system DPMC-water, where they find $\lambda_p \sim 10^{-30}$ cm^2/poise. If their result is related to the fact than $D_{//} << D_\perp$, as they claim, it must be explained why λ_p is so large in the other compounds. This can only be related to a large effective permeation mechanism through pores and screw dislocation cores (Fig. 1). A simple calculation applying Darcy law (6) to the channels of the screw lines gives :

$$\lambda_p \sim c\, \xi^4/\mu \, , \tag{7}$$

where c is the density (per unit surface of layer) of screw dislocations and ξ a core size. With $c \sim 10^{10}/cm^2$, $\xi \sim 30$ Å and $\mu \sim 0.1$ poise, we obtain the right order of magnitude. It is not yet understood why there are no screw dislocations in the samples of ref. 1, while they are so numerous in all the microscopic measurements of their densities in freeze-fractured specimens (2,23,24).

Let us now consider glide, which requires layers to break and fuse again. This process is difficult in most compounds, but probably easy in $C_{12}E_5$. The reason is that, since λ/d is so small, the core of edge dislocations is very anisotropic and extends on a distance $\xi_z \sim d^3/\lambda$ in a direction perpendicular to the layers ($\xi_x \sim$ b). The core size is indeed the distance z at which the parabolas $x^2 = \pm - \lambda z$ which bound the regions of strong layer distortions (25) have a width of the order of the layers' thickness d (26). A possible model of such an extended core is as in Fig. 5 ; this is akin to what is observed in 2-d layered systems formed by the roll electrohydrodynamical instabilities of nematics (27). The quantity analogous to λ is the frequency of the a.c. electric voltage which drives the instability : at small frequencies dislocations of the rolls system have been observed to glide easily. In $C_{12}E_5$ or $C_{12}E_6$, one expects that easy glide will result from activated nucleation of "kinks" formed of screw

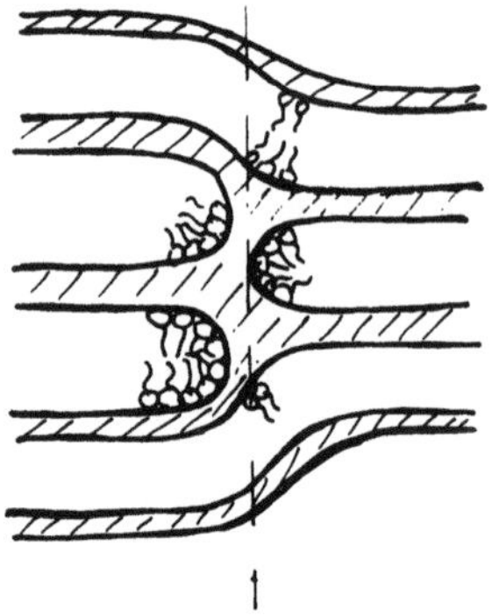

Fig. 5 Extended core of an edge dislocation in a small λ/d material.

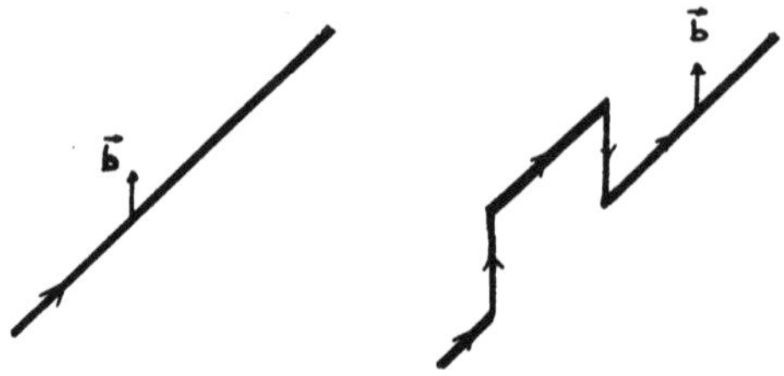

Fig. 6 A kink along an edge dislocation line.

segments of opposite sign (Fig. 6) along the edge line, and of the easy displacement of the screw segments under the action of the σ_{13} or σ_{23} stress produced by the edge line : this stress is large since it is its value in the near vicinity of the core which is relevant here. Therefore, the leading factor of easy glide is the low activation energy for the formation of pores (which form the cores of the screw segments) made easier by the stretching of the layers in the extended core region.

3. SCREW DISLOCATIONS.

They move parallel to themselves under the action of configurational forces $f_y = b\sigma_{13}$ or $f_x = b\sigma_{23}$, or can change shape under σ stresses which provoke a helical instability above a certain displacement threshold of the layers.

A screw dislocation moving parallel to itself with velocity v provokes a backflow very akin to the backflow set up by the motion of a cylinder of radius ξ in a liquid, except that an important v_z component appears whose role is to avoid permeation, which as a result is nonexistent outside the core (28). There are therefore two contributions to the drag force which acts on the dislocation line, a classical Stokes force proportional to $\mu_2 v$ and a (larger) force due to the v_z component, which reads (28) :

$$f_d = \mu_3 \frac{b^2}{4\pi\xi^2} v . \tag{8}$$

To produce such a large force by the action of mechanical stresses requires a large σ_{13} ($f_d = b\sigma_{13}$) which, as mentioned above, could exist near a kink on an edge line. However, if the dislocation is pinned, and the sample sheared, the velocity field which is set up can be large enough to couple to the screw dislocation. This produces an anomalous flow behaviour and an increase of viscosity which has been demonstrated experimentally and analyzed (29).

Helical instabilities of screw lines have been predicted by Bourdon et al. (30) and have up to now been observed only in thermotropic smectics (31). The mechanism is described in fig. 7 : the σ_{33} stress is relaxed (above some displacement threshold of the order of d) by the production of a helical turn which is equivalent to the formation of an edge dislocation loop (Fig.7) ; a layer is missing inside the loop. Successive helical instabilities show up when the stress is increased ; they correspond to smaller relaxation times. It would be of interest to look for these instabilities in lyotropic specimens ; they would provide for a direct measurement of the mobility of screw dislocations.

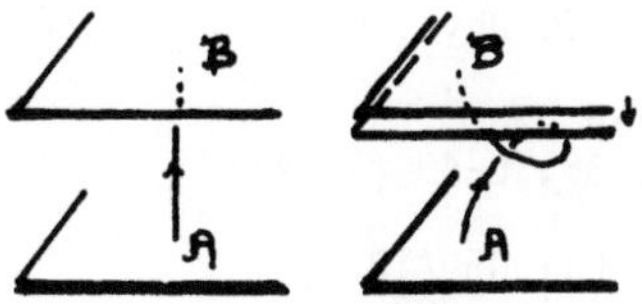

Fig. 7 Mechanism of the helical instability of a screw dislocation line pinned in two points A and B (here on the boundaries of the sample).

4. STRUCTURAL DEFECTS IN NON-IONIC SURFACTANTS.

The L_α phases of $C_{12}E_5$ and $C_{12}E_6$ have two types of 'defects'.

At low temperature (17) there are fluctuations of the layers which have been seen by E.P.R. measurements (7,17) ; they seem to correspond to the layers corrugations observed on replicas of freeze-fractured specimens by electron microscopy (2,23) ; these corrugations, which repeat with a periodicity of $\Lambda \sim 4000$ Å, have negative gaussian curvature. They might therefore well be precursors of the cubic phase which exists at lower temperatures. These corrugations are also revealed in viscosity measurements of homeotropic specimens (16) : viscosity is constant as long as the total thickness of the specimen is smaller than 15 µm, then begins to increase. This is explained by assuming that the flat surface of the boundary kills the corrugations or a distance $D \sim \Lambda^2/\lambda$, which is about 15 µm (16).

At higher temperatures, there is measured by E.P.R. an increase (activation energy $\sim$ 0.5 eV) of 'curvature defects' with T. These defects were shown, also by freeze-fracture observations, to consist, at least in part, of vertical dislocation loops (2 edge and 2 screw segments). An analysis of viscosity measurements relate correctly the increase of viscosity with temperature in this region with an increase of the number of vertical loops (16), which gather in clusters of 10 to 20 loops. These loops have a very small elastic energy ($\sim$ 0.1 eV), which makes their role comparable to the role of vacancies in metals despite their size. Since their activation energy ($\sim$ 0.5 eV) is larger than the energy that the theory (26) attributes to them, their nucleation refers to a mechanism which is most probably the creation of a pore (0.5 eV is the calculated energy of a pore). Furthermore if the size that these thermodynamical defects reach is their optimal size, the theory shows that the core free energy of the screw segments is negative, which can be described phenomenologically by a positive $\overline{K}$ saddle-splay coefficient for the monolayer : the current model (as in fig. 1b) assumes indeed that the surface of the monolayer in the core has a negative gaussian curvature.

REFERENCES

1. Chan W.K. and Webb W.W., J. Physique 42, 1007 (1981).
2. Allain M., Thesis, Orsay, 1987 and references hereunder.
3. Holmes M.C. and Charvolin J., J. Phys. Chem. 88, 810 (1984).
4. Oswald P., Thesis, Orsay, 1985 and references hereunder.
5. Nallet J., these proceedings.
6. Oswald P. and Allain M., J. Physique 46, 831 (1985).
7. di Meglio J.M., these proceedings.
8. Benton W.J., Toor E.W., Miller C.A. and Fort T. Jr., J. Phys. 40, 107 (1979).
9. Oswald P., Behar J. and Kleman M., Phil. Mag. A46, 899 (1982).
10. Horn R. and Kleman M., Ann. Phys. 3, 229 (1978).
11. Ben-Abraham S.I. and Oswald P., Mol. Cryst. Liq. Cryst. 94, 383 (1983).
Ben-Abraham S.I., Mol. Cryst. Liq. Cryst. 123, 77 (1985).
12. Williams C.E. and Kleman M., J. Phys. Colloq. 36, C1-315 (1975).
13. Marignan J., Thesis, Montpellier, 1981.
Marignan J., Parodi O. and Dubois-Violette E., J. Physique 44, 263 (1983).
14. Durand G., J. Chimie Phys. 80, 119 (1983).
15. Bartolino R. and Durand G., Phys. Rev. Lett. 39, 1346 (1977).
16. Oswald P. and Allain M., submitted to J. Coll. Interf. Science.

17. di Meglio J.M., Paz L., Dvolaitzky M. and Taupin C., J. Phys. Chem. 88, 6036 (1984).
18. Friedel J., Dislocations, Pergamon Press, 1964 ; chap. 2.
19. Kleman M. and Williams C., J. de Physique, Lettres, 35, L-49 (1974).
20. Helfrich W., Phys. Rev. Lett. 23, 372 (1969).
21. Orsay Liquid Crystal Group, J. Physique Colloq. 36, C1-305 (1975).
22. Oswald P. and Kleman M., J. Physique Lettres, 43, L-411 (1982).
23. Allain M., J. Physique 46, 225 (1985).
24. Costello M.J., Gulik-Krzywicki T., Kleman M. and Williams C.E., Phil. Mag. 35, 33 (1977).
25. Kleman M., Points, Lines and Walls, Wiley Pub. Cy, (1983), chapt. 5.
26. Allain M. and Kleman M., Thermodynamic defects, instabilities and mobility processes in the lamellar phase of a non-ionic surfactant, submitted to J. de Physique.
27. Ribotta R. and Joets A., in "Cellular Structures in Instabilities", ed. by Wesfreid J.E. and Zaleski S., Springer (1984)
28. Pleinert H., Phil. Mag. A54, 421 (1986).
29. Oswald P., J. Physique 47, 1091 (1986).
30. Bourdon L., Kleman M., Lycek L. and Taupin D., J. Physique 42, 261 (1981).
31. Oswald P. and Kleman M., J. de Physique Lettres, 45, L-319 (1984).
32. Durand G., Comptes Rendus Acad. Sci., Paris, 275B, 629 (1972).

On the Coexistence of Two Lamellar Phases

H. Wennerström

Division of Physical Chemistry 1, Chemical Center,
P.O. Box 124, S-22100 Lund, Sweden

1. INTRODUCTION

It is now experimentally well established that two lamellar phases with quite different water contents can coexist in amphiphile water systems. Examples of this behaviour can be found in the early work by Vincent and Skoulios on gel phases of potassium soaps [1,2]. Similar observations have been made for gels of acylglycerols [3]. More recent studies have revealed similar effects for a number of lamellar liquid crystalline systems with charged amphiphiles containing both divalent [4,5] and monovalent [6] counterions. It is tempting to conclude that the phase separation of the two lamellar systems is caused by some important molecular difference between the two phases. However, both NMR [4,6,7] and X-ray experiments [1,2,6] provide little support for this conclusion.

Studies of the electrostatic double layer interactions for ionic systems have shown that the phase separation should be expected under certain circumstances simply from the form of the interaction between the lamellar [8,9]. In this paper we generalize the analysis of the relation between force-distance curves and phase behaviour presented shortly in refs [8] and [9]. The aim is to obtain a more systematic description which will enable us to make some predictions for new systems and possibly resolve a standing controversy in the literature on lecithin systems.

2. THERMODYNAMIC MODEL

Figure 1 shows a model description of a lamellar phase. The well defined and measurable repeat distance d can, with some approximation, be decomposed into a thickness d_a of the apolar and a thickness d_w of the aqueous layer. In Fig. 1 interfaces are represented by their average flat configuration, and the effect of the fluctuations of the surfaces is only considered indirectly through the interaction. Such a 'mean field' description precludes a quantitative description of the critical region.

Consider a two-component system water (subscript w) - amphiphile (subscript a). Two phases (α and β) are in equilibrium when the chemical potentials μ_i of both components in both phases are equal; i.e.

$$\mu_w(\alpha) = \mu_w(\beta), \tag{1a}$$

$$\mu_a(\alpha) = \mu_a(\beta). \tag{1b}$$

In a binary system the concentration dependence of the chemical potentials are linked by the Gibbs-Duhem relation

$$n_w d\mu_w + n_a d\mu_a = 0 \tag{2}$$

at constant temperature and pressure. Here n_i denotes the molar amount of component i.

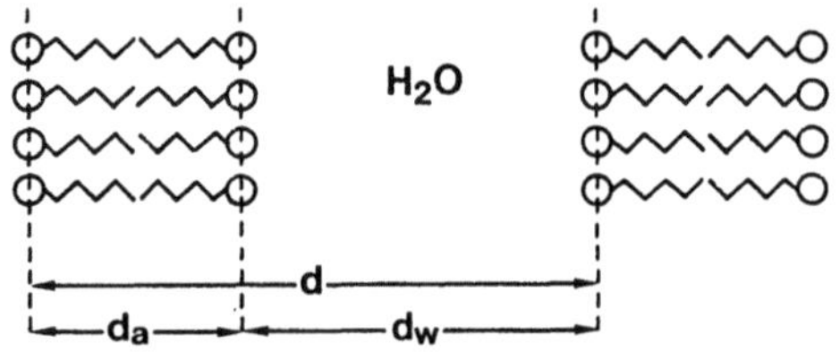

Fig. 1. Schematic illustration of a lamellar system with a repeat distance d, a bilayer thickness d_a and a thickness d_w of the aqueous layer.

The eqs. (1) and (2) are of a formal thermodynamic nature and to proceed it is necessary to introduce molecular concepts and relate these to the model of Fig. 1. It is a classical problem in colloid science to evaluate the force between two particles. These forces are often discussed in terms of the behaviour of the force per unit area between two parallel infinite surfaces [10]. There is clearly a direct relation between these forces and the thermodynamics of lamellar phases, a fact that has been elegantly exploited by Parsegian, Rand and coworkers [11].

3. FORCE-DISTANCE RELATIONS and PHASE EQUILIBRIA

It is customary to represent the force between two interfaces in terms of a force/unit area - distance curve. For the examples considered in the introduction it would be relevant to change the water content and the schematic force - distance curve shown in Fig. 2 then relates the osmotic pressure, or water chemical potential, to the thickness of the aqueous layer.

A possible equilibrium between two lamellar phases can only occur when the osmotic pressure is the same at two different separations. This is a necessary but not sufficient condition,as is evident from eq. (1). To also satisfy eq. (1b) an additional condition applies, which we can investigate using the Gibbs-Duhem eq. (2). Neglecting the solubility of the amphiphile in the water and the solubility in the water of the amphiphile we have

$$n_w = d_w \cdot A/V_w , \qquad (3a)$$

$$n_a = d_a \cdot A/V_a , \qquad (3b)$$

where A is the area and V_i is the molar volume of component i. The eq. (2) can now be written

$$d\mu_a = -\frac{V_a}{V_w}\frac{d_w}{d_a}\, d\mu_w \,. \qquad (4)$$

Integrating this between two compositions α and β chosen so that $\mu_w(\alpha) = \mu_w(\beta)$ leads to

$$\mu_a(\alpha) - \mu_a(\beta) = \frac{V_a}{V_w}\int_\alpha^\beta \frac{d_w}{d_a}\, d\mu_w \,. \qquad (5)$$

To satisfy eq. (2) the integral on the r.h.s. should vanish. The force distance curve of Fig. 2 gives a relation between d_w and μ_w, through the osmotic pressure. If the thickness of the amphiphilic layer is constant in the relavant range the vanishing of the integral in eq. (5) is equivalent to the well-known Maxwell equal area construction in the van der Waals' theory of liquid-vapour equilibria as illustrated in Fig. 3. The same method can be applied for the case of a varying thickness d_a of the amphiphilic layer if the distance d_w is scaled relative to d_a.

Having found a solution to eq. (1) this represents an extremum and it remains to verify that the solution represents a global minimum in free energy. A stable two-phase equilibrium can only be obtained for a positive osmotic pressure π and at d_w values such that $\frac{d\pi}{dd_w} < 0$. There should thus be a non-monotonic behaviour of the

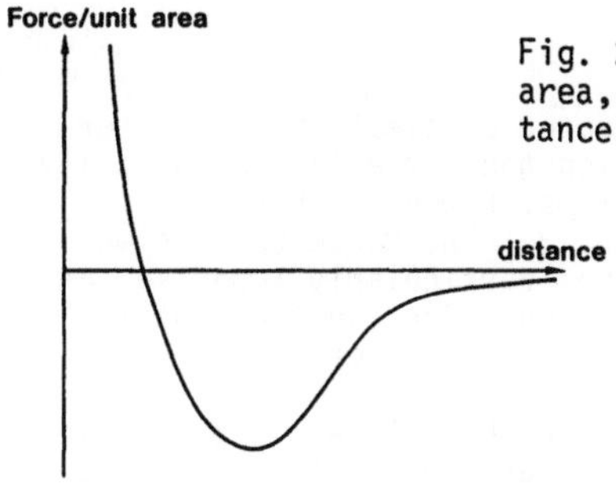

Fig. 2. A schematic illustration of the force per unit area, sometimes called disjoining pressure, versus distance. A repulsive force is defined as positive.

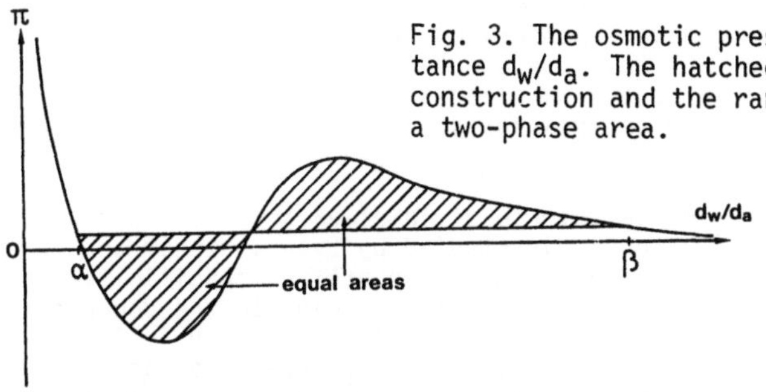

Fig. 3. The osmotic pressure π versus the scaled distance d_w/d_a. The hatched areas illustrate the Maxwell construction and the range between α and β represents a two-phase area.

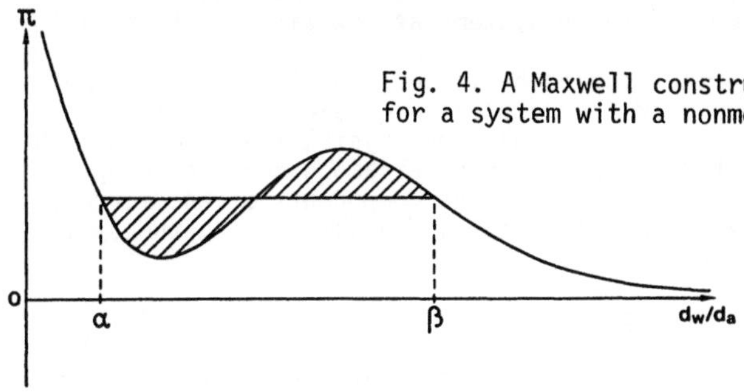

Fig. 4. A Maxwell construction and two-phase behaviour for a system with a nonmonotonic purely repulsive force.

force in the repulsive regime ($\pi > 0$). Two-phase equilibria can under these circumstances be observed both for a purely repulsive non-monotonic force (Fig.4) or in the presence of both attractive and repulsive regimes as illustrated in Fig. 3.

For a three- or multi-component system the thermodynamic relations become in general more involved. If one can assume that the composition of the amphiphilic layer is constant,the concepts developed for the two-component case still apply for the three-component one. However, with a varying distribution of one component between the two types of environments one has to invoke a more complex analysis,as demonstrated in ref. [9].

In the discussion above it has been assumed that the intrinsic state of the amphiphilic layer is independent of distance d_w except for a possible elastic deformation. The situation is quite different when one has the possibility of two different states as, for example, a liquid crystalline and a gel (or crystalline) state of the amphiphiles. In such a case one can obtain an equilibrium between two lamellar structures simply due to the interplay betweeen a repulsive interlayer interaction, specific for each state, and the intralamellar free energies,as has been explicitly demonstrated for the lecithin water system [12].

4. APPLICATIONS

For ionic amphiphiles the interlamellar force is dominated by electrostatic interactions. In the mean-field Poisson-Boltzmann description the force is monotonically repulsive and one expects highly swollen lamellar systems. However, there is also an attractive component due to ion-ion correlations [8,13]. The Monte Carlo simulations [8], show that this attractive component becomes particularly important at short separations and for high valencies of the counterions. One can then obtain the typical non-monotonic behaviour of the force distance curve (see insert Fig. 3 of ref. 8) that result in the separation of two lamellar phases. This effect is clearly illustrated by the phase behaviour of the H_2O-(Na)AOT - $Ca(AOT)_2$ system [4]. In a narrow range in the Na/Ca ratio two lamellar phases are in equilibrium. This finding is also in a semiquantitative way consistent with the MC simulations. At lower Na/Ca ratios only one lamellar phase is observed. From the phase behaviour it is clear that this is caused by the fact that the short-range attractive minimum gets deeper, which pushed the possible second lamellar phase to higher water contents. However, the surfactant is sufficiently water soluble that an isotropic phase is more stable in this range.

Another recent example of two-phase coexistence of lamellar is found in the system water - $(C_{12}H_{25})_2N^+(CH_3)_2Br^-$ [6] extensively studied by Evans, Ninham and coworkers [14]. Here the counterion is monovalent and the counterion specificity of the effect shows that it is not of a purely electrostatic origin. In the concentration range 15% to 70% water there is a lamellar phase coexistence which implies that the interlayer force behaves in a nonmonotonic way. The occurrence of an attractive short-range component should be of considerable importance for understanding the structure of the microemulsions formed in these systems at low water contents [14].

The gel phases of potassium stearate studied by Vincent and Skoulios [1,2] provides an additional interesting example of the complexity of electrostatic interactions at short range. In the gel phase the chains show crystalline order as in a plastic crystal. This order in the chains imposes an order among charged headgroups, which will have the tendency to increase the attractive components of the force [15]. This effect will be enhanced the lower the temperature, finally driving the system into its minimum energy state i.e. the crystal.

By adding small amounts of an ionic amphiphile with low aqueous solubility to a zwitterionic lamellar lecithin system one can induce a high swelling [16]. The ionic amphiphile contributes a far-reaching repulsive tail in the inter-lamellar interaction and at small additions one expects an equilibrium between a highly swollen and a moderately swollen lamellar phase [9]. At higher concentrations of the charged amphiphile the force distance curve will be monotonically repulsive, resulting in a continuous swelling of the lamellar phase.

The phase behaviour of the pure lecithin systems is an intriguing and much debated question. The prevailing point of view is that truly zwitterionic lecithins swell in excess water to incorporate approximately 30 to 40% water (w/w) (see for example ref. 11). In contrast to this Harbich and Helfrich [17] have found that egg-lecithin swells in a seemingly unlimited way in excess water. The discrepancy between these findings might of course be due to chemical subtleties, but here we want to point out another possible solution, which has also in a more general context important implications.

For lecithins one has identified three components of the interbilayer force; i) a short range, approximately exponentially decaying, strong repulsive force [11]; ii) a van der Waals' attraction which at short range decays as d_w^{-3} while at longer range ($d_w > d_a$) it goes over to a d_w^{-5} decay [11]; iii) a repulsive undulation force [18] which decays as d_w^{-3}. In the vicinity of a critical region such a decomposition of the forces is not valid [19] and the conclusion based on such a decomposition can only be valid away from a critical point.

Most estimates of the relative magnitude of the van der Waals' attraction and the undulation repulsion show that the former dominates at short range. It is commonly assumed that the observations by Harbich and Helfrich [17] on the swelling of lecithins are inconsistent with such a relation between the van der Waals' and the undulation repulsion. This is not necessarily the case.

The transition region from a d_w^{-3} to a d_w^{-5} dependence of the van der Waals' force starts at approximately 50% water and for a lecithin system with a far-reaching repulsive force there is only a narrow region where the system exhibits a strong van der Waals' attraction. The undulation repulsion, even though weaker, have a long reaching tail. We can anticipate a force distance curve as in Fig. 4 and a crucial question is whether in a Maxwell construction the area of the repulsive tail can balance the attractive minimum. Since the intregral $\int_{d_0}^{\infty} 1/X^3\,dx$ is finite a Maxwell construction will be possible for some values of the parameters while not for others. If we assume that the relative magnitude of the van der Waals' attraction and the undulation repulsion change with, for example, temperature, the qualitative phase bahaviour is shown in Fig. 5. There is a region with a coexistence between two lamellar phases, one highly swollen, one at a degree of swelling determined by the short-range repulsion. The observations of Rand and Parsegian and others on one hand and those of Harbich and Helfrich on the other would be consistent if the egg lecithin system had characteristic parameters that led to this coexistence of two lamellar phases. It is commonly stated in the literature that the two-phase region found above 'maximum swelling' of zwitterionic phospholipids is pure water with minute amounts of monomeric or oligomeric lecithin. This is correct in some cases but clearly not in others and one should try to carefully determine in which regime the particular system is.

For the highly swollen lecithin system there is the possibility that another phase with a different symmetry is even more stable. Lecithins have a very low aqueous solubility so molecular solution is only stable at extreme dilutions. An isotropic vesicular phase is a more likely alternative. An intriguing possibility suggested by Helfrich is that a bicontinuous cubic phase can be formed for a lamellar system that prefers a negative gaussian curvature.

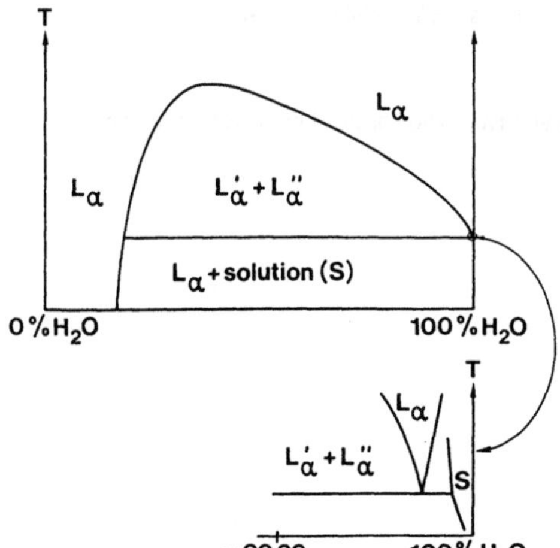

Fig. 5. A schematic phase diagram of non-charged bilayer-forming amphiphile. As a function of temperature, or some other proper field variable, one can see 3 regimes i) below the three-phase line a moderately swollen lamellar phase (L_α) is in equilibrium with a dilute isotropic solution (s); ii) Between the three-phase line and the critical point moderately swollen (L'_α) and highly swollen (L_β") lamellar phases are in equilibrium; iii) above the critical point there is a continous swelling of the lamellar phase.

REFERENCES

1. J.M. Vincent, A. Skoulios: Acta Cryst. 20, 432 (1966)
2. J.M. Vincent, A. Skoulios: Acta Cryst. 20, 447 (1966)
3. K. Larsson, N. Krog: Chem. Phys. Lipids. 10, 177 (1973)
4. A. Khan, B. Jönsson, H. Wennerström: J. Phys. Chem. 89, 5180 (1985)
5. A. Khan: To be publ.
6. K. Fonlell, A. Ceglie, B. Lindman, B. Ninham: Acta Chem. Scand. A40, 247 (1986)
7. B. Mely, J. Charvolin: Chem. Phys. Lipids. 19, 43 (1977)
8. L. Guldbrand, B. Jönsson, H. Wennerström, P. Linse: J. Chem. Phys. 80, 2221 (1984)
9. B. Jönsson, P. Persson: J. Colloid Interface Sci. In press.
10. See for example; J. Israelachvili: Intermolecular and surface forces (Academic Press, New York, 1985)
11. See; R.P. Rand: Ann. Rev. Biophys. Bioeng. 10, 277 (1981)
12. L. Guldbrand, B. Jönsson, H. Wennerström: J. Colloid Interface Sci. 89, 532 (1982)
13. R. Kjellander, S. Marcelja: Chem. Phys. Letters. 112, 49 (1984)
14. S.J. Chen, D.F. Evans, B.W. Ninham, D.J. Mitchell, F.D. Blum, S. Pickup: J. Phys. Chem. 90, 482 (1986)
15. B. Jönsson, H. Wennerström: J. Chem. Soc: Faraday Trans. 2 79, 19 (1983)
16. L. Rydhag, T. Gabran: Chem. Phys. Lipids. 30, 309 (1982)
17. W. Harbich, W. Helfrich: Chem. Phys. Lipids. 36, 39 (1984)
18. W. Helfrich: Z. Naturforsch. 33a, 305 (1978)
19. R. Lipowsky, S. Leibler: Phys. Rev. Letters. 56, 2541 (1986)

Discussions with Bengt Jönsson and Jean Charvolin, who made the author aware of refs 1 and 2, are gratefully acknowledged.

Part IV

Vesicles

Long Range Interactions Between Lipid Bilayers in Salt Solutions and Solutions of Non-Adsorbant Polymers: Comparison of Mean-Field Theory with Direct Measurements

E. Evans[1] *and D. Needham*[2]

[1]Pathology and Physics, University of British Columbia, Vancouver, B.C. V6T 1W5, Canada
[2]Pathology, University of British Columbia, Vancouver, B.C. V6T 1W5, Canada

Micromechanical experimentation has made possible direct measurements of weak adhesive interactions between surfactant bilayers in aqueous media. Interaction properties - free energy potentials (per unit area) for adhesion - are tested by micro-assembly of two vesicles with control of bilayer tensions to regulate contact formation/separation. Adhesion energies of phospholipids (neutral and charged) and a neutral "sugar" lipid have been studied - in salt solutions to examine classical prescriptions for colloidal attraction and repulsion - and in solutions of non-adsorbant polymers to evaluate mean-field theories for attraction induced by polymers in good solvents.

1. INTRODUCTION

Surfactant bilayer membranes in aqueous media interact non-specifically via long-range electrostatic, electrodynamic, and solvation forces.[1-5] These colloidal forces are commonly recognized as van der Waals' attraction, electric double-layer repulsion, and hydration repulsion. Other steric and structural interactions exist that are not yet clearly defined. Even though the magnitude of each interaction can be very large, the free energy minimum at stable contact is extremely low. Addition of large polymers to the aqueous environment can greatly augment the weak attraction between lipid bilayers.[6-7] For neutral polymers like polyglucose (dextran), this augmented attraction exhibits similar dependence on polymer concentration for several types of lipid head groups. Hence, the interaction between surfaces is again non-specific. For such neutral polymers, colloidal forces simply superpose on interactions peculiar to the suspended polymer. Because of micromechanical technology for the study of weak bilayer interactions, it is possible to quantitate adhesion energies - test reversibility - and critically evaluate disparate theories for interactions in concentrated polymer solutions as well as in simple buffers.[7-10]

Here, we discuss experiments that show how different lipids (neutral and charged) affect the intrinsic colloidal attraction between bilayer membranes. We examine the efficacy of classical prescriptions for van der Waals' attraction and electric double-layer repulsion to correlate measured free energy potentials for adhesion. Then, we will describe direct measurements of the free energy potential for adhesion of neutral-and charged-lipid bilayers in solutions of dextran over a wide range of volume fraction (0-0.1) and molecular weight (10000-150000). Since attempts to measure the concentration of polymer in the gap between adherent bilayers by fluorescence assay showed a significant reduction compared to the exterior bulk solution, we examine the results in the context of a self-consistent, mean-field theory for adhesion induced by polymer depletion in the gap. First, the experimental approach will be outlined.

2. MICROMECHANICAL METHOD

The experimental approach is to create large bilayer capsules that can be manipulated to test surface-surface adhesivity. Lipids from two classes have been used: synthetic phospholipids (1-stearoyl-2-oleoyl-phosphatidylcholine SOPC, 1-palmitoyl-2-oleoyl-phosphatidylserine POPS, and dioleoyl phosphatidylglycerol DOPG)) and a natural "sugar" lipid (digalactosyl diacylglycerol DGDG). These lipids were chosen to represent two types of neutral head group structures and two electrically charged species. Diacyl lipids exhibit the useful feature that they spontaneously form closed vesicular capsules when hydrated from an anhydrous state. Although small in number, a few of these capsules are of sufficient size (10-20 x 10^{-4}cm) that they can be subjected to direct micromechanical adhesion tests. Since the vesicles are formed in non-ionic (sucrose or other small sugars) buffers, very small levels of surface charge are sufficient to separate lamillae and thus vesicles are usually single bilayer structures.[11] When resuspended in iso-osmotic salt solutions, the small difference in index of refraction between the interior and exterior of the vesicle greatly enhances the optical image as shown in Fig. 1. Because of the extremely low solubility of lipid molecules in aqueous media, single-walled vesicles form cohesive surfaces with limited area compressibility.[11-12] Further, because of the osmotic strength of solutes trapped inside the vesicles and the small pressure differentials applied to vesicles, vesicle volumes remain essentially constant during mechanical tests. Hence, vesicles deform as liquid-filled bags with nearly constant surface area and volume; spherical vesicles appear rigid and undeformable. Provided the surface is not forced to dilate, bilayers only oppose deformation with a small bending rigidity.[11,13] For big vesicles, the bending stiffness is so small that when the vesicle is slightly deflated by osmotic increases in the exterior solution, the vesicle becomes completely flaccid and deformable. As consequences of these mechanical properties, spherical vesicles are too rigid to form adhesive contacts whereas flaccid,

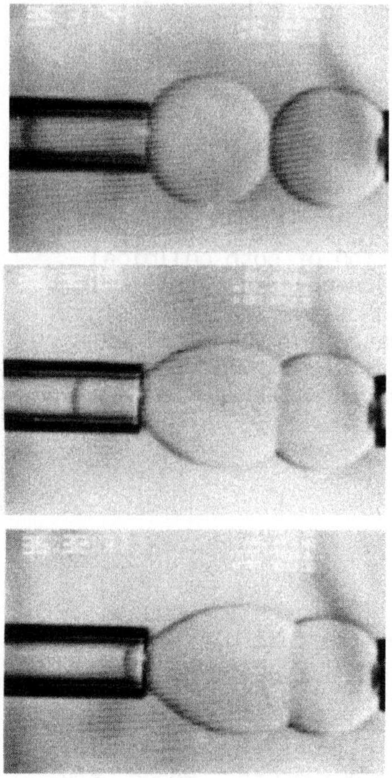

Fig. 1.

Video micrographs of a vesicle adhesion test. (a) Vesicles maneuvered into position for contact. (b,c) Adhesion configurations at equilibrium controlled by the suction pressure applied to the left-hand vesicle. (Pipet calibre ~ 5 x 10^{-4}cm; vesicle diameters ~ 2 x 10^{-3}cm.)

non-spherical vesicles easily form adhesive contacts with essentially no resistance to deformation until the surfaces are pressurized into spherical segments.

We have taken advantage of the general deformability properties of bilayer vesicles to establish a sensitive method for measurement of adhesion energy between vesicle surfaces.[6-10] First, two spherical vesicles are selected and transferred from the initial suspension in a chamber on a microscope stage to an adjacent chamber which contains a slightly more concentrated buffer (0.1 M pH 7.0 phosphate buffered NaCl-PBS) or PBS plus polymer at the desired concentration. There, the vesicles rapidly deflate to new equilibrium volumes. One vesicle is aspirated by a small micropipet and held with sufficient suction to form a rigid spherical segment outside of the pipet; this forms the "test" surface for adhesion. The second vesicle is aspirated by another pipet with low suction controlled to regulate the adhesion process. The second vesicle is then maneuvered into close proximity of the test vesicle surface (Fig. 1a) and the adhesion process is allowed to proceed in discrete (equilibrium) steps established by pipet suction (Fig. 1b,c). This experimental procedure yields the tension in the adherent vesicle bilayer as a function of the extent of coverage of the test vesicle, both for the forward process of adhesion and the reverse process of separation. The extent of coverage x_c of the test vesicle is measured by the polar height z_c of the adhesion "cap" divided by the diameter $2R_s$ of the spherical test surface ($x_c = z_c/2R_s$). In the tests to be reported here, the adhesion appeared to be totally reversible as illustrated in Fig. 2 (i.e. suction pressures that allowed contact formation were equivalent to pressures required to separate contact).

The distance scale for forces that act between bilayer surfaces is much smaller than the macroscopic dimensions observable in these experiments; thus, it is not possible to directly determine the bilayer forces. However, cumulation of the forces into an integral over distance is measurable. This integral is the negative work or free energy reduction per unit area for assembly of the bilayers from infinite separation to stable contact where the force between the surfaces is zero,

$$\gamma \equiv - \int_{\infty}^{z_g} \sigma_n \cdot dz \,. \tag{1}$$

The stress σ_n normal to the surfaces is the total action of both colloidal forces and the force induced by polymer in solution. Mechanical equilibrium in the adhesion process is established when small reductions in free energy due to formation of contact just balance small increases in mechanical work required to deform the vesicle.[8-9] This variational statement leads to a direct relation between the free energy potential for adhesion and the suction applied to the adherent vesicle,

$$\frac{\gamma}{P \cdot R_p} = f(\text{geom}) \,. \tag{2}$$

When the product of pipet suction P and radius R_p is converted to tension for the adherent vesicle bilayer, this equation takes the form of the classic Young equation where the geometric factor is $(1-\cos\,\theta_c)$; θ_c is the included angle between the bilayer surfaces. Because of the special mechanical properties of vesicles, the vesicle area and volume are constrained to remain fixed throughout contact formation and the bilayer region exterior to the pipet conforms to a surface with constant mean curvature. Hence, the contact angle can be derived from measurements of the extent of encapsulation

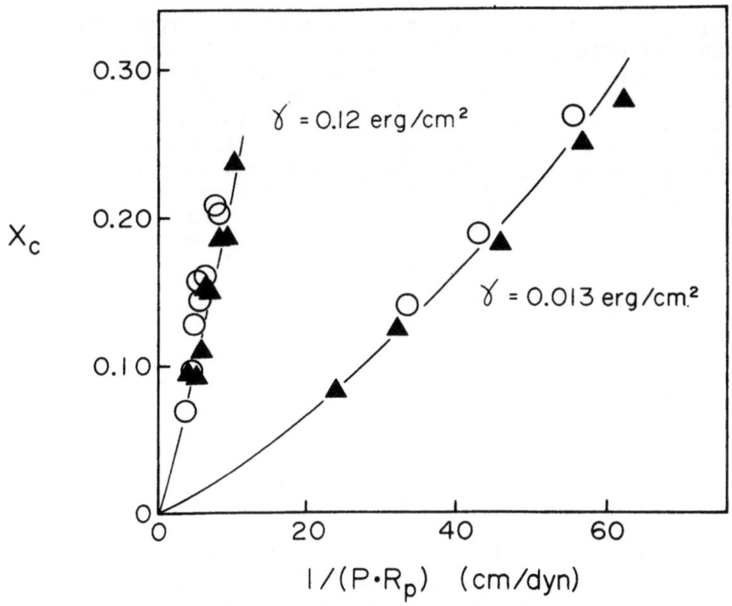

Fig. 2. Data from two vesicle adhesion tests: fractional area x_c of the rigid vesicle covered by the adherent vesicle versus the reciprocal of suction pressure P multiplied by pipet radius R_p. Triangles (▲) represent contact formation and the open circles (○) represent separation of the contact. The solid curves are predictions from mechanical analysis for uniform values of adhesion energy - listed on the figure. The lower curve is for neutral SOPC vesicles in salt buffer only. The upper curve is for neutral SOPC vesicles but in salt buffer plus 9.3 g/100cm^3 dextran - 36500 MW polymer.

of the spherical test surface with a computational algorithm.[8] If the free energy potential for adhesion is uniform over the adhesion zone, then a single curve is predicted for the relationship between pipet suction and the fractional extent of coverage of the test vesicle (as shown in Fig. 2).

3. INTRINSIC COLLOIDAL ATTRACTION AND REPULSION BETWEEN BILAYERS IN SALT SOLUTIONS

Separated by a medium with different polarizability properties, bilayers are drawn together by long-range van der Waals' forces described by Lifshitz's theory for continuous media.[14] Electrically charged lipids in salt solutions oppose this attraction with double-layer forces at large distances; but at short distances, solvation and structure forces associated with the intervening liquid and surfactant head groups dominate repulsion for all lipids.[1-5] Reversible adhesion will occur if the attraction is sufficient to overcome long-range repulsion but not sufficient to overwhelm the short-range forces that prevent collapse and phase instability. X-ray diffraction studies of forced dehydration of multilamellar lipid-water phases show that the short-range repulsion follows a steep exponential-like decay.[2,15] Hence, the energetics of adhesion for electrically neutral bilayers can be conceptually viewed as approach along a soft van der Waals' attraction to a limit determined by the magnitude and decay of the strong repulsion. The free energy potential for assembly of neutral bilayers to stable contact essentially represents the attractive van der Waals' potential at the final separation distance. If the bilayers are unsupported, repulsion may be enhanced by secondary effects due to thermo-mechanical excitation of bilayer undulations.[16-17] Bilayer undulations appear not to greatly alter the close-range repulsion; but there is predicted to be significant attenuation of the free energy potential at the minimum and an increase in the equilibrium separation because of the slower decay of the "fluctuation-enhanced" repulsion.[17] For lipids with electrically charged head groups, the long-range repulsion follows an exponential decay determined by the characteristic screening distance of the ions in solution as described by Guoy-Chapman theory.[1] At a specific separation z_g, the total free energy per unit area of contact formation is given approximately by,[2,17]

$$\tilde{F} = \lambda_{sr}\cdot P_{sr}\cdot e^{-z_g/\lambda_{sr}} - A_H\cdot f(z_\ell/z_g)/(12\cdot\pi\cdot z_g^{2}) + \lambda_{es}\cdot P_{es}\cdot e^{-z_g/\lambda_{es}} + \tilde{F}_{f\ell}. \qquad (3)$$

P_{sr} is the coefficient for the strong short-range repulsion and λ_{sr} is the decay length; A_H is the Hamaker coefficient for the van der Waals' attraction and $f(z_l/z_g)$ is a weak retardation function of the ratio of the bilayer thickness z_l to the distance between bilayers; P_{es} is the coefficient for the electric double-layer repulsion and λ_{es} is the decay length. The last term in the free energy expression $\tilde{F}_{fl}$ is the free energy excess due to thermally-excited mechanical undulations of the bilayers. The stress applied normal to the bilayer surfaces is given by the derivative of Eq. (3) as,

$$\sigma_n^c = -P_{sr}\cdot e^{-z_g/\lambda_{sr}} + \frac{A_H\cdot f(z_g/z_\ell)}{6\pi\cdot z_g^{3}} - P_{es}\cdot e^{-z_g/\lambda_{es}} - \sigma_{f\ell}. \qquad (4)$$

Equilibrium contact is specified by the condition at the secondary minimum where the stress σ_n is equal to zero.

Even though the interactions are given as functions of the same separation distance, the actual origin for each interaction is difficult to define because of molecular motion and architecture. The "mass-average" water gap between bilayers seems to be the best functional definition of distance because compositional and colligative properties are based on water content and work to remove water from lipid dispersions.[2,18] Compositional and x-ray diffraction studies of lamellar arrays naturally combine to yield the mean water gap with only the assumption of a layered structure.[19] This choice of common distance creates difficulty in the critical evaluation of van der Waals' attraction and electric double-layer repulsion when the bilayers are in close proximity (z_g 3 x 10^{-7} cm). To circumvent this difficulty, the usual approach is to introduce arbitrary displacements for the origins of these interactions which represent the location of head groups and charges relative to the mass-average water interface.

3.1 Free Energy Potentials for Neutral Bilayer Adhesion

Adhesion tests for bilayers composed of neutral PC's (egg PC, DMPC, SOPC) in the liquid state yield comparable values (0.01 - 0.015 erg/cm^2) of free energy reduction per unit area of contact.[10,20] Vesicles made from less-hydrated phosphatidylethanolamine POPE exhibit much stronger adhesion with energies that are an order of magnitude greater (0.12-0.15 erg/cm^2). Even larger free energy potentials (0.22 erg/cm^2) are measured for adhesion with DGDG (sugar-lipid) bilayers.[13,20] X-ray diffraction studies of forced dehydration (e.g. by osmotic or mechanical stress) of multilamellar arrays provide the essential measures of separation distance and the interlamellar stress at close range dominated by $P_{sr}EXP(-z_w/\lambda_{sr})$.[2,15,21] Results for these neutral lipids in the L_α (liquid) phase are given in Table 1.[21] Obviously, larger values of adhesion energy are characteristic of smaller separations at full hydration. Note, however, DGDG and POPE form adhesive contact at similar separations but the adhesion energy for DGDG bilayers is nearly twice as strong as that for POPE bilayers. Since the phospholipids and the "sugar"-lipid possess similar acyl chain cores, the conclusion is that the augmented attraction is due to attraction between sugar head groups of DGDG across the water gap. Thus, a layered-model must be introduced to represent the differential attraction properties of head groups and acyl chains across the water gap between bilayers.

Table 1 "Mass-Average Water Separation and Parameters for Short-Range Repulsion Derived from X-Ray Diffraction Studies[21]"

Neutral Lipid	z_w ($\times 10^{-8}$cm)	P_{sr} (dyn/cm^2)	λ_{sr} ($\times 10^{-8}$cm)
SOPC	27	1.0×10^{10}	2.45
POPE	11.5	1.1×10^{12}	0.95
DGDG	14	1.3×10^{11}	1.50

Theoretical prescriptions for attraction between layered composites are well established and given approximately by,[22-23]

$$\tilde{F}_A \cong -A_1\cdot f_1/\left[12\pi\cdot(z_g-z_p)^2\right] - A_2\cdot f_2/\left[12\pi\cdot z_g^{\,2}\right] - A_3\cdot f_3/\left[12\pi(z_g+z_p)^2\right],$$

where again z_g is the "mass-average" water separation; $z_p/2$ is the head group projection beyond the mass-average water interface. The retardation functions (f_1, f_2, f_3) depend on the ratio of acyl chain thickness to water gap separation and polar head group thickness to water gap separation and are on the order of unity at close range.[22] The Hamaker coefficients (A_1, A_2, A_3) are related to correlations of differences in polarizability between adjacent layers integrated over the electromagnetic spectrum: A_1 ~ (head group-water)^2; A_2 ~ (head group-water) (acyl chains-head group); A_3 ~ (acyl chains-head group)^2. Simple calculations of head group geometry (supported by comprehensive studies) indicate projections beyond the mass-average water interface of about 3 Å for the phospholipids and 4-5 Å for DGDG.[24-26] For bilayers in close proximity (e.g. POPE and DGDG), head group attraction across the water gap will be the major effect even for small Hamaker coefficients (A_1). At large separations away from equilibrium contact, attraction is characterized by the cumulative coefficient (A_1 + A_2 + A_3). Three coefficients for attraction plus geometric properties of the head group allow great freedom in selection of (A_1, A_2, A_3) consistent with measurements of adhesion energy. However, theoretical calculations indicate that the cross-coefficient A_2 is very small ($< 10^{-15}$ erg) and that the coefficient A_3 for hydrocarbon attraction across the head group region remains essentially constant ($\sim 3 \times 10^{-14}$ erg) for a wide range of head group properties.[22,27] Thus, only the coefficient A_1 is used to correlate the experimental results. For the phospholipids SOPC and POPE, A_1 is estimated to be about 1.2×10^{-14} erg and about 4×10^{-14} erg for the sugar-lipid DGDG. The significant observation is that head groups appear to play an important role in attraction at close range.

How are estimates for attraction coefficients affected by inclusion of the free energy for the short-range repulsion? If we accept an exponential form for the short-range repulsion, attenuation of the adhesion energy by repulsion is about 10-40% based on the parameters (P_{sr}, λ_{sr}) in Table 1; coefficients for attraction must be increased by about 50% to match the measured adhesion energies. However, it is not certain that the exponential form observed for repulsion at close range is appropriate for larger distances nor that the energy contribution can be obtained by direct integration of a single exponential relation. If soft repulsive effects exist further out, the energy contribution could be much greater than deduced from $\lambda_{sr}P_{sr}EXP(-z_w/\lambda_{sr})$. For example, softer repulsion is predicted from analysis of thermally-excited bending undulations in unsupported bilayers.[16-17] Repulsion is enhanced by the work required to drive heat out of the system as the bilayers are forced together. For separations less than equilibrium contact, the free energy excess due to bilayer undulations is determined by the short-range repulsion,[17]

$$\tilde{F}_{f\ell} \sim (\pi \cdot k \cdot T/16)\sqrt{P_{sr}/B \cdot \lambda_{sr}} \cdot e^{-z_g/2 \cdot \lambda_{sr}},$$

where k is Boltzmann's constant and B is the bilayer curvature or bending elastic modulus. In this range, the primary effect of the added repulsion is to increase the observed decay length λ_{sr} and expand the separation at full hydration. For unsupported SOPC multilayers, analysis shows that the measured 2.45 Å decay would reduce to 2.1 Å if the layers were rigidly fixed to solid substrates. An auxiliary prediction is that the adhesion energy for immobilized PC bilayers should be about four to five times larger than the value measured for unsupported PC vesicle membranes. This may account for the stronger adhesion energies measured for PC bilayers adsorbed to rigid mica sheets.[28] At distances well beyond equilibrium separation, the free energy excess from thermo-mechanical excitations is predicted to follow a weak steric repulsion given by,

$$\tilde{F}_{f\ell} \sim c \cdot (\pi \cdot k \cdot T/16)^2 / (B \cdot z_g^2),$$

where $c \approx 5$.[16-17] For neutral bilayers, this weak steric repulsion should overcome the retarded van der Waals' attraction at large separations ($\sim 10^{-6}$ cm). However, the repulsive stress is predicted to be so small (~ 10 dyn/cm^2) that the barrier is insignificant.

3.2 Electric Double-Layer Repulsion Between Charged Bilayers in Salt Solutions

Adhesion energies for mixtures of electrically charged (POPS) and neutral (SOPC) lipids in 0.1 M NaCl are shown in Fig. 3. A few percent of charged lipid overcomes the weak van der Waals' attraction.[13] Based on area per lipid molecule derived from x-ray diffraction studies (e.g. A_{SOPC} = 66-68 Å^2)[21] and one electrical charge per PS, adhesion energies are calculated by inclusion of double-layer repulsion in the analysis for neutral bilayers.[52] As indicated by molecular structure, the charge may be located 1-2 Å beyond the mass-average water interface into the water gap. Therefore, two curves are shown in Fig. 3 to illustrate the effect of charge position. Conceptually, the results can be viewed as reduction of the adhesion energy at essentially fixed separation-as the charge density is increased-followed by a "disjoining" transition where the bilayers separate to large distances. Ideal double-layer theory works well at this separation of about 3 Debye lengths (λ_{es} = 9.6 Å in 0.1 M NaCl). Furthermore, the outward displacement of the bilayers (implicit in the "disjoining" transition) is consistent with the combined distance dependence for attraction and double-layer repulsion.

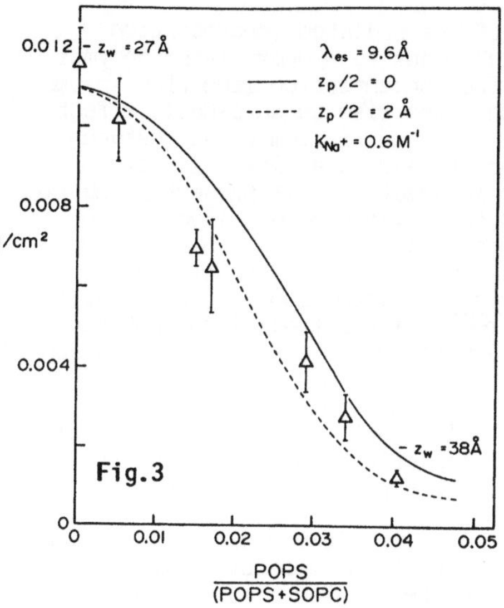

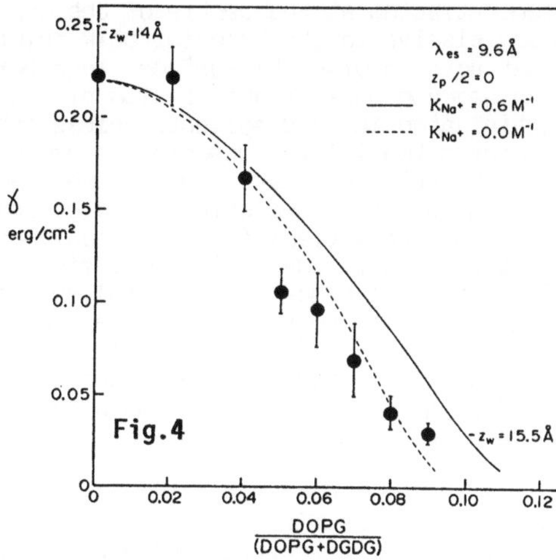

Fig. 3. Attenuation of adhesion energy by double-layer repulsion in 0.1M NaCl for mixed (POPS+SOPC) vesicles versus composition:POPS/(POPS+SOPC). Theoretical curves are predictions of electric double-layer theory.

Fig. 4. Attenuation of adhesion energy by electric double-layer repulsion in 0.1M NaCl for mixed (DOPG+DGDG) vesicles versus composition: DOPG/(DOPG+DGDG). Theoretical curves are predictions of electric double-layer theory.

To test the double-layer theory at even smaller separations, adhesion energies in 0.1 M NaCl (Fig. 4) have been measured with vesicles made from mixtures of DGDG and DOPG which start out at separations of about 1.5 decay lengths.[13] The solid curve is the prediction for adhesion energy calculated with superposition of the double-layer repulsion on the neutral bilayer interactions. Here, an area per molecule of 72 $Å^2$ is used for DGDG and the charge on DOPG is assumed to be located at the mass-average water interface because of molecular structure. Since the equilibrium constant for Na^+ binding to PG in sugar-lipid bilayers is not known, two curves are plotted in Fig. 4 to illustrate the effect of ion binding. Again, ideal theory appears to work well.

4. POLYMER-INDUCED ATTRACTION BETWEEN BILAYERS IN SALT SOLUTIONS

Two diverse conceptual views of non-specific adhesion processes form the bases for contemporary theories that have been introduced to rationalize observations of colloidal stability and flocculation in polymer solutions.[30-32] The first view is based on adsorption and cross-bridging of the polymers between surfaces. The interaction of polymer with the surfaces is assumed to be a short-range attraction proportional to area of direct contact. Theories derived from this concept usually indicate a rapid initial rise in surface adsorption for infinitesimal volume fractions[31,33]; this is followed by a plateau with gradual attentuation of attraction because of excluded volume effects in the gap between surfaces at the larger volume fractions.[31,34-35] The second - completely disparate - view of non-specific adhesion is based on the concept that there is exclusion or depletion of polymer in the vicinity of the surface, i.e. no adsorption to the surfaces.[36-37] Here, theory shows that attraction is caused by interaction of the depleted concentration profiles associated with each surface which leads to a depreciated polymer concentration at the centre of the gap. The concentration reduction in the gap relative to the exterior bulk solution gives rise to an osmotic effect that acts to draw the surfaces together. When equilibrium exists between the gap and bulk, stabilization or approach to a plateau level is not anticipated in good solvents unless other interactions are present. The free energy potential for adhesion is expected to increase progressively with concentration even at large volume fractions. Thus, adsorption-based and (non-adsorption) depletion-based concepts predict distinctly different adhesion properties: (i) excess polymer versus a reduction in concentration in the contact zone; (ii) rapid rise in free energy potential for adhesion at infinitesimal concentrations followed by a plateau and eventual attenuation of the attraction versus adhesion energy that progressively increases with concentration and without stabilization. Also, adsorption-based phenomena should exhibit specific dependencies on chemical attributes of the polymer and surface molecules whereas (non-adsorption) depletion-based processes should depend only on the colligative properties of the polymer in the solution.

If polymer does not interact with the electric field between bilayers nor contribute significantly to the polarizability of the aqueous region between the surfaces, then these colloidal forces can be treated simply as external stresses that superpose on the action of the polymer, i.e.

$$\sigma_n = \sigma_n^c + \sigma_n^p \,. \tag{5}$$

Neutral dextran at volume fractions below 0.1-0.2 can be considered as an ideal, non-interacting polymer for the following reasons. Our measurements of osmotic pressure in salt buffers and distilled water show that the polymer does not interact with the monovalent salt (NaCl). Thus, there is only a negligibly small effect on the decay length of double-layer repulsion (due to the reduced volume fraction of water) when the polymer is present in 0.1M NaCl solutions. Likewise, theoretical calculations show that sugar concentrations of 30% (wt:wt) do not significantly alter the van der Waals' attraction between lipid bilayers at long range.[38] Hence, we assume that the intrinsic colloidal interactions between the bilayers act as a separable external field as expressed in Eq. (5).

4.1 Bilayer Adhesion in Dextran Solutions

Sharp-cut dextran fractions were provided by a gracious gift from Dr. K. Granath (Pharmacia, Sweden) through our colleague, Dr. D.E. Brooks (Univ. of British Columbia). Three fractions were used, labelled as 11200, 36500, and 147500 weight-average molecular weights with ratios of weight-average to number-average molecular weights of 2.07, 1.34, and 1.56 respectively. For solutions of these fractions up to 10% by volume, osmotic pressures were measured by freezing point depression and checked by vapour pressure osmometry at high concentration. Results are plotted in Fig. 5. The first and second virial coefficients (listed in Table 2) were derived from the osmotic pressure data which show that the number-average molecular weights were 3100, 27200, and 94300[53] for these fractions. Also, the Flory interaction parameter which characterizes the quality of the solvent was derived from the second virial coefficient and found to be 0.43-0.44, i.e. in the "good" solvent range.[39]

Micromechanical adhesion tests were carried out on 5-10 neutral (SOPC) vesicle pairs at specific concentrations of these polymers in the range of 0-15 g/100cm^3 (wt/vol). These results are plotted in Fig. 6 as a function of volume fraction (v = 0.611 x wt/vol). Even for these concentrated solutions, adhesion energies increased progressively (from 0.01 to > 0.2

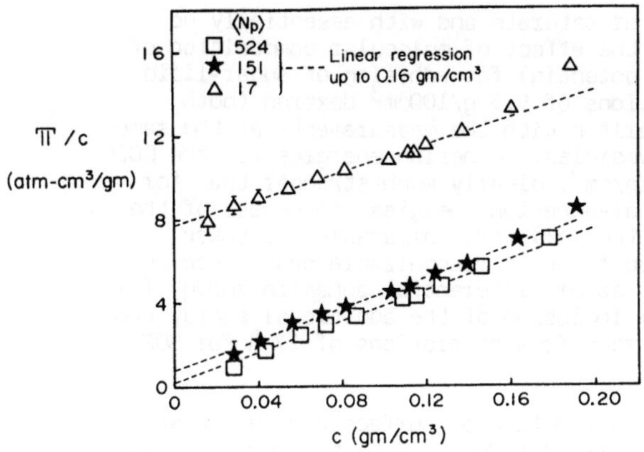

Fig. 5. Measurements of osmotic pressure divided by polymer concentration (wt/vol) plotted versus concentration. The intercept and slope provide measures of the first and second virial coefficients for the three dextran fractions (△-11200, ★-36500, □-147500 MW).

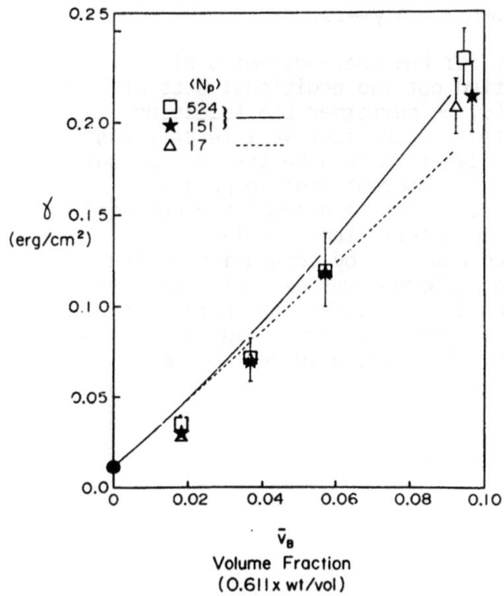

Fig. 6.
Adhesion energies (erg/cm^2) for neutral SOPC bilayers in 0.1M salt (PBS) plus dextran polymer at volume fractions up to 0.1. Results are given for three dextran fractions represented by number-average polymer indices, N_p (i.e. equivalent number of glucose monomers). The curves are predictions obtained from Self-Consistent, Mean-Field theory (outlined in the text).

Table 2 "First and Second Virial Coefficients, Interaction Parameter, and Number Average Molecular Weight for Three Dextran Fractions from Osmometry"

Dextran Fraction (MW)	First Virial Coefficient (atm-cm^2/g)	Second Virial Coefficient (atm-cm^6/g^2)	Number Average Molecular Weight (M_n)	Flory Interaction Parameter (χ)
11,200	7.739	31.90	3,139	0.437
36,500	0.8930 [53]	36.29	27,200 [53]	0.428
147,500	0.2576 [53]	36.29	94,300 [53]	0.428

erg/cm^2) with no tendency to plateau or saturate and with essentially no dependence on polymer size. To test the effect of molecular composition of the neutral surface, the free energy potential for adhesion of sugar-lipid (DGDG) vesicles was measured in solutions of 9.3 g/100cm^3 dextran (both 147500 and 36500 fractions) for comparison with the measurements at the same concentration with the neutral SOPC vesicles. Adhesion energies for the DGDG vesicles in both dextrans were 0.4 erg/cm^2, clearly much stronger than for the SOPC vesicles; but the differential-adhesion energies (in excess of the baseline van der Waals' attraction potential) were comparable for these neutral lipids. Hence, there appeared to be no recognizable dependence on surface composition. [The slightly greater differential-adhesion energy for DGDG vesicles can be accounted for by inclusion of the additional energy due to displacement of polymer-induced forces from separations of 27 Å for SOPC to 14 Å for DGDG as discussed later.]

Attenuation of adhesion energy as a function of surface charge (shown in Fig. 7) was measured by assembly of mixed (SOPC + POPS) bilayer vesicles in dextran solutions fixed at 9.3 g/100cm^3 ($\bar{v}_B$ = 0.057). Unlike the behaviour of the adhesion energy for neutral lipid bilayers, there was an obvious difference between the small and higher molecular weight polymers. Adhesion could be prevented by surface charge contents greater than 15 mol% POPS in solutions of the small molecular weight polymer whereas even for POPS concentrations in excess of 30 mol%, there was a measurable level of adhesion between vesicles with the higher molecular weight polymers.

In order to select an appropriate theory for the thermodynamics of adhesion in these polymer solutions, we carried out two additional sets of experiments. In the first set of experiments, we performed the following sequence of vesicle adhesion-separation tests: an adherent vesicle pair was first assembled in the salt buffer without polymer where adhesive contact was established by van der Waals' attraction. This adherent vesicle pair was transferred within a few seconds (with care so as not to disrupt the contact) to an adjacent chamber that contained a high concentration of polymer ($\geq$ 5 g/100cm^3). The adhesion energy was then measured by separation of the contact in the final solution. Similarly, an adherent vesicle pair was first assembled in a solution with polymer and then transferred to the pure salt buffer without disruption of the contact; again the adhesion energy was measured by separation in the final solution. The rationale behind these

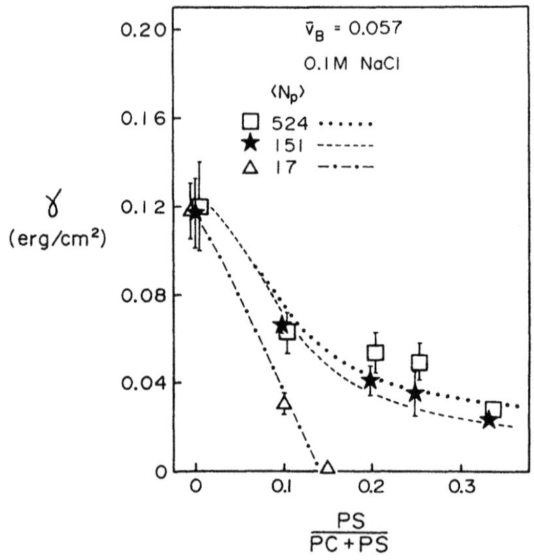

Fig. 7.

Adhesion energies (erg/cm^2) for charged bilayers in 0.1M salt (PBS) plus dextran polymer fixed at a volume fraction of 0.057 (9.3 g/100cm^3) <u>versus</u> charge content given by composition: POPS/(POPS+SOPC). Again, curves are predictions obtained from Self - Consistent, Mean-Field theory (outlined in text) in conjunction with electric double-layer repulsion.

tests was the expectation that if polymer cross-bridges were involved in adhesion, the measured values for energy should have been determined primarily by the composition of the initial solution in which adhesion was first established. However, the results were exactly opposite.

In the second set of experiments, we attempted to quantitate the number of polymer molecules captured in the gap between adherent surfaces. The procedure involved formation of an adherent vesicle complex in a solution that contained fluorescently labelled dextran polymer (150000 MW), followed by transfer of the adherent vesicle pair to an adjacent chamber that contained the same concentration of polymer but without fluorescent label. Since the vesicles did not separate, it was expected that fluorescent polymer trapped in the adhesion zone would be detectable over a long time period until diminished by exchange diffusion with the exterior solution. The results were negative; we could not detect any fluorescence in the contact zone. Based on molecular dimensions ($\sim 10^{-6}$ cm) as estimates of the minimum gap thickness and the bulk concentration of polymer, fluorescence should have easily been detected with our photometric system.

4.2 Theoretical Implications and Methods of Analysis

Adhesion tests for lipid bilayer vesicles in dextran polymer solutions are consistent with the (non-adsorption) depletion type of interaction. This conclusion is based on: (i) the null observation that fluorescently labelled polymer could not be detected in the gap between bilayers; (ii) the progressive increase in adhesion energy for neutral bilayers as the polymer concentration was increased to large volume fractions; and (iii) the transfer of adherent vesicle pairs with subsequent separation which showed that adhesion energy depended only on composition of the medium exterior to the contact zone. Hence, we have chosen to carefully examine depletion-based theories in conjunction with these experiments. First, we outline a simple thermodynamic approach that provides a formalism for derivation of physical stresses from free energy of mixing and polymer configuration. Then, we use the approach to establish a mean-field prediction for attraction induced by non-adsorbant polymers in good solvents.

Variations in total free energy associated with adhesion must include both the gap and exterior (bulk) regions,

$$\delta F = \delta F_g + \delta F_B .$$

These changes are subject to conservation requirements for the total number of solute molecules and the total volume of gap and exterior regions (which implies conservation of solvent),

$$\delta(\bar{v}_g \cdot V_g + \bar{v}_B \cdot V_B) \equiv 0 ,$$

$$\delta(V_g + V_B) \equiv 0 ,$$

where V_g, V_B are the volumes of the gap and bulk regions respectively; $\bar{v}_g$, $\bar{v}_B$ are the mean volume fractions of the solute molecules in the gap and bulk regions. The free energy depends on the following variables: (i) the spatial distribution $\phi = v/\bar{v}_g$ and configuration of polymer molecules in the gap; (ii) the mean concentration or volume fraction $\bar{v}_g$ of polymer in the gap; (iii) the gap thickness z_g; and (iv) the contact area A. First of all, we assume that the polymer configuratons and distribution in the gap are at local equilibrium for fixed gap dimensions and number of molecules in the gap, i.e.

$$\delta F_g \Big|_{(\bar{v}_g, \bar{z}_g, A)} = 0 \cdot$$

The variation implies an optimum profile ϕ for the local volume fraction in the gap. The variation can also be written with the requirement for fixed number of polymer molecules included explicitly by,

$$\delta F_g \Big|_{(\bar{v}_g, z_g, A)} = \delta F_g \Big|_{(z_g, A)} - \lambda_c \cdot A \cdot \int_0^{z_g} \delta v \Big|_{(z_g, A)} \cdot dz = 0 . \tag{6}$$

The Lagrange multiplier λ_c represents a "pressure" which ensures that the constraint,

$$\bar{v}_g \equiv \frac{1}{z_g} \cdot \int_0^{z_g} v \cdot dz = \text{constant}$$

is satisfied appropriate to conditions for exchange of polymer between gap and bulk regions. For local equilibrium in the gap, the total free energy variation is given by the sum of variations with respect to mean concentration v_g, separation z_g, and contact area A,

$$\delta F = \frac{\partial F}{\partial \bar{v}_g} \Big|_{(z_g, A)} \cdot \delta v_g + \frac{\partial F}{\partial z_g} \Big|_{(\bar{v}_g, A)} \cdot \delta z_g + \frac{\partial F}{\partial A} \Big|_{(\bar{v}_g, z_g)} \cdot \delta A . \tag{7}$$

For reversible assembly processes, the mechanical work to displace stresses at the surfaces is equal to the free energy change at <u>constant</u> contact area,

$$(\sigma_n^p \cdot A) \cdot \delta z_g = \delta F \Big|_A . \tag{8}$$

Likewise, the free energy potential to create contact area at constant separation is given by,

$$\gamma \cdot \delta A = \delta F \Big|_{z_g = \infty} - \delta F \Big|_{z_g} , \tag{9}$$

which shows the expected result equivalent to Eq. (1),

$$\frac{\partial}{\partial A} (\sigma_n^p \cdot A) \Big|_{z_g} = - \frac{\partial \gamma}{\partial z_g} \Big|_A .$$

A great deal of theoretical development over the past decade or so has been designed to predict the configurations and distribution of polymer segments in the vicinity of a solid surface or between surfaces and thereby to predict the deviation of the free energy of mixing from that in the adjacent bulk region.[40-46] The simplest approach is to introduce a "constitutive" relation for the free energy as a sum of the free energy for uniform concentration plus a term that represents the free energy excess due to configurational entropy gradients.[34,36,42,47,48] For linear flexible polymers, the relation can be deduced from a statistical equation for pair-correlation functions of segment distribution to give a self-consistent, mean-field (SCMF) approximation.[36,42,45] Here, the free energy density is expressed by,

$$F_g = A \cdot \int_0^{z_g} \left(\tilde{F} + \frac{a_m^2}{6} \cdot \left|\frac{d\psi}{dz}\right|^2\right) \cdot dz \,, \tag{10}$$

where $\psi^2 \equiv v/\upsilon_m$ is the expectation value for local segment concentration; a_m is the effective length of a rigid segment of the flexible polymer. $\tilde{F}$ is the free energy density evaluated at the local concentration in the absence of gradients and given in terms of chemical potentials as,

$$\tilde{F} = \frac{v \cdot \mu_p}{N_p \cdot \upsilon_m} + \frac{(1 - v) \cdot \mu_s}{\upsilon_s}$$

or,

$$\tilde{F} = \frac{v \cdot \mu_p}{N_p \cdot \upsilon_m} - (1 - v) \cdot \Pi \,, \tag{11}$$

where μ_p and μ_s are chemical potentials for the polymer and solvent; υ_m and υ_s are the molecular volumes of a monomer segment and solvent; N_p is the polymer index or number of segments per chain; Π is the osmotic pressure which is proportional to the chemical potential of the solvent.[54] A more useful expression for the free energy density is in terms of the chemical potential $\bar{\mu}_p$ for the polymer adjusted by the osmotic pressure work for displacement of solvent,

$$\frac{\bar{\mu}_p}{N_p \cdot \upsilon_m} \equiv \frac{\mu_p}{N_p \cdot \upsilon_m} + \Pi = \frac{\partial \tilde{F}}{\partial v} \,,$$

$$\tilde{F} = \frac{v \cdot \bar{\mu}_p}{N_p \cdot \upsilon_m} - \Pi \,. \tag{12}$$

For bulk regions of large extent, the free energy F_B for the polymer solution is simply,

$$F_B = \left(\frac{\bar{v}_B \cdot \bar{\mu}_p^B}{N_p \cdot \upsilon_m} - \Pi_B\right) \cdot V_B \,.$$

Procedures for evaluation of the variations in Eq. (6) and (7) (consistent with conservation requirements) are given in detail elsewhere[56] with particular expressions in terms of adjusted chemical potential, osmotic pressure, and gradients. Here, we will assume that the assembly (adhesion) process is "complete" equilibrium for polymer exchange between gap and bulk regions as well as local to the gap. This restricts the variation with respect to mean concentration $\bar{v}_g$ in the gap (at fixed dimensions) to be identically zero as given by,

$$\int_0^{z_g} \left[\frac{v\cdot(\bar{\mu}_p - \bar{\mu}_p^B)}{N_p \cdot \upsilon_m} + \frac{a_m^2}{6} \left|\frac{d\psi}{dz}\right|^2 \right] dz \equiv 0 \,, \tag{13}$$

i.e. deviations of $\bar{\mu}_p$ from the adjusted chemical potential for the bulk solution act as a "field" to offset free energy excess due to gradients. This embodies the essence of the self-consistent, mean-field (SCMF) approximation. Eq. (13) for complete equilibrium is an important auxilliary equation and further simplifies the other variations to,

$$\delta F_g \Big|_{(\bar{v}_g, z_g, A)} = \delta \left\{ \int_0^{z_g} \left[\frac{v\cdot\bar{\mu}_p}{N_p \cdot \upsilon_m} - \Pi + \frac{a_m^2}{6} \left|\frac{d\psi}{dz}\right|^2 - \lambda_c \cdot v \right] dz \right\} \Bigg|_{(z_g, A)} \equiv 0 \,, \tag{14}$$

$$\delta F \Big|_{(\bar{v}_g, A)} = \frac{\partial}{\partial z_g} \left\{ \int_0^{z_g} \left[\frac{v\cdot(\bar{\mu}_p - \bar{\mu}_p^B)}{N_p \cdot \upsilon_m} + (\Pi_B - \Pi) + \frac{a_m^2}{6} \left|\frac{d\psi}{dz}\right|^2 \right] \cdot dz \right\} \Bigg|_{(\bar{v}_g, A)} A\cdot\delta z_g \,, \tag{15}$$

$$\delta F \Big|_{(\bar{v}_g, z_g)} = \left\{ \int_0^{z_g} \left[\frac{v\cdot(\bar{\mu}_p - \bar{\mu}_p^B)}{N_p \cdot \upsilon_m} + (\Pi_B - \Pi) + \frac{a_m^2}{6} \left|\frac{d\psi}{dz}\right|^2 \right] \cdot dz \right\} \cdot \delta A \,, \tag{16}$$

where the Lagrange multiplier λ_c is $\bar{\mu}_p^B / N_p \cdot \upsilon_m$ for complete equilibrium.

Based on Eq. (13)-(14), the polymer-induced stress is given by Eq. (15) which reduces to the osmotic pressure difference between the bulk solution and the mid-point of the gap,

$$\sigma_n^p = \frac{\partial}{\partial z_g} \left\{ \int_0^{z_g} (\Pi_B - \Pi)\cdot dz \right\} \Bigg|_{(\bar{v}_g, A)} = \Pi_B - \Pi_{z_g/2} \,, \tag{17}$$

because to first order,

$$\int_0^{z_g} \left\{ \frac{\partial \Pi}{\partial z_g} \right\} \Bigg|_{(\bar{v}_g, A)} dz \Rightarrow 0 .$$

In order to determine the volume fraction at the mid-point of the gap, Eq. (14) is solved to yield the optimum profile for the volume fraction in the gap,

$$\frac{a_m^2}{6} \cdot \left|\frac{d\psi}{dz}\right|^2 = \Delta\tilde{F} - \Delta\tilde{F}_{z_g/2} \,,$$

where

$$\Delta\tilde{F} = \frac{v\cdot(\bar{\mu}_p - \bar{\mu}_p^B)}{N_p\cdot\upsilon_m} + (\Pi_B - \Pi)\,,$$

which is integrated to give the mid-point relation,

$$z_g/2 = \frac{a_m}{\sqrt{6}}\int_0^{\psi_{z_g/2}} \frac{d\psi}{\left[\Delta\tilde{F} - \Delta\tilde{F}_{z_g/2}\right]^{1/2}} \tag{18}$$

Chemical potentials for uniform mixing of polymer and solvent molecules are described classically by Flory-Huggins theory[39] or by more esoteric scaling theories.[45] Here, we use the classical Flory equations,

$$\mu_p = \ell n(v) + \frac{\chi\cdot N_p\cdot\upsilon_m}{\upsilon_s}\cdot(1 - v)^2 + (1 - \frac{N_p\cdot\upsilon_m}{\upsilon_s})(1 - v),$$

$$-\mu_s = \upsilon_s\cdot\Pi = -\ell n(1 - v) + v\cdot(\frac{\upsilon_s}{N_p\cdot\upsilon_m} - 1) - \chi\cdot v^2, \tag{19}$$

where χ is the Flory interaction parameter. In addition, a free energy contribution (from overall elastic extension - compression - of a polymer chain to fit into a gap less than twice the radius of gyration) must be included because the free energy excess due to monomer concentration gradients $-a_m^2|d\psi/dz|^2$- does not represent long-range changes in chain conformation. Hence, we add the elastic free energy $\tilde{F}_{el}$ of deformation[39] to the free energy of uniform mixing,

$$\tilde{F}_{e\ell} = \frac{\bar{v}_{B'}}{N_p\cdot\upsilon_m}\left[(\frac{2\cdot R_g}{z_g}) + \frac{1}{2}\cdot(\frac{z_g}{2\cdot R_g})^2 - \frac{3}{2}\right] \quad ; \quad z_g \lesssim 2\cdot R_g\,. \tag{20}$$

Differences exist between classical Flory[39] and scaling[45] theories for deformation energies and radii of gyration; but again, we use Flory's equations, e.g. $R_g = a_m N_p^{1/2}$. The effect of this ad-hoc inclusion of polymer deformation energy is to slightly change the reference state for the polymer in the gap (defined by the polymer configuration without monomer gradients). Hence, the adjusted chemical potential $\bar{\mu}_p$ for the reference state becomes $\bar{\mu}_p^{B'}$ given by,

$$\bar{\mu}_p^{B'} = \bar{\mu}_p^B - \tilde{F}_{e\ell}\cdot\upsilon_s/\bar{v}_{B'}\,, \tag{21}$$

which defines a reference volume fraction $\bar{v}_B{}'$ in the absence of monomer gradients. In other words, $\bar{v}_B{}'$ would be the volume fraction of polymer in the gap for equilibrium exchange with the bulk region but without perturbation of the monomer distribution by the proximity of solid surfaces.

We will particularize the rest of the analysis to the range of polymer concentrations dominated by the second virial coefficient, i.e. the "semi-dilute" range in good solvents (< 0.5). The approach is clearly reasonable for solutions of the two larger dextran polymers used in our adhesion tests over volume fractions of 0.01 to 0.1. For this range, the following approximations for chemical potential differences (evaluated at a local volume fraction in the gap minus the reference value) are derived from Flory equations,

$$\frac{v\cdot(\bar{\mu}_p - \bar{\mu}_p^B)}{N_p\cdot \upsilon_m} \cong \left(\frac{C_\chi}{\upsilon_s}\right)\cdot v\cdot(v - \bar{v}_{B'}) \,,$$

$$(\Pi_{z_g/2} - \Pi) \cong \left(\frac{C_\chi}{2\cdot \upsilon_s}\right)\cdot(v^2_{z_g/2} - v^2) \,, \tag{22}$$

where $C_\chi \equiv 1 - 2\cdot\chi + \upsilon_s/(N_p\cdot\upsilon_m\cdot\bar{v}_{B'})$. If we assume complete equilibrium, Eq. (18) determines the concentration at the midpoint of the gap and is given by an elliptic integral with the use of Eq. (22) and $\psi^2 = v/\upsilon_m$,

$$z_g/2 = \frac{\zeta_m}{\sqrt{A_v}} \int_0^1 \frac{d\tilde{\psi}}{\left[(1-\tilde{\psi}^2)(1 - B_v\cdot\tilde{\psi}^2)\right]^{1/2}} \,. \tag{23}$$

This equation has been scaled to give a dimensionless form where the parameters are,

$$A_v \equiv 1 - v_{z_g/2}/(2\cdot\bar{v}_{B'}) \,, \tag{24}$$

$$B_v \equiv v_{z_g/2}/(2\cdot\bar{v}_{B'}\cdot A_v) \,, \tag{25}$$

$$\zeta_m \equiv a_m \left[\frac{\upsilon_s}{6\cdot\upsilon_m\cdot\bar{v}_{B'}\cdot C_\chi}\right]^{1/2} \quad ; \quad \tilde{\psi} \equiv \psi/\psi_{z_g/2} \,. \tag{26}$$

The distance scale ζ_m represents the depletion zone near the surface and the decay of depletion with distance. Hence, the volume fraction $v_{z_g/2}$ at the centre of the gap is specified along with the normal stress σ_n^p induced by the polymer,

$$\sigma_n^p = \Pi_B - \Pi_{z_g/2} \,.$$

Figure 8 shows the predictions[55] of stress versus separation for SOPC bilayers in dextran solutions of 0.01 and 0.1 volume fractions; plotted separately are the polymer-induced stresses and the natural colloidal interaction.[56] Equilibrium separation is defined by zero total stress; hence it is apparent that the gap is reduced slightly from 27 to 24 Å at the higher polymer concentration. Integration of the total stress from infinity to equilibrium separation yields the free energy reduction per unit area - or adhesion energy. Predictions for adhesion energy from our theoretical development are plotted in Fig. 6 along with the direct measurements. Even

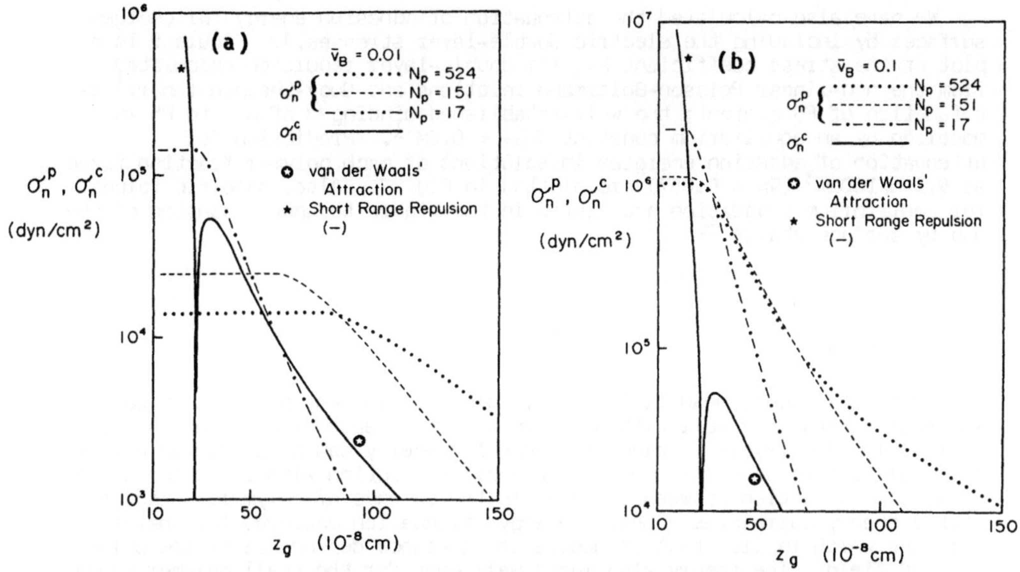

Fig. 8. Predictions of attractive stresses induced by non-adsorbant polymers in good solvents and the stress for intrinsic colloidal attraction/ repulsion between bilayers versus separation. a) Polymer volume fraction of 0.01; b) volume fraction of 0.1 (note the order of magnitude change in scale).

though the approximations embodied in Eq. (22) are not expected to work, we have calculated the interaction potential for the small dextran polymer and plotted it in Fig. 6. Similarly, adhesion between DGDG bilayers in polymer solutions can be represented by superposition of the colloidal attraction in the absence of polymer, i.e. 0.22 erg/cm^2, plus the polymer-induced attraction. The polymer-induced attraction is predicted to include the same free energy reduction per unit area as for SOPC bilayers plus an added reduction given by the osmotic pressure multiplied by the gap displacement from 24 Å down to 14 Å (since theory shows that no polymer is present in the gap below 30 Å), i.e. $\gamma \cong 0.22 + 0.12 + (4 \times 10^5)(10^{-7}) \cong 0.4$ erg/cm^2.

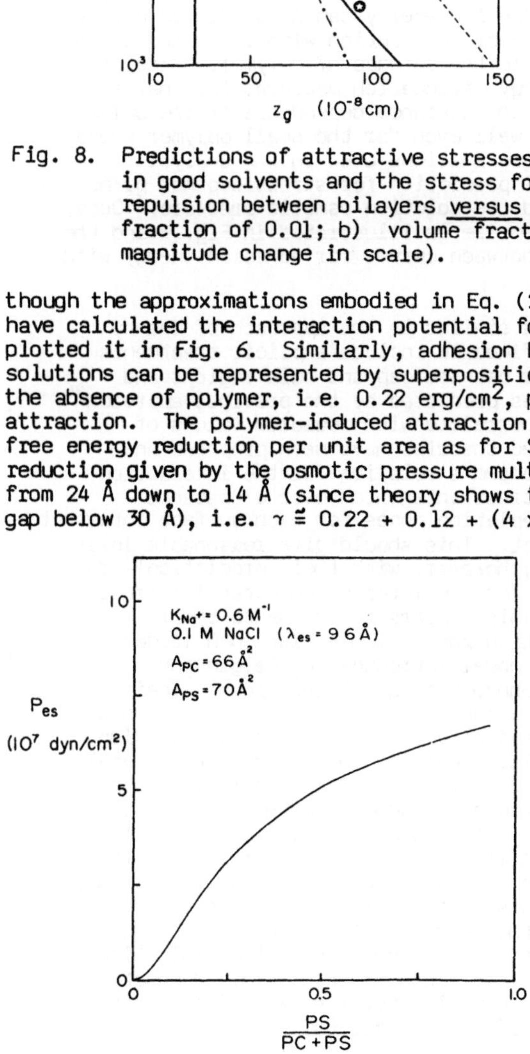

Fig. 9.
Stress coefficient P_{es} for electric double-layer repulsion (at separations $z_g > \lambda_{es}$) in 0.1M NaCl versus charge content given by composition:POPS/(POPS+SOPC).

We have also calculated the attenuation of adhesion energy for charged surfaces by including the electric double-layer stresses.[13] Figure 9 is a plot of the stress coefficient P_{es} for double-layer repulsion calculated from the non-linear Poisson-Boltzmann relations and Guoy-Chapman theory[1] as a function of PS content; the well-established binding[49] of Na^+ to PS was modelled by an equilibrium constant $K_{Na^+} = 0.6M^{-1}$. Prediction for attenuation of adhesion energies in solutions of each polymer fraction fixed at 9.3 g/100cm^3 ($\bar{v}_B = 0.057$) are plotted in Fig. 7. Also, specific values for equilibrium separation are listed in the figure to show expansion of the gap by surface charge.[56]

5. CONCLUSIONS

The self-consistent, mean-field theory for non-adsorbent polymers in good solvents agrees very well with our direct measurements of adhesion energy for neutral lipid bilayers. Values for adhesion energy can be predicted directly from colligative properties of the polymer in solution with inclusion of the intrinsic attraction between bilayers in the absence of polymer. Also, the (SCMF) theory correlates adhesion energy attenuation measured for charged surfaces which implies that it models the distance-dependence of the polymer-induced field. The theory also works well even for the small polymer whose characteristics do not satisfy the approximations used in the analysis. This is most likely because the theoretical prediction for stress depends primarily on the scale of the depletion zone and the solution osmotic pressure. Decay of the depletion zone for each polymer size is illustrated in Fig. 8 and the effect is apparent in the separation between curves for adhesion energy with charged surfaces - Fig. 7.

The (SCMF) model presented here is similar to the approach of Joanny et al.[36] but the analysis has been refined to include explicit treatments of the equilibrium exchange of polymer between gap and bulk regions and derivation of a work potential. Values predicted by the previous approach[36] for the interaction potential do not correlate with measured values of adhesion energy. Other depletion-based theories have been proposed and examined.[37,50,51] Fleer et al.[37] proposed a relation for the interaction potential given by the bulk osmotic stress integrated over a distance - beginning at a separation where the depletion zones for each surface intersect - down to the final equilibrium contact. This should give reasonable lower-bound estimates for adhesion energies; however, with their prediction[13] for depletion-zone dimensions, the values for the interaction potential are too large by a factor of three or more. This appears to be due to their expression for the scale of the depletion zone which yields much larger distances than our analysis. Another model introduced by Feigin and Napper[50] predicts entirely different phenomenological behaviour. Their model predicts a repulsive barrier at large separations then attraction at closer range. We have found no evidence of any repulsion in our adhesion measurements. The values of the repulsion barrier indicated in their analysis would be on the order of 0.04 erg/cm^2. If this repulsive barrier existed, vesicles charged with only 5 mol% PS would not adhere in the polymer solution because they do not adhere in the salt buffer alone. This is clearly not the case (Fig. 7). Finally, Luckham and Klein[51] proposed an empirical relation for the interaction potential which is similar to the result derived in our analysis. In their elegant measurements of the force as a function of distance between curved mica sheets, they were unable to verify this relation in a non-adsorbent polymer (polyethylene oxide) because of the limitation on force detection and volume fraction in their experiments. This demonstrates a very useful aspect of vesicle adhesion experiments, i.e. weak interactions can be measured in concentrated solutions of macromolecules.

In general, it is not expected that mean-field theories will provide accurate predictions. However, these approaches are very useful because they provide reasonable approximations - and good insight - to the physics of the process. The approach outlined in this paper is readily extended to situations of "restricted" exchange of polymer between the gap and bulk regions; non-specific adsorption of polymer to the surfaces; inclusion of additional surface-fixed polymer groups; interactions in two-phase polymer solutions; etc. The primary advantage over sophisticated computational analyses of polymer configurations in the vicinity of surfaces is that the approach basically gives a simple analytic solution which can be evaluated at minimal cost. Further (based on the results outlined in the previous paragraph), it is not clear that more sophisticated methods give more accurate results. Careful comparison of both approaches is necessary to identify the breakdown in the continuous mean-field approximation and, ultimately, all approaches should be examined by direct experimental test.

6. ACKNOWLEDGEMENT

This work was supported by the Medical Research Council of Canada through Grant MT 7477. EAE is grateful to the Alexander von Humboldt Foundation, Federal Republic of Germany, for a Senior Scientist Award.

7. REFERENCES

1. E.J.W. Verwey and J.Th.G. Overbeek: Theory of the Stability of Lyophobic Colloids (Elsevier, Amsterdam, 1948).
2. V.A. Parsegian, N. Fuller and R.P. Rand: Proc. Natl. Acad. Sci. USA, 76, 2750 (1979).
3. V.A. Parsegian: Ann. Rev. Biophys. Bioeng., 2, 221 (1973).
4. J.N. Israelachvili and R.M. Pashley: In Biophysics of Water, F. Franks, Ed. (John Wiley and Sons, 1982) p. 183.
5. R.P. Rand: Ann. Rev. Biophys. Bioeng., 10, 277 (1981).
6. E. Evans: Colloids and Surfaces 1984, 10, 133.
7. E. Evans and M. Metcalfe: Biophys. J., 45, 715 (1984).
8. E. Evans: Biophys. J. 1980, 31, 425.
9. E. Evans and V.A. Parsegian: In Surface Phenomena in Hemorheology: Theoretical, Experimental, and Clinical Aspects; A.L. Copley and G.V.F. Seaman, Eds. (N.Y. Acad. Sci., 1983) p. 13.
10. E. Evans and M. Metcalfe: Biophys. J., 46, 423 (1984).
11. R. Kwok and E. Evans: Biophys. J., 35, 637 (1981).
12. E. Evans and D. Needham: Faraday Discuss. Chem. Soc., No. 81 (1987, in-press).
13. E. Evans and D. Needham: J. Phys. Chem. (submitted).
14. E.M. Lifshitz: J. Exp. Theor. Phys. USSR, 29, 94 (1955); Sov. Phys. JETP, 2, 73 (1956).
15. L.J. Lis, M. McAlister, N. Fuller and R.P. Rand: Biophys. J., 37, 657 (1982).
16. W. Helfrich: Z. Naturforsch, 33a, 305 (1978).
17. E. Evans and V.A. Parsegian: Proc. Natl. Acad. Sci. USA, 83, 7132 (1986).
18. E. Evans and R. Skalak: Mechanics and Thermodynamics of Biomembranes (CRC Press, Boca Raton, Fla., 1980).
19. V. Luzzati: In Biological Membranes, D. Chapman, Ed. (Academic Press, New York, 1968) p. 71.
20. E. Evans, D. Needham and R.P. Rand: Colloids and Surfaces (to be submitted).
21. R.P. Rand and V.A. Parsegian: (to be submitted).
22. D. Le Neveu, R.P. Rand, D. Gingell and V.A. Parsegian: Biophys. J., 18, 209 (1977).

23. B.W. Ninham and V.A. Parsegian: J. Chem. Phys., 53, 3398 (1970).
24. J. Marra: J. Colloid Interface Sci., 107, 446 (1985).
25. M. Hauser, I. Pascher, R.M. Pearson and S. Sundell: Biochim et Biophys. Acta, 650, 21 (1981).
26. T.J. McIntosh and S.A. Simon: Biochem., 25, 4058 (1986).
27. V.A. Parsegian: Private Communication.
28. J. Marra and J. Israelachvili: Biochem., 24, 4608 (1985).
29. M. Eisenberg, T. Gresalfi, T. Riccio and S. McLaughlin: Biochem., 18, 3213 (1979).
30. Th.F. Tadros: In Polymer Colloids, R. Buscall, T. Corner and J.F. Stageman, Eds. (Elsevier Applied Sci., London, 1985) p. 105.
31. B. Vincent: In Polymer Adsorption and Dispersion Stability, E.D. Goddard, B. Vincent, Eds. (Am. Chem. Soc., Washington, 1984) p. 1.
32. D.H. Napper: Polymeric Stabilization of Colloidal Dispersions, Academic Press, London, 1983.
33. P-G. de Gennes: J. Phys. (Paris) 1976, 37, 1445.
34. P-G. de Gennes: Macromolecules 1982, 15, 492.
35. J.M.H.M. Scheutjens and G.J. Fleer: Macromolecules 1985, 18, 182.
36. J.F. Joanny, L. Leibler and P-G. de Gennes: J. Polym. Sci. Polym. Phys. Ed. 1979, 17, 1073.
37. G.J. Fleer, J.M.H.M. Scheutjens and B. Vincent: In Polymer Adsorption and Dispersion Stability, E.D. Goddard, B. Vincent, Eds. (Am. Chem. Soc., Washington, 1984) p. 245.
38. V.A. Parsegian: Ann. Rev. Biophys. Bioeng. 1973, 2, 221.
39. P.J. Flory: Principles of Polymer Chemistry (Cornell University Press, Ithaca, 1953)
40. E. Mackor, and J. van der Waals: J. Colloid Sci. 1952, 7, 535.
41. S. Ash and G. Findenegg: Trans. Faraday Soc. 1971, 67, 2122.
42. S.F. Edwards: Proc. Phys. Soc. (London) 1965, 85, 613.
43. A. Silberberg: J. Chem. Phys. 1967, 46, 1105; 1968, 48, 2835.
44. R.J. Roe: J. Chem. Phys. 1974, 60, 4192.
45. P-G. de Gennes: Scaling Concepts in Polymer Physics (Cornell University Press, Ithaca, 1979).
46. J.M.H.M. Scheutjens and G.F. Fleer: Adv. Colloid Interface Sci. 1982, 16, 361; 1983, 18, 309.
47. J.W. Cahn and J.E. Hilliard: J. Chem. Phys. 1958, 28, 258.
48. G. Widom: Physica 1979, 95A, 1.
49. M. Eisenberg, T. Gresalfi, T. Riccio and S. McLaughlin: Biochem. 1979, 18, 3213.
50. R.I. Feigin and D.H. Napper: J. Colloid Interface Sci. 1980, 75, 525.
51. P.F. Luckham and J. Klein: J. Macromolecules 1985, 18, 721.
52. Double-layer repulsion is modelled by the full-nonlinear Poisson-Boltzmann equations and Guoy-Chapman theory.[1] For these small charge contents, linearized equations are completely acceptable. Also, Na^+ binding to the (PS + PC) bilayers is well established and is characterized by an equilibrium constant of 0.6 M^{-1}.[29]
53. The number-average molecular weights for the larger 36500 and 147500 fractions could not be accurately measured by osmotic pressure methods; hence the values given in Table 1 for the first virial coefficient are those measured by Dr. Granath and were used to provide the intercepts at zero concentration in Fig. 5.
54. Note: all chemical potentials and free energies are assumed to be in units of thermal energy; i.e. normalized by kT.
55. Note: The data required for these predictions are $\nu_s = 29.9 \times 10^{-24}$ cm^3; $\nu_m = 182.2 \times 10^{-24}$ cm^3; $a_m = \nu_m^{1/3} = 5.67 \times 10^{-8}$cm; $N_p = M_n/180$; and the interaction parameter for each polymer fraction.
56. E. Evans and D. Needham: Macromolecules (submitted).

Niosomes: A Case in the Design of Lipid Vesicles

G. Vanlerberghe

Laboratoires de Recherche – L'OREAL – 1 Av. E. Schueller,
F-93601 Aulnay-sous-Bois, France

INTRODUCTION

Experimental evidence as well as theoretical considerations concurred for a long time to support the view that single-chained lipids form micelles whereas those containing at least two hydrocarbon chains self-assemble into vesicles.
Some twelve years ago [1] we found that mono-alkyl non-ionic amphiphiles can be dispersed in water and form closed lamellar bodies for which we coined the term "Niosomes". Although this behaviour is not a prerogative of a single group of compounds, the fact still remains that the polyglycerol ethers that will be presented here offer a unique combination of physical and chemical properties. The flexibility of their design is a valuable asset in order to scan the influence of molecular structure on interfacial and colloidal phenomena.

I - PREPARATION AND PHYSICO-CHEMICAL PROPERTIES OF POLYGLYCEROL ETHERS

1. Synthesis

Owing to their trifunctional character, glycerol molecules can be linked one to another through primary, secondary or primary-secondary ether bonds. The compounds we have studied more extensively are represented by the general formula :

$$R\,[OCH_2\underset{\displaystyle CH_2OH}{\underset{|}{CH}}]_n\,OH \qquad (I)$$

wherein R designates an alkyl group having 12 to 18 carbon atoms.
They have been prepared according to a general procedure that has been described previously [2]. Homologues with a specified degree of polymerization have been separated by conventional methods.

Some linear polyglycerol derivatives have also been synthetized in order to compare their properties with those of the branched isomers represented by formula I.

2. Monolayer properties

Insoluble polyglycerol ethers of formula (I) can be spread in monolayers at the surface of water. These experiments were possible when $n \leq 4$ provided that the hydrocarbon chain contains at least 14 carbon atoms. The salient feature of the compression isotherms is that all of them reveal an expanded state of the monolayers, even with C_{14}-C_{16} derivatives, as soon as $n = 2$.

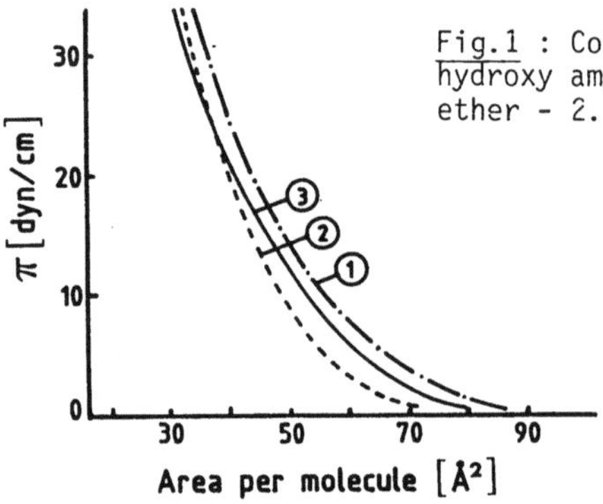

Fig.1 : Compression isotherms on water of three polyhydroxy amphiphiles at 20°C. 1. Diglycerol hexadecyl ether - 2. Hexadecyl glucoside - 3. Oleylglucoside.

On this condition, no transition by compression is observable.
Branching of the polyglycerol groups causes no significant expansion of the monolayers. As it can be seen on Fig.1, the hexadecyl ether of diglycerol occupies a larger area at the water/air interface that its glycosidic counterpart [3].
The bidimensional compressibility module of both is comparable.

In spread monolayers, dodecyl-polyglycerol ethers (n = 2) were found to be significantly less compressible that $C_{12}E_4$ (E representing an ethylene oxide unit).

3. Three-dimensional states

While the correspondence between monolayers and bilayers is rather uncertain, the three-dimensional state of amphiphiles is closely related to their capability of forming vesicles.
Whereas the monolayers of the compounds covered in the foregoing are liquid, a solid three-dimensional state is observed at room temperature when the chain-length of their alkyl group is elongated.
However, the liquefying influence of the polyglycerol group is quite evident and still more pronounced by branching. In this context, it is noteworthy that this influence is only slightly enhanced in the presence of water. This behaviour is strongly contrasting with that of glucosides [3].
Both types of amphiphiles undergo a transition from the solid to the lamellar state. X -rays diffraction patterns have been obtained by P.LEMAIRE [4]. They show that the molecular area is roughly constant when water is added to polyglycerol ethers in the neat (lamellar) phase. The latter is the only one appearing in presence of increasing water amounts, as long as n < 3.
A distinctive characteristic of the moieties $- OCH_2\underset{|}{C}H -$ (with CH_2OH branch) and $- OCH_2CHOHCH_2 -$ is their strong hydrophilicity which is further substantiated by the high cloud points of the amphiphilic compounds comprising them. This property, which is obtained with no excessive occupation of interfaces, is one of their main assets for the formation of vesicles.

II - PREPARATION AND PROPERTIES OF NIOSOMES

Owing to their physical properties, polyglycerol ethers can be hydrated in the bulk at moderate temperatures and then dispersed in water. Under these conditions multi-

lamellar bodies are formed. Size reduction is obtained by conventional methods.
As we discussed it in a previous communication [5], the length and configuration of the hydrocarbon chain are strong determinants of stability.
Specifically, Niosomes based on dodecyl compounds are stable but leaky. Those formed from hexadecyl derivatives are the least permeable to glucose used as a permeability probe. However they are metastable at room temperature, since the gel - liquid crystal transition takes place at about 40°C. The bilayer fluidity of Niosomes has been investigated by differential polarized phase fluorometry [6]. Whereas polyglycerol and polyethyleneglycol ethers having comparable amphiphilic properties display the same chain order in the gel stage, the liquid-crystalline of the latter is less ordered. The lowering of the phase transition temperature can be obtained by branching or dissymmetry of the hydrocarbon group. Cis double bonds in oleyl derivatives have the same effect. However, in this case, reversal to the planar lamellar phase has been observed.
The addition of a second liquid such as cholesterol to the non-ionic amphiphile is one of the most powerful means to obtain the liquid crystalline state at room temperature and to reduce the permeability of Niosomes.
In this context, it is noteworthy that the orientational effect of cholesterol is less pronounced with a polyethoxylated amphiphile than with its polyglycerol counterpart.
Finally, it is worthwhile mentioning that Niosomes prepared from polyglycerol derivatives are unexpectedly stable in the presence of detergents and oils. It has been found that they can stabilize oil in water emulsions.

CONCLUSION

The existence of Niosomes can now be rationalized in terms of the critical packing factors calculated by ISRAELACHVILI et al. [7].
The work presented here shows that the polar head contribution plays an evident role mainly on the dynamic properties of vesicles. This observation is most important for practical purposes and, all the same, it confirms the possibility of mimicking biological membranes. A better quantitative understanding of these phenomena is still one of the exciting challenges propounded to physicists.

Acknowledgements
I am grateful to the Management of L'OREAL for authorizing the presentation of this work and deeply indebted to the co-workers mentioned in the references.

REFERENCES

1. L'OREAL FP. N° 2.315.991 filed 30.06.75
2. G.VANLERBERGHE, R.M.HANDJANI-VILA, C.BERTHELOT, H.SEBAG
   Ber.vom VI Int.Kongr. GrenzPI. Stoffe (1972) Bd I p.140-155
3. G.VANLERBERGHE, R.M.HANDJANI, B.RONDOT
   Rev.Fr.C.Gras 27e A. N°5 p.237-242 (1980)
4. P.LEMAIRE Thesis Univ. Paris VI
5. G.VANLERBERGHE, R.M.HANDJANI-VILA, A.RIBIER
   Colloq. Nat. CNRS N°938 Physicochimie des Amphiphiles p.303-311 (1978)
6. A.RIBIER, R.M.HANDJANI-VILA, E.BARDEZ, B.VALEUR
   Coll.and Surf. 10 (1984) p.155-161
7. J.N.ISRAELACHVILI, S.MARCELJA, R.G.HORN
   Quat. Rev. Biphysics 13, 2 (1980) p.121-200

Direct Visualization of Amphiphilic Phases by Video Enhanced Microscopy and Cryo-Transmission Electron Microscopy

D.D. Miller[1], *J.R. Bellare*[1], *D.F. Evans*[1], *Y. Talmon*[2], *and B.W. Ninham*[3]

[1]Dept. of Chemical Engineering and Materials Science, University of Minnesota, Minneapolis, MN 55455, USA
[2]Department of Chemical Engineering, Technion – Israel Institute of Technology, Haifa 32000, Israel
[3]Dept. of Applied Mathematics, Research School of Physical Sciences, Institute of Advanced Studies, Australian National University, Canberra A.C.T., 2601 Australia

1. INTRODUCTION

Establishing the phase diagram of a surfactant-water system is a relatively simple matter; visual observations between crossed polarizers [1] are combined with data obtained from polarizing microscopy [2] and other techniques to generate diagrams such as shown in Fig. 1. The determination of the structure within these phases is more complicated. Commonly used techniques such as NMR, and X-ray, light and neutron scattering go a long way towards elucidating average structural tendencies in surfactant phases. However, since these non-invasive, though indirect, techniques rely on mathematical models for interpretation — models with a priori assumptions of simple geometric form (spheres, cylinders, planes, etc.) for surfactant aggregates — they cannot provide detailed information on systems with complicated topologies. For example, paired disclinations in lamellar phases [3] and self-similar structures, which scale over many orders of magnitude, of a kind to be illustrated below, are difficult, if not impossible to study by scattering techniques.

In this paper we describe two new experimental techniques, video-enhanced microscopy and cryo-transmission electron microscopy, which allow direct, model-independent visualization of microstructure in amphiphilic phases. These techniques reveal astonishing and counter-intuitive aggregation patterns in ionic double-chained surfactants and demonstrate heretofore unpostulated subtlety in the interplay of counterion, amphiphile concentration and temperature in surfactant aggregation. For the first time, some of the complexity of self-assembly in biological systems is shown to be mimicked by simple, non-living analogues.

2. EXPERIMENTAL

The major obstacles to direct visualization of surfactant microstructures by optical microscopy methods are 1) contrast limitations place a lower bound on the size of aggregate that can be selected for study while 2) resolution limitations restrict the amount of structural information that can be extracted from microscopic observations.

Many surfactant aggregates, vesicles and microtubules in particular, are of inherently low contrast and cannot be distinguished from the background solution with ordinary light microscopy techniques. With video-enhanced microscopy, contrast enhancement is obtained in three ways. 1) Optically, image contrast is boosted by using rectified differential interference contrast (DIC). This technique gives high contrast light or dark bands at regions in the sample where there are sharp refractive index gradients (such as at surfactant aggregate-water interfaces). 2) Electronically, image contrast is increased through the use of a video camera instead of the human eye as the detection device [4]. Unlike the eye, which is a quasi-logarithmic device and saturates at high light levels, a television camera is a linear device; it responds equally well to small differences in intensity no matter what the background light level is. Thus, a television camera linked to a differential interference contrast microscope (VEM) improves contrast by responding linearly to contrast at all light levels. 3) Digitally, image contrast is increased by using a real-time digital image processor to perform background subtraction [4] and gray-scale transformations [5]. As a result of these manipulations the background mottled pattern created by inaccessible dirt and lens imperfections is subtracted frame by frame in real time from the video image, and the narrow region in gray-scale space occupied by the image (e.g. from gray level 120 to 140) is expanded to cover the full gray scale range of the digitizing equipment (i.e., 0 to 255).

As a result of this contrast enhancement, small, isolated colloidal particles with diameters as small as 0.05 μm (such as unilamellar vesicles or polystyrene latex spheres) can be clearly and dynamically visualized. This is an important result, since it means that VEM can be used as a detection device in the study of spontaneous, unilamellar vesicle formation in double-chained surfactants [6].

While there is no theoretical limit to the size of an isolated particle that can be detected by video-enhanced microscopy (given sufficient constrast to distinguish the particle from the background), VEM, like all microscopy techniques, is limited in resolution by the wave nature of light (the resolution limit of VEM is ~ 0.1 to 0.25 μm). Higher resolution, though static, images of surfactant aggregates are obtained with a second new technique, cryo-transmission electron microscopy (cryo-TEM).

If a thin sample of a surfactant dispersion is plunged into liquid ethane at its melting point, the cooling rate is fast enough to solidify the water in the specimen without crystallization, forming a vitreous ice (I_v) matrix. As a result the original microstructure of the system is preserved, and, when the specimen is observed with a transmission electron microscope, the contrast between surfactant aggregates and the I_v matrix is vastly improved over specimens frozen in hexagonal (I_h) ice [7]. As will be seen below, surfactant bilayers and other structures as small as 1 nm are clearly resolved.

3. RESULTS

We now present some video-enhanced and cryo-transmission electron micrographs of microstructures formed by double-chained surfactants in water. A typical phase diagram for such a system is shown in Fig. 1. The photographs clearly demonstrate that aggregation in the "water + liquid crystal" region and the isotropic, "micellar solution" phase is more complex than once thought.

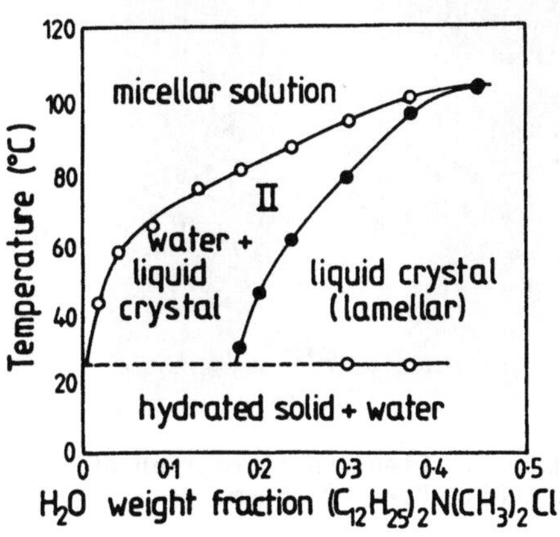

Figure 1. Phase diagram of didodecyldimenthyl-ammonium chloride (from ref. [8])

3.1 Biphasic "Water + Liquid Crystal" Region

In Fig. 2 we see a VEM picture of a 6-month old sample of sodium-8-phenyl-n-hexadecyl-ρ-sulfonate (SHBS). At 1.7 wt%, this sample is located well within the biphasic "water + liquid crystal" region of the phase diagram. While the frequently observed birefringent, liquid crystalline liposomes that are thought to characterize this phase region are certainly observed here (structure A), many other, non-birefringent structures with a wide variety of topologies are present as well (structures B, C and D). In structure B we see a large vesicle of ~ 15 μm diameter enclosing many smaller vesicles. The caged motion of these smaller vesicles can be followed in real time on video tape. Structure C is a vesicle enclosing a tightly coiled microtubule. Cryo-TEM pictures of the same 1.7 wt% SHBS sample reveal similar structures, though on a smaller size scale. For instance the single-walled vesicle in Fig. 3 (structure A) is of diameter 0.5 μm but in every way similar in structure to the larger structure (B) of Fig. 2. Similarly, the small vesicle encasing a coiled tubule (structure B, Fig. 3) is topologically similar to the much larger structure in Fig. 2 (structure C). It is difficult to determine whether these large, complex, non-birefringent aggregates belong to the isotropic or liquid crystalline phase. A wide variety of other double-chained surfactants have been observed by the authors with VEM and cryo-TEM, and, in all cases, the biphasic region contains similar menageries of microstructures.

3.2 Aggregation in Isotropic Phases — Vesicles and Micelles

At room temperature, most surfactants with two long alkyl chains form clear isotropic phases which are very dilute in surfactant (see Fig. 1). Consequently, it is difficult, if not impossible, to determine aggregate structure by the usual scattering or fluorescence techniques. Suitable strategies can be devised, however, to increase the surfactant content of the isotropic phase. These strategies involve decreasing the degree of counterion binding to the surfactant; high binding, such as seen at room temperatures in those surfactants most studied (e.g. didodecyldimethylammonium halides and SHBS), favors bilayer and liquid crystalline phases whereas low counterion binding favors smaller aggregates (such as micelles and unilamellar vesicles) and isotropic phases.

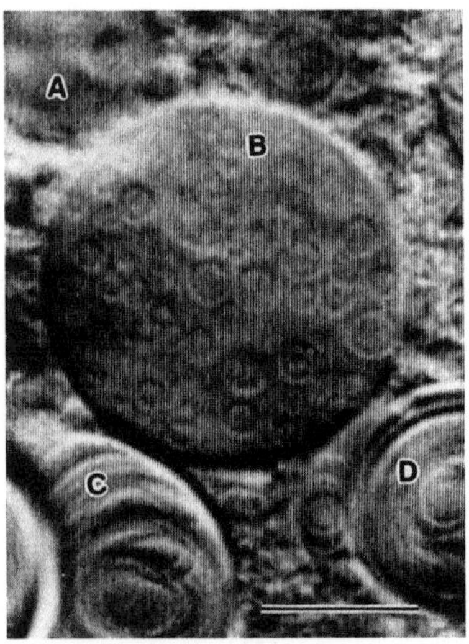

Figure 2. VEM micrograph of 1.7 wt% SHBS. Bar = 10 μm

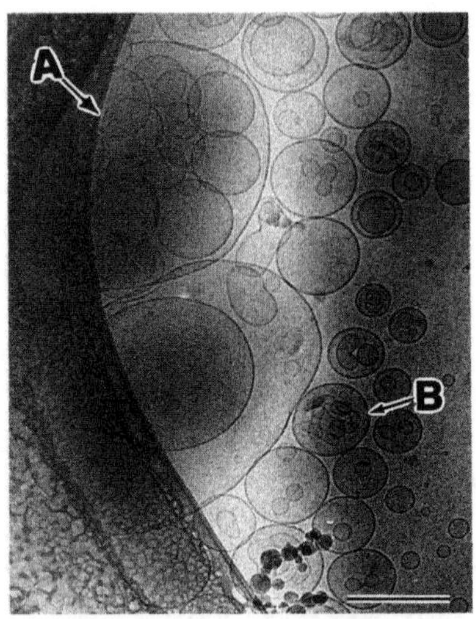

Figure 3. Electron micrograph of 1.7 wt% SHBS. Bar = 0.25 μm

Low counterion binding can be achieved by employing counterions with large hydrated radii. For example, the isotropic phase of aqueous didodecyldimethylammonium bromide extends to only 10^{-4} M whereas the highly hydrated acetate counterion gives solutions that remain clear up to and beyond 0.5 M [9]. Similarly, the solubility of SHBS in water is remarkably enhanced by the addition of macrocyclic polyethers and cryptates that complex the sodium counterion [10] and thereby increase its effective radius. From Fig. 1 we see that the isotropic phase can also be extended into an experimentally accessible range by increasing the temperature. Apparently, the mechanism that is operating here is the desorption of counterions with increasing temperature [11].

Video-enhanced microscopy and cryo-transmission electron microscopy of isotropic phases of didodecyldimethylammonium acetate, SHBS plus the cryptate C222, and didecyldimethylammonium bromide at 70°C reveal an unexpected aggregation sequence; in dilute solution (i.e. 10^{-3} M) these surfactant systems form unilamellar vesicles with diameters between 0.02 μm and 0.1 μm [9]. With increasing concentration, the vesicles gradually shrink in size and eventually disappear at 0.1 M, presumably collapsing into small micelles [9].

The transformation from an isotropic solution of vesicles and micelles to a biphasic dispersion of liposomes and other complex structures was followed in real time with VEM. 0.01 M solutions of didodecyldimethylammonium acetate and hydrobromic acid were pumped through a vortex mixer and then rapidly injected into the microscope flow cell described by MILLER et al. [9]. After stopping the flow, the dynamics of the transformation from the acetate to the bromide surfactant were recorded on video tape. After approximately one minute, the vesicles and micelles had reassembled into quite thin microtubules (see Fig. 4). These tubules thicken and elongate into "worm-like" aggregates, examples of which are shown 10 minutes into the reaction in Fig. 5. After six hours the reaction is complete, and the aggregates resemble those shown in Fig. 2. The dynamic sequence vesicles and micelles → "worms" → liposomes has been observed in other systems, notably in the contacting of didodecyldimethylammonium hydroxide with HBr [9] and in the rapid cooling of an isotropic solution of SHBS at 90°C to room temperature [12].

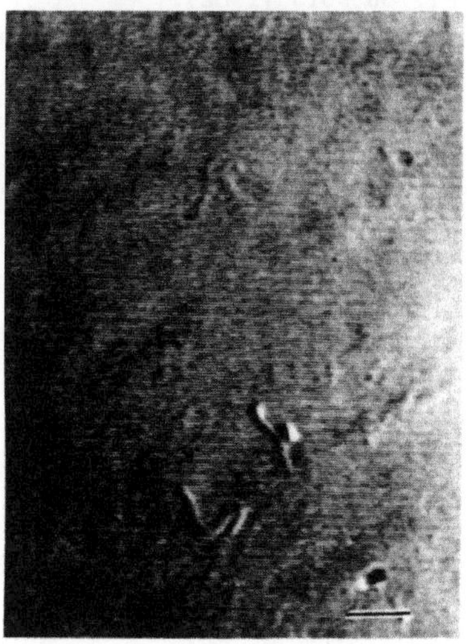

Figure 4. Structures formed upon neutralization of didodecyldimethylammonium acetate by HBr one minute into the reaction. Bar = 10 μm

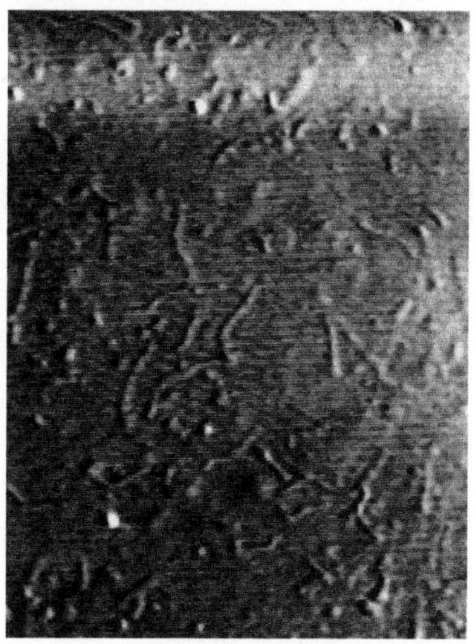

Figure 5. Structures formed upon neutralization of didodecyldimethylammonium acetate by HBr ten minutes into the reaction. Magnification same as Fig. 4

4. CONCLUSIONS

The use of VEM and cryo-TEM reveals the true complexity of aggregation in double-chained surfactants. The topologically complex structures pictured in Figures 2 and 3 are unobserved and unobservable with other, indirect techniques. These direct techniques also allow us to observe transient intermediates in the transformation of one type of aggregate into another. The existence of such complexity in simple two-component surfactant plus water systems gives us hope that at least some of the bewildering variety of pictures presented by cell biology can be interpreted in the language of association colloid science.

5. ACKNOWLEDGEMENTS

D.D.M. acknowledges funding from the National Science Foundation. J.R.B. thanks his advisors, Profs. L.E. Scriven and H.T. Davis. This work was funded by U.S. Army Grant DAAG29-85-K-0169, NIH Grant IROI-Gm 34341-01, NSF equipment Grant NSF/CPE-8312539, and NSF Grant NSF/CPE-8215342.

6. REFERENCES

1. R.G. Laughlin: J. Colloid Interface Sci. 55, 239 (1976)
2. F.B. Rosevear: J. Amer. Oil Chem. Soc. 31, 628 (1954)
3. M.B. Schneider, W.W. Webb: J. Physique 45, 273 (1984)
4. S. Inoué: Video Microscopy (Plenum Press, New York 1986)
5. W. Niblack: An Introduction to Digital Image Processing (Prentice-Hall International (UK) Ltd., London 1986)
6. S. Hashimoto, et al.: J. Colloid Interface Sci. 95, 594 (1983)
7. J.R. Bellare, et al.: J. Colloid Interface Sci. (in preparation)
8. H. Kunieda, K. Shinoda: J. Phys. Chem. 82, 1710 (1978)
9. D.D. Miller, et al.: J. Phys. Chem. 91, 674 (1987)
10. D.D. Miller, et al.: J. Colloid Interface Sci. 116, 598 (1987)
11. D.F. Evans, P.J. Wightman: J. Colloid Interface Sci. 86, 515 (1982).
12. G. Javadi, private communication

Spontaneous Formation of Vesicular Structures from a Swollen Lamellar Phase by Dilution and Control of Surface Charge Density

*W.J. Benton**

Dept. of Chemical Engineering, Case-Western Reserve University, Cleveland, OH 44106, USA

Over the past several years there has been increased interest in the phase behavior and morphology of the dilute region of amphiphilic systems. A general sequence of phases has been shown to exist when a single variable is changed, such as alcohol, salinity or temperature of an amphiphilic surfactant solution. The surfactant can be anionic, cationic, zwitterionic or nonionic. The sequence of phase transformations represents a spontaneous change in the curvature and self-association of the amphiphile whereby an isotopic micellar solution transforms into a highly swollen homogeneous lamellar phase made up of sheets of bilayers, to an optically isotropic phase which scatters light and exhibits birefringence when sheared (1). These phases are designated L_1, L_α and L_3 respectively.

The research area that encompasses vesicles, their topology, formation mechanisms and polymerization overlaps the area of phase behavior and morphology of dilute amphiphilic systems. But in general the relationship between these two areas of research has not been correlated. It is clear, however, that vesicles form, whether spontaneously or by extensive mechanical energy techniques, in the $L_1 + L_\alpha$ two-phase region, in close proximity to the L_1 phase boundary. Some recent studies indicate that vesicles form spontaneously on crossing the L_1 into the $L_1 + L_\alpha$ two-phase region (2) or by varying the pH (3). For the most part vesicles are formed by inducing high levels of mechanical energy, such as with the French press or ultrasonication techniques.

The purpose of this lecture is to demonstrate that the broad two-phase region of $L_1 + L_\alpha$ that exists between the L_1 and L_α phases is in certain regions of phase space kinetically stable and macroscopically homogeneous. These kinetically stable solutions can exist for extended periods of time, several years in some cases. The topology and size of the aggregate morphology varies from system to system but is dependant upon such variables as order of mixing and thermal history. With a judicious choice of the formulation variables both the aggregate size and in certain cases the morphology can be controlled. This poses several interesting questions about the physics involved as well as having pronounced implications on the application of these systems. Some recent theoretical considerations indicate that single bilayered vesicles may be thermodynamically stable when they reach their own preferred radius of curvature (4).

Typically a solution is formulated by addition of the components and a series of heating, mixing and cooling cycles followed by equilibration of the solution at constant temperature. For the case of interest here, i.e. the $L_1 + L_\alpha$ two phase region, the resultant solutions exhibit a macroscopically homogeneous, but kinetically stable dispersion of lamellar aggregates. When viewed by polarized light microscopy the lamellar aggregates show the well-known Maltese cross. Some of these aggregates have been shown to have the two-dimensional topology of sheets of bilayers spontaneously curved into the limiting case of the focal conics, the Dupin cyclide, with a singular line and a circle or loop defects.

---

* also at The Standard Oil Company, Research Center, Cleveland, Ohio, USA

Other less well known or understood topologies also occur and in several instances also exhibit birefringent textures similar to the maltese cross. Solutions of $L_1 + L_\alpha$ formed in this manner can remain kinetically stable for extended periods of time, several years in some cases.

However in certain instances the $L_1 + L_\alpha$ dispersions flocculate out of solution. Observations with enhanced video microscopy (EVM) indicate a change in topology takes place from the original Dupin cyclide geometry to that of torii, which are clustered together. A simple explanation of this transformation is that the outer bilayers of the Dupin cyclide have been stripped off retaining the bilayers surrounding the circle defect of the Dupin cyclide - i.e. torii. The stripping of bilayers destabilizes the aggregates and the initial kinetic stability is lost,with the result that the aggregates flocculate.

Another approach is to change the order of mixing process to a two-step process, eliminating the heating and cooling cycles of the final solution state. This was accomplished by first formulating a system as a homogeneous lamellar phase as described above. A specific example consisted of a 9.0 vol.% of a 63:37 ratio of the monoethanolomine salt of dodecyl orthoxylene sulfonate and tert-amyl alcohol in 4.8% brine. The solution exhibited the well-known lamellar oily streak and parabolic focal conic textures by polarized light microscopy. Two experiments were then conducted to study the topology of the aggregates, a) serial dilution of the lamellar phase with H_2O to a final composition of 2.25 vol.% of the amphiphile/alcohol mixture and 1.2% brine and b) contacting the lamellar phase with H_2O and following the interfacial phenomena by EVM.

The lamellar phase diluted with H_2O was mixed gently by hand. The solution quickly lost all birefringence but retained some translucency when observed between crossed polars. With continued mixing the solution became less translucent. When observed with modulation contrast optics on the EVM the solution showed small aggregates with an average size of about one micrometer or less. Each successive mixing step produced on average smaller aggregates. Even after several months storage no change in the average size of aggregates or coalescence was observed. A few seconds of ultrasonication produced transparent solutions.

The second approach involved the contacting of water or brine with the lamellar phase and recording the subsequent interfacial phenomena by EVM. Within a few minutes of contact closed shell structures grew out from the lamellar phase and ejected after necking off into the water region. The process of spontaneous formation and growth of vesicles can be seen in Figure 1, where the lamellar phase is on the left-hand side and pure water on the right. Along the interface from top to bottom the transformation process to vesicular forms is in progress. Several vesicles can be seen in the process of necking-off and ejecting into the water region. These observations show the unequivocal spontaneous formation of vesicles under conditions where convection and mixing are minimized. The average size of the vesicles can be large, some with diameters of 50 micrometers. From the thickness of the interface it would appear that the vesicles are made up of perhaps fifteen to twenty bilayers. The fluctuations of the vesicle walls are easily observed and have an average motion from the local median of about two micrometers.

The dilution process described here has two effects on solution behavior. First by simply diluting the solution so that the phase boundary is traversed from a homogenous lamellar phase into the $L_1 + L_\alpha$ two-phase region. Second that the dilution of total ionic strength of the solution decreases the head group repulsion and changes the surface charge density in the polar head region of the bilayers. This produces a change in the osmotic pressure difference at the interior of the vesicle and the continuous phase with the overall result of an increased stability of the vesicles. Further details will appear elsewhere.

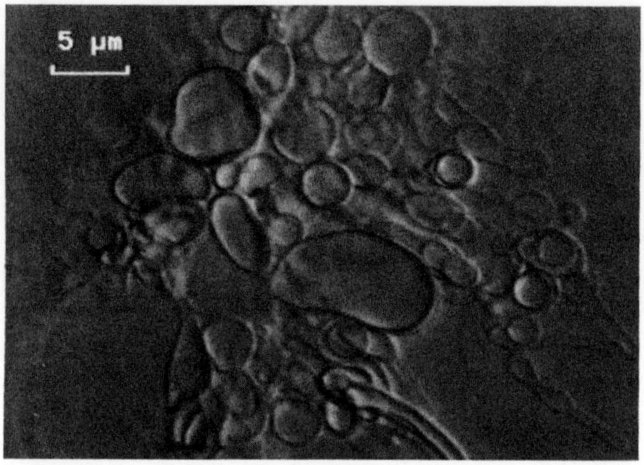

Figure 1. A digital enhancement of a video frame showing vesicle walls formed spontaneously after contacting a lamellar phase with H_2O. After 30 minutes of contact. Modulation contrast optics. See text for explanation.

References:

[1] C.A. Miller, O. Ghosh and W.J. Benton, Colloids and Surfaces, 19, 197-223 (1986)
[2] Y. Talmon, D.F. Evans and B.W. Ninham, Science, 221, 1047-1048 (1983)
[3] N. Gains and H. Hauser, J. Membrane Sci., 22, 225-234 (1985)
[4] Y. Suezaki and N. Tsuji, J. Colloid and Interface Sci., 114, 131-139 (1986)

Part V

Micelles

Scattering of Concentrated Dispersions of Colloidal Particles: Microemulsion Droplets, Polydisperse Hard Spheres and Charged Hard Spheres

A. Vrij

Van't Hoff Laboratory, University of Utrecht,
Padualaan 8, NL-3584 CH Utrecht, The Netherlands

One of the ways in which microemulsion systems manifest themselves is in the form of colloidal spheres [1] (Schulman-model). In the microemulsion literature one speaks of the Droplet Model of a microemulsion. Small water droplets, stabilized by a protecting surface layer of long-chain surfactant molecules, are dispersed in "oil".

Some ten years ago we introduced this microemulsion system in light scattering studies of interparticle forces [2]: a "hard sphere repulsion" supplemented with an "attractive interaction" [3], treated as a perturbation [4] according to PONCE and RENON [5]. The attractive forces cannot be explained by the macroscopic London-van der Waals forces between the water cores. They must be sought in local van der Waals forces and local entropy terms on the level of "segments" of long-chain and solvent molecules [6,7] at the surface layer of the particles. Their quantitative description must be described in terms of local "equation of state" or "liquid structure" properties.

A complication - always found in practice - is the polydispersity effect. It is caused by the inherent distribution in the size of the particles. It is not appropriate to allow for the influence of the distribution of the scattering power of the particles and to neglect the influence on the interparticle interactions [8] . We proved this by a theoretical analysis based on hard sphere interactions [9,10]. This was corroborated by Monte Carlo simulations on hard sphere mixtures [10].

The results were applied to small-angle X-ray and neutron studies of stearyl-coated silica model-particles in cyclohexane which behave as hard spheres [11].

Finally, recent results were shown of surface-coated, charged silica spheres in polar, non-aqueous solvents in which very large repulsive interactions could be measured [12].

1. J.H. Schulman and J.A.Friend, Kolloid Z. 115, 67 (1949)
2. A. Vrij, E.A. Nieuwenhuis, H.M. Fijnaut and W.G.M. Agterof, Faraday Discussions of the Chem. Soc. 65, 101 (1978)
3. W.G.M. Agterof, J.A.J. van Zomeren and A. Vrij, Chem. Phys. Lett., 43, 363 (1976)
4. A.A. Caljé, W.G.M. Agterof and A. Vrij, In Micellization, solubilization and microemulsions, ed. by K.L. Mittal, Vol. 2 (Plenum Press, New York, 1979) p.779
5. L. Ponce and H. Renon, J. Chem. Phys., 64, 638 (1976)
6. P.J. Flory, R.A. Orwoll and A. Vrij, J. Am. Chem. Soc., 86, 3515 (1964)
7. P.J. Flory, Discussions of the Faraday Soc., 49, 7 (1970)
8. M. Kortlarchyk and S-H. Chen, J. Chem. Phys., 79, 2461 (1983)
9. P. van Beurten and A. Vrij, J. Chem. Phys., 74, 2744 (1981)
10. D. Frenkel, R.J. Vos, C.G. de Kruif and A. Vrij, J. Chem. Phys., 84, 4625 (1986)
11. J. Moonen, C.G. de Kruif and A. Vrij, to be published.
12. A. Philipse and A. Vrij, to be published.

Self-diffusion of Globules

R. Klein, U. Genz, and J.K.G. Dhont

Fakultät für Physik, Universität Konstanz,
D-7750 Konstanz, Fed. Rep. of Germany

1. INTRODUCTION

A major tool to characterize microemulsions and other supramolecular fluids is the application of scattering techniques. Static (angle-dependent) scattering provides a form factor and the static structure factor S(k), whereas dynamic experiments like photon correlation spectroscopy and neutron spin echo experiments lead to the dynamic structure factor S(k,t) (also called the intermediate scattering function). S(k,t) contains interesting information, which reflects the time development on various time scales of the system. In most cases only the short-time behavior is investigated which determines essentially the collective diffusion coefficient. The self-diffusion coefficient, on the other hand, can be determined by forced Rayleigh scattering or by fluorescence recovery after fringe-pattern photobleaching.

These scattering experiments are very useful tests of models of the internal structure of certain microemulsion phases. A quantitative interpretation is, however, only possible for some rather simple models such as the globular phase. For more complicated cases a thorough theoretical description for the above mentioned static and dynamic quantities is essentially lacking.

In this contribution we will therefore concentrate on the consequences of a droplet model for water-in-oil microemulsion /1/. This model consists of a hard core of radius b which contains the water and the head groups of the surfactant molecules and a spherical shell of thickness h , which represents the surfactant "tails", which are immersed in the continuous oil part of the system. Therefore, the droplet is an object of radius a = b + h. Two droplets are allowed to approach each other such that their respective surfactant tails overlap, giving rise to an attractive interaction /1/. Therefore, the effective interaction of the model consists of

$$V(r) = \begin{cases} \infty & ; \quad r < 2b, \\ -V_0 & ; \quad 2b < r < 2a, \\ 0 & ; \quad r > 2a . \end{cases} \tag{1}$$

Since this interaction is short range, hydrodynamic interactions are expected to be of considerable importance for dynamical properties. But it should be realized that this expected importance of the hydrodynamic interactions leads to severe limitations for the applicability of theoretical results, since the precise form of these interactions are only known at relatively small concentrations. Many-particle hydrodynamic interactions have to be taken into account for more concentrated systems /2/. These are sometimes approximated in a semi-empirical way by introducing "effective" two-particle hydrodynamics /3/. Here, we restrict ourselves to lowest order in the volume fraction of droplets in order to avoid the complications from many-body effects.

The model for the droplet interactions has been used to calculate the static structure factor /4/ and the collective diffusion coefficient. To lowest order in volume fraction ϕ the static structure factor at very small values of the scattering wave vector k can be written as $S(k\approx 0;\phi) = 1 - k_s\phi$, where the virial coefficient k_s is a function of the two independent parameters of the model interaction /5/, $k_s(a/b,V_o)$. Using hard-sphere hydrodynamics in the Felderhof approximation /6/, the collective diffusion coefficient has also been calculated /5,7/ in a low-order virial form, $D_c=D_o(1 + k_D\phi)$ with $k_D=k_D(a/b,V_o)$. Although in principle, two independent experiments should be sufficient to determine the two parameters of the model, one should aim to calculate as many physical properties as possible in terms of a model and to compare with as many experiments as possible to really establish the validity of the model. It is in this spirit that the self-diffusion coefficient D_s will be calculated here.

2. THE DYNAMICAL EVOLUTION EQUATION

The basis for the dynamic description of the interacting droplet system will be taken to be the Smoluchowski equation

$$\frac{\partial}{\partial t} P_N(\{R^N\},\{R_o^N\},t) = \hat{\Omega}_N P_N(\{R^N\},\{R_o^N\},t) \tag{2}$$

for the conditional probability density of finding the N droplets at time t at the positions $\{R^N\} \equiv \{\underline{r}_1,\ldots,\underline{r}_N\}$, given that they had been at $\{R_o^N\} \equiv \{\underline{r}_1^o,\ldots,\underline{r}_N^o\}$ at time zero. The Smoluchowski operator is

$$\hat{\Omega}_N = \sum_{i,j=1}^{N} \underline{\nabla}_i \cdot \underline{\underline{D}}_{ij}^N \cdot \left(\underline{\nabla}_j + (\underline{\nabla}_j \beta U^N)\right) \ , \quad \beta = (k_B T)^{-1} \ , \tag{3}$$

where U^N is the N-particle potential interaction and $\underline{\underline{D}}_{ij}^N$ are diffusion tensors, connecting the velocity of particle i with a force $\underline{F}_j$ on particle j :

$$\underline{v}_i = \beta \sum_j \underline{\underline{D}}_{ij}^N (\{R^N\}) \cdot \underline{F}_j \ . \tag{4}$$

Here, it is indicated explicitly that the hydrodynamic tensors in general depend on the positions of all droplets.

In general, eq. (2) cannot be solved. Since one is not interested in P_N directly, but rather in certain correlation functions, projection operator techniques could be used. Since we want to restrict ourselves here to first order effects in ϕ, however, it is sufficient to solve the two-particle Smoluchowski equation. Introducing the relative coordinates $\underline{r} = \underline{r}_1 - \underline{r}_2$ of two particles, denoted by 1 and 2, eq. (2) can be reduced to

$$\frac{\partial}{\partial t} \rho(\underline{r},\underline{r}_o,t) = \hat{\Omega}_{12}\, \rho(\underline{r},\underline{r}_o,t) \ , \tag{5}$$

where ρ is the conditional probability for the relative vector to be $\underline{r}$ at time t under the condition that it was $\underline{r}_o$ at time zero. Furthermore,

$$\hat{\Omega}_{12} = \underline{\nabla} \cdot \underline{\underline{D}} \cdot \left(\underline{\nabla} + (\underline{\nabla}\, \beta V)\right) \ , \tag{6}$$

where V is the two-droplet potential (1) and $\underline{\underline{D}} = 2(\underline{\underline{D}}_{11} - \underline{\underline{D}}_{12})$ denotes the relative tensor, which can be expressed as

$$\underline{\underline{D}} = A(r)\, \underline{\underline{P}} + B(r)\, (1 - \underline{\underline{P}}) \ . \tag{7}$$

$\underline{\underline{P}} = \hat{\underline{r}} : \hat{\underline{r}}$ is the projector on $\underline{r}$, so that the first term describes motion of the two particles parallel to the line connecting their centers and the second one describes perpendicular motion. The functions A(r) and B(r) are known for hard spheres /8/. They can be obtained by a reflection method as a power series in (a/r) or in the form of expansion for small interparticle separations.

To illustrate the effects of hydrodynamic interactions it is quite instructive to recall the results for D_s for the case of pure hard-sphere interactions (V_o=0 in (1)). Writing $D_s = D_o(1 - \alpha\phi)$, the coefficient α is equal to 2 for the case without hydrodynamic interactions /9,10/. Using the most elementary way to include the latter is to employ the Oseen approximation; in this case /9,11/ $\alpha = 0.09$, which is a rather large effect as compared to the case of hard-sphere interactions only. If one uses for A(r) and B(r) the results obtained by Felderhof to the order $(a/r)^7$, one finds /9/ $\alpha = 1.89$. This result is much closer to that without hydrodynamic interactions, so that one may conclude that neglecting hydrodynamic interactions completely, is better than to employ the Oseen approximation, at least for short-range potential interactions. Finally, Batchelor /12/ has found $\alpha = 2.10$, taking the correct behavior of the hydrodynamic functions at small separations into account.

3. ONE-PARTICLE PROPERTIES IN CORRELATED SYSTEMS

There are various equivalent methods to calculate the self-diffusion coefficient D_s. The so-called relaxation method /13/ gives a rather direct expression for D_s. From a theoretical point of view it is, however, of interest to know more about the single-particle properties of a strongly correlated system. D_s is a transport coefficient which is a number, if the volume fraction and the potential are given. On the other hand, transport coefficients can be represented as integrals of time-dependent correlation functions, which contain much more information about the dynamical behavior of the system over the whole time regime, from the short-time behavior to long times and the cross-over between these two limits. This type of information is contained in the one-particle propagator (or self-intermediate scattering function)

$$G(k,t) = \langle e^{-i\underline{k}\cdot\underline{r}_1(0)}\, e^{i\underline{k}\cdot\underline{r}_1(t)} \rangle , \tag{8}$$

where $\underline{r}_1(t)$ is the position of the tagged particle at time t and the bracket denotes an ensemble average. If $\hat{\Omega}^+$ is the Hermitean adjoint of the operator appearing in the basic transport equation (which in our case is eq. (2), but the treatment can be extended to more general equations such as the Fokker-Planck equation), then the Laplace transform of (8) can be shown /14/ to be

$$\begin{aligned}\tilde{G}(k,z) &\equiv \int_0^\infty dt\, e^{-zt}\, G(k,t)\\ &= \langle e^{-i\underline{k}\cdot\underline{r}_1} \left[z - \hat{\Omega}^+\right]^{-1} e^{i\underline{k}\cdot\underline{r}_1} \rangle\\ &\equiv \frac{1}{z + D_s(k,z)k^2} \ .\end{aligned} \tag{9}$$

Here, $\tilde{D}_s(k,z)$ is a generalized (wavevector and frequency-dependent) self-diffusion function, whose hydrodynamic limit is the self-diffusion coefficient:

$$D_s = \lim_{\substack{k,z\to 0\\ z/k^2\ \mathrm{const}}} \tilde{D}_s(k,z) \ . \tag{10}$$

It is easy to show that other relevant one-particle properties are given in terms of the one-particle propagator. The mean-squared displacement of the tagged particle is

$$W(t) \equiv \frac{1}{6} \langle \left(\underline{r}_1(t) - \underline{r}_1(0)\right)^2 \rangle = -\frac{1}{2}\frac{\partial^2}{\partial k^2} G(k,t)\Big|_{k=0} \ , \tag{11}$$

the velocity auto-correlation function is

$$V(t) \equiv \frac{1}{3} \langle \underline{v}_1(t) \cdot \underline{v}_1(0) \rangle = -\lim_{k\to 0} \frac{\partial^2}{\partial t^2} G(k,t) \tag{12}$$

and the self-diffusion coefficient can be expressed by these quantities as

$$D_s = \lim_{t\to\infty} \frac{W(t)}{t} = \int_0^\infty dt\ V(t) \quad . \tag{13}$$

This program of calculating time-correlation functions can be developed quite generally in close analogy to what is known as molecular hydrodynamics /15/ in the theory of dynamical properties of simple liquids (in which case $\hat{\Omega}$ is the Liouville operator). One can show /14/ that $\tilde{D}_s(k,z)$ is related to a generalized self-friction function $\tilde{\zeta}_s(k,z)$ by an Einstein-like relation, generalized to finite k and z,

$$\tilde{D}_s(k,z) = \frac{k_BT/m}{z + \tilde{\zeta}_s(k,z)/m} \quad . \tag{14}$$

Using a Mori-Zwanzig projection operator technique, explicit expressions can be derived for $\tilde{\zeta}_s(k,z)$ in the form of time-correlation functions. They also contain the full hydrodynamic tensors, which, for reasons mentioned in the Introduction, complicate the problem tremendously. There are, however, cases in the field of complex fluids, where further progress along these lines can be made, namely highly charged colloidal particles. These systems are rather dilute but at the same time strongly correlated because of strong screened Coulomb interactions. It is well established that hydrodynamic interactions are unimportant in these systems as long as the ionic strength is sufficiently low. In this case, the friction function has been calculated /14/ with the result

$$\zeta_s(k,t) = \zeta_o\ \delta(t) + \frac{c\ k_BT}{(2\pi)^3}\int d^3k' \left(k_z'\ c_D(k')\right)^2\ G(\underline{k}-\underline{k}',t)\ S(k',t) \ . \tag{15}$$

Here, ζ_o is the friction coefficient at infinite dilution, c is the concentration, S(k,t) is the dynamic structure factor of the system under consideration and $c_D(k) = [S(k) - 1]/c\ S(k)$ denotes the direct correlation function. If the static structure factor S(k) is known (either from static experiments or theoretically /16/) and if S(k,t) is approximated by its mean-field approximation $S(k,t) \simeq S(k) \cdot \exp[-D_0 k^2 t/S(k)]$, eqs. (9), (14) and (15) form a closed set of equations, from which D_s , V(t) and W(t) can be obtained.

Here we will only mention some general results to illustrate what can be achieved in these systems; details can be found elsewhere /14,17/. Writing the generalized self-diffusion function as $\tilde{D}_s(k,z) = D_o - \Delta\tilde{D}(k,z)$, the mean-squared displacement is

$$W(t) = D_o t - \int_0^t dt'\ (t-t')\ \Delta D(0,t') \ , \tag{16}$$

which shows that $W(t) \approx D_o t$ for short times and $W(t) = D_s t$ for $t\to\infty$, where in general $D_s < D_o$. At short times the tagged particle starts to diffuse without taking notice of its neighbors, but as soon as the integral in (16) begins to contribute the particle feels the repulsive interaction with other particles, which leads to a slowing-down of its diffusive motion. The long-time behavior of W(t), which is related to D_s, is the result of many such interactions. It is clear from this short discussion that the full knowledge of W(t) contains much more information about the dynamics of the system than D_s does. Such information could in principle be obtained experimentally, if it were possible to measure the "time-dependent self-diffusion coefficient" $D_s(t) = W(t)/t$ on different time-scales relevant for the system.

The dynamic part $\Delta D(0,t)$ of the self-diffusion function determines also the velocity-autocorrelation function V(t). Finally, it should be noted that a nonlinear closed equation for D_s can be obtained from (10), (14) and (15), when $G(k,t) = \exp[-D_s\ k^2 t]$ is used in the integral in (15). The first approximation to this equation is /14/

$$D_s = D_o \left[1 + \frac{c}{6\pi^2} \int_0^\infty dk \, \frac{h^2(k)\, k^2}{1 + S(k)}\right]^{-1} \quad ; \quad h(k) = \big(S(k) - 1\big)/c \, , \tag{17}$$

which expresses D_s entirely in terms of the static structure factor S(k). It is found that results obtained on the basis of eq. (17) are in very good agreement with computer simulations for charged polystyrene spheres /18/.

4. SELF-DIFFUSION OF WATER-IN-OIL MICROEMULSION DROPLETS

Because of the complexity of the hydrodynamic problem, the calculation of D_s in the cases of short-ranged potential interactions cannot be carried to the same level as indicated in the last section for highly charged systems. Therefore, we return to the limit of calculating D_s to first order in ϕ from the solution of the two-particle Smoluchowski equation. The result can be expressed as /19/

$$D_s = \langle \hat{\underline{k}} \cdot \underline{\underline{D}}_{11} \cdot \hat{\underline{k}} \rangle + D_L \quad ; \quad \hat{\underline{k}} = \underline{k}/k \, . \tag{18}$$

The first term is a mean-field contribution expressing short-time self-diffusion. It is the equivalent of the first term in (16), but now including hydrodynamic interactions,

$$\underline{\underline{D}}_{11} = a_{11}(r)\, \underline{\underline{P}} + b_{11}(r)\, (1 - \underline{\underline{P}}) \, , \tag{19}$$

where $a_{11}(r)$ and $b_{11}(r)$ are again known for hard spheres /8/. The second term in (18) describes memory effects and can be expressed as

$$D_L = - \frac{N}{V} \int d^3r \int d^3r_o \; F(r)\, F(r_o)\, \tilde{\rho}(\underline{r},\underline{r}_o,z{=}0)\, \exp\left[-\beta V_R\right] \, . \tag{20}$$

Here, F(r) is a known function /19/ of the hydrodynamic functions A(r) and B(r) and of the potential interaction $V(r) = V_{\text{Hard sphere}} + V_R$. $\tilde{\rho}(\underline{r},\underline{r}_o,z{=}0)$ is the Laplace transform at z=0 of the solution of the two-particle Smoluchowski equation with appropriate boundary conditions.

Since the hydrodynamic functions are not known for our droplet model, some approximations have to be made to include the presence of the hydrocarbon tails within the spherical shell of thickness h = a - b. For r > 2a the tails do not overlap and the hydrodynamic interaction is taken as among hard spheres of radius b according to a b/r expansion given by Jones and Burfield up to order $(b/r)^{20}$. In the overlap region $2b \leqslant r \leqslant 2a$ the perpendicular motion (described by B(r) and $b_{11}(r)$) will be strongly hindered. Therefore, this part is treated like touching spheres having stick boundary conditions /20/:

$$B(r) \cong B = 0.802\, D_o \quad ; \quad b_{11}(r) \cong b_{11} = 0.891\, D_o \, . \tag{21}$$

The parallel motion is not expected to be as strongly influenced by the presence of the surfactant tails. Therefore we employ the b/r expansion for $a_{11}(r)$. Since the 1/r expansion for A(r) does not give the correct result for (nearly) touching hard spheres, we approximate in the overlap region

$$A(r) \cong A_- = f \, \frac{1}{2h} \int_{2b}^{2a} dr \, A_{HS}(r) = f \, \bar{A}_{HS}(a/b) \, . \tag{22}$$

Here, $A_{HS}(r)$ has been taken from the known results for hard spheres and f < 1, representing additional resistance due to the presence of the surfactant tails.

Using these approximations for the hydrodynamic interactions the result for D_s can be expressed as /19/

$$D_s = D_o + \phi_b \frac{d_s^{(0)} + d_s^{(1)} \exp[\beta V_o] + d_s^{(2)} \exp[2\beta V_o]}{1 + C \exp[\beta V_o]}$$

$$\equiv D_o + d_s \phi_b \quad ; \quad \phi_b = (4\pi/3)\, b^3\, (N/V) \; , \qquad (23)$$

where d_s is the desired virial coefficient of D_s. The quantities $d_s^{(i)}$, i=1,2,3, and C are entirely expressed by a/b and f, whereas the dependence on the depth of the attractive potential well is explicit.

5. RESULTS AND DISCUSSION

According to the model, βV_o is proportional to the maximum penetration volume:

$$\beta V_o \sim \frac{\pi h^2}{6} (3a - h/2) \xrightarrow[h \ll a]{} \frac{\pi h^2 a}{2} \quad ,$$

assuming that the length h is small as compared to the outer radius a. Therefore, we write $\beta V_o = Pa$, where P is the constant of proportionality.

To illustrate the result (23) for D_s or d_s we consider an H_2O/AOT/decane microemulsion, for which P = 0.0613 $Å^{-1}$ and h = 2.4 Å have been estimated on the basis of analyzing the static structure factor /21/. In fig. 1 the result for d_s in units of D_o is depicted as a function of the radius a. Comparing with curve (c), which is the result neglecting hydrodynamic interactions /22/, it is clear that there is a large effect of hydrodynamic interactions. The latter are long-ranged compared to the range of the potential interactions. Therefore, the tagged particle feels the presence of the other droplets long before the attraction can come into play. The strong decrease of d_s in the case without hydrodynamic interactions arises from the direct forces acting at r = 2b and r = 2a in the long-time (or memory) term only (the mean-field term being just D_o). In the presence of hydrodynamic interactions there are two effects: The first one is in the short-time contribution, which represents the decrease of the mean mobility of the tagged particle in the presence of other particles. It depends only on the equilibrium distribution, and therefore describes D_s on the time scale on which the particle positions do not change appreciably. This is the main contribution as can be seen by comparing curves (a) and (b) in the figure. The second part is the additional long-time term D_L. Due to the presence of hydrodynamic interactions the direct forces at r = 2b and r = 2a are now being weighted by the appropriate hydrodynamic functions which are rather small factors at these small separations. This reduces the importance of short-ranged forces substantially, which is also reflected in the weak dependence of curve (a) on the factor f.

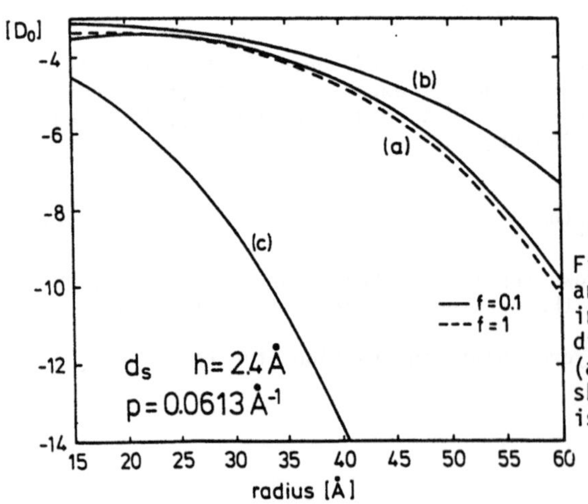

Fig. 1. The slope d_s of D_s versus ϕ_b for an AOT system (h = 2.4 Å, P = 0.0.613 $Å^{-1}$) in units of D_o as a function of the radius a. Note that D_o itself depends on a. (a) is the long-time result, (b) the short-time contribution only and (c) is d_s without hydrodynamic interactions

The result (23) has also been used to determine d_s for fixed a and h as a function of P, the strength of the potential. One finds /19/ d_s to increase in magnitude with increasing P, as expected. This could be used to determine P experimentally.

Very few systematic measurements of D_s have so far been performed. The results obtained by CHATENAY et al. /23/ are in qualitative agreement with conclusions of the present theory: the initial slope of D_s versus ϕ increases with increasing attractive interaction.

This work was partially supported by Deutsche Forschungsgemeinschaft, SFB 306.

6. LITERATURE

1. B. Lemaire, P. Bothorel, D. Roux: J. Phys. Chem. 87, 1023 (1983)
2. P. Mazur, W. van Saarloos: Physica 115A, 21 (1982); C.W.J. Beenakker, P.Mazur: Physica 120A, 388 (1983); 131A, 311 (1985)
3. W. van Megen, I. Snook, P.N. Pusey: J. Chem. Phys. 78, 931 (1983); I. Snook, W. van Megen, R.J.A. Tough: ibid. 78, 5825 (1983); S. Walrand, L. Belloni, M. Drifford: J. Physique, to appear
4. J.S. Huang, S.A. Safran, M.W. Kim, G.S. Grest, M. Kotlarchyk, N. Quirke: Phys. Rev. Letters 53, 592 (1984)
5. R. Finsay, A. Devriese, H. Lekkerkerker: J. Chem. Soc. Faraday Trans.2 76, 767 (1980)
6. B.U. Felderhof: J. Phys. A 11, 929 (1978)
7. M.W. Kim, W.D. Dozier, R. Klein: J. Chem. Phys. 84, 5919 (1986)
8. B.U. Felderhof: Physica 89A, 373 (1977); R. Schmitz, B.U. Felderhof: ibid. 116A, 163 (1982); R.B. Jones, G.S. Burfield: ibid. 133A, 152 (1985); D.J. Jeffrey, Y. Onishi: J. Fluid Mech. 139, 261 (1984)
9. S. Hanna, W. Hess, R. Klein: Physica 111A, 181 (1982)
10. B.J. Ackerson, L. Fleishmann: J. Chem. Phys. 76, 2675 (1982)
11. J.A. Marqusee, J.M. Deutch: J. Chem. Phys. 73, 5396 (1980)
12. G.K. Batchelor: J. Fluid Mech. 131, 155 (1983); 137, 467 (1983)
13. G.K. Batchelor: J. Fluid Mech. 74, 1 (1976); H.N.W. Lekkerkerker, J.K.G.Dhont: J. Chem. Phys. 80, 5790 (1984)
14. W. Hess, R. Klein: Adv. Phys. 32, 173 (1983)
15. J.P. Boon, S. Yip: Molecular Hydrodynamics (McGraw-Hill, New York 1980)
16. J.B. Hayter, J. Penfold: Mol. Phys. 42, 109 (1981); J.P. Hansen, J.B. Hayter: ibid. 46, 651 (1982)
17. G. Nägele, M. Medina-Noyola, R. Klein, J.L. Arauz-Lara: Progr. Colloid Polym. Sci., to appear
18. G. Nägele, M. Medina-Noyola, R. Klein: to be published
19. U. Genz, J.K.G. Dhont, R. Klein: J. Chem. Phys., to be published
20. A. Nir, A. Acrivos: J. Fluid Mech. 59, 209 (1973); see also D.J. Jeffrey, Y. Onishi, ref. 8
21. J.S. Huang: J. Chem. Phys. 82, 480 (1985)
22. C. van den Broeck: J. Chem. Phys. 82, 4248 (1985)
23. D. Chatenay, W. Urbach, A.M. Cazabat, D. Langevin: Phys. Rev. Letters 54, 2253 (1985)

Dynamics of Charged Systems

L. Belloni

CEA-IRDI, Département de Physico-Chimie, CEN-Saclay,
F-91191 Gif-sur-Yvette Cedex, France

I INTRODUCTION

The dynamics of colloidal solutions is classically studied in terms of the short-time diffusion coefficient D(q) which is extracted from the first cumulant of the auto-correlation function of the scattered amplitude I(q,t):

$$D(q) = -\frac{1}{q^2}\frac{\partial}{\partial t}\log I(q,t=0). \tag{1}$$

The function I(q,t) is measured in dynamic light scattering or in neutron scattering with the spin-echo technique. For monodisperse systems of spherical particles, the theory is well developed. Its main result is the following relation which relates the diffusion coefficient D(q) to equilibrium factors [1,2]:

$$D(q) = \frac{H(q)}{S(q)}\ ; \tag{2}$$

S(q) is the static structure factor and is the Fourier-transform of the pair distribution function g(r). The factor H(q) contains the so-called hydrodynamic interactions. Its exact expression is:

$$H(q) = \frac{kT}{N}\left\langle \sum_{i,j}^{N} \hat{q}\cdot\mu_{ij}\cdot\hat{q}\ e^{i\vec{q}\vec{r}_{ij}} \right\rangle\ ; \tag{3}$$

μ_{ij} is the generalized mobility tensor.

In part II, we extend the theory to the case of charged systems where there is a strong long-distance electrostatic coupling between the motion of the polyions and that of the small ions. In the part III, we discuss how to calculate the hydrodynamic factor H(q) in concentrated solutions.

II MULTI-COMPONENT DIFFUSION IN CHARGED SOLUTIONS

The starting point of the theory is the Smoluchowski equation and the Mori-Zwanzig projection technic [3-5]. In the short-time limit, we neglect the relaxation or memory effects and the time-evolution of the dynamic structure factors S(q,t) is governed by:

$$\frac{\partial S(q,t)}{\partial t} = -\,\Omega(q)\ S(q,t)\ . \tag{4}$$

For polydisperse systems as charged systems, eq.4 is a matricial equation. The frequency matrix $\Omega(q)$ is:

$$\Omega(q) = q^2 H(q) S^{-1}(q). \tag{5}$$

In the hydrodynamic limit ($q \to 0$), the elements of the inverse matrix $S^{-1}(q)$ are nothing more than the partial derivatives of the chemical potentials with respect to the ionic concentrations $\left(\frac{\partial \mu_i}{\partial \rho_j}\right)_{\rho_k}$ [6]. The solution of eq.4 is:

$$S(q,t) = A_1 e^{-\Gamma_1 t} + A_2 e^{-\Gamma_2 t} + \ldots, \tag{6}$$

where the normal modes Γ_i are the eigenvalues of $\Omega(q)$ and the A_i are its eigenmatrices.

The long-distance behavior of the unscreened coulombic potential leads to some problems in the limit $q=0$: the partial derivatives of the chemical potentials have no sense since it is impossible to add a polyion to the solution without adding the corresponding counterions. In another words, it is necessary to use the electrochemical potentials $\tilde{\mu}_i$ instead of the chemical potentials μ_i:

$$\tilde{\mu}_i = \mu_i^0 + kT\ln\rho_i + kT\ln\gamma_i + Z_i e\psi ; \tag{7}$$

γ_i is the activity coefficient of species i, ψ is the electrostatic potential.
In modern language, the matrix $S^{-1}(q)$ is related to the matrix $c(q)$ of the direct correlation functions through the Ornstein-Zernike equation and, then, presents a known divergence at $q=0$ [7-8]:

$$S^{-1}(q) = 1 - c(q) = 1 + \frac{Q}{q^2} - c'(q). \tag{8}$$

The matrix Q contains the elements $q_{ij} = Z_i Z_j \sqrt{\rho_i \rho_j}\, \frac{e^2}{\epsilon_0 \epsilon kT}$. The square of the Debye screening constant κ is equal to the trace of Q.
At this stage, it is important to note that the present normal modes theory has been often used without non-ideality terms γ_i, which is identical to neglecting the correction $c'(q)$ and to identify $c(q)$ to its divergent part [7-8]. This is exactly the Debye-Hückel approximation, which is correct for simple electrolytes but fails for polyelectrolytes. Its use has led to erroneous results in the literature [9-12].

Contrary to the uncharged case, the matrix $\Omega(q)$ does not behave as q^2 when $q \to 0$ but tends to a non-zero matrix. Thus, there is one relaxation mode, the plasmon or Debye mode $\Gamma_1 \equiv \Gamma_D = \mathrm{Trace}(H(0)Q)$, which is independent of q, and (n-1) diffusional modes $\Gamma_i \approx d_i q^2$. In quasielastic light scattering, the experimental time $\tau \approx 1\mu s$ is not short enough to allow the detection of the Debye mode $\Gamma_D^{-1} \approx 1ns$.
The measured mutual diffusion coefficient $D_m \equiv D(0)$ corresponds to an apparent first cumulant. When the light is essentially scattered by the large and heavy polyions (p), its theoretical expression is [7-8]:

$$D_m = \frac{H_{pp}(0) - \dfrac{H(0)QH(0)|_{pp}}{\Gamma_D}}{S_{pp}(0)} . \tag{9}$$

In the time scale $\tau \gg \Gamma_D^{-1}$, the different ions must diffuse together, due to the local electroneutrality condition. The mutual diffusion coefficient is then smaller than the classical expression (2), $\frac{H_{pp}(0)}{S_{pp}(0)}$, which is obtained in the one-component approach [1-2]. Note that the corrective factor in eq.9 disappears in the limit of point-like ions.

Expression 9 has been successfully applied to fit various experimental data for micellar or biological solutions [8,13-15]. This theory improves the classical one-component model, especially for charged systems at low ionic strength, for which the effect of the non-instantaneous diffusion of the small ions is emphasized.

II HYDRODYNAMIC INTERACTIONS IN CONCENTRATED SOLUTIONS

The main problem in evaluating the expression (3) is that the mobility tensor μ_{ij} depends on the position of all particles and not only on $\vec{r}_{ij}$. Thus, contrary to S(q), H(q) cannot be expressed as an integral of g(r). The total mobility tensor is a sum of one, two, three... particles mobility tensors:

$$\mu_{ij} = \frac{1}{6\pi\eta a}\delta_{ij} + \mu_{ij}^{(2)} + \mu_{ij}^{(3)} + \mu_{ij}^{(4)} + \ldots \tag{10}$$

The first one is the Stokes's result (a is the radius of the particles). The pair mobility tensor is given by Felderhof [16]. Mazur and Van Saarloos have explicitly calculated the following tensors [17] but their use is limited since the average (3) needs the knowledge of the 3, 4 ... particle distribution functions and time-consuming calculations of n-bodies integrals. Except at low volume fraction ($\lesssim$5%), the hydrodynamic interactions cannot be reduced to the pair interactions. Thus, we have chosen a semi-phenomenological way to take account of the multi-particles interactions, which consists in replacing eq.10 by:

$$\mu_{ij} = \frac{1}{6\pi\eta a}\delta_{ij} + \mu_{ij}^{(2)^{\text{effective}}}. \tag{11}$$

The hydrodynamic factor is then given by a simple integral:

$$H(q) = \frac{kT}{6\pi\eta a} + kT\rho \int \left(\mu_{11}^{(2)^{\text{eff}}} + \mu_{12}^{(2)^{\text{eff}}} e^{i\vec{q}\vec{r}_{12}}\right) g(r_{12})\, d\vec{r}_{12}\,. \tag{12}$$

The effective pair mobility tensors are solutions of the screened Navier-Stokes equation [18-19]:

$$-\vec{\nabla}P + \eta\Delta\vec{v} - \eta\kappa^2\vec{v} = \vec{0}\,. \tag{13}$$

The last term in eq.13 is an empirical representation of the hydrodynamic interactions of particles 3,4... on particles 1 and 2. With the induced forces formalism of Mazur and Van Saarloos, Stephan Walrand has calculated the self- and cross- effective mobility tensors up to the "screened" r^{-7} term [20-21]. We have investigated two choices for the screening constant κ:

$$\kappa a = K_0\sqrt{\Phi} \quad \text{or} \quad \kappa a = K_0'\Phi\,; \tag{14}$$

Φ is the volume fraction of the solution. K_0 and K_0' are fitting parameters which are common to all systems and are independent of the concentration, charge, salinity...

This model has been applied to different micellar and colloidal systems. First, we have reproduced experimental curves $D_m(\rho)$ obtained in dynamic light scattering on cationic micellar solutions at five different salinities [22]. The fitting values are $K_0 \approx 0.45$ or $K'_0 \approx 0.98$. Similar curves corresponding to smaller and less charged micelles have been reproduced with the same constants. The concentration-dependence of the mutual diffusion coefficient in uncharged colloidal systems is correctly taken into account, up to $\Phi \approx 50\%$, with the second choice for κ. Lastly, this model has been used to predict a curious and characteristic curve $D_m(\Phi)$ in the case of hard-spheres with an attractive interaction [21,23].

The present theory is not an exact treatment of the many-bodies hydrodynamic interactions. Its interest is to show the relative universality of these interactions, which depend only on the volume fraction at large concentration, and to allow to fit various experimental data corresponding to very different systems with a single constant.

IV CONCLUSION

We have shown that the classical expression (2) for the short-time diffusion coefficient must be used with caution: at low concentration and low salinity, it must be corrected so as to take the finite ionic diffusion into account. The ionic distortion around a moving particle is missing in the present theory and will be studied in future works [8]. At large concentration, the calculation of the hydrodynamic factor is very difficult or impossible except by using the notion of effective or "screened" pair hydrodynamic interactions. This empirical model leads to good agreements with simple calculations.

REFERENCES

1. P.N. Pusey, J. Phys. A8, 664 and 1433 (1975)
2. B.J. Ackerson, J. Chem. Phys. 64, 242 (1976)
3. A.Z. Akcasu and H. Gurol, J. Polymer Science 14, 1 (1976)
4. B.J. Ackerson, J. Chem. Phys. 69, 684 (1978)
5. A.Z. Akcasu, M. Benmouna and B. Hammouda, J. Chem. Phys. 80, 2762 (1984)
6. B. Berne and R. Pecora, Dynamic Light Scattering, (Wiley) 1976
7. L. Belloni, M. Drifford and P. Turq, J. Physique Lett. 46, L207 (1985)
8. L. Belloni, Thèse d'Etat, Paris (1987)
9. M.J. Stephen, J. Chem. Phys. 55, 3878 (1971)
10. P. Doherty and G.B. Benedek, J. Chem. Phys. 61, 5426 (1974)
11. A. Rhode and E. Sackmann, J. Coll. Interf. Sci. 78, 330 (1980)
12. P. Tivant, P. Turq, M. Drifford, H. Magdelenat and R. Menez, Biop. 22, 643 (1984)
13. L. Belloni and M. Drifford, J. Physique Lett. 46, L1183 (1985)
14. M. Drifford, L. Belloni, J.P. Dalbiez and A. Chattopadhyay, J. Colloid Interf. Sci. 105, 587 (1985)
15. L. Cantu, M. Corti and V. Degiorgio, Europhys. Lett. 2, 673 (1986)
16. B.U. Felderhof, Physica 89A, 373 (1977)
17. P. Mazur and W. Van Saarloos, Physica 115A, 21 (1982)
18. S.A. Adelman, J. Chem. Phys. 68, 49 (1978)
19. I. Snook, W. Van Megen and R.J.A. Tough, J. Chem. Phys. 78, 5825 (1983)
20. S. Walrand, L. Belloni and M. Drifford, Physics Lett. A118, 422 (1986)
21. S. Walrand, Thèse de l'Université, Paris (1986)
22. S. Walrand, L. Belloni and M. Drifford, J. Physique 47, 1565 (1986)
23. S. Walrand, L. Belloni and M. Drifford, Europhys. Lett. (1987)

Counterion Complexation. Ion Specificity in the Diffuse Double Layer of Surfactant and Classical Colloids

D.F. Evans, J.B. Evans, R. Sen, and G.G. Warr

Dept. of Chemical Engineering and Materials Science,
University of Minnesota, Minneapolis, MN 55455, USA

The effect of complexation of alkali metal ions by macrocyclic ligands in dispersions of anionic surfactants and clay colloids has been studied as a route to understanding counterion specific effects in these classes of colloidal dispersions. The rigidity of the lattice of charge in the solid leads to vastly different results than are observed in micellar solutions, where the aggregates rearrange to minimise their free energy.

1. Introduction

The role of counterion specificity in determining the properties of charged surfaces remains an important issue. For colloidal sols, where stability is the key concern, counterion effects are usually described in terms of the Debye screening length as embodied in the DLVO theory. Once the role of potential determining ions and the valency dependence are accounted for, specific ion effects are minimal. For surfactant aggregates the situation is rather different. The main issues here being size, shape, and polydispersity of the colloidal particles. Here counterion effects are frequently described in terms of an ion binding parameter, and they can dramatically alter the structure and reactivity of these surfactant aggregates. The differences in language used to describe the counterion properties of these two kinds of colloidal systems reflect their very different properties.

In two previous publications[1,2] we have shown how the properties of sodium dodecylsulfate micelles and dispersions of Texas I could be changed by modulation of counterion binding *via* complexation with macrocyclic ligands. In this paper we extend this study of counterion-cryptate complexes to include a colloidal clay, mica. Addition of cryptates permits the potassium counterion to be complexed and the properties of the system are determined using the surface forces apparatus. Critical micelle concentrations (cmc), aggregation numbers, and the effect of added salt upon cryptate plus surfactant solutions containing LiDS, SDS, KDS and tetramethylammonium (NMe_4) DS are also reported. These two parallel studies provide the opportunity to compare how the charged interfaces of a classical colloidal material and a surfactant aggregate respond to perturbations induced by cryptate-counterion complexation.

2. Experimental Section

Sodium dodecyl sulfate (SDS) and lithium dodecyl sulfate (LiDS) were obtained from BDH (extra pure) and were used as received. Potassium dodecyl sulfate (KDS), silver dodecyl sulfate (AgDS) and tetramethylammonium dodecyl sulfate (NMe_4DS) were prepared as described previously[3,4]. SDS showed a slight minimum in surface tension at the cmc, which was determined to be 6.3 mM. LiDS showed no minimum.

The Krafft temperature of KDS was 32°C which compares well with literature results[3]. 4,7,13,16,21,24-hexaoxa-1,10-diazabicyclo[8.8.8]hexacosane (C222) and 4,7,13,16, 21- pentaoxa-1,10,diazabicyclo[8.8.5]tricosane (C221) were purchased from Merck and Sigma at 98% purity and were used as received. Octa(ethylene glycol) -mono-n-dodecyl ether, $C_{12}E_8$, was obtained from Nikkol Co., Japan, and was used as received. All inorganic salts were AR grade. NMe_4Br (Eastman) was recrystallized from methanol before use. Surfactant solutions were prepared in water which was once distilled and filtered through a Millipore® Milli-Q system.

Brown muscovite mica for the surface forces measurement was obtained from United Mineral. Water was distilled, Milli-Q filtered and redistilled from a Pyrex still under an oxygen atmosphere containing a platinum wire heater. The water was recycled over the heater wire for five hours to oxidize any remaining organic contaminants.

The surface forces apparatus employed in this study is identical in principle to the apparatus used by ISRAELACHVILI and ADAMS[5]. It has been described in some detail elsewhere[4]. Some of the measurements were made by filling the entire box with solution, while others utilised a small (30ml) stainless steel bath.

Critical micelle concentrations (cmc), and areas per molecule at the air-solution interface were determined from surface tension measurements taken on a Du Nuoy ring apparatus. The apparatus used for conductivity measurements has been described elsewhere[6]. Average micelle aggregation numbers, <i>, were determined by fluorescence quenching of micelle-bound ruthenium tris(bipyridyl)$^{2+}$ (ICN Pharmaceuticals) with 9-methyl anthracene (Eastman, recrystallized), measured using a Spex Fluorolog 222 spectrofluorometer.

3. Results and Discussion

3.1 Surfactant Solutions

The micellar and interfacial properties of SDS/C222 solutions, with added NaCl, are shown in Table 1. The degree of dissociation was estimated by two methods. The first involved taking the ratio of slopes of conductivity versus concentration above and below the cmc. The degrees of binding obtained in this way agree well with the result obtained from the variation of ln(cmc) *vs* ln(cmc +[NaCl]) in the presence of C222. Although the cmc of SDS depends upon C222/Na^+, the variation is small above a ratio of about 0.5, so we may legitimately obtain degrees of micellar dissociation from the salt dependence of the cmc (see below).

As the ratio of C222 to Na^+ is increased, the aggregation number decreases while both the degree of dissociation and the area per head group increase. This is consistent with the removal of the counterion from the micellar surface upon encapsulation into the cryptate cage, resulting in increased repulsion between surfactant head groups. At C222>> Na^+ asymptotic limits of <i>=40 and a molecular area of ~140Å^2 are reached, suggesting that the micelle surface has become saturated with cryptate complex. The effect of added C222 on SDS micelles is perhaps most dramatic with salt also added . The usual increase in aggregation number is completely suppressed, and <i> remains constant at 40 up to 0.3M NaCl with equimolar C222.

Table 1: Micelle and surface parameters of SDS in the presence of C222.

$[C222]/[Na^+]$	[NaCl] [M]	cmc [mM]	Degree of dissociation	Area per mol. [$Å^2$]	Agg. no. <i>
0.0	0.0	6.3	0.29	65	65
0.20	0.0	2.5	-	118	65
0.40	0.0	1.8	0.48	130	50
2.00	0.0	1.3	0.49	136	40
0.83	0.1	0.28	0.55 ± 0.02^a	172	44
1.0	0.2	-	-	-	40
1.0	0.3	-	-	-	38

^{a} from a plot of log(cmc) *vs* log(cmc + [NaCl])

Table 2: Micelle and surface parameters of dodecyl sulfate solutions with C222 and C221.

Surfactant	Ligand, L	$[L]/[M^+]$	cmc [mM]	Area/mol. [$Å^2$]	Agg. No. <i>	$\log_{10}K_{st}{}^a$ H_2O	MeOH
LiDS	-	0.0	8.7	73±1	66	-	-
	C222	1.0	2.1	80±6	49	$< 2^b$	1.8
	C221	1.0	1.8	120±6	44	2.5	4.2
SDS	-	0.0	6.3	65±3	65	-	-
	C222	2.0	1.3	136±6	40	4.7	7.1
KDS	-	0.0	-	68±2	-	-	-
	C222	1.1	1.3	123±3	44	5.2	9.4
NMe_4DS	-	0.0	5.6	70±3	88	-	-
	C222	0.9	1.1	78±5	-	-	-
$C_{12}E_8$	-	-	0.081[8]	64[8]	90[9]	-	-
	C222	50.0^c	0.025	44	-	-	-

^{a} From ref. 7, MeOH refers to a 95:5 methanol:water mixture.

^{b} The technique used[7] is unreliable at $K_{st}<2$. We can safely regard the value in water as being lower than that in MeOH/H_2O.

^{c}at the cmc. The concentration of C222 was fixed at 1.3 mM, which approximates [C222] at the cmc of SDS/C222 systems.

The cmc, area per molecule, and aggregation numbers of LiDS, SDS, KDS, NMe_4DS and $C_{12}E_8$ with and without added C222 and C221 are summarized in Table 2. The complex stability constants in water and in 95:5 methanol:water are included also, and the effects of complex stability on DS micelles can clearly be seen. The extreme cases of NMe_4DS and $C_{12}E_8$ are almost unaffected by addition of complexing ligand, except for the lowering of their cmcs. Tetramethylammonium ion is not complexed by C222, and $C_{12}E_8$ has no counterion to interact with the ligand. The observed cmc reduction must thus be due to nucleation of micelles by the ligand alone, and is unrelated to the effect of the cryptate complex. Similarly C222, which only weakly complexes lithium, has a marginal effect on the properties of LiDS micelles, whereas C221 has a more dramatic effect, comparable with C222/Na^+ and C222/K^+ systems.

3.2 Surface Forces Measurements

Surface forces measurements were carried out on KCl solutions in the presence of varying amounts of C222 (Fig. 1)[4]. A dilute KOH solution was used initially to obtain a pH of 10, both because of the dibasic nature of the ligand (pK_1=7.28, pK_2=9.60[7]) and also to expel adsorbed protons from the mica surface[10]. To this was added C222 and KCl to give varying ratios of K^+/C222 and varying total ion concentrations. The pH was checked periodically during the experiment by withdrawing aliquots of solution, and was found to remain constant. Data were analysed according to DLVO theory using a nonretarded Hamaker constant of 2.2 $\times 10^{-20}$J[4], and the best fits obtained are shown as solid lines in Fig.1. In these fits the surface which determines the potential was taken to coincide with the vertical portion of the force curve.

In the presence of only KOH a repulsive barrier due to adsorbed, hydrated potassium ions was encountered at 10Å separation of surfaces, and the surface potential here was 115mV. The Debye length of 240Å corresponds to an ionic strength of 0.16mM, which can be partially attributed to the presence of dissolved carbonates. Addition of C222 to 0.54mM (>99% K^+ complexed) moved the repulsive barrier out to 23Å separation with the potential at the wall remaining 155mV, and decreased the Debye length to 200Å. (This latter effect can be attributed to hydrolysis of the excess ligand.) Movement of the hard, repulsive wall must be a result of adsorbed potassium cryptate on the mica surface. Addition of 1 mM KCl led to the recovery of the 10Å barrier, which increased again to 23Å with added C222 at 1.6mM. Thus the potassium cryptate can be moved on and off the mica surface by ion exchange.

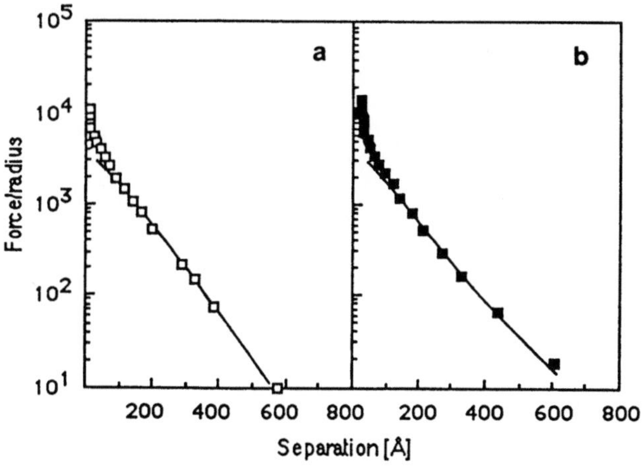

Figure 1: Force *vs* distance curves (at pH=10) for 0.08mM KOH in the absence (a) and presence (b) of complexing C222.

With the exception of the 13Å displacement, the force curves obtained with and without added C222 are identical at constant potassium concentration (see Fig. 1)[4]. At higher potassium concentration, where excess C222 is not necessary to induce complexation, the Debye length in the system is unchanged when C222 is added, and the surface potential is also constant at the plane of hard wall repulsion.

Complexation of counterions by macrocyclic ligands affects surfactant and clay colloids in very different ways. Clearly the main influence of cryptate formation exists very near to the anionic surface, and the colloid responds according to its ability to rearrange in response to changes at such intimate distances. With mica the charges are immobile and cryptate formation acts only to displace the diffuse double-layer away from the solid surface. In surfactant systems, counterion complexation changes head group interactions, leading to a rearrangement of the aggregate structure with a new minimum free energy state.

References

1. D.F. Evans, R. Sen and G.G. Warr, J. Phys. Chem., 90, 5500, (1986)
2. D.D. Miller, D.F. Evans, G.G. Warr, J.R. Bellare and B.W. Ninham, J. Colloid Interface Sci., 116, 598 (1987)
3. Y. Moroi, S. Motomura and R. Matuura, J. Colloid Interface Sci., 81, 486, (1981)
4. D.F. Evans, J.B. Evans, R. Sen and G.G. Warr, submitted to J. Phys. Chem.
5. J.N. Israelachvili and G.E. Adams, J. Chem. Soc., Faraday Trans. 1, 74, 975 (1978)
6. S.J. Chen, D.F. Evans and B.W. Ninham, J. Phys. Chem., 88, 1631 (1984)
7. J.M. Lehn and J.P Sauvage, J. Amer. Chem. Soc., 97, 6700 (1975)
8. C.J. Drummond, G.G. Warr, F. Grieser, D.F. Evans and B.W. Ninham, J. Phys. Chem., 89, 2103, (1985)
9. G.G. Warr, C.J. Drummond, F. Grieser, D.F. Evans and B.W. Ninham, J. Phys. Chem., 90, 4581, (1986)
10. R.M. Pashley, J. Colloid Interface Sci., 83, 531 (1981)

Light Scattering Experiments on Interacting Micelles

M. Corti[1], *V. Degiorgio*[1], *and L. Cantu*[2]

[1]Dipartimento di Elettronica, Sezione di Fisica Applicata,
Università di Pavia, I-27100 Pavia, Italy

[2]Dipartimento di Chimica e Biochimica Medica,
Universitá di Milano, I-20133 Milano, Italy

1. Introduction

A micellar solution is a system of polydisperse and interacting aggregates which are in multiple chemical equilibrium. Such a system is very hard to describe in its full generality, so that some approximation is needed to interpret experimental results. In this paper we describe two experiments performed on solutions of strongly interacting micelles. The first experiment describes a system, ionic biological micelles in a low ionic strength aqueous solution, which shows little structural changes with temperature, amphiphile concentration or ionic strength, so that it represents an almost ideal candidate to study in detail the effect of interactions on both static and dynamic properties. The second experiment refers to nonionic micelles with added ionic impurities. The phase diagram of this latter system is explained only partially and qualitatively by a model which considers interactions and neglects structural changes.

2. Ganglioside micelles in low ionic strength solutions

We have performed light scattering measurements on ionic micelles of biological glycolipids (gangliosides) in aqueous solution at very low ionic strength /1/. Gangliosides are anionic glycolipids occurring in neuronal plasma membranes. The material used in the present investigation was prepared as a sodium salt by Tettamanti and coworkers. Detailed information about nomenclature, properties and preparation procedure can be found in Ref./2/. We have used the ganglioside GM1 which has a critical micelle concentration smaller than 10^{-8} M. The solutions were prepared with doubly distilled and degassed water. The cell was accurately precleaned, and was flushed with large amounts of pure water before each measurement in order to eliminate possible ionic impurities. All measurements were performed at 25°C. The light scattering apparatus includes a 514.5 nm argon laser, and a Langley-Ford digital correlator. Both scattered intensity and correlation function measurements were made at a scattering angle of 90°.

The effect of NaCl addition to the ganglioside solution was studied at fixed ganglioside concentration. We show in Fig. 1 the behavior of the scattered intensity as a function of the ionic strength at the GM1 concentrations 0.5 mM and 1 mM. The reported values are normalized by the intensity scattered from the ideal solution which was derived by measuring the scattered intensity at various GM1 concentrations in 30 mM NaCl solutions and extrapolating to zero micelle concentration. The ionic strength is calculated as $c_s + \frac{1}{2}Qc/m$ /1/, where c_s is the NaCl molarity, c is the amphiphile molar concentration, Q is the electric charge of the micelle in electronic units and m is the micelle aggregation number. In order to check whether the aggregation number depends on the ionic strength, we have measured the intensity of scattered light I_s as a function of the ganglioside concentration at two distinct salt concentrations, 1 and 30 mM NaCl. The extrapolated value of I_s/c at zero concentration, which is proportional to m, is the same for the two solutions. The obtained aggregation number is $m = 302 \pm 30$.

The measured correlation functions were exponential in all cases, except for very low ionic strengths. The correlation functions obtained for various NaCl concentrations were analyzed by the standard cumulant fit which gives the diffusion coefficient D and the relative variance v. The ratio D_0/D is reported in Fig. 2 as a function of the ionic strength I for two distinct GM1 concentrations, D_0 being the diffusion coefficient of the individual GM1 micelle. D_0 is derived by measuring D at various GM1 concentrations and extrapolating to zero micellar concentrations. From D_0 we derive the hydrodynamic radius which is 5.87 nm for the GM1 micelle.

As an example of the observed nonexponentiality at low ionic strength, we show in Fig. 3 the intensity correlation functions measured at two distinct scattering angles, 22° and 162°, for a 1 mM GM1 solution with about 1 mM NaCl. It is interesting to note that the deviation from exponential behavior is more marked for the low-angle measurement. For sake of comparison, we present in Fig. 3 also a correlation function measured at 90° in a 30 mM NaCl solution.

Since the micelle size and shape do not change with the ionic strength /1/, we can say that the quantity plotted in Fig. 1 is indeed the static structure factor S(k) evaluated at $k = 2k_0 \sin\theta/2$, where $k_0 = 2\pi n/\lambda$, n is the index of the refraction solution and $\theta = 90°$ in our experiment. We have calculated the theoretical structure factor by using the HNC approximation for g(r). Our computer program is a copy of that used by Cannell et al. /3/. We have considered the GM1 micelles as spherical particles interacting through a potential consisting of a hard core repulsion plus a screened Coulomb potential. As discussed in detail elsewhere /4/, the aggregation number of GM1 micelles is too large to be compatible with a spherical shape, but the micelles are still globular with an axial ratio about 2. The continuous curves in Fig. 1 show the theoretical structure factor calculated by using the best fit value of the micellar charge, Q = 48, which corresponds to a fractional ionization $\alpha = Q/m = 0.16$.

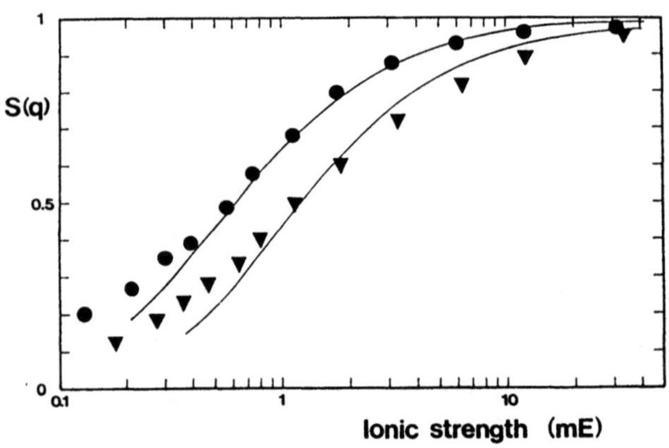

Fig. 1. Structure factor S of 0.5 mN (●) and 1 mM (▼) GM1 solutions measured as functions of the ionic strength I at 25°C and scattering angle of 90°. The full curves respresent theoretical results calculated for a micellar charge of 48 electronic units.

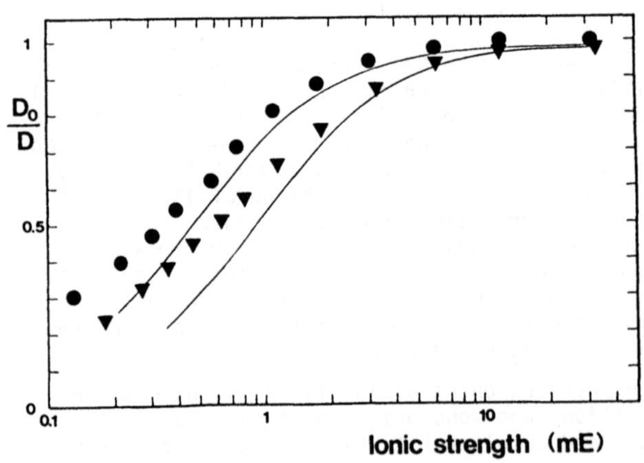

Fig. 2. The quantity D_0/D versus the ionic strength I. All the parameters are the same as in Fig. 1.

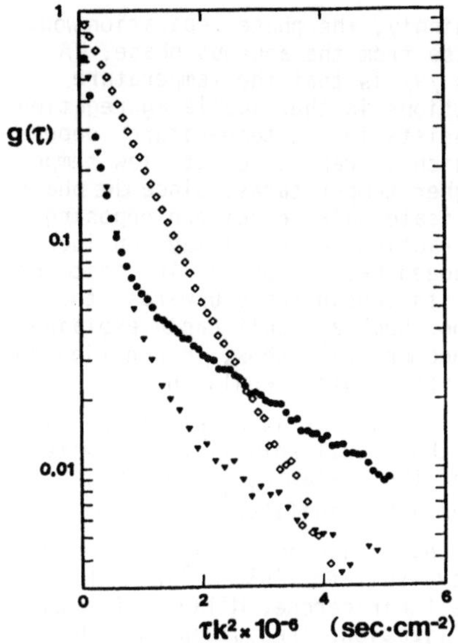

Fig. 3. Normalized time-dependent part of the intensity correlation function measured for 1 mM GM1 solutions. (•) 1 mM NaCl, θ = 22°; (▼) 1 mM NaCl, θ = 162°; (◇) 30 mM NaCl, θ = 90°

As shown in Fig. 1, the fit between experiment and theory with Q = 48 becomes less good at low ionic strength (below 0.5 mM NaCl). In order to improve the fit one has to assume that Q is decreasing with the ionic strength I at very low I. A plausible explanation for such a dependence could be that strongly repulsive intermicellar forces might enhance ion condensation on the micelle surface. Fig. 1 shows that deviations are larger when the micelle concentration is larger. This would also be consistent with an explanation in terms of interactions. Unfortunately all the theoretical treatments of counterion condensation neglect interparticle interactions, so that we cannot compare our data with any model. As discussed in detail in Ref. 1, one should not neglect the possibility that the apparent change of Q with I is an artifact due, for instance, to the fact that the theoretical model we have used does not take into account the finite size of the small ions /5/.

We discuss now the dynamic data. The full curves in Fig. 3 are derived by inserting the calculated g(r) into the expression derived by Ackerson /6/ which considers hydrodynamic interactions at the Oseen level, by using for Q the best fit value derived from the static data, Q = 48, and by applying the correction factor proposed by Belloni and Drifford /7/ to take into account the non-negligible size of small ions. As discussed in more detail in Ref./1/,the trend is correct, but the agreement is only qualitative. Finally, we briefly comment on the data presented in Fig. 3. Since explicit theoretical results concerning the full shape of the intensity correlation function are not available, we do not know whether our results are compatible with the theory of strongly interacting monodisperse particles. The fact that exponential correlation functions were obtained by Cannell et al. /3/ with a monodisperse protein solution in a similar range of ionic strengths, could suggest that the observed nonexponentiality is due to the intrinsic polydispersity of the micellar solution /8/. Further theoretical and experimental results are needed to clarify this point.

3. Nonionic micelles in the cloud point region: effect of the addition of an ionic amphiphile

Aqueous solutions of nonionic amphiphiles (in particular, the alkylpolyoxyethylenes C_iE_j) have been actively investigated in the last few years with the aim of clarifying the microscopic structure of the solution and the mechanism of the cloud point transition. The prevailing view in the early literature was that the approach to the cloud point is associated with a shape instability of the individual micelle which would grow, as T is increased toward the critical temperature T_c, from a globular shape to a long flexible cylinder. Accordingly, the phase separation would consist in a segregation of an amphiphile-rich phase from the aqueous phase. A completely different view, discussed in our papers /9/ is that the temperature change does not produce necessarily dramatic variations in the micelle aggregation number and that the phase transition mechanism consists in the temperature dependence of the intermicellar interaction potential which is repulsive at low temperature but becomes predominantly attractive at higher temperatures. Since the phase transition temperature is determined by a very delicate balance between opposing effects, a slight change in the intermicellar interaction potential may considerably affect the cloud point temperature. It has indeed been reported /10/ that the addition of a small amount of ionic amphiphile shifts considerably upwards the cloud temperature of a nonionic amphiphile solution. Such a result can be explained by noting that the micelles containing ionic monomers will show, in addition to the attraction typical of nonionic micelles, an electrostatic repulsion.

Before describing our experiment, we must mention that Firman et al. /11/ have reported a study on the effect of electrolytes on the mutual solubility of water and C_iE_j in which they present the phase diagram of the system C_4E_1-H_2O (which does not form micelles) with the addition of sodium alkyl sulfates.

We have performed an experimental study of the phase diagram of $C_{12}E_6$ solutions with the addition of small amounts of sodium dodecyl sulfate (SDS) /12/. High purity $C_{12}E_6$ was prepared by Platone and coworkers (Eniricerche, Milan). SDS was obtained from BDH and purified by repeated crystallization. The solvent was D_2O (99% purity) obtained from Carlo Erba, Milan. We have used heavy water instead of normal water in order to increase the density mismatch between solvent and solute, thus increasing the speed of the phase separation process and allowing a systematic study of the phase diagram in a reasonable amount of time. The cloud curves were obtained by a turbidimetric technique developed previously in our laboratory /13/. Eight sealed cells, each filled at a different amphiphile concentration, are placed in a water bath whose temperature is increased at constant rate, typically a few °C per hour. The phase transition temperature is determined by monitoring with a laser beam the variation of the cell turbidity on approaching the phase separation point. The accuracy of the instrument is about 0.1°C. Each cloud curve is derived by fixing the molar ratio r between SDS and $C_{12}E_6$ and measuring the transition temperature as a function of the total amphiphile concentration. We found it rather difficult to obtain reproducible results and the reported data should be considered as preliminary. The cloud curves measured with pure $C_{12}E_6$ and with four distinct SDS/$C_{12}E_6$ molar ratios are shown in Fig.4. The system, besides showing an upward shift of the lower consolute point, also shows an upper transition point which is downshifted by the addition of amphiphile. The two-phase region shrinks progressively as r increases and disappears when r is above ≈ 0.01. In Ref. 12 we have interpreted the upper transition as an upper consolute point above which the system presents a single-phase isotropic solution. However, according to a very recent work by Carvell et al. /14/, the phase appearing above the upper transition presents a weak birefringence and should be identified as a lamellar phase, whereas the upper consolute point appears at much higher temperature. This new assignment of phases would limit the validity of the interpretation discussed in Ref. 12 only to the lower branch of the curves shown in Fig. 4.

In order to explain qualitatively the upward shift of the cloud curve, we can make reference to the lattice model for critical binary mixtures (a more accurate description could be achieved by using recent theories of phase separation in micellar solutions /15, 16/) which says that the phase behavior of the system is controlled by the value of the effective interaction parameter χ. If χ is below a critical value, χ_c, the system consists of a single-phase solution. When χ is above χ_c, the solution splits into two phases. The existence of a lower consolute point implies that there is a temperature range over which χ is increasing with T. Eventually χ will reach a maximum and after decrease monotonically at high temperatures. If the maximum of χ is larger than χ_c, the system will present both a lower and an upper critical point (see, for instance, Ref. 15). The addition of a repulsive contribution to the micelle interaction potential produces es-

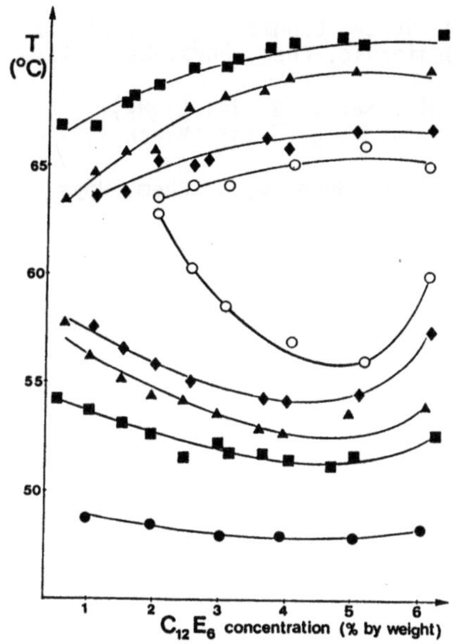

Fig. 4. Phase diagram of the system $C_{12}E_6$/SDS/D_2O at various molar ratios r. (●) r = 0, (■) r = 0.0045, (▲) r = 0.0056, (♦) r = 0.0090

sentially a downward shift of the χ versus T plot. By increasing the SDS fraction we increase the average electric charge of the micelle and, consequently, the amplitude of the repulsive contribution. Clearly such an increase will produce an upward shift of the lower critical temperature, in agreement with the trend reported in Fig. 4.

We thank G. Tettamanti and E. Platone for the preparation and gift of the amphiphiles used in this work. We acknowledge support from the Italian Ministry of Public Education (MPI 40% funds).

References

1. L. Cantù, M. Corti and V. Degiorgio, Europhys. Lett. 2, 672 (1986); L. Cantù, M. Corti and V. Degiorgio, Faraday Disc. 83, (1987)
2. G. Tettamanti et al., in Physics of Amphiphiles: Micelles, Vesicles and Microemulsions, edited by V. Degiorgio and M. Corti (North-Holland, Amsterdam, 1985), p. 607
3. D. S. Cannell, in Physics of Amphiphiles: Micelles, Vesicles and Microemulsions, edited by V. Degiorgio and M. Corti (North-Holland, Amsterdam, 1985), p. 202; D. G. Neal, D. Purich and D. S. Cannell, J. Chem. Phys. 80, 3469 (1984)
4. L. Cantù, M. Corti, S. Sonnino and G. Tettamanti, Chem. Phys. Lipids 41, 315 (1986)
5. G. Nägele, R. Klein and M. Medina-Noyola, J. Chem. Phys. 83, 2560 (1985); L. Belloni, J. Chem. Phys. 85, 519 (1986)
6. B. J. Ackerson, J. Chem. Phys. 64, 242 (1976); 69, 684 (1978)
7. L. Belloni and M. Drifford, J. Physique Lett. 46, 1183 (1985)
8. M. B. Weissman, J. Chem. Phys. 72, 231 (1980); P. N. Pusey in Light Scattering in Liquids and Macromolecular Solutions (Plenum Press, New York, 1980)
9. V. Degiorgio, in Physics of Amphiphiles: Micelles, Vesicles and Microemulsions, edited by V. Degiorgio and M. Corti (North-Holland, Amsterdam, 1985), p. 303; M. Corti, C. Minero and V. Degiorgio, J. Phys. Chem. 88, 309 (1984); V. Degiorgio, R. Piazza, M. Corti and C. Minero, J. Chem. Phys. 82, 1025(1985); M. Corti and V. Degiorgio, Phys. Rev. Lett. 55, 2005 (1985)
10. H. Hoffmann et al., J. Colloid Interface Sci. 80, 237 (1981); P.-G. Nilsson and B. Lindman, J. Phys. Chem. 88, 5391 (1984); B. S. Valaulikar and C. Manohar, J. Colloid Interface Sci. 108, 403 (1985)

11. P. Firman, D. Haase, J. Jen, M. Kahlweit and R. Strey, Langmuir 1, 718 (1985)
12. L. De Salvo Souza, M. Corti, L. Cantù and V. Degiorgio, Chem. Phys. Lett. 131, 160 (1986)
13. M. Corti, C. Minero, L. Cantù and R. Piazza, Coll. Surf. 12, 341 (1984)
14. M. Carvell, C. A. Leng, F. J. Leng and G. J. T. Tiddy, preprint (March 1987)
15. R.E. Goldstein, J. Chem. Phys. 84, 3367 (1986)
16. G.M. Thurston, D. Blankschtein, M. R. Fish and G. B. Benedek, J. Chem. Phys. 84, 4558 (1986)

Nonionic Micelles

B. Lindman and M. Jonströmer
Physical Chemistry 1, Chemical Center, Lund University,
P.O. Box 124, S-22100 Lund, Sweden

1. INTRODUCTION

Systems with nonionic surfactants of the oligo (ethylene oxide) variety have attracted a strikingly great interest in the last few years, both from theoreticians and experimentalists. There are a number of reasons for this: the unusual property of the aqueous solutions to exhibit phase separation and a lower consolute curve; the rich and diverse phase behaviour with several isotropic solution and liquid crystalline phases also for two-component aqueous systems; the efficiency of these surfactants in mixing large amounts of water and oil in a homogeneous phase (l.c. or microemulsion); the availability of homologue-pure surfactants and the possibility of varying the head-group size systematically; the large and rapidly growing technical use of these and related compounds.

LANG and MORGAN /1/ investigated in great detail the phase behaviour of the $C_{10}E_4$-H_2O system (we use the common notation C_xE_y for $CH_3(CH_2)_{x-1}(OCH_2CH_2)_yOH$) and demonstrated the closed loop appearance and the presence of an additional isotropic solution phase (termed anomalous phase in ref. 1 and L_3 phase by others). MITCHELL et al. /2/ in another extensive investigation established the phase behaviour for a large number of aqueous nonionic surfactant systems and could inter alia demonstrate the occurrence of a number of different l.c. phases. CORTI and DEGIORGIO (e.g. ref. 3) provided lower consolute curves and investigated critical effects for a number of systems and DEGIORGIO /4/ has tabulated much significant data (cmc's, critical temperatures, critical concentrations etc.) The authors cited were also concerned with theoretical analyses of different aspects of their observations and there are additional important papers devoted entirely to theoretical analyses of the phase separation and the lower consolute curves, see in particular KJELLANDER /5/ and BLANKSCHTEIN et al. /6,7/. It is natural in such work to have the poly(ethylene glycol) - water system as a reference system as it shows in fundamental respects an analogous (but much less complex) phase behaiour /8-12/.

Depending on the level of understanding desired, different types of experimental information on the system are required. Fundamental information is micelle size, shape and intermicellar interactions at different concentrations and numerous studies using different techniques - static and dynamic light scattering, neutron scattering, self-diffusion, fluorescence quenching, NMR relaxation - have been performed in recent years (see inter alia other papers in this volume). On another level we wish to understand the mechanism of the observed micellar sizes and shapes and we should then investigate molecular packing, order and dynamics in the micelles and micelle hydration all as a function of temperature and concentration. On a further level we wish to assign mechanisms in terms of molecular interactions which can explain these characteristics.

In this paper we will discuss some of these matters and try to give a brief account of our present understanding. In so doing we will inter alia review our earlier studies using mainly self-diffusion and NMR relaxation (work done in collaboration with P.G. NILSSON, B. FAUCOMPRE, T. AHLNÄS, G. KARLSTRÖM AND H. WENNERSTRÖM; see reference list). We will also present a number of previously unpublished observations

(M. JONSTRÖMER and B. LINDMAN, in preparation). Various experimental aspects are to be found in the original publications. We will just note that for the determination of the different self-diffusion coefficients by NMR the FT version of the pulsed-gradient spin-echo technique /13/ has dramatically improved the situation. We continue to investigate ^1H NMR relaxation as the transverse relaxation rate is very sensitive to micelle growth /14/; care was taken to faithfully reproduce the natural line-shape without distortions. In part of the recent studies we have devoted special interest to solutions close to phase boundaries; special care was taken to minimize temperature variations (both spatially and temporally) but we should emphasize that in our experiments temperature control achieved is far from that in some experiments by other techniques.

2. MICELLIZATION

Since the translational mobilities over large distances are very different for free, non-associated, molecules and for molecules confined to micelles, multicomponent self-diffusion studies (as performed most conveniently by the FT NMR technique) are well suited for investigating the partitioning of different species between micelles and the bulk solution (surfactant molecules, solubilized molecules, counterions and water molecules) /15/. Studies of surfactant self-diffusion in ionic surfactant systems emphasize the large cooperativity of micellization and give evidence for a pronounced maximum in free surfactant concentration around the cmc. The marked decrease in free surfactant concentration above the cmc which is well understood from Poisson-Boltzmann calculations is associated with an unfavourable entropy in the counterion distribution /16/. Apparently, such an effect being absent for surfactant molecules with no net charge, a corresponding finding is not expected for nonionic and zwitterionic systems. Indeed, BERNARD FAUCOMPRÉ in investigations of a number of surfactants in these classes /17/ found either an approximately constant (for surfactants with long alkyl chain) or a slowly increasing free surfactant concentration above the cmc. (It was also noted that an even relatively small admixture of ionic surfactant, may, in the absence of added salt, have an important effect). As an example, for C_8E_4 cmc was obtained to be $7.1 \cdot 10^{-3}$ molal and the free surfactant concentration increases to ca. $9 \cdot 10^{-3}$ molal at 10 x cmc.

3. MICELLE SELF-DIFFUSION AND MICELLE SIZE

Micelle self-diffusion coefficients (D_m) can be obtained either from the molecular D value of the surfactant (at concentrations >> cmc and in the absence of direct intermicellar exchange; cf. below) or from D of a probe confined to the micelle. For a number of zwitterionic and nonionic micelles /17/ the concentration dependence of D_m follows

$$D_m = D_m^o (1 - k\emptyset)$$

with k = 1.7 within 10%. (The micelle volume fraction $\emptyset$ is taken to include the water of hydration). Such a behaviour is close to theoretical predictions for hard spheres with hydrodynamic interactions. (Cf. lecture by R. KLEIN at this meeting).

From D_m^o we obtain using the Stokes-Einstein equation, hydrodynamic radii of the micelles that are close to or somewhat below the lengths of the extended surfactant molecules.

For the C_xE_y surfactants investigated, spherical micelles are found at low concentrations /17-19/. For $C_{12}E_8$ the rate of decrease of D_m with increasing concentration gives strong evidence for spherical micelles (or closely spherical) up to very high concentrations while with a shorter EO-chain, like $C_{12}E_5$, there is a very pronounced growth with increasing concentration /18, 19/. This difference in behaviour is not unexpected in view of the occurrence at higher concentrations of a cubic l.c. phase built up of small globular units for $C_{12}E_8$ and a hexagonal l.c. phase of very long rod-shaped aggregates for $C_{12}E_5$.

4. MICELLE SIZE AT HIGHER TEMPERATURES

In our first study /18/, we demonstrated from the D_m values obtained at low concentrations, where intermicellar interactions are negligible, that $C_{12}E_5$ micelles grow dramatically with increasing temperature while there is little growth for $C_{12}E_8$. Later, self-diffusion studies /20,21/ have documented a major growth with temperature for $C_{12}E_6$; these investigations were also concerned with some other aspects such as the distinction between micellar growth effects and critical fluctuations through a combination of self-diffusion and mutual diffusion (dynamic light scattering).

The general pattern of micelle growth illustrated in Fig. 1 has also been observed in studies by other techniques, for example, NMR relaxation /18/, fluorescence quenching /22/ and neutron spin-echo (DEGIORGIO's lecture at this symposium).

The growth of the minimal spherical micelles into larger units can be envisaged to occur in different ways, mainly in a "real" growth involving a growth of the hydrocarbon cores or an aggregation of small micelles which retain their identity. The former mechanism has been established to be the general one for micelle growth to rods in many classes of surfactants. There is direct evidence for a growth of the hydrocarbon cores also for systems like $C_{12}E_5$ and $C_{12}E_6$ from NMR relaxation /18/ and fluorescence quenching /22/. Especially for surfactants with large EO groups, the other mechanism is also probably significant, but this is as yet little investigated.

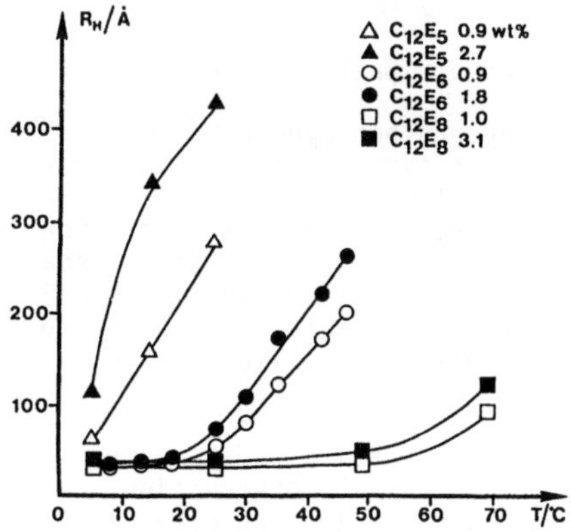

Fig. 1. Temperature dependence of hydrodynamic radii of nonionic micelles from self-diffusion data. Self-diffusion data are from Ref. 18 ($C_{12}E_5$ and $C_{12}E_8$) and Ref. 21 ($C_{12}E_6$, cf. also Ref. 20)

5. MICELLE SHAPE

It is straightforward to investigate if there is micellar growth or not, but much more difficult to establish which nonspherical shape the micelles have in case of growth. However, as shown by JÖNSSON et al. /23/ the self-diffusion of small molecules in colloidal systems is strongly dependent on the shape of the colloidal particles; in particular, the obstruction effect of large oblate or disc-shaped particles is much more important than that of large prolate or rod-shaped particles. These principles were used to quite strongly support the notion of (large and flexible) disc-shaped micelles in the L_3 phase and rod-shaped micelles (in case of growth) in the normal micellar or L_1 phase /24,19/. In recent work we have been interested in further investigating this problem and in particular to study aggregate properties close to two-phase regions.

Using the principles outlined previously /24/ water self-diffusion data have been evaluated to obtain the obstruction factor, A (ratio between D_{water} in the presence and absence of obstructing particles after accounting for hydration), for the $C_{12}E_5$ system. While micelles are large according to surfactant self-diffusion we do not observe in the vicinity of the lower consolute boundary any deviations from the obstruction factor expected for rod micelles. Furthermore, we do not find any changes

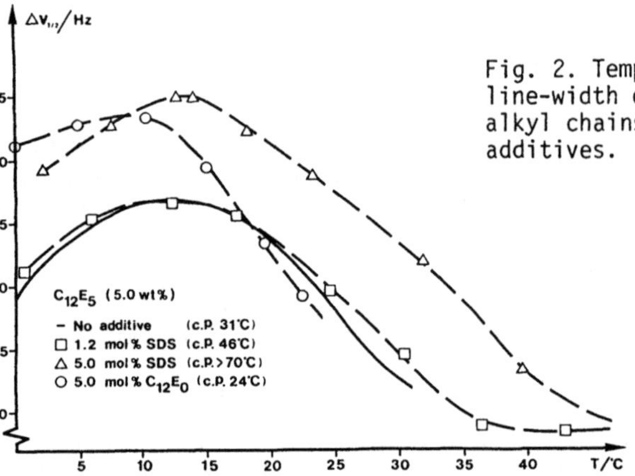

Fig. 2. Temperature dependence of the ^1H NMR line-width of the main methylene peak of the alkyl chains for $C_{12}E_5$ in the presence of additives.

in addition to the general temperature effect in either surfactant self-diffusion or ^1H NMR transverse relaxation of surfactant methylenes close to the lower consolute boundary for the $C_{12}E_5$, C_8E_4 and $C_{12}E_8$ systems. There is thus no indication of any particular change in either micelle size and shape due to the closeness to the phase separation curve. Furthermore, we prepared samples above the lower consolute boundary in the two-phase region and let them phase separate. Measurements on these phases did not indicate any change in character of the solutions as one passes the cloud point.

According to these experiments, micelle size and shape do not appear to be directly related to the distance to the cloud point. Another type of studies indicating that micelle growth is rather primarily determined by temperature involves displacing the cloud point by adding small amounts of additive. For example, replacing 1% of $C_{12}E_5$ by SDS (by mole) for a 5% $C_{12}E_5$ solution, which displaces the phase separation from 32 to 46°C, has no detectable influence on either self-diffusion or NMR relaxation (Fig. 2) at any temperatures investigated (0-50°C). (Refs. 25 and 19 and unpublished work). At higher SDS additions there are changes in micelle size and flexibility.

For the $C_{12}E_5$ system we also investigated the micellar properties on approach of the lamellar l.c. region. It is striking that no changes in the obstruction factor that could indicate the formation of large disc micelles were observed, at least at 25°C. This was so even for micellar solutions in equilibrium with the lamellar phase. (At 40°C a minor decrease in the obstruction factor was observed, but it was too small to be indicative of large disc micelles). On the other hand, large rod micelles are found on the approach of the hexagonal phase for this and other systems.

6. INTERACTIONS AND MECHANISMS

The change over from net repulsive to attractive intermicellar interactions with increasing temperature implied by the phase behaviour has also been deduced in surfactant self-diffusion /18,19/ and light scattering /20,21/ studies. CLAESSON et al. /26/ have in direct force measurements between surfaces onto which $C_{12}E_5$ was adsorbed demonstrated a change over from repulsive to attractive interactions as temperature is increased.

This change in intermicellar interactions has frequently been referred to a dehydration of the EO groups at higher temperatures, but it seems that this notion has seldom been documented by direct experiments. Water self-diffusion measurements /24/ have shown that the number of water molecules diffusion with the micelles decreases with increasing temperature (ca. 5.5 water per EO at 5°C, 4.5 at 25°C and 4 at 50°C). However, the decrease is slow and progressive and not dramatic and concentrated to the region of the cloud point, as has been suggested.

A decreased hydration allows a denser packing of the polar head groups, thus favouring micellar growth and a change in aggregate shape sphere → rod → disc with increasing temperature. A decreased hydration favours both by direct and indirect mechanisms phase separation. The direct mechanism concerns the decreased intermicellar repulsion that results from dehydration, while the indirect mechanism concerns the fact that phase separation increases in importance with micelle size.

A more fundamental question relates to the mechanism of the decreased hydration. Based on the observation /27/ that a gauche orientation around the C-C bonds together with anti orientation around the C-O bonds in a poly(ethylene oxide) chain has a lower energy in a polar solvent than other orientations, KARLSTRÖM /28/ has presented a model relevant to this observation. Thus, the particular conformation mentioned creates a segment of the chain which has a rather large dipole moment and therefore would be expected to interact favourably with water. With increasing temperature also other less polar conformations become populated. Such a simple notion of an oligo(ethyleneoxide) chain that changes conformation and becomes less polar with increasing temperature /28/ can also explain a large number of other observations for these systems /12/, but certainly it needs further examination.

An attempt to provide some further insight into chain conformation and packing is provided by a recent ^{13}C NMR chemical shift and relaxation study /29/. The chemical shifts are mainly determined by chain conformation and can be used to extract at least qualitative information on trans-gauche equilibria around the C-C bonds. For the alkyl chains in $C_{12}E_5$ higher trans population is found for micelles both relative to the aqueous monomer and the neat liquid surfactant. With increasing temperature an increased gauche population is found, a general result for many micellar and nonmicellar situations. For the EO chains an opposite temperature dependence was found, as predicted by theory. Micelle formation from monomers increases the fraction of trans conformers in the EO chains and a further increase is noted from micelles to neat liquid. Both these observations agree with the notion that conformation is significantly influenced by the medium polarity.

7. ACKNOWLEDGEMENTS

Håkan Wennerström is thanked for useful comments on the manuscript. The collaboration on nonionic surfactants with Per-Gunnar Nilsson, Håkan Wennerström, Bernard Faucompré, Thomas Ahlnäs, Gunnar Karlström and others is gratefully acknowledged.

8. LITERATURE

1. J.C. Lang, R.D. Morgan: J. Chem. Phys. 73, 5849 (1980)
2. D.J. Mitchell, G.J.T. Tiddy, L. Waring, T. Bostock, M.P. McDonald: J. Chem. Soc. Faraday 1, 79, 975 (1983)
3. M. Corti, C. Minero, V. Degiorgio: J. Phys. Chem. 88, 309 (1984)
4. V. Degiorgio: In Physics of Amphiphiles, ed. by V. Degiorgio and M. Corti (North-Holland, Amsterdam, 1985) p. 303.
5. R. Kjellander: J. Chem. Soc. Faraday 2, 78, 2025 (1982)
6. G.M. Thurston, D. Blankschtein, M.R. Fisch, G.B. Benedek: J. Chem. Phys. 84, 4558 (1986)
7. D. Blankschtein, G.M. Thurston, G.B. Benedek: J. Chem. Phys. 85, 7268 (1986)
8. F.E. Bailey, Jr, J.V. Koleske: In Polyethyleneoxide, (Academic Press, New York, 1976)
9. S. Saeki, N. Kuwahara, M. Nakara, M. Kaneko: Polymer 17, 685 (1976)
10. R. Kjellander, E. Florin: J. Chem. Soc. Faraday 1, 77, 2053 (1981)
11. G. Karlström: J. Phys. Chem. 87, 4762 (1985)
12. B. Lindman, G. Karlström: Z. Phys. Chem. in press.
13. P. Stilbs: Progr. NMR Spectroscopy 19, 1 (1987)
14. J. Ulmius, H. Wennerström: J. Magn. Resonance 28, 309 (1977)
15. B. Lindman, O. Söderman, H. Wennerström: In Surfactant Solutions, ed. by R. Zana (Marcel Dekker, New York, 1987) p. 295

16. G. Gunnarsson, B. Jönsson, H. Wennerström: J. Phys. Chem. 84, 3114 (1980)
17. B. Faucompré, B. Lindman: J. Phys. Chem. 91, 383 (1987)
18. P.G. Nilsson, H. Wennerström, B. Lindman: J. Phys. Chem. 87, 4548 (1983)
19. P.G. Nilsson, H. Wennerström, B. Lindman, Chem. Scripta 25, 67 (1985)
20. W. Brown, R. Johnson, P. Stilbs, B. Lindman: J. Phys. Chem. 87, 4548 (1983)
21. T. Kato, T. Seimiya: J. Phys. Chem. 90, 3159 (1986)
22. J.E. Löfroth, M. Almgren: In Surfactants in Solution, ed. by K.L. Mittal and B. Lindman, Vol. 1 (Plenum, New York, 1984) p. 627. See also more recent work by the groups of Zana, Grieser and Turro.
23. B. Jönsson, H. Wennerström, P.G. Nilsson, P. Linse: Colloid & Polymer Sci. 264, 77 (1986)
24. P.G. Nilsson, B. Lindman: J. Phys. Chem. 88, 4764 (1984)
25. P.G. Nilsson, B. Lindman: J. Phys. Chem. 88, 5391 (1984)
26. P.M. Claesson, R. Kjellander, P. Stenius, H.K. Christenson: J. Chem. Soc. Faraday 1, 82, 2735 (1986)
27. M. Andersson, G. Karlström: J. Phys. Chem. 87, 4757 (1985)
28. G. Karlström: J. Phys. Chem. 87, 4762 (1985)
29. T. Ahlnäs, G. Karlström, B. Lindman: J. Phys. Chem. in press.

SANS Study of Structure, Growth, and Polydispersity of Short-Chain Lecithin Micellar Systems. A Ladder Model Analysis

S.H. Chen, T.L. Lin, and C.F. Wu*

Nuclear Engineering Dept. and Center for Material Science and Engineering, 24-211, Massachusetts Institute of Technology, Cambridge, MA 02139, USA

A method for analyzing small angle neutron scattering data from polydispersed rod-like micelles is presented and applied to micellar systems formed by short-chain lecithins in aqueous solutions. The thermodynamic theory developed for these micellar aggregates allows us to extract quantitatively the minimum micelle size, the size distribution, and the growth of micelles as a function of lecithin concentration. The criterion for growth at a given concentration depends on the minimum micelle size and the difference between the chemical potentials of a monomer in the end caps and in the straight section of the cylindrical micelle.

I. Introduction

Short-chain lecithins (6-8 carbons per fatty acyl chain) are useful synthetic phospholipids which form micelles in aqueous solutions whose average size depends strongly on the fatty acid chain lengths[1-3]. This is in contrast to the long-chain phospholipids ($\geq$ 12 carbons per fatty acyl chain) which form bilayers and thus are important as model systems for biological membranes.

The shortest of the micellar lecithins, dihexanoylphosphatidylcholine (diC_6PC), has been studied extensively by high resolution NMR[4] and by small angle neutron scattering (SANS)[5] techniques. The diC_6PC micelles exhibit no appreciable growth with increasing lecithin concentration in the range of 27 mM to 360 mM[5]. The SANS data were fitted by a prolate ellipsoid structure of 19 ± 1 molecules with the major and minor axes for the hydrophobic core of fatty acyl chains equal to 24 Å and 7.8 Å, respectively. The next homologue, diheptanoylphosphatidylcholine (diC_7PC), behaves quite differently from diC_6PC[6]. The addition of an extra methylene group to each chain causes the micelles to become very polydispersed and to grow rapidly with increasing diC_7PC concentrations[2,3]. Analogous comparisons can also be made between diC_6PC, diC_7PC, and other asymmetric short-chain lecithin micellar systems such as

---

*Present address: Nuclear Engineering Department, National Tsing-Hua University, Hsinchu, Taiwan, ROC.

$1\text{-}C_6\text{-}2\text{-}C_8PC$ and $1\text{-}C_8\text{-}2\text{-}C_6PC$, both with 14 carbons, and $1\text{-}C_7\text{-}2\text{-}C_8PC$ and $1\text{-}C_8\text{-}2\text{-}C_7PC$, both with 15 carbons[7]. The growth and polydispersity at a given concentration increase rapidly with the total carbon number. But within the same homologous group (same carbon number) one can observe differences among asymmetric lecithins with different distributions of carbons in the two inequivalent chains[7]. Therefore a detailed comparative structural and polydispersity study for a series of short-chain lecithin micelles would greatly increase our understanding of the geometry of the packing of the hydrophobic tails in the micellar core and its relation to the hydrophobic free energy of micellization. In this article we shall discuss (i) an experimental technique to determine the rod-like micellar structure and, (ii) a thermodynamic model which quantitatively accounts for the observed micellar growth and size distribution for any of the short-chain lecithin homologues mentioned above.

II. A Thermodynamic Theory of Micellization and Micellar Growth

In a micellar solution we imagine that there are chemical equilibria between the monomers and the aggregates of various sizes denoted by the aggregation numbers n and m. The thermodynamic conditions for these multiple chemical equilibria can be expressed as

$$\tilde{\mu}_n = \tilde{\mu}_m \,, \qquad \text{for } n, m = 1, 2, \ldots \tag{1}$$

where $\tilde{\mu}_n$ and $\tilde{\mu}_m$ are the chemical potentials of a monomer in aggregates of sizes n and m, respectively. We can now write the standard expressions for these chemical potentials as

$$\tilde{\mu}_n = \tilde{\mu}^o_n + \frac{kT}{n}\ell n X_n + \chi_n \,, \tag{2}$$

$$\tilde{\mu}_m = \tilde{\mu}^o_m + \frac{kT}{m}\ell n X_m + \chi_m \,, \tag{3}$$

where k is the Boltzmann constant, the first term on the right hand side expresses the standard chemical potential, the second term is the contribution from the entropy of dispersion, and the third is from the interaction. Using eqs. (2) and (3) in eq. (1) we obtain the ratio of molar fractions given by

$$\frac{(X_n)^m}{(X_m)^n} = \exp[\frac{nm}{kT}(\tilde{\mu}^o_n - \tilde{\mu}^o_m)]\exp[-\frac{nm}{kT}(\chi_n - \chi_m)] \,. \tag{4}$$

Since lecithins are zwitterionic the inter-aggregate interaction is expected to be weakly attractive. Furthermore, the volume fractions of lecithins in our experiments are all below 2%. A simple virial expansion of the interaction term is then appropriate. Following Ben-Shual and Gelbart[8] we write

$$\chi_n = -\frac{kT}{n}\ell n\, \gamma_n = -\frac{kT}{n}\sum_s \beta_1(n,s)\rho_s \,, \tag{5}$$

where γ_n is the activity coefficient of the monomer incorporated in an n-mer, $\beta_1(n,s)$ is the first order virial coefficient, and ρ_s is the number density of the monomer in an aggregate of size s. According to Onsager[9] the first order virial coefficient of rod-like particles is dominated by the overlapping volume of the two cylinders of lengths ℓ_n and ℓ_s, namely,

$$\beta_1(n,s) \sim \ell_n \, \ell_s \sim n\, s \; . \tag{6}$$

The length ℓ_n of the micelle is proportional to its aggregation number n because it has a compact core. Use of eq. (6) in eq. (5) would result in χ_n which is independent of n, this in turn implies that the last exponential factor in eq. (4) is unity. This conclusion is significant in the sense that the micellar distribution function is insensitive to the inter-aggregate interaction at least to the first order in the density. For a further application to SANS data analysis we are interested in the size distribution of micelles in terms of the mole fraction of the monomer X_1. We therefore set m = 1 in eq.(4) to get

$$X_n = (X_1)^n \exp\left[\frac{n}{kT}\,(\tilde{\mu}_n^o - \tilde{\mu}_1^o)\right] \; . \tag{7}$$

Equation (7) means that the probability of having a micelle of size n is proportional to two factors: the first is the probability of finding n monomers in the same spatial location, and the second is a Boltzmann factor which expresses the free energy advantage of packing n monomers together to form a micelle of size n. This energy advantage is essentially derived from the hydrophobic interaction of the hydrocarbon tails in an aqueous environment. We shall call $E_n \equiv \tilde{\mu}_n^o - \tilde{\mu}_1^o$ a micellization energy, which is typically about 10 kT per monomer for lecithin micelles.

The Ladder Model

The ladder model was formulated originally by Tausk and Overbeek[10] for application to lecithin micellar systems. It was later developed in great detail by Mazer et al.[11] and by Missel et al.[12] for SDS micellar systems with high salt. We shall briefly summarize the results here in a form convenient for SANS analysis. The basic assumptions of the model are: (i) the aggregation number of a micelle can only be equal to or greater than a minimum number n_0 which is largely determined by the geometrical packing condition of the tails in the minimum micelle, and (ii) the energy of micellization takes the form[13]

$$n(\tilde{\mu}_n^o - \tilde{\mu}_1^o) = \Delta + (n - n_0)\delta, \qquad \text{for } n \geq n_0 \; , \tag{8}$$

where $\Delta < 0$ is the step change (lowering) in chemical potential when n_0 monomers come together to form a minimum micelle. For any further addition of a monomer to the minimum micelle, the lowering in the chemical potential of a monomer in the micelle is a constant, $\delta < 0$. Obviously the condition for forming larger micelles is that $-\delta$ is larger than $-\Delta/n_0$ or, equivalently, $(\Delta - n_0\delta)/kT$ is greater than zero. Applying eq. (8) to eq. (7) and defining

$$\beta \equiv X_1 \, e^{-\delta/kT} \tag{9}$$

we get the micelle distribution function as

$$X_n = \beta^n \exp[-(\Delta - n_0\delta)/kT] \equiv \beta^n/K, \text{ for } n \geq n_0 . \tag{10}$$

It is interesting to note from eq. (10) that the micelle size distribution function is a geometrical distribution.

In order to determine β as a function of lecithin concentration we now impose a material conservation relation

$$X = X_1 + \sum_{n=n_0}^{\infty} n X_n , \tag{11}$$

which can easily be summed to obtain

$$X = \beta \, e^{\delta/kT} + \frac{1}{K} \{ n_0 \beta^{n_0} [\frac{1}{1-\beta} + \frac{\beta}{n_0(1-\beta)^2}] \} , \tag{12}$$

where X is the total mole fraction of lecithin in the solution. It is clear from eq. (12) that in order to determine β, and hence the size distribution function, at each lecithin concentration we need three input model parameters, namely, n_0, K, and δ/kT. Using eq. (10) we can easily calculate the weight-averaged aggregation number, n_w,

$$n_w \equiv \langle n^2 \rangle / \langle n \rangle = n_0 + \frac{\beta}{1-\beta} [1 + \frac{1}{n_0(1-\beta) + \beta}] , \tag{13}$$

and the number-averaged aggregation number n_n,

$$n_n \equiv \langle n \rangle = n_0 + \frac{\beta}{1-\beta} . \tag{14}$$

From these two mean aggregation numbers, the polydispersity index p can be conveniently defined as

$$p \equiv \frac{\sqrt{\langle n^2 \rangle - \langle n \rangle^2}}{\langle n \rangle} = \sqrt{\frac{n_w}{n_n} - 1} = \frac{\sqrt{\beta}}{n_0(1-\beta) + \beta} . \tag{15}$$

We shall give numerical values for these parameters later in Table I.

III. SANS Cross Section of Polydispersed Rod-like Micelles

The scattering cross section per unit volume denoted by I(Q), with a dimension 1/cm, can be written as[(6)]

$$I(Q) = \sum_{n=n_0}^{\infty} \frac{C_n}{n} n^2 (b_m - v_m \rho_s)^2 \tilde{P}_n(Q) , \tag{16}$$

where

C_n = number density of monomers which form n-mers,
b_m = scattering length of a monomer,
v_m = dry volume of a monomer,
ρ_s = scattering length density of the solvent,
$\tilde{P}_n(Q)$ = normalized form factor of the n-mer.

Using D_2O as a solvent we have $X_n = 20\ C_n/(N_A n)$ and $\sum_{n=n_0}^{\infty} C_n = C - \text{cmc}$, where N_A is Avogadro's number, C is the total monomer concentration, and cmc is the critical micelle concentration, we can rewrite eq. (16) as

$$I(Q) = I(0) \langle \tilde{P}_n(Q) \rangle \tag{17}$$

with

$$I(0) = n_w\ (C - \text{cmc})\ (b_m - v_m\rho_s)^2 \tag{18}$$

and

$$\langle \tilde{P}_n(Q) \rangle \equiv [\sum_{n=n_0}^{\infty} n^2 \beta^n \tilde{P}_n(Q)] / [\sum_{n=n_0}^{\infty} n^2 \beta^n] \ . \tag{19}$$

The function $\tilde{P}_n(Q)$ is the normalized form factor for a randomly oriented cylinder of length L and cross sectional area πR^2, it can be shown[6] that

$$\tilde{P}_n(Q) = \frac{1}{2}\int_{-1}^{1} d\mu \left[\frac{\sin(\frac{1}{2} QL\mu)}{\frac{1}{2} QL\mu} \right]^2 \left[\frac{2J_1(QR(1-\mu^2))}{QR(1-\mu^2)} \right]^2 . \tag{20}$$

The aggregation number n is related to L and R by $nv_m = \pi R^2 L$. At a given Q, for a sufficiently long cylinder such that $QL > 2\pi$, one has, to a good approximation,

$$\left[\frac{\sin(\frac{1}{2} QL\mu)}{\frac{1}{2} QL\mu} \right]^2 \xrightarrow{QL > 2\pi} \frac{2\pi}{QL}\, \delta(\mu) \ , \tag{21}$$

therefore for large QL and small QR we have the approximation

$$\tilde{P}_n(Q) \xrightarrow{QL > 2\pi} \frac{\pi}{QL} \left[\frac{2J_1(QR)}{QR} \right]^2 \xrightarrow{QR < 1} \frac{\pi}{QL}\, e^{-\frac{1}{4}Q^2R^2} \ . \tag{22}$$

This is illustrated in Fig. 1 for the L = 100Å and R = 25Å case. It is seen from the figure that the Guinier approximation is reasonable for X = QR < 1. Thus a practical method for testing whether the shape of the micelles is cylindrical or not is to plot $\ell n[QI(Q)]$ vs. Q^2, according to the equation

$$\ell n[QI(Q)] = \ell n[\pi(b_m - v_m\rho_s)^2 \frac{n_w}{L}] - \frac{1}{4} Q^2R^2 \ . \tag{23}$$

The slope of the straight line should then give the value of R and the Q = 0 intercept the mean aggregation number per unit length n_w/L. One can always refine these parameters by using the exact formula, eq. (20), to fit the curve for the entire Q range.

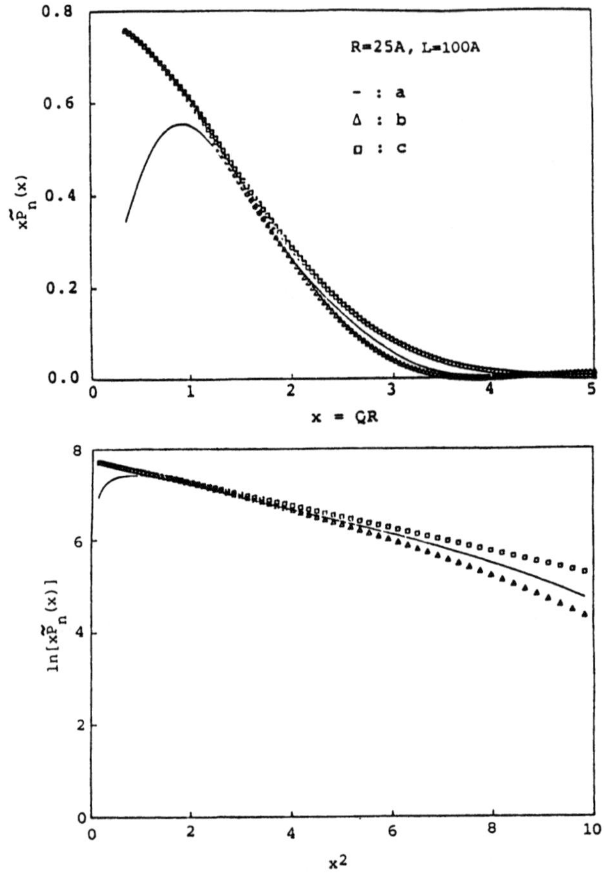

Fig. 1 Curves showing $x\tilde{P}_n(x)$ vs. x and $\ln[x\tilde{P}_n(x)]$ vs. x^2 for a cylindrical micelle with R=25A and L=100A. In both figures, curve a represents the exact expression for $\tilde{P}_n(x)$ as in eq.(20), curve b is the 1st approximation, $\tilde{P}_n(x)=(\pi/QL)[2J_1(x)/x]^2$, and curve c is the Guinier approximation of b, $\tilde{P}_n(x)=(\pi/QL)\exp(-x^2/4)$.

IV. Discussion of the Results from SANS Data Analysis

The basic structural model of diC_6PC we arrived at by the SANS technique is depicted in Fig. 2[(5)]. It has a constant average aggregation number n_w = 19 in the whole observed concentration range from 27 mM to 361 mM. The micelle has a globular shape which is consistent with the packing of monomers with a tail volume v_T = 324 Å^3 and a fully stretched tail length of 7.83 Å, the head with a volume v_h = 346 Å^3 extends in a direction roughly perpendicular to the tail and occupies an area a_H = 102 Å^2. The neutron scattering intensity distributions as a function of Q and monomer concentration are shown in Fig. 3. It should be noted that the depression of the intensities in the low Q region as exemplified in curves 1-5 is

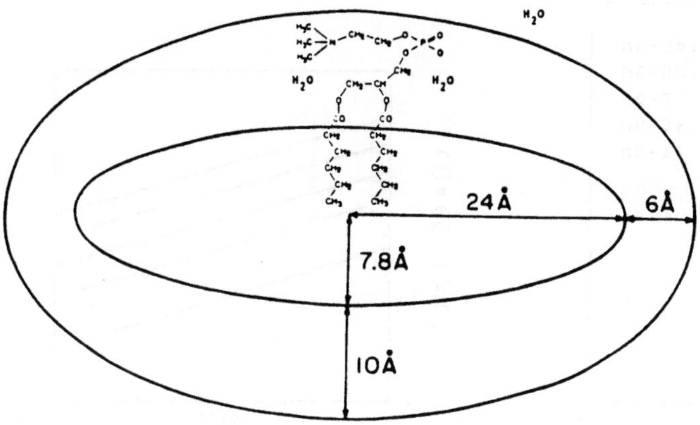

Fig.2 The proposed micellar structure for dihexanoyl-PC. The fatty acyl chains form a close-packed ellipsoidal core and the head groups stick out in the solvent. This two-dimensional figure shows the general position of the tails and head group of the monomer but not their exact conformation in the micellar state.

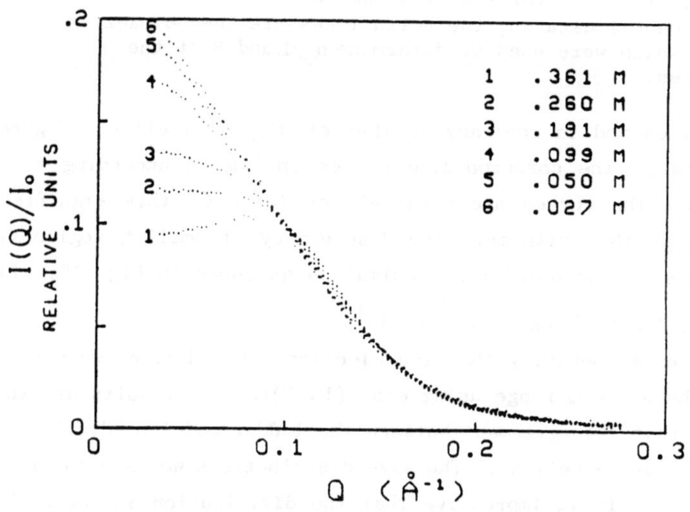

Fig.3 The scattering intensity profiles of diC_6PC at various concentrations are scaled by their scattering amplitude. These curves are seen to coincide very well for $Q>0.1Å^{-1}$, implying that the micellar size and the structure do not change with lecithin concentration.

due to the intermicellar interaction[5]. It is therefore concluded that for lecithin concentration less than about 30 mM the intermicellar interaction effect can be neglected in the intensity distribution.

In Fig. 4 we show the intensity distributions for diC_7PC micelles in D_2O in the concentration range 2.2 - 35 mM[7]. One striking feature observed is the bending upward of all the curves in the small Q region, this is due to the

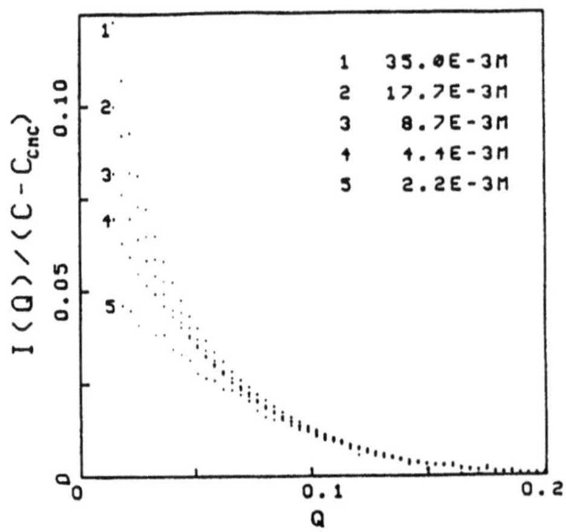

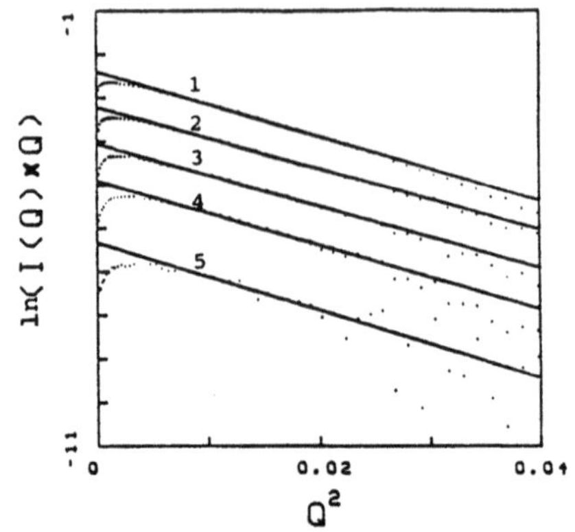

Fig.4 SANS intensity profiles normalized by micellar concentration for samples containing 2.2, 4.4, 8.7, 17.7, and 35.0 mM diC_7 lecithin are plotted against Q.

Fig.5 Plots of ln[QI(Q)] vs. Q^2 for the same samples as in Fig.4. Dots are experimental data and the solid lines are the Guinier approximations which were used to determine n_w/L and R of the rod-like micelles.

polydispersity of micelles and is contrary to that of diC_6PC micelles. Figure 5 shows the Guinier plots of the corresponding curves in Fig. 4 according to eq. (23). We note that all the curves are parallel for large Q, this signifies a constant cross section of the cylinder. The discrepancy at small Q region from the straight line is due to the Guinier approximation as shown in Fig. 1B. From these plots we obtained R = 17 Å and nλ = 0.76 $Å^{-1}$.

After identifying experimentally that the micelles are rod-like, we can then fit the SANS data at the whole Q range using eqs. (18-20). The results are shown in Fig. 6. It is seen that the agreement between the ladder model predictions and the experimental results are excellent. The size distributions we used to fit the data are given in Fig. 7. It is impressive that the distribution at the highest concentration, which is 25 times cmc, is so broad as to cover a range from n = 27 to 400. The computed parameters are given in Table I.

From the ladder model one can derive the criterion for an appreciable growth beyond a certain concentration X as given by the equation

$$X - X_{cmc} > 0.8\, n_0^2 / K . \tag{24}$$

From data in Table I we get the RHS of eq. (24) to be 4.1 x 10^{-5} which means that one would see appreciable growth above 3.5 mM, in agreement with the experimental observations as shown in Table I.

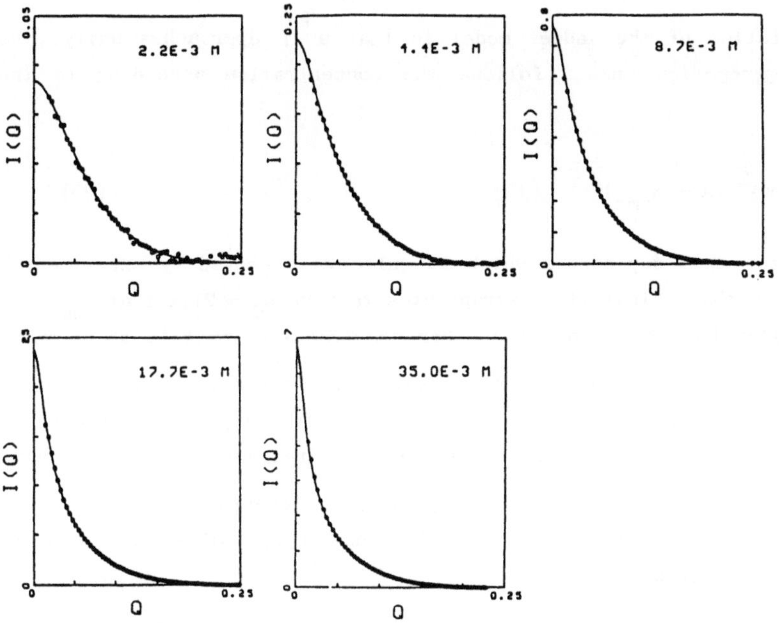

Fig.6 Theoretical(solid curves) and experimental SANS intensity data are plotted against Q for diC_7 at different concentrations.

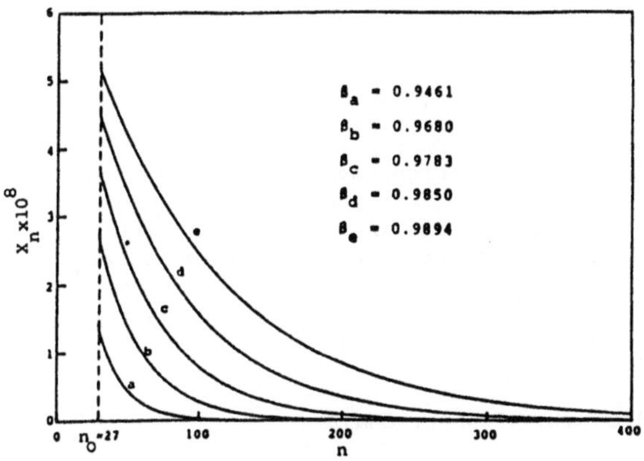

Fig.7 Curves of size distribution X_n vs. aggregation number n for diC_7PC lecithin micelles at different concentrations. The corresponding β values taken from Table I are given.

Table I. Ladder Model Parameters for diC_7PC Micelles.

C(mM)	β	p(%)	n_w
2.2	0.9461	40	48
4.4	0.9680	54	74
8.7	0.9783	63	101
17.7	0.9850	71	140
35.0	0.9894	78	193

cmc=1.4mM, R=17Å, $n_w/L=0.75\text{Å}^{-1}$, $n_0=27$, $(\Delta-n_0\delta)/kT=16.46$, and $\delta/kT=-10.48$ were used to obtain this table.

Another prediction of the ladder model is that as β approaches unity, the weight-averaged aggregation number follows the concentration according to the relation

$$n_w = n_0 + 2\,(K)^{\frac{1}{2}}\,(X - X_{cmc})^{\frac{1}{2}}\,. \qquad (25)$$

Figure 8 illustrates this dependence which is quite well obeyed by experimental data. Note that the theoretical line extrapolates to give $n_0 = 27 \pm 1$ at X_{cmc}.

Similar analyses have been done for other homologous asymmetric short-chain lecithins(7). Here we shall only quote the results of the analyses in Fig. 9 and Table II. From Fig. 9, for example, we can infer that for diC_8PC, n_0 would be 55 and $(\Delta - n_0\delta)/kT = 28$. These predictions are in agreement with the results obtained by Blankschtein et al.(14) from analysis of the coexistence curve. From these numerical values, we expect from eq. (24) that diC_8PC micellar system is already very polydispersed at cmc.

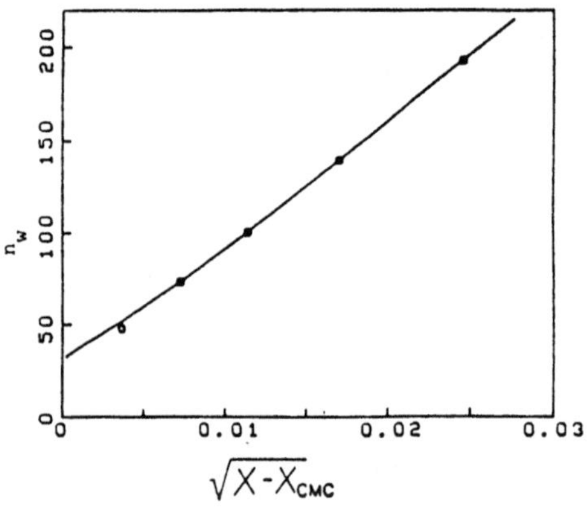

Fig.8 The weight-averaged aggregation number is plotted against $(X-X_{cmc})^{1/2}$. The solid line is theoretical curve and the points are experimental data.

Table II. Parameters Derived from SANS Data of Short-Chain Lecithin Micelles.

Lecithin	n_0	δ/kT	$(\Delta-n_0\delta)/kT$	R(Å)	n_w/L(Å^{-1})
diC_6PC	16		7		
$1-C_6-2-C_7PC$	25±2	-10.65±.02	12.09±.02	16.8	0.75±.07
diC_7PC	27±1	-10.48±.02	16.46±.02	18.0	0.75±.02
$1-C_6-2-C_8PC$	32±5	-10.57±.02	16.08±.03	17.8	0.85±.06
$1-C_8-2-C_6PC$	27±5	-10.68±.03	15.17±.02	18.0	0.82±.04
$1-C_7-2-C_8PC$	48±10	-11.20±.01	23.52±.05	21.5	1.06±.10
$1-C_8-2-C_7PC$	45±10	-11.34±.01	22.71±.05	19.5	0.94±.06

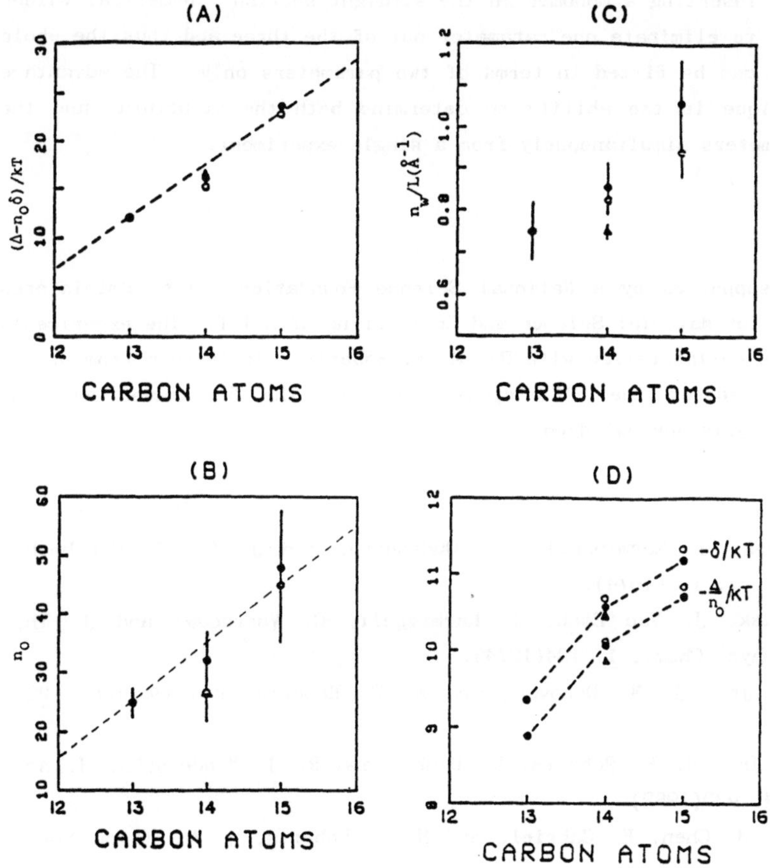

Fig.9 The values of $(\Delta-n_0\delta)/kT$, n_0, n_w/L, and $-\delta/kT$ and $-(\frac{\Delta}{n_0})/kT$ for short-chain lecithin micelles determined by SANS data fitting are plotted vs. the total number of carbon atoms of the two fatty acyl chains. Numerical data are also given in Table II. Asymmetric lecithins with the S1 chain shorter than its S2 chain are represented by filled circles while those with longer S1 chain are shown by open circles. Triangles represent the results of diheptanoyl-PC.

V. Conclusion

We have demonstrated experimentally that short-chain lecithins form rod-like micelles in aqueous solutions. The growth and polydispersity of a specific lecithin system depend strongly on the total carbon number and the distribution of the carbons in the two inequivalent chains as illustrated in Fig. 9. One can formulate a thermodynamic model to account for the growth and polydispersity quantatively. The basic parameters of the theory are n_0, the minimum micelle size, $(\Delta - n_0\delta)/kT$, the free energy advantage of putting n_0 monomers in the straight section of the cylinder as compared to that in the end caps, and δ/kT, the free energy of inserting a monomer in the straight section. Numerical values of cmc can be used to eliminate one parameter out of the three and thus the whole range of SANS data can be fitted in terms of two parameters only. The advantage of the SANS technique is the ability to determine both the structural and the thermodynamic parameters simultaneously from a single experiment.

Acknowledgements

This research is supported by a National Science Foundation grant administered through the Center for Material Science and Engineering of M.I.T. The experiments were carried out in collaboration with Dr. M. F. Roberts. Assistance from Dr. D. Schneider in the use of the SANS spectrometer at the Brookhaven National Laboratory is gratefully acknowledged.

REFERENCES

1. R. J. M. Tausk, J. Karminggelt, C. Oudshoorn, and J. Th. G. Overbeek, Biophys. Chem., 1, 175(1974).
2. R. J. M. Tausk, J. Van Esch, J. Karmiggelt, G. Voordouw, and J. Th. Overbeek, Biophys. Chem., 1, 184(1974).
3. R. A. Burns, Jr., J. M. Donovan, and M. F. Roberts, Biochemistry, 22, 964(1983).
4. R. A. Burns, Jr., M. F. Roberts, R. Dluhy, and R. J. Mendelsohn, J. Am. Chem. Soc., 104, 430(1982).
5. T. L. Lin, S. H. Chen, E. Gabriel, and M. F. Roberts, J. Am. Chem. Soc., 108, 3499(1986).
6. T. L. Lin, S. H. Chen, E. Gabriel, and M. F. Roberts, J. Phys. Chem., 91, 406(1987).
7. T. L. Lin, S. H. Chen, and M. F. Roberts, J. Am. Chem. Soc., 109, 2321(1987).
8. A. Ben-Shaul and W. M. Gelbart, J. Phys. Chem., 86, 316(1982).
9. L. Onsager, Am. N. Y. Acad. Sci., 51, 627(1949).
10. R. J. M. Tausk and J. Th. G. Overbeek, Colloid Interface Sci., 2, 379(1976).
11. N. A. Mazer, M. C. Garey, and G. B. Benedek, in Micellization, Solubilization, and Microemulsions, ed. by K. L. Mittal(Plenum, New York 1977).
12. P. J. Missel, N. A. Mazer, G. B. Benedek, C. Y. Yong, and M. C. Garey, J. Phys. Chem., 84, 1044(1980).
13. J. N. Israelachvili, "Physics of Amphiphiles: Micelles, Vesicles, and Microemulsions", International School of Physics 'Enrico Fermi', Varenna, 1983.
14. D. Blankschtein, G. M. Thurston, and G. B. Benedek, J. Chem. Phys., 85, 7268(1986).

Theory of Thermodynamic Properties and Phase Separation of Micellar Solutions with Upper and Lower Consolute Points

D. Blankschtein[1], *G.M. Thurston*[2], *and G.B. Benedek*[2]

[1]Dept. of Chemical Engineering, Massachusetts Institute of Technology, Cambridge, MA 02139, USA

[2]Dept. of Physics and Center for Materials Science and Engineering, Massachusetts Institute of Technology, Cambridge, MA 02139, USA

1. Introduction

In this paper we review our work on the thermodynamic properties and phase separation of micellar solutions. A detailed exposition may be found in Refs.[1], [2] and [3].

Micelles are noncovalently bonded macromolecular aggregates which are continually and reversibly exchanging amphiphiles with one another and with the amphiphilic monomers in the solution [4]. As a result the individual micelles do not maintain a distinct, unchanging identity. Instead, a description of the state of the micellar solution requires a specification of the distribution of micellar sizes which exists on average in the solution and is governed by the thermodynamic principle of multiple chemical equilibrium [5]. It is essential to recognize, in distinction to other multicomponent mixtures, that the micellar size distribution is sensitively dependent upon solution conditions such as amphiphile concentration, temperature, pressure and salt concentration. In this respect micellar solutions are fundamentally different from multicomponent mixtures in which the polydispersity is a fixed and unchanging feature determined solely by the initial concentrations of the non-associating solute molecules.

At amphiphile volume fractions below 20% micellar solutions often exist as homogeneous isotropic liquid phases [6]. These phases are composed of micelles which are randomly dispersed in the solvent, and do not exhibit long-range positional or orientational ordering. Phase separation and critical phenomena can often be induced in this concentration range by changing temperature, pressure, salt concentration and other solution conditions [7]. In many such phase separations, a single isotropic micellar phase separates into two isotropic phases, both of which contain micelles and solvent, but which differ in total amphiphile concentration [6, 7]. In particular, phase separations of this type can be induced by lowering the temperature, as in solutions of the zwitterionic amphiphile dioctanoyl phosphatidylcholine (C_8-lecithin) and water [1, 8], and by raising the temperature, as in solutions of the nonionic amphiphile n-dodecyl hexaoxyethylene glycol monoether ($C_{12}E_6$) and water [7, 9, 10]. These cases lead to coexistence curves which exhibit upper and lower consolute (critical) points, respectively. In addition, micellar solutions can exhibit both types of consolute points as the temperature is varied monotonically over a finite range, leading to closed-loop coexistence curves. An example of the latter behavior is found in solutions of n-decyl pentaoxyethylene glycol monoether ($C_{10}E_5$) and water [11]. Typically, consolute points in these phase transitions occur at very dilute amphiphile concentrations, for example, in C_8-lecithin at about 2% volume fraction. The coexistence curves usually show a pronounced asymmetry between the dilute and concentrated branches [6, 1]. Although polymers and other macromolecules in solution can exhibit similar behavior [12], micellar solutions are fundamentally different as emphasized above.

In order to formulate a theory of the equilibrium thermodynamic properties of micellar solutions including the phase separation phenomena, it is necessary to incorporate the unique characteristics of micellar aggregates and their size distribution, which distinguish micellar solutions from previously considered multicomponent mixtures[1-3,13]. Previous theoretical investigations of phase separation in micellar solutions[6,14-17] did not include these unique features, which we have described above.

The present theory[1-3] represents an effort to include in a single, unified theoretical framework, the effect of both intermicellar interactions and multiple chemical equilibrium on the micellar size distribution and on the equilibrium thermodynamic properties in both the single-phase and two-phase regions of micellar solutions. As usual, the fundamental mechanism leading to phase separation involves a competition between the entropy of the solution, which favors random mixing of micelles and water molecules, and net attractive micelle-micelle and water-water interactions, which favor demixing into micelle-poor and micelle-rich phases. In addition, as the system is brought towards phase separation, the micellar size distribution can change according to the principle of multiple chemical equilibrium, and this is the essential new physical feature which we have included in our theory. We have adopted a thermodynamic approach in which we introduce a phenomenological Gibbs free energy which incorporates the essential physical ingredients of the micellar solution. All the equilibrium properties of the solution, including phase separation, can be calculated from this free energy. As a result, we have put the theory of micellar solutions in the same category as existing mean-field type theories of simple binary mixtures or polymer solutions.

2. Theoretical and Experimental Results

Consider a solution of N_s solute molecules (amphiphiles) and N_w solvent molecules (water) in thermodynamic equilibrium at temperature T and pressure p. The self-association of the amphiphiles produces a distribution $\{N_n\}$ of micellar sizes, where N_n is the number of micelles having n amphiphiles. We provisionally consider micelles of different sizes as distinct chemical species, and model[1] the Gibbs free energy G of the solution as consisting of three additive parts, G_f, G_m and G_{int}. These parts are chosen so as to provide a heuristically appealing identification of the factors which are responsible for micellar formation and growth, on the one hand, and for phase separation, on the other.

Our model for G_f has the form

$$G_f = N_w\mu_w^o + \sum_n N_n\bar{\mu}_n^o\,, \qquad (1a)$$

where $\mu_w^o(T,p)$ is the free energy change of the solution when a water molecule is added to pure water, and $\bar{\mu}_n^o(T,p)$ reflects the free-energy change of the solution when a single aggregate of size n is placed at a given position in pure water. G_f summarizes the many complex physical factors that are responsible for the formation of micelles. These contributions are evaluated for a dilute reference solution that lacks intermicellar interactions, as is reflected in the definitions of μ_w^o and $\bar{\mu}_n^o$. These factors include hydrophobic, hydrogen-bonding, electrostatic, steric and van der Waals interactions, as well as other subtle considerations that are the topics of active experimental and theoretical research[6,18-21].

Our model for G_m has the form

$$G_m = k_BT\,[N_w \ln(X_w) + \sum_n N_n \ln(X_n)]\,, \qquad (1b)$$

where $X_w = N_w/(N_w + N_s)$, $X_n = N_n/(N_w + N_s)$, and k_B is the Boltzmann constant. $-G_m/T$ models the entropy of mixing of the formed aggregates, the monomeric amphiphiles and the solvent. This entropy of mixing reflects the number of geometric configurations that describe the possible positions of the micelles, the monomeric amphiphiles, and the water molecules in the solution, as a function of the relative proportions of each of these constituents. These relative proportions are represented by the mole fractions X_w and $\{X_n\}$.

A fundamental calculation of the entropy of mixing is available only for a very few simple models [12, 22]. There is as yet no first principles calculation of the mixing entropy which includes the effects of size, shape, flexibility and polydispersity of micellar solutions. In the absence of such information we have adopted in (1b) an *ad hoc* expression to represent the entropy of mixing. The present model for G_m has a form which is similar to that for ideal solutions, but its justification must be *a posteriori*. The consequences of adopting a Flory-Huggins model for the entropy of mixing, instead of (1b), are presented in Ref.[3].

Our model for G_{int} reflects the interactions between the formed micellar aggregates, the monomeric amphiphiles and the water molecules. Like the free energy of mixing, G_{int} would be very difficult to calculate from first principles. Our choice of G_{int} results from a simple mean-field type approximation, in which interactions are averaged uniformly over spatial and orientational configurations, thus neglecting correlations. Since the aim of the present theory is to describe isotropic micellar phases, which lack both positional and orientational long-range order, we believe that this approximation preserves the essential physical ingredients of the interactions.

G_{int} has the form

$$G_{int} = -(1/2)CN_s\phi\,, \tag{1c}$$

where the parameter $C(T,p)$ represents an effective interaction free energy, mediated by the solvent, between pairs of amphiphiles on different micelles, and ϕ is the total volume fraction of amphiphile. Denoting the volume of an individual amphiphile by Ω_1, that of a water molecule by Ω_w, and their ratio by $\gamma = \Omega_1/\Omega_w$, the volume fraction of amphiphile is given by $\phi = \gamma N_s/(N_w + \gamma N_s)$. Physically, (1c) reflects the fact that in the actual solution, a certain fraction ϕ of the environment of each micelle consists of amphiphilic material found in other micelles, rather than water. Our choice for G_{int} reflects a simple mean-field approach, whose justification can also be properly made *a posteriori*. It is instructive to recognize that our choice for G_{int} has precisely the quadratic form widely used in the description of polymer solutions and binary mixtures [12, 22].

We next review the thermodynamic consequences of this model [1-3]. The self-association equilibrium and the phase separation are both governed by the chemical potentials of the water, μ_w, and each n-mer, μ_n. These are calculated from G and are given by

$$\mu_w = \mu_w^o + k_BT[\ln(1-X) + X - M_0] + (C/2)\gamma X^2/[1+(\gamma-1)X]^2\,, \tag{2a}$$

$$\mu_n = \mu_n^o + k_BT[\ln(X_n) + n(X-1-M_0)] + (C/2)n\{(1-X)^2/[1+(\gamma-1)X]^2 - 1\}\,, \tag{2b}$$

where $\mu_n^o(T,p) = \bar{\mu}_n^o(T,p) + k_BT$, $X = N_s/(N_w+N_s)$ is the total amphiphile mole fraction and $X_n = N_n/(N_w+N_s)$ is the mole fraction of n-mers. M_0 is the 0th moment of the distribution of micellar sizes, where the kth moment of this distribution is given by $M_k = \sum_m m^k X_m$.

Until this point we have treated micellar aggregates of different sizes as independent chemical species in the solution. However, as emphasized in the Introduction, the aggregates continually and reversibly exchange amphiphiles with one another. These reversible material exchanges can be conveniently described using the thermodynamic principles which govern multiple chemical equilibrium. This implies that the chemical potential per amphiphile must be the same in all the aggregates: $(\mu_n/n) = \mu_1$, independently of n. Using (2b) in this condition yields the distribution of micellar sizes

$$X_n = (X_1)^n \exp(-\beta[\mu_n^\circ - n\mu_1^\circ]) \,, \tag{3}$$

where $\beta = 1/k_BT$. Equation (3), combined with the conservation of amphiphilic monomers in solution, $X = \sum_n nX_n$, determines the micellar size distribution as a function of total amphiphile concentration X, provided that the sequence of chemical potentials $\{\mu_n^\circ\}$ is known as a function of T, p and other solution conditions.

A particularly important consequence of our choice of G_{int} is that intermicellar interactions do not affect the micellar size distribution. This can be immediately seen from (1c), since G_{int} depends only on the total amount of amphiphile N_s, and not on the specific way that amphiphiles are distributed. Nevertheless, since the micellar size distribution $\{X_n\}$ depends explicitly on X, it follows that after phase separation the dilute and concentrated phases, having mole fractions Y and Z, respectively, will have different micellar size distributions $\{X_n(Y,T,p)\}$ and $\{X_n(Z,T,p)\}$. Clearly, other models of G_{int} may lead to distributions which will depend explicitly on intermicellar interactions [23]. In that case, certain aggregate structures and sizes may be preferred over others and stabilized by these interactions [24]. These interesting possibilities are the subject of current theoretical research.

Another important consequence of the form of (3) is that, in conjunction with the definition of the moments M_k of the distribution of micellar sizes, it implies[1,25] that all these moments are related to the second moment M_2, through $M_{k+1} = M_2(dM_k/dX)$. This conclusion is valid regardless of the form of the $\{\mu_n^\circ\}$. As a result, all the influence of the micellar size distribution on equilibrium properties of the solution, such as the osmotic pressure $\pi = (\mu_w^\circ - \mu_w)/\Omega_w$, the osmotic compressibility $(\partial\pi/\partial X)^{-1}_{T,p}$, the coexistence curve and the spinodal line, is determined by the dependence of M_2 on X, T, p and other solution conditions [1-3].

To compare the theoretical predictions with experiments in a real micellar solution, it is first necessary to evaluate $M_2(X,T,p)$ for that system. We have done so for two experimental systems [1-3]. The first [1,3] is solutions of dioctanoyl phosphatidylcholine (C_8-lecithin) and water, which show an upper consolute point, and the second [2,3] is solutions of n-dodecyl hexaoxyethylene glycol monoether ($C_{12}E_6$) and water, which show a lower consolute point. For both systems we have shown that the predictions of the theory may be completely summarized by the temperature dependence of two parameters: (i) $\Delta\mu$, a chemical potential difference which governs the extent of micellar growth in these systems, and (ii) C, from (1c) above, which reflects the effective attractive intermicellar interactions which lead to phase separation.

For the C_8-lecithin and water solutions, using the experimental coexistence curve concentrations we found that the values of $\Delta\mu$ and C are approximately temperature independent. These values can then be determined from, for example, the experimental critical concentration X_c and critical temperature T_c. The resulting values of C and $\Delta\mu$ were used to make a theoretical prediction for the shape of the coexistence curve,which is in excellent agreement with experiment. Furthermore, the value of $\Delta\mu$ determined by this procedure is in good agreement with the value of $\Delta\mu$ that was found by using independent measurements to characterize the micellar size distribution in the single-phase region. Details may be found in Refs.[1] and [3].

For the $C_{12}E_6$ and water system, we have shown that a linear dependence of $\Delta\mu$ and C on T is consistent with (a) single-phase measurements of the weight-average association number, $\langle n\rangle_w$, as a function of X and of the osmotic compressibility along the critical isochore, and (b) measurements of the overall amphiphile concentrations on the coexistence curve and the location of the critical point. Figure 1 shows the theoretical (full curve) and experimental osmotic compressibility along the critical isochore for this system. Note that for this comparison we have used the concentration variable c, the weight of $C_{12}E_6$ per unit volume of the solution, instead of X. The experimental

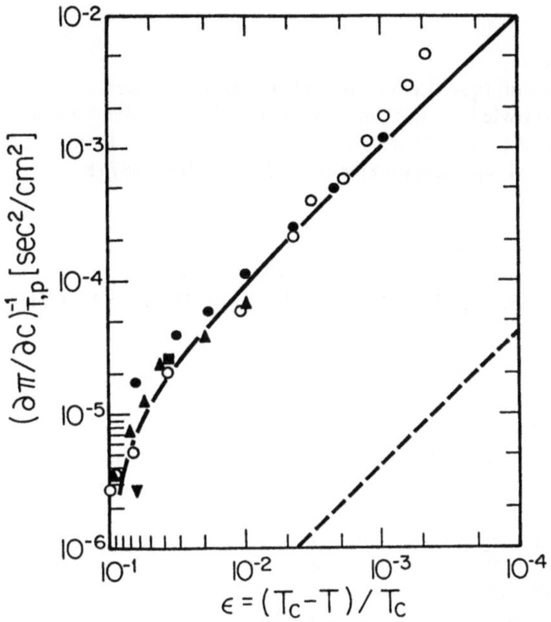

Figure 1: Osmotic compressibility vs. reduced temperature along the critical isochore for the $C_{12}E_6$ and water system. The experimental points are: (●)-Ref.[26], (▲)-Ref.[9], (▼)-Ref.[27], (■)-Ref.[28], (○)-Ref.[29]. The solid curve represents our theoretical prediction using the model described in this paper and the dashed line represents the prediction of the Flory-Huggins model described in Ref.[3]

determinations made by five groups [9,26-29] are in excellent agreement with the theoretical prediction over a three-decade range of osmotic compressibility and a two-decade range of reduced temperature. Details may be found in Refs.[2] and [3].

3. Conclusions

The theoretical analysis presented above permits us to describe the physically important features of (i) interactions between micellar aggregates, (ii) entropy of mixing of the micellar solution, and (iii) the micellar multiple chemical equilibrium, with the aid of two phenomenological parameters, $\Delta\mu$ and C. The parameter $\Delta\mu$ describes the free-energy advantage associated with micellar growth, and the parameter C describes the magnitude of the effective attractive intermicellar free energy. The theory in its present form enables us to provide an analytic representation of the following equilibrium properties of the micellar solution: the shape and location of the coexistence curve and spinodal line, the osmotic compressibility, and the concentration and temperature dependence of the micellar size distribution. This representation has proven to be an accurate description of experimental data in two distinct systems, one showing an upper consolute point (C_8-lecithin and water) and the other a lower consolute point ($C_{12}E_6$ and water). We believe that this advance provides researchers with a mean-field description of micellar systems which places this field on the same theoretical footing as has previously existed for binary mixtures, polymer solutions and the gas-liquid phase transition.

4. Acknowledgements

Daniel Blankschtein would like to thank Dominique Langevin and Jacques Meunier for the invitation to present this work. He also acknowledges the support by the Texaco-Mangelsdorf Career Development Professorship at M.I.T. and by the Alfred P. Sloan Fund for Basic Research. This work was supported by the National Science Foundation under Grant No. DMR84-18718.

5. References

1. D. Blankschtein, G.M. Thurston, G.B. Benedek: Phys. Rev. Lett. 54, 955 (1985)
2. G.M. Thurston, D. Blankschtein, M.R. Fisch, G.B. Benedek: J. Chem. Phys. 84, 4558 (1986)
3. D. Blankschtein, G.M. Thurston, G.B. Benedek: J. Chem. Phys. 85, 7268 (1986)
4. C. Tanford: The Hydrophobic Effect, 2nd.ed., (Wiley, New York 1980), and J. Phys. Chem. 78, 2469 (1974)
5. E. Ruckenstein, R. Nagarajan: J. Phys. Chem. 79, 2622 (1975)
6. For comprehensive experimental and theoretical surveys of the field of micellar systems see:
   a. K.L. Mittal, B. Lindman (eds.): Surfactants in Solution, Vols.1-3 (Plenum, New York 1984)
   b. V. Degiorgio, M. Corti (eds.): Proceedings of the International School of Physics Enrico Fermi - Physics of Amphiphiles: Micelles, Vesicles and Microemulsions (North-Holland, The Netherlands 1985)
7. V. Degiorgio, R. Piazza, M. Corti, C. Minero: J. Phys. Chem. 82, 1025 (1985)
8. R.J.M. Tausk, C. Oudshoorn, J.Th.G. Overbeek: Biophysical Chem. 2, 53 (1974)
9. R.R. Balmbra, J.S. Clunie, J.M. Corkill, J.F. Goodman: Trans. Faraday Soc. 58, 1661 (1962)
10. R. Strey, A. Pakusch: In Proceedings of the Fifth International Symposium on Surfactants in Solution (Bordeaux, France 1985)
11. J.C. Lang, R.D. Morgan: J. Chem. Phys. 73, 5849 (1980)
12. P.J. Flory: Principles of Polymer Chemistry (Cornell University Press, Ithaca 1953) Chaps. 12 and 13
13. R.E. Goldstein: J. Chem. Phys. 84, 3367 (1986)
14. R. Kjellander: J. Chem. Soc., Faraday Trans. 2 78, 2025 (1982)
15. L. Reatto, M. Tau: Chem. Phys. Lett. 108, 292 (1984), and in Ref.[6b], p. 448
16. J.C. Lang: in Ref.[6a], Vol. 1, p. 35
17. C.A. Leng: J. Chem. Soc., Faraday Trans. 2 81, 145 (1985), and in Ref.[6b], p. 469
18. K.A. Dill, P.J. Flory: Proc. Natl. Acad. Sci. U.S.A. 77, 3115 (1980), and Proc. Natl. Acad. Sci. U.S.A. 78, 676 (1981)
19. A. Ben-Shaul, I. Szleifer, W.M. Gelbart: Proc. Natl. Acad. Sci. U.S.A. 81, 4601 (1984)
20. D.W.R. Gruen: J. Phys. Chem. 89, 146 (1985), and J. Phys. Chem. 89, 153 (1985)
21. B. Owenson and L.R. Pratt: J. Phys. Chem. 88, 2905 (1984)
22. For a review see, for example, E.A. Guggenheim: Mixtures: The Theory of the Equilibrium Properties of Some Simple Classes of Mixtures, Solutions and Alloys (Clarendon Press, Oxford 1952)
23. A. Ben-Shaul, W.M. Gelbart: J. Phys. Chem. 86, 316 (1982)
24. W.M. Gelbart, A. Ben-Shaul, W.E. McMullen, A. Masters: J. Phys. Chem. 88, 861 (1984)
25. J.M. Corkill, J.F. Goodman, T. Walker, J. Wyer: Proc. Roy. Soc. A312, 243 (1969)
26. M. Corti, C. Minero, V. Degiorgio: J. Phys. Chem. 88, 309 (1984)
27. J.M. Corkill, J.F. Goodman, R.H. Ottewill: Trans. Faraday Soc. 57, 1627 (1961)
28. D. Attwood, P.H. Elworthy, S.B. Kane: J. Phys. Chem. 74, 3529 (1970)
29. J.P. Wilcoxon, E.W. Kaler: J. Chem. Phys. (in press)

Electric Birefringence of Nonionic Micellar Solutions Near the Cloud Point

V. Degiorgio and R. Piazza

Dipartimento di Elettronica, Sezione di Fisica Applicata,
Università di Pavia, I-27100 Pavia, Italy

We have measured the Kerr coefficient B of aqueous solutions of the nonionic amphiphiles C_6E_3, C_8E_4, $C_{10}E_5$ and $C_{12}E_6$ as a function of the temperature T in a temperature region close to the cloud point and at the concentration corresponding to the minimum of the cloud curve. The used nonionic amphiphiles are high-purity products prepared by the group of Dr. Platone (Eniricerche, S. Donato, Milano, Italy). In order to lower the sample conductivity, the compounds were further purified by means of repeated extraction with organic solvents to reduce the amount of residual ionic impurities. A detailed description of our experimental apparatus can be found in Ref./1/. The setup used in this work includes a quarter-wave plate inserted between the Kerr cell and the analyzer. The cell temperature was controlled within 0.01°C. Voltage pulses had height of 0.3-1 kV and duration of 10-300 μs. Some of the results are shown in Figs. 1-2. A full report of our work will appear elsewhere /2/. Earlier data on the C_iE_j-H_2O system, both static and dynamic, have been published two years ago /3/. All the investigated solutions present, far from the cloud point, a very low Kerr coefficient, comparable to the value expected for pure water, $B \approx 3\times10^{-13} m/V^2$. All systems show a considerable increase of B as the temperature is raised toward the critical point. In a double logarithmic plot B is seen to depend linearly on the temperature distance $T_c - T$ when the data are taken sufficiently close to T_c. This corresponds to a power-law behavior, $B \approx (T_c - T)^{-\psi}$, with an exponent $\psi \approx 0.9$. Far from T_c the behavior of B may be more complex, as shown in Fig. 2 for the system $C_{12}E_6$ - H_2O which was studied in a wide range of temperatures (8-50°C). The power-law divergence of B near T_c is connected with critical concentration fluctuations, and has nothing to do with changes of the micelle shape. In the case of the system $C_{12}E_6$ - H_2O the growth of B in the region 15-45°C very likely reflects a moderate micellar growth as function of T: if one completely neglects the effect of interactions, the change of B can be explained by a change of the aggregation number by a factor of three.

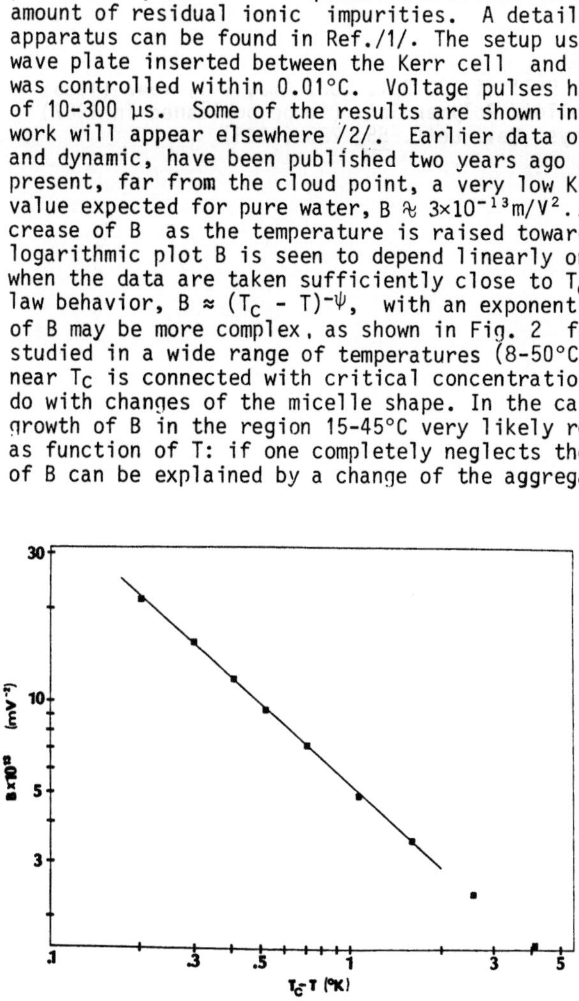

Fig. 1. Kerr coefficient B versus the temperature distance from the critical point for a 13% solution of C_6E_3 in H_2O.

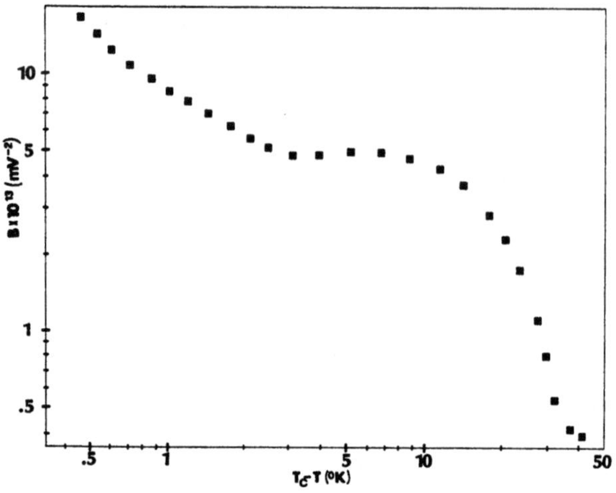

Fig. 2. Kerr coefficient B versus the temperature distance from the critical point for a 2.2% solution of $C_{12}E_6$ in H_2O.

References

1. R. Piazza, V. Degiorgio and T. Bellini, J. Opt. Soc. Ame. B 3, 1642 (1986)
2. V. Degiorgio and R. Piazza, Colloid Polymer Sci. (to be published in 1987)
3. V. Degiorgio and R. Piazza, Phys. Rev. Lett. 55, 288 (1985)

Aspects of the Statistical Thermodynamics of Amphiphilic Solutions

R.E. Goldstein

Laboratory of Atomic and Solid State Physics and Materials Science Center, Clark Hall, Cornell University, Ithaca, NY 14853, USA

I. Introduction

Perhaps the most distinctive feature of solutions of amphiphilic molecules is that their microstructure depends sensitively on the bulk thermodynamic variables which specify the state of the system: Dilute solutions of surfactants are isotropic and homogeneous molecular dispersions, while micellar aggregates and a variety of liquid crystalline mesophases are encountered at progressively higher concentrations. Of particular interest in the present discussion is the interplay of self-association and phase equilibria, especially those involving the isotropic micellar phases which have been discussed extensively at this conference. In the following, we indicate some of the important conclusions of recent work in this area and highlight some fundamental open questions. The discussion below is divided into two sections, the first dealing with general aspects of phase equilibria in associating solutions, the second specializing to the neighborhood of second order phase transitions. A common theme here is the concept of *effective interactions* in micellar solutions, especially near critical points or points of rapidly changing microstructure.

II. Self-Association and Phase Transitions

The basic principles of the statistical thermodynamics of chemical equilibria which have been applied quite successfully[1] to the study of dilute micellar solutions, at concentrations near the critical micelle concentration (*cmc*), have recently been extended in the framework of phenomenological models[2,3] to describe the interplay of phase equilibria and self-association in concentrated surfactant systems. These mixtures differ from conventional heterogeneous solutions in that the distribution of particle sizes and shapes varies with thermodynamic variables, rather than being imposed by the nominal composition. While the details of aggregation are seemingly beyond present theory, it has emerged that there is a great simplification possible in describing phase equilibria through a link between properties of two-phase coexistence and certain *moments* of the micelle size distribution function. This connection has been invaluable in the discussion of self-associating systems, and promises to provide a unifying mean field description of amphiphilic solutions. We summarize here key conclusions and unresolved theoretical issues, particularly with regard to the global phase diagram of such systems.

The thermodynamic models[2,3] start by adopting the view that polydisperse micellar solutions are equivalent to *non-associating multicomponent* mixtures, with each of the N_n aggregates of n amphiphiles comprising a different species. The total free energy F of the solution then consists of three basic contributions arising from (i) the internal free energy μ_n^0 of each of the n-mers in solution relative to some dilute, noninteracting reference system; (ii) the entropy of mixing of the distribution of micelles and solvent, and (iii) the mutual interactions between all species in solution. The chemical potential of an n-mer $\mu_n = (\partial F/\partial N_n)_{T,p,N_{m\neq n}}$ then has the generic form $\mu_n = \mu_n^0 + \mu_n^{mix} + \mu_n^{int}$, where the second and third terms are, in general, *functionals* of the *entire* distribution function $\{N_n\}$. Since a binary mixture of surfactant and water is like any other two-component system in being described by only two independent chemical potentials, we know that the function $\{N_n\}$ must be specified fully by these *thermodynamic fields* and temperature, and this is achieved by invoking the minimization of F with respect to a normalized distribution N_n, giving the law of mass action, $\mu_n = n\mu_1$ for all n.

It is significant to note that this is a *self-consistent* integral equation for the distribution in the sense that the various contributions to the μ_n depend on distribution moments of the form $M_s = \sum_{n=1}^{\infty} n^s N_n$. It is self-consistent in that the distribution both determines and is determined by the mixing entropy and interaggregate interactions. One may view the description of a surfactant system in terms of micelles as a kind of "coordinate transformation" analogous to those which abound in condensed matter physics, in which the strongest interactions in a system (here the hydrophobic energy) are accounted for by defining new basic "particles" which are (hopefully) less strongly interacting than those in the bare Hamiltonian. Certainly there are many circumstances in which this is a reasonable change of variables, yet we wish to emphasize that the relationship between the basic surfactant-surfactant interactions and the form of the effective interactions between *micelles* is still far from clear (see below).

In the case of simple dilute-solution models of the mixing entropy, and random-mixing type interactions, one finds that the densities ρ_n of n-mers follow a relation of the general form

$$\rho_n = (\rho_1)^n e^{-\beta\Delta_n} g(\{\rho_n\}), \tag{1}$$

where $\beta = 1/k_B T$, $\Delta_n = \mu_n^0 - n\mu_1^0$, and the function g summarizes micellar interactions. This demonstrates that the monomer density ρ_1 plays the role of the surfactant fugacity, and given the total surfactant density $\rho = \sum n\rho_n$, we may in principle solve for $\rho_1(\rho)$ and hence the complete distribution function. In describing two-phase coexistence, the equality of chemical potentials of each species separately in the two phases is guaranteed by that equality holding for μ_1, and this in turn leads to the spinodal and critical-point conditions involving the compositional derivatives of the moments M_s. These derivatives often obey recursion relations,[2,3] thus relating the coexistence-curve properties ultimately to certain low-s moments, namely the mean and variance.

From these considerations, several important points emerge:[3]

(1) For intermicellar interactions described by a mean field or van der Waals-like free energy weighted by the number density of species, $g \equiv 1$, and the form of ρ_n is explicitly independent of overall composition, although depending implicitly on ρ through ρ_1. This simplification arises from the fact that such mean field interactions are equivalent to exact treatments of infinite-range potentials, for which the free energy is clearly independent of the overall aggregation state of the system.

(2) The inversion of the normalization condition on ρ_1 to obtain $\rho_1(\rho)$ demonstrates,[4,3] at least for sharply peaked distributions of compact micelles, that the critical micelle concentration has aspects in common with a true thermodynamic first order phase transition, albeit one rounded by the finite aggregate size. This appears as singularities of the osmotic pressure in the plane of *complex fugacity* (complex ρ_1), in much the same way as in the Lee-Yang[5] theory of phase transitions. The importance of dealing with polydispersity in this region of the phase diagram is highlighted in a reinterpretation[6] of Widom's continuum model[7] for microemulsions which demonstrates a formal equivalence to that of a micellar solution.

(3) Again for compact objects, and with $g = 1$, there is a particular kind of fractionation of the surfactant between coexisting phases in which, while the total number density may differ dramatically between them, the characteristic micelle size may be nearly identical,[3] leading to what has been called an *effective thermodynamic field variable*,[8] a quantity like a chemical potential or temperature which assumes a common value in coexisting phases. Such effective fields have also been detected in microemulsions.[8,9] Their existence is linked to the relation of the *cmc* to a first-order transition (see (2) above) in that the similarity in micelle distributions arises from the near invariance of the monomer density ρ_1 (the fugacity) with composition, itself analogous to the rigorous invariance of the chemical potential at a true first-order transition in the region of two-phase coexistence.

(4) The critical point at the extremum of two-phase coexistence between isotropic surfactant solutions generally involves the coalescence of two *micellar* solutions, since the *cmc* is typically far below the critical composition. In this case, one may view the solution point as analogous to the liquid-vapor transition of a simple fluid, with the solvent as some kind of spectator species. Clearly, this picture is intimately related to the existence of an effective field variable related to the aggregate geometry. The extreme asymmetry[10] of the coexistence curve with respect to the critical composition suggests however that the analogy with a simple fluid breaks down, and indeed these systems are more in keeping with the Flory-Huggins type polymer solution thermodynamics, even those systems which appear to have spherical aggregates.[3] The breakdown of a law of corresponding states with respect to simpler solutions of spherical particles is puzzling. It may be due to the nature of the intermicellar interactions, or to the existence of two disparate length scales, namely the particle size and interaction range, quantities normally quite similar in conventional fluids. The issue of the range of interactions also enters the discussion of critical phenomena below.

The central point which is bypassed in present discussions concerns the proper formulation of the statistical mechanics of a fluid whose internal structure may change dramatically with thermodynamic variables, particularly in the neighborhood of phase transitions. It is known from experiment and described in the theoretical work[3] that, far away from the consolute point, the phase equilibrium in nonionic surfactant solutions involves a dilute *nonmicellar* phase and a dense micellar phase, whereas closer to criticality the dilute phase concentration passes through the *cmc*: The effective entities in solution change with concentration (or better, with surfactant chemical potential), and so does the effective Hamiltonian which governs their interactions. The kind of self-consistent approach to the thermodynamics which appears in the phenomenological models is surely only a crude approximation to the more rigorous approach in which effective interactions are functions of the governing thermodynamic fields, chemical potentials and temperature. Only this kind of approach is capable of dealing with the neighborhood of the critical point. We remark that a proper treatment of effective potentials in fluids with many-body *dispersion* forces has recently been developed,[11] and it highlights the importance of a full accounting of fluctuations. For instance, it is known that the extent of a chemical reaction, of which micelle formation is an example, exhibits an anomaly at critical points,[12,13] and one might ask about the observable consequences of this behavior in amphiphilic systems.

In micellar solutions, we suggest that rigorous study of this issue may be possible within the context of the type of lattice model introduced by Widom,[14] for which the statistical mechanics is more tractable than for continuum fluids. It is also interesting to note that coupled chemical (aggregation) and phase equilibria are also believed to occur in *metallic* fluids like mercury, in which ionization and clustering equilibria vary rapidly as the dilute insulating vapor is compressed toward the critical density, and then through a metal-nonmetal transition.[15] There, again, the effective Hamiltonian for the fluid varies dramatically as a function of thermodynamic variables, leading to certain dramatic effects at criticality.[16]

III. Critical Phenomena in Nonionic Micellar Solutions

Some of the most puzzling experimental findings in recent years in the study of phase transitions of amphiphilic solutions have been the observations of apparent nonuniversality in the critical behavior near the lower critical solution temperatures T_c^L of aqueous nonionic surfactant systems.[17] The "nonuniversality" manifests itself in the amphiphile- and solvent-dependence of critical exponents which, in more conventional binary liquid mixtures, are unique and in agreement with those of the three-dimensional Ising model or lattice gas. In particular, Corti, Degiorgio, and collaborators have found from light scattering measurements that the osmotic compressibility $\chi \equiv (\partial\Pi/\partial c)_T^{-1}$ and correlation length ξ have critical divergences of the usual form

$$\chi \sim \chi_0 t^{-\gamma}, \qquad \xi \sim \xi_0 t^{-\nu},$$

with $t \equiv (T_c^L - T)/T_c$ the reduced temperature, χ_0 and ξ_0 system-dependent amplitudes, and γ and ν *also* varying. The most studied system is the nonionic $C_{12}E_8$, for which the exponents vary from near the Ising values $\gamma_I \simeq 2\nu_I \simeq 1.24$ for $C_{12}E_8$-D_2O ($\xi_0 = 9Å$) down to the classical mean field or van der Waals values $\gamma_{MF} = 2\nu_{MF} = 1$ for C_12E_8-(H_2O/D_2O) ($\xi_0 = 15Å$), and even significantly below those, to $\gamma \simeq 2\nu \simeq 0.88$ for C_12E_8-H_2O ($\xi_0 = 23Å$). Now, although it has been suggested[18] that a new type of critical phenomenon may play a role in these results, a more likely explanation has been suggested by Fisher,[19] namely that what is being probed is a "crossover" phenomenon *controlled by the range of interactions*, and interpolating between the van der Waals behavior of infinite-range potentials and the pure Ising behavior for short-range potentials. We briefly discuss these conclusions and point to possible alternative explanations, tentative new results, and further directions for theory and experiment.

From the classical Ginzburg criterion for the validity of mean field theory (MFT), that is, the neglect of fluctuations, it is known that a crossover temperature t_X may be defined above which MFT is valid, and below which the mean field exponents give way smoothly to the asymptotic nonclassical values. In terms of the bare correlation length ξ_0 and the "lattice constant" or inverse momentum cutoff a,

$$t_X \sim (\xi_0/a)^{-2d},$$

in d spatial dimensions, so that in three dimensions even a relatively minor change of a factor of two or more in the relative range ξ_0/a can move the crossover region by several decades. From renormalization group studies[20] it is expected that the *effective* exponents such as $\gamma_{eff} \equiv -(\partial \ln\chi/\partial \ln t)$ will be smooth functions of t_{exp}/t_X, with t_{exp} say the midpoint of the temperature range in which γ_{eff} is determined. Fisher identified a with the micelle radius, which was measured in separate light-scattering experiments, and was able to show that indeed the experimental $\gamma_{eff}(t_{exp}/t_X)$ defined a smooth crossover, suggesting that different experiments, while performed at roughly comparable values of t_{exp}, simply probed different regions of the crossover function by the variation of ξ_0.

The existence of this smooth function strongly supports a crossover controlled by the range of the interactions, but there are many puzzling aspects of the results which demand further study. First of all, the crossover function appears to be *nonmonotonic* in the range in falling *below* the mean field value $\gamma_{MFT} = 1$. A monotonic crossover is predicted from ϵ-expansion results, and in more sophosticated work recently reported by

Bagnuls and Bervillier[21] for the standard ϕ^4 theory of critical phenomena. These studies have been in the context of what could be called "simple" Ising models, those without any of the microscopic information needed to describe polydispersity of molecular sizes, or to describe the existence of reentrant transitions. From previous theoretical studies of such transitions,[22] it is known that independent of the range of the interactions the asymptotic critical region is significantly compressed compared to that at normal upper consolute points. It was pointed out by Robledo, *et al.*,[23] that the peculiarities of the crossover may ultimately be due solely to the temperature-dependent interactions which give rise to the LCST. In binary mixtures of comparable size molecules, especially aqueous organic solutions, a variety of studies has suggested that hydrogen bonding interactions between the two components are responsible for the reentrant solubility (lower critical solution point), and several lattice models have been advanced which explain the phenomenon by invoking the orientational degrees of freedom which are frozen out in the formation of the hydrogen bond. By a suitable statistical trace over those orientational degrees of freedom, it is possible to derive a temperature-dependent effective interaction J_{eff} for an underlying lattice-gas model of such solutions.[22,24,25] In fact, it has been shown[26] that the form of this coupling is identical in all such models, and may be derived by a general prescription from basic statistical mechanics. In a lattice model of these systems, the general functional form of the reduced interaction $K_{\text{eff}} \equiv -J_{\text{eff}}/T$ is simply the free energy of a two-level system representing the hydrogen-bonded and non-bonded states of two neighboring particles:

$$K_{\text{eff}}(T) = \frac{1}{T} - \ln\left(1 + \omega\left(\exp\left((R+1)/T\right)\right)\right),$$

with $\omega \ll 1$ representing the degeneracy ratio between the two states, and $R > 1$ the ratio of bonded to non-bonded energies.

The crucial point of the effective interaction which was pointed out by Robledo, *et al.* is that at low T, K_{eff} switches sign from being ferromagnetic to antiferromagnetic, and vanishes at a *decoupling* point T_D slightly below the lower consolute temperature. T_D is thus analogous to the Boyle point of a gas or the theta point of a polymer solution, where the second virial coefficient vanishes and the system behaves very much like an ideal solution. The decoupling point of nonionic surfactant solutions has been measured by light scattering determinations of the osmotic compressibility in dilute solutions, and it is known to lie within about $30K$ of T_L.[17] This implies a highly compressed critical region, and may indeed be responsible for some of the peculiarities of the exponents. In addition, the rapid variation of the effective interaction suggests that the *range* ξ_0 may itself be a strong function of temperature, and we[27] are currently investigating the consequences of this for the exponent crossover.

Many questions immediately come to mind in interpreting these results. What is the relationship between the micelle size and the range of interactions as measured by ξ_0? What physical features control ξ_0, and why does it appear to change so dramatically with the substitution of D_2O for H_2O? We remark that a previous study[28] of isotopic effects in systems with lower critical points suggests strong modifications of the scaling variables with isotopic substitution. We remark here that although there exist heuristic derivations of simple analytic expressions for the effective interaction referred to above, the true underlying physical mechanism is not well understood. It is certain to have important implications in the study of the so-called "hydration repulsion" which acts at short range between neutral phospholipid bilayers,[29] and is believed to arise from the particularly strong structuring tendencies of water.

I am grateful to S. Leibler and R. Lipowsky for their valuable insights and continued encouragement, and to B. Widom and M.E. Fisher for a number of enlightening discussions. This work was supported in part by NSF, through Grant Nos. DMR84-15669 and 81-17011 at Cornell University, and by a graduate fellowship from the Fannie and John Hertz Foundation.

References

1. C. Tanford, *The Hydrophobic Effect*, 2nd ed. (Wiley, New York, 1980), and references therein.
2. D. Blankschtein, G.M. Thurston, and G.B. Benedek, J. Chem. Phys. **85**, 7268 (1986), and references therein.
3. R.E. Goldstein, *J. Chem. Phys.* **84**, 3367 (1986).
4. F.H. Stillinger and A. Ben-Naim, *J. Chem. Phys.* **74**, 2510 (1981).
5. C.N. Yang and T.D. Lee, *Phys. Rev.* **87**, 404 (1952).
6. R.E. Goldstein and S. Leibler, unpublished.
7. B. Widom, *J. Chem. Phys.* **81**, 1030 (1984).
8. D. Roux and A.M. Bellocq, *Phys. Rev. Lett.* **52**, 1895 (1984).
9. M. Kotlarchyk, S.-H. Chen, J.S. Huang, and M.W. Kim, *Phys. Rev. A* **29**, 2054 (1984).
10. See, e.g., D.J. Mitchell, G.J.T. Tiddy, L. Waring, T. Bostock, and M.P. McDonald, *J. Chem. Soc. Faraday Trans. 1* **79**, 975 (1983).
11. R.E. Goldstein, A. Parola, N.W. Ashcroft, M.W. Pestak, M.H.W. Chan, J.R. de Bruyn, and D.A. Balzarini, *Phys. Rev. Lett.* **58**, 41 (1987).
12. I. Procaccia and M. Gitterman, *Phys. Rev. A* **27**, 555 (1983); J.C. Wheeler and R.G. Petschek, *Phys. Rev. A* **28**, 2442 (1983).
13. J.L. Tveekrem, R.H. Cohn, and S.C. Greer, *J. Chem. Phys.* **86**, 3602 (1987).
14. B. Widom, *J. Chem. Phys.* **84**, 6943 (1986).
15. W. Hefner and F. Hensel, *Phys. Rev. Lett.* **48**, 1026 (1982).
16. S. Jüngst, B. Knuth, and F. Hensel, *Phys. Rev. Lett.* **55**, 2160 (1985); R.E. Goldstein and N.W. Ashcroft, *ibid.*, **55**, 2164 (1985).
17. V. Degiorgio, R. Piazza, M. Corti, and C. Minero, *J. Chem. Phys.* **82**, 1025 (1985), and references therein. Variable exponents also appear in microemulsions; A.M. Bellocq, P. Honorat, and D. Roux, *J. Phys.* **46**, 743 (1985).
18. Y. Shnidman, *Phys. Rev. Lett.* **56**, 201 (1986). For comments on this work, see R.G. Caflisch, M. Kaufman, and J.R. Banavar, *Phys. Rev. Lett.* **56**, 2545 (1986); L. Reatto, *ibid.* **58**, 620 (1987).
19. M.E. Fisher, *Phys. Rev. Lett.* **57**, 1914 (1986).
20. P. Seglar and M.E. Fisher, *J. Phys. C* **13**, 6613 (1980).
21. C. Bagnuls and C. Bervillier, *Phys. Rev. Lett.* **58**, 435 (1987).
22. See e.g. R.E. Goldstein and J.S. Walker, *J. Chem. Phys.* **78**, 1492 (1983).
23. A. Robledo, G.F. A.-Noaimi, and G. Martinez-Mekler, preprint (1987).
24. G.R. Anderson and J.C. Wheeler, *J. Chem. Phys.* **69**, 2082, 3403 (1978).

25. J.S. Walker and C.A. Vause, *Phys. Lett. A* **79**, 421 (1980).
26. R.E. Goldstein, *J. Chem. Phys.* **83**, 1246 (1985).
27. R.E. Goldstein, A. Parola, and J.J. Rehr, unpublished.
28. R.E. Goldstein, *J. Chem. Phys.* **79**, 4439 (1983).
29. R.P. Rand, *Ann. Rev. Biophys. Bioeng.* **10**, 277 (1981).

Rheological Properties of Semi-Dilute Micellar Systems

S.J. Candau[1], *E. Hirsch*[1], *R. Zana*[2], *and M. Adam*[3]

[1]Laboratoire de Spectrométrie et d'Imagerie Ultrasonores,
Unité Associée au CNRS, Université Louis Pasteur,
4 Rue Blaise Pascal, F-67070 Strasbourg Cedex, France
[2]Institut Charles Sadron, (C.R.M.-E.A.H.P.), CNRS/U.L.P.,
6 Rue Boussingault, F-67083 Strasbourg Cedex, France
[3]C.E.N. Saclay, F-91191 Gif-sur Yvette Cedex, France

1. INTRODUCTION

Aqueous micelles of ionic surfactants adopt an anisodiametric shape at high ionic strength /1-9/. An increase of surfactant concentration induces a micellar growth, leading to very elongated aggregates for most surfactants with a single linear alkyl chain /10-12/. Several studies have shown that these giant micelles are flexible and behave in many respects like polymers /7-9/ /11/ /12/. For instance, the scattered intensity and the cooperative diffusion coefficient measured by elastic and quasi-elastic light scattering respectively, obey power laws of the surfactant volume fraction with exponents close to those predicted for semi-dilute polymer solutions /11/ /12/. These results indicate that the micelles form a transient network with lifetime much longer than the time scale probed in light scattering experiments.

In this paper we investigate the rheological properties of such networks of entangled flexible micelles. We measured the zero shear viscosity η_s and the longest viscoelastic relaxation time T_R of a 0,7 M aqueous cetyltrimethylammonium bromide (CTAB) solution in the presence of 0.1 M KBr at temperatures from 20 to 50°C by means of a magnetorheometer /13/.

2. THEORY

The existence of flexible cylindrical micelles with a persistence length varying as 1/T has been predicted by Safran et al from a microscopic model for the elasticity of the interface in dilute microemulsion /14/.

The average surfactant aggregation number, in the case where the statistics of the broad distribution of chain length is included is given by :

$$N = 2\,\Phi^{0.5}\,\exp\frac{1}{2}\left(\frac{K'}{T}+1\right), \tag{1}$$

where Φ is the surfactant volume fraction and K' is the energy difference per surfactant between the hemispherical endcap and the cylindrical region of a micelle. The $\Phi^{0.5}$ dependence of N has been observed for several systems /15/ /16/.

The viscoelastic behaviour of semi-dilute micellar solutions has been recently investigated theoretically by Cates. This work deals with the dynamics of the stress relaxation through reptation in a system of living polymers. The viscoelastic behaviour depends on the relative values of the reptation time τ_{rep} of a micelle and the average time τ_{Br} taken by a micelle to break into two parts /17/.

If $\tau_{rep} \ll \tau_{Br}$, then the stress relaxation function should decay according to a broad distribution of exponentials because of the size polydispersity of the micelles. The average relaxation time T_R and the relative zero shear viscosity η_r are given by :

$$T_R \simeq \tau_{rep} \propto N^3\,\Phi^{3.1}, \tag{2}$$

$$\eta_r = \eta_s/\eta_o \propto N^3 \Phi^{3.92}. \quad (3)$$

Combining Eq.(1) with Eqs.(2) and (3) leads to the following temperature dependence of T_R and η_r for a given surfactant volume fraction :

$$T_R \sim \eta_R \propto \exp(\frac{3}{2} K'/T). \quad (4)$$

In the opposite limit $\tau_{rep} \gg \tau_{Br}$, the Cates theory predicts a single exponential decay with a relaxation time:

$$T_R \simeq (\tau_{Br}\,\tau_{rep})^{1/2} \propto C_1^{-1/2}\, N\, \Phi^{1.3}, \quad (5)$$

where C_1 is the rate constant for the breaking process. Assuming for C_1 an Arrhenius behaviour with an activation energy W and using Eq.(1) leads to the following relation :

$$T_R \sim \eta_r \propto \exp(K'/2 + W)/T. \quad (6)$$

Turning now to the network properties, they can be characterized through the shear modulus $G = \eta_s/T_R$ which should be, as a first approximation, independent of temperature.

3. EXPERIMENTAL RESULTS AND DISCUSSION

The stress relaxation function provided by the magnetorheometer is described with a very good accuracy by a single exponential line shape at all temperatures investigated. This suggests that, in the viscous flow, the motion of the micelles is controlled by their breaking time τ_{Br} which then would be shorter than the viscous relaxation time τ_{rep}.

Figure 1 shows the semi-log plots of T_R and η_r respectively as a function of 1/T. The variation of log T_R with 1/T is seen to be linear ; from the slope of the corresponding straight line and taking into account the change of the solvent viscosity with temperature, one obtains an apparent activation energy $W_{app} \simeq 12000$°K. On the assumption that $\tau_{rep} \ll \tau_{Br}$, that is $W_{app} = 3/2\ K'$, one obtains $K'/T \simeq 27$ at 25°C. This value is large, compared to that obtained for the growth in the di-

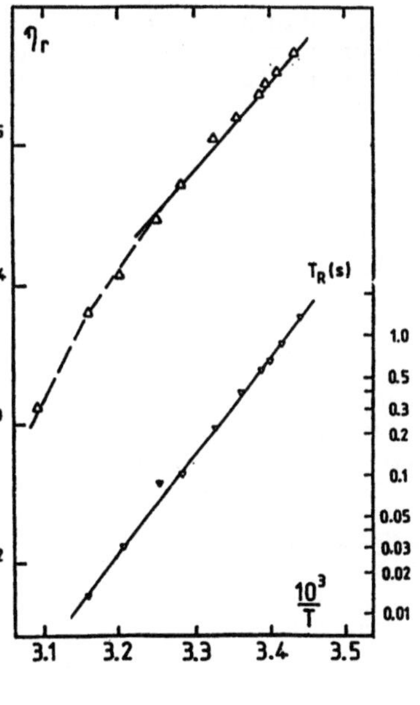

Fig.1

Semi-logarithmic variations of the viscoelastic relaxation time T_R and the relative zero shear viscosity η_r with 1/T for a 0.7 M CTAB solution in H_2O - 0.1 M KBr.

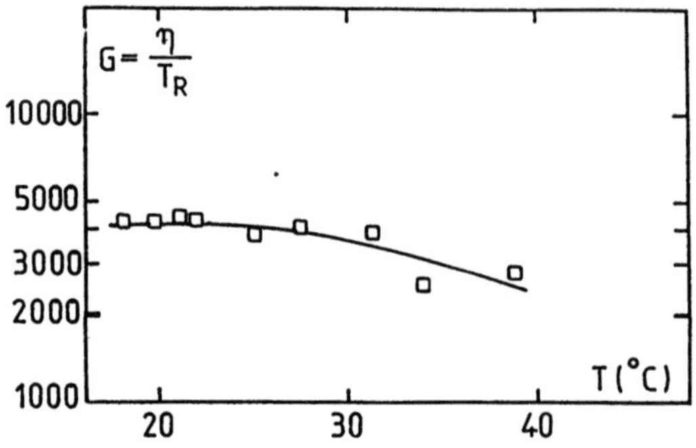

Fig.2

Temperature dependence of the shear modulus for a 0.7 M CTAB solution in H_2O - 0.1 M KBr.

lute range K'/T ≃ 15 - 19 in light scattering studies /1/ /7/ /8/ and leads to physically unsound value of N.

On the other hand, if one assumes $\tau_{rep} \gg \tau_{Br}$, which is consistent with the observation of a single exponential decay of the stress-relaxation function,then: W_{app} = 1/2 (K'+W). Taking for K'/T the values 15-19 obtained in the dilute range leads to the following estimate of the activation energy for the rate constant associated with the breaking process : W ∿ 18000 - 20000°K. The same analysis applied to the linear part of the variation of η_r with 1/T yields values of W in the same range. This rather large activation energy is in good agreement with the results obtained by means of T-jump, p-jump and stopped-flow techniques /18/.

Figure 2 shows the variation of the shear modulus G. It can be seen that G decreases slightly as T increases. This behaviour is similar to that observed for semi-dilute polymer solutions and can be attributed to a decrease of the number and the efficiency of the entanglements.

In conclusion, the rheological data reported here are consistent with a model of reptation troncated by the kinetics of fragmentation-coagulation of the micelles. These kinetics effects cannot be probed by light scattering in the dilute range because of the time scale investigated which is shorter than τ_{Br}. A confirmatory check of the validity of the model would be given by the volume fraction dependence of T_R and η_r. The results obtained on CTAB aqueous solutions led to larger variations of T_R and η_r with Φ than predicted by the Cates model. However in these experiments, the micellar growth was likely enhanced by a simultaneous increase of volume fraction and ionic strength upon increasing surfactant concentration. Also, the theory of micellar growth derived for dilute systems might not apply to semi-dilute ones. Clearly more experiments are needed to clarify this point.

REFERENCES

1. C. Young, P. Missel, N. Mazer, G. Benedek, M. Carey: J. Phys. Chem. 82, 1375 (1978) ; P. Missel, N.Mazer, G. Benedek, M. Carey: J. Phys. Chem. 87, 1264 (1983) and references therein.
2. S. Ikeda, S. Ozeki, M.A. Tsunoda: J. Colloid Interface Sci. 73, 27 (1980) ; S. Ozeki, S. Ikeda: ibid 77, 219 (1980).
3. S. Hayashi, S. Ikeda: J. Phys. Chem. 84, 744 (1980).
4. W. Schorr, H. Hoffmann: J. Phys. Chem. 85, 3160 (1981).
5. H. Hoffmann, G. Platz, W. Ulbricht: J. Phys. Chem. 85, 3160 (1981).
6. (a) H. Hoffmann, G. Platz, H. Rehage, W. Schorr: Ber. Bunsenges. Phys. Chem. 25, 877 (1981); (b) Adv. Colloid Interface Sci. 17, 275 (1982).
7. G. Porte, J. Appel, Y. Poggi: J. Phys. Chem. 84, 3105 (1980).
8. G. Porte, J. Appel: J. Phys. Chem. 85, 2511 (1981).
9. J. Appell, G. Porte, Y. Poggi: J. Colloid Interface Sci. 87, 492 (1982).
10. F. Reiss-Husson, V. Luzzati: J. Phys. Chem. 68, 3504 (1964).
11. S.J. Candau, E. Hirsch, R. Zana: J. Colloid Interface Sci. 105, 521 (1985).

12. S.J. Candau, E. Hirsch, R. Zana: J. Phys. 45, 1263 (1984).
13. M. Adam, M. Delsanti: J. Phys. 44, 1185 (1985).
14. S. Safran, L. Turkevich, P. Pincus: J. Phys. Lett. 45, L-19 (1984).
15. D. Blankshtein, G. Thurston, G. Benedek: J. Chem. Phys., to be published.
16. W. Brown, R. Johnsen, P. Stilbs, B. Lindman: J. Phys. Chem. 87, 4548 (1983).
17. M. Cates: Macromolecules, submitted for publication.
18. J. Lang, R.Zana: In Surfactant Solutions New Methods of Investigation, R.Zana ed. (Dekker, New York 1987) p.405 and references therein.

Shear Induced Micellar Structures

H. Hoffmann, H. Rehage, and I. Wunderlich

Institut für Physikalische Chemie der Universität Bayreuth,
Universitätsstr. 30, D-8580 Bayreuth, Fed. Rep. of Germany

Abstract: Aqueous solutions of cationic surfactants with strongly binding counterions show the striking phenomenon of flow induced phase transitions. In this paper we discuss new results which have been obtained from rheological measurements and from flow birefringence data.We examine the stability of the shear induced state as a function of temperature and surfactant concentration and we analyse the effect of solubilisation of hydrocarbons.The results are interpreted in terms of a kinetic model which accounts for the observed behavior.

1. INTRODUCTION

Surfactants of a particular kind form supermolecular structures during flow [1-5].The new structures are not present in the solutions at rest but they are formed under the orienting forces of a velocity gradient. On grounds of this unusual behavior we have called these supermolecular phases "shear- induced structures "(SIS).The new phase can only be formed when the shear rate exceeds a well- defined threshold value.The critical shear rate depends very much on the specific conditions of the system like the temperature, chain-length or the ionic strength of the solution.The appearence and the build-up of the SIS can easily be monitored by rheological and flow birefringence measurements.

When the flow process is stopped ,the SIS decays and the solution relaxes to its equilibrium state.The relaxation time for this process depends on the specific conditions of the surfactant system.In practical applications the surfactant solution can come into contact with different additives and it is therefore of interest to know how the SIS is influenced by the addition of other additives.In the present paper we make an effort to investigate this problem. As a test system we used Tetradecyltrimethylammoniumsalicylate (TTAS) which shows the striking phenomenon of (SIS). As experimental techniques we used flow birefringence and rheological measurements ,which are well suited for the investigation of these phenomena.

2. EXPERIMENTAL

Tetradecyltrimethylammoniumsalicylate (TTAS) was a gift from Hoechst Company. The surfactant was prepared by an ion exchange procedure from TTACl-solutions or by dissolving TTAS which had been synthesized before[1-5].Both methods gave identical results.The solutions were left standing for two days in order to reach equilibrium.The rheological properties of the surfactant solutions were measured with the Rheometrics Fluid Rheometer RFR 7800 (cone and plate geometry). The experimental equipment for flow birefringence measurements consists of a concentric cylinder apparatus which has been described in detail in Refs.[6-8].

3. RESULTS

The phenomenon of shear-induced structures can be observed only in the highly dilute concentration regime ,where the zero shear viscosity is very near the one of water.Above the transition point where the viscosity starts to rise ,the solutions are elastic already at rest and exhibit strong gel properties.The transient behavior of the shear-induced phase transition can be investigated by measuring the

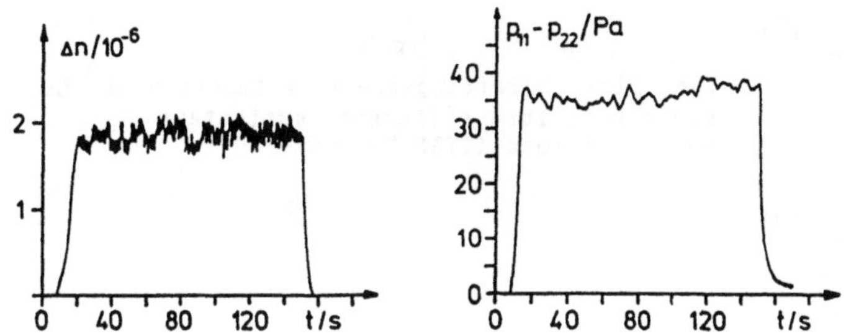

Fig.1
The first normal stress difference $p_{11}-p_{22}$ and the flow birefringence Δn as a function of the shear time after a step function shear rate (TTAS $c=2.5$ mmol, $T=20^{\circ}C$, $\dot{\gamma}=200 s^{-1}$).

time-dependent rheological properties. In this type of experiment ,a step function shear rate is suddenly applied to the solution and the flow birefringence and the first normal stress difference are measured as a function of time. In former investigations we have pointed out that the formation of the flow induced structure is accompanied by only a small increase of the shear stress ,but there are large effects concerning the first normal stress difference and the flow birefringence[4]. Typical results of these measurements are represented in Fig.1.

In this experiment the step function shear rate has been applied at t=0.Immediately after the onset of flow ,the first normal stress difference is equal to zero, indicating that the solution still behaves as a Newtonian fluid.After a short time of shearing, however,the rheological properties change dramatically and the first normal stress difference rises steeply to a plateau value. The occurrence of normal forces is usually attributed to the presence of finite strains in viscoelastic materials [9]. This definition implies that only the flow-induced state is viscoelastic whereas the solution at rest behaves like a Newtonian liquid.Below the critical shear time of 10s only viscous properties can be detected.

In analogy to the transient rheological experiments , time dependent flow birefringence measurements can be performed. We observe a curve similar to that of the first normal stress difference (Fig. 1). After the critical time of shearing, the flow birefringence increases steeply and after about 20s a steady state value is reached.In this range the new structure ,the shear induced phase, is formed and strong birefringence appears.It is evident that the flow-induced structure has anisotropic optical properties.

It is interesting to note that the extinction angle ,which describes the dynamic orientation process of particles under shear conditions, is equal to zero degrees for the shear induced structure[4]. That means: The flow - induced phase is completely aligned in the direction of flow. Under these conditions there exists a simple relationship between the flow birefringence and the first normal stress difference [4]:

$$\Delta n=C(p_{11}-p_{22}). \qquad (1)$$

The formation of the shear-induced state depends upon the actual surfactant concentration. Typical results of flow birefringence measurements are summarized in Fig. 2.

It is interesting to note that for all concentrations the optical anisotropy effect can be observed at the same characteristic threshold value of the shear rate. At high values of the velocity gradient ,the curves approach a saturation value. The plateau values of Δn increase more or less linearly with the total

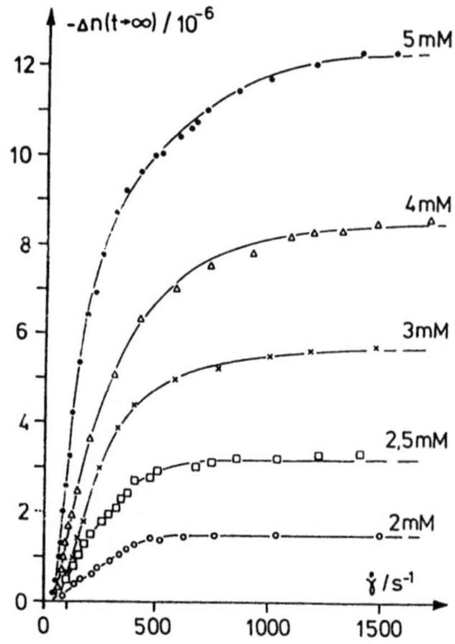

Fig.2

The flow birefringence as a function of the shear rate for different surfactant concentrations (TTAS, T=20° C).

surfactant concentration. The experimental results indicate that there are no shear-induced states below a surfactant concentration of 1.5 mmol. This phenomenon could be caused by a sphere/rod transition of the micellar aggregates and we could argue that the globular micelles do not form a SIS. The formation of the supermolecular structure depends on the temperature of the surfactant solution.This is clearly seen in Figure 3, where the flow birefringence is given as a function of the shear rate for several temperatures.

With increasing temperature , the critical shear rate where the SIS can be detected is shifted to higher values.At temperatures above 40° C the flow-induced state disappears. This indicates a rod-sphere transition at elevated temperatures. Globular micelles, which are formed under these conditions ,are not able to form shear-induced phases.

Micellar solutions can always solubilize a certain amount of hydrocarbons [10]. The amount which can be enriched in the hydrophobic core of the micellar aggregates depends very much upon the particular conditions and may vary for the

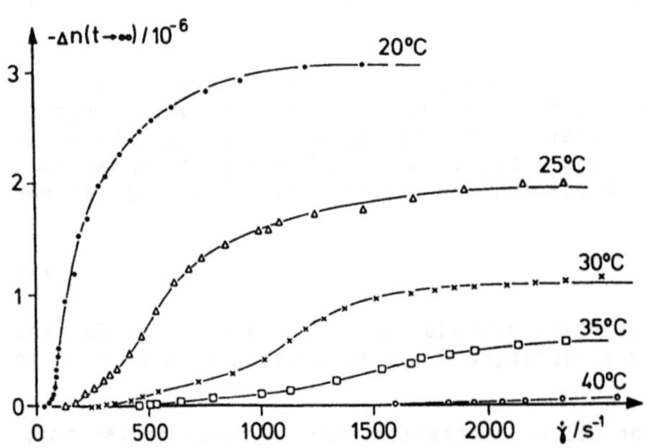

Fig.3

The flow birefringence as a function of the shear rate at different temperatures (TTAS, c=2.5mmol).

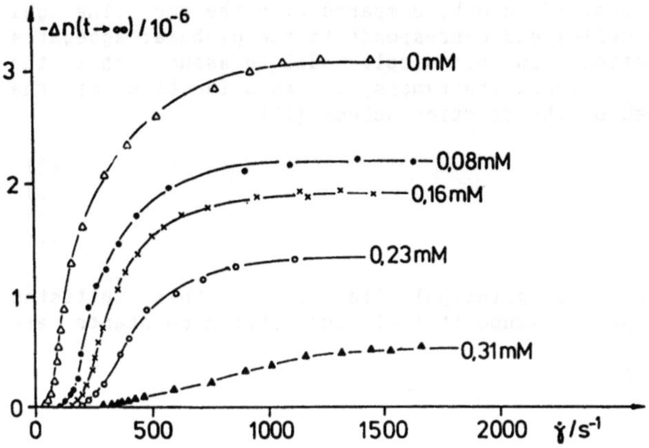

Fig.4
The flow birefringence of TTAS-solutions as a function of the shear rate for different concentrations of decane (c=2.5mmol, T=20°C).

same surfactant ion by more than one order of magnitude as a function of salt concentration or temperature . In the case of globular micelles, the amount of single chain hydrocarbons which can be solubilized by a single chain surfactant of about equal length is very limited. Usually the molar ratio between the hydrocarbon and the surfactant is below 0.1 . Rod-like micelles have a rather high solubilization capacity for single chain hydrocarbons . In such systems solubilisation ratios of the order of 1 can be observed. In order to investigate the effect of solubilization on the SIS ,we have carried out measurements of the flow birefringence of these solutions.Typical results of our measurements are shown in Fig. 4.

With increasing solubilization the critical shear rates are shifted to higher values while the plateau level of the flow birefringence decreases. It is obvious that the solubilization affects the SIS in the same way as an increase of the temperature.This suggests a decrease of the axial ratio of the anisometric particles with increasing oil concentration. The SIS disappears at conditions where the rods are transformed into globular micelles.From geometrical constraint it follows that the available area for the surfactant headgroup at the micellar interface is about twice as large as that of the cross section of the chain for the rod-like micelles and about three times for the globular micelles. The strongly adsorbed counterions need at least as much area as the headgroups themselves. It is thus likely that the whole interface of the cylinders is completely covered with the headgroups and the counter-ions while at the end caps there is some free area available in which the hydrocarbon core is in direct contact with water. It is conceivable that these sites are responsible for the contacts between different rods. In the globular particles ,which form when the hydrocarbon is solubilized in the interior , these areas disappear because the surfactant molecules can now be packed more tightly at the micellar interface.

4.DISCUSSIONS

The main objective in this part is to explain the formation of the shear-induced structures in a qualitative way and to form a basis for a more quantitative treatment which will follow in a forthcoming paper. Micellar aggregates are always formed when the chemical potential of the monomers is continuously increased until the value of the monomers in the micelle is reached. This is usually achieved by increasing the surfactant concentration. Close to the critical micelle concentration (cmc), micelle formation can already be induced by a small change in temperature or pressure or by a change of the salt concentration. In the shear induced structures ,the sol/gel transition can be compared with the cmc .The gel state is formed from clusters of micelles and corresponds to the globular aggegates in the normal process of micellisation. In this simple model we assume that the rod-like micelles can form dimers, trimers ,tetramers,... as a function of the shear rate.This process is described by the reaction scheme [11]

$$A_1 + A_1 = A_2, \tag{2}$$
$$A_2 + A_1 = A_3, \tag{3}$$
$$A_3 + A_1 = A_4, \tag{4}$$
$$A_{s-1} + A_1 = A_s. \tag{5}$$

In order to show qualitatively the principal features of this multistep equilibrium system ,it is sufficient to assume that all equilibrium constants are identical and have the value K_1
The concentration A_s is then given by

$$A_s = (K_1 * A_1)^{s-1} A_1. \tag{6}$$

This equation shows that the concentration A_s is small when s is in a large number and the product $K_1 A_1 < 1$. With this condition only a few higher aggregates are present in the quiescent state. This situation has its analogue in the formation of micelles from monomers when we are below the value of the critical micelle concentration (cmc). In the thermodynamics of micelle formation A_s would correspond to a micelle with the aggregation number s. This same kinetic model can be applied to the formation of SIS by introducing the shear rate dependence into this theory. Supermolecular structures and higher aggregates can be formed by collisions which are due to the Brownian motion or due to the influence of a velocity gradient. We thus have two contributions for the formation of the oligomers and we can write

$$dA_2/dt = k^+ A_1^2 + \dot{\gamma}\xi A_1^2 - k^- A_2. \tag{7}$$

In these equations k^+ denotes the chemical rate constant for the formation of dimers. k^- is the dissociation rate constant and ξ is a constant with the dimensions of a reciprocal concentration.The value of ξ depends upon the effective cross-section of the rod-like micelles.
At stationary state conditions for the dimers and oligomers we obtain

$$(k^+/k^-)A_1^2 + (\dot{\gamma}\xi)/k^- A_1^2 = A_2, \tag{8}$$

$$(k^+/k^-) + (\dot{\gamma}\xi)/k^- = A_2/A_1^2 = K_2, \tag{9}$$

$$\text{or} \quad K_2 = K_2^0 + (\dot{\gamma}\xi)/k^-. \tag{10}$$

The equilibration constant K_2 is thus given by a shear-rate-independent term K_2^0 and a shear-rate-dependent term $(\dot{\gamma}\xi)/k^-$, which is equal to $(\dot{\gamma}\xi/\tau_D)$ when $\tau_D = 1/k^-$. From equations (10) and (6) we finally obtain

$$A_s = ((K_2^0 + \dot{\gamma}\xi\tau_D)A_1)^{s-1} A_1. \tag{11}$$

The equation shows that the SIS will be formed when the product $(K_2^0 + \dot{\gamma}\xi\tau_D)A_1$ approaches unity. The shear rate at which this occurs is the critical threshold value. The equation furthermore shows that this threshold value will depend very much on the value of τ_D. At short relaxation times τ_D the critical shear rate has to be high enough in order to shift the equilibrium towards the SIS. This simple kinetic model is also capabel of explaining another characteristic feature of the shear induced state. All the experiments show that the product of the relaxation time of the SIS and the critical shear rate is constant and a large number of the order 100. The decay of the SIS state is ,in analogy to its formation ,a multistep process. After the cluster of rods has been formed ,the new structure can also disappear by a succession of steps. The situation is similar like to micellar kinetics where all steps do not automatically go in the same direction. In analogy to the fast process in micellar kinetics we can therefore write

$$1/\tau_{SIS} = k^- / \sigma^2 . \quad (12)$$

In this equation σ describes the variance of the size distribution of the different clusters. We thus obtain

$$\tau_{SIS} \approx (\sigma^2 / k^-) \approx \sigma^2 \tau_D \quad (13)$$

and $$\tau_{SIS} \dot{\gamma} \approx \sigma^2 \tau_D \dot{\gamma} \approx \sigma^2 . \quad (14)$$

It furthermore follows that the SIS is only stable at conditions where $\tau_{SIS} \gg 1$. At such high velocity gradients any network structure will be completely stretched. It is not surprising that under these drastic conditions the aggregates of the SIS are completely aligned in the direction of flow. The angle of extinction for a dilute solution of rod-like particles is 45 ° for $\dot{\gamma}\tau \ll 1$ and approaches the angle zero for $\dot{\gamma}\tau \gg 1$.

5.CONCLUSIONS

The above observations suggest that the shearing process results in the reversible formation of a supermolecular structure. The flow-induced structure has gel properties whereas the quiescent state behaves as a simple sol. The formation and the decay of the SIS can qualitatively be described by a kinetic model which is based on the flow-induced aggregation of the rod-like micelles .

6.ACKNOWLEDGEMENT

Financial support of this work by a grant from the " Deutsche Forschungsgemeinschaft (SFB 213, Project C1)" is gratefully acknowledged.

7.REFERENCES

1. D.Ohlendorf, W.Interthal, H.Hoffmann : Proc.IX Int.Congress on Rheology , Mexico , 41(1984)
2. D.Ohlendorf ,W.Interthal ,H.Hoffmann : Rheol.Acta 25, 468 (1986)
3. H.Hoffmann ,M.Löbl ,H.Rehage : Physics of Amphiphiles:Micelles Vesicles and Microemulsions, XC Corso,Soc.Italiana di Fiscia,Bologna, Ital. (1985)
4. H.Rehage ,I.Wunderlich ,H.Hoffmann :Prog. Colloid & Polymer Sci.72, 51 (1986)
5. H.Rehage ,H.Hoffmann :Rheol.Acta 21 ,561 (1982)
6. M Löbl ,PhD-Thesis ,Universität Bayreuth (1985)
7. D. Koemann , H.Janeschitz-Kriegl :Prog. Colloid Polym. Sci.65, 265 (1978)
8. H.Janeschitz-Kriegl : Polymer Melt Rheology and Flow Birefringence , Springer-Verlag, Berlin , Heidelberg , New York (1983)
9. R. Darby ,L.F. Albright ,McKetta Maddox JJ(eds): Viscoelastic Fluids: An Introduction to Their Properties and Behavior , Marcel Dekker Inc. New York, Basel (1976)
10. H. Hoffmann ,W. Ulbricht :Tenside ,Surfactants ,Detergents 24 (1987)
11. E.A.G. Aniansson , S.N. Wall :J.Phys.Chem 79, 857 (1975)

Dynamics of Intermicellar Exchanges

R. Zana and J. Lang

Institut Charles Sadron, CNRS,
6 Rue Boussingault, F-67000 Strasbourg, France

Micelles are dynamic assemblies and a number of processes involving micelles occur spontaneously in micellar solutions /1-3/.The two most important processes are the intermicellar exchanges through which material (surfactant and any kind of solubilized compound: alcohol, oil, fluorescent probe, reactant, etc...) is exchanged between micelles and the process of micelle formation-breakdown.The time scale for these processes stretches from second to nanosecond.This is why their study was performed using fast kinetics methods, chemical relaxation /4/ and time-resolved fluorescence probing /5/ being the most extensively used.In the following the term micelle refers to simple or mixed micelles as well as micelles containing solubilized molecules and droplets of oil or water in oil in water or water in oil microemulsions.

Our purpose in this short review is to discuss the various types of intermicellar exchanges and to complement the current picture of micellar dynamics.At this stage it is noteworthy that studies of micellar dynamics are important because

(i) they provide an understanding of micellar systems, per se ;

(ii) they may help in the understanding of other properties of these systems such as diffusion or viscosity (see S.J. CANDAU et al, this volume) ;

(iii) there appears to be a relationship between dynamics, polydispersity, interactions and critical behavior ;

(iv) the rate of chemical reactions performed in micellar systems may be strongly coupled to micellar dynamics.

Figure 1 gives a schematic representation of the three types of intermicellar exchanges which have been evidenced thus far.Process A involves the exit of a micellar component, its diffusion in the bulk phase and its association to another micelle.In process B the exchange of material takes place following a collision between two micelles, during the time they remain merged.Process C is intermediate between process A and B.It involves the detachment of a small micellar fragment (fragmentation) which diffuses through the bulk and attaches to another micelle (coagulation).These fragments carry material from micelle to micelle.Depending on the system and on the material one or the other of these processes predominates. Clearly, process A will be negligible for micellar components insoluble in the bulk.The three processes are reviewed successively.

1. INTERMICELLAR EXCHANGE OF MICELLAR COMPONENTS

The most relevant results concerning this process have been obtained essentially with aqueous micellar solutions.The exchange of the surfactant has been mainly investigated by the ultrasonic absorption and the shock-tube techniques /4/.Time-resolved fluorescence and phosphorescence probing /5/, EPR, chemical relaxation /4/ and flash photolysis were used to study the exchange of the other micellar components.

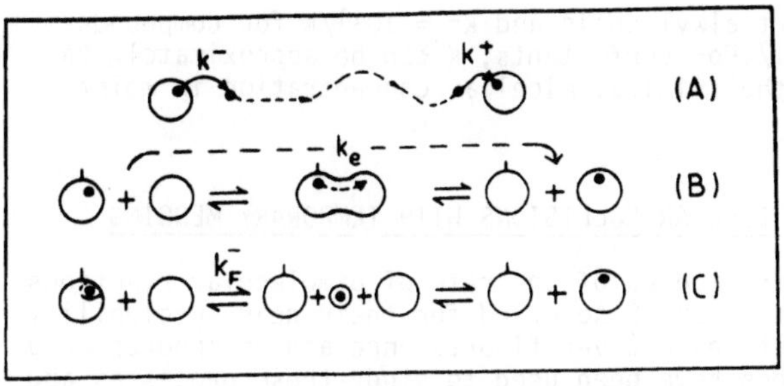

Figure 1. Intermicellar exchange processes. For the sake of clarity, the micelles are represented as spherical, whereas they may be anisodiametric.

The association rate constant k^+ (Fig.1) is usually close to, or only slightly lower than, the value which can be calculated for a diffusion-controlled process ($\sim 10^9$-10^{10} $M^{-1}s^{-1}$), whereas the dissociation rate constant k^- strongly depends on the hydrophobicity of the micellar component leaving the micelle /6,7/. Thus for surfactants k^- decreases from 10^9 to 5×10^4 s^{-1} when the surfactant chain length is increased from 6 to 16 carbon atoms. The residence time of surfactants in micelles, which is given by $T_R = N/k^-$ (N = micelle aggregation number), undergoes the same large variation. Table 1 lists the values of k^+ and k^- for typical compounds which can be solubilized in micelles /1/. It can be seen that k^- depends little on the nature of the surfactant. The k^+ values for the arenes are seen to be larger than for compounds with a medium alkyl chain. The difference reflects the fact that the former are solubilized in the micelle palisade layer whereas the alkyl chains of the latter must penetrate in the micelles.

The rate constant k^- can be evaluated from the partition coefficient K of any micellar component, within a factor 2 to 3, using the relationships $k^- \simeq 5\times10^{11}/K$

Table 1. Values of k^+ and k^- for selected compounds at 25°C (Adapted from reference 1 and references therein)

Compound	$k^+(M^{-1}s^{-1})$	$k^-(s^{-1})$
1-bromonaphthalene	4×10^{10}	2.5×10^4(a)
	-	4×10^4(a)
1-chloronaphthalene	-	4.3×10^4(a)
	-	1.9×10^3(b)
Biphenyle	-	1.2×10^5(b)
m-dicyanobenzene	10^{10}	6×10^6(a)
Toluene	-	1.4×10^6(a)
Acetophenone	$\sim 2\times10^{10}$	7.8×10^6(a)
Propiophenone	$\sim 2\times10^{10}$	3×10^6(a)
Benzophenone	$\sim 2\times10^{10}$	2×10^6(a)
Duroquinone	5×10^{10}	6×10^5(b)
Acetone	$> 10^{10}$	1-4×10^8(a,b)
Ethyliodide	2×10^{10}	5×10^6(a)
cis 1,3-pentadiene	1.2×10^9	8.9×10^6(a)
1,3-hexadiene	8.3×10^8	2.3×10^6(a)
1,3-octadiene	-	1.3×10^5(a)
1,3-cyclooctadiene	-	3.5×10^5(a)
t-butyl-(1,1-dimethylpentyl)nitroxide	-	7.5×10^5(c)
	-	5.9×10^5(d)
	-	3.7×10^5(a)
	-	2.2×10^5(e)
Di-t-butylnitroxide	1.5×10^9	2.8×10^5(a)
1-pentanol	$> 10^{10}$	$\sim 10^9$(b)

(a) Values in sodium dodecylsulfate micelles; (b) Values in cetyltrimethylammonium bromide micelles; (c), (d) and (e) Values in sodium, octyl, decyl and tetradecylsulfates, respectively.

for arenes and compounds with a short alkyl chain and $k^- \approx 10^{11}/K$ for compounds with a medium to long alkyl chain /1/.For surfactants, K can be approximately taken as: $K \approx 55.5/CMC$, where CMC is the critical micellar concentration in mole/liter.

2. INTERMICELLAR EXCHANGES THROUGH MICELLAR COLLISIONS WITH TEMPORARY MERGING

Such processes have been evidenced in studies of the rate of bimolecular reactions between micelle-solubilized reactants A and B selected for their near insolubility in the bulk phase.The stopped-flow, time-resolved fluorescence and phosphorescence quenching and flash photolysis methods have been used to study these processes and to obtain the second order overall rate constant k_e (Fig.1) for micellar collision with temporary merging and exchange (see references 66 and 69-74 in reference 3).

2.1 Aqueous Micelles

Both a time-resolved fluorescence quenching study of aqueous ionic micelles /8/ and a comparative stopped-flow study of the rate of dissolution of oil and water in microemulsions /9/ led to the conclusion that the rate of collisions between electrically charged micelles is very low.Clearly this is due to the strongly repulsive intermicellar interactions arising from the micelle electrical charge.

On the contrary, time-resolved fluorescence probing studies showed that collisions with temporary merging can take place in solutions of nonionic micelles /10/. The processes become detectable some thirty degrees below the cloud temperature T_c and its rate constant increases rapidly with T to reach near T_c, values close to that for a diffusion-controlled process (10^9-10^{10} $M^{-1}s^{-1}$).

2.2 Reversed Micelles and Water in Oil Microemulsions

The most extensive results have been reported for AOT/water/n-alkane ternary systems (see Table 2 and ref.11).The value of k_e has been found to increase considerably with the temperature, with the alkane chain length and with the value of the molar ratio $w = |water|/|AOT|$.This increase of k_e which correlates with an increase of micelle size, reflects increasing attractive intermicellar interactions.Some typical results are represented in Fig.2 and listed in Table 2.It can be seen that the values of k_e lie between 1.6×10^7 $M^{-1}s^{-1}$ in n-heptane and up to 7×10^9 $M^{-1}s^{-1}$ in n-decane.

A study of cationic surfactants /water/chlorobenzene ternary systems /12/ has shown that an increase of the surfactant chain length brings about a decrease of k_e and of micelle size.These two effects are opposite to that of the alkane chain on k_e and the micelle size in the case of AOT-based ternary systems.However the effects of T and w on both k_e and N are qualitatively similar in the two types of systems.Following HUANG et al./13/, the above results can be explained in terms of oil penetration and packing of the alkyl chains.

Values of k_e for surfactant/water/alcohol/oil quaternary systems are also listed in Table 2.These values can be quite large (up to 10^9 $M^{-1}s^{-1}$).However systematic studies of the dependence of k_e on the nature of the alcohol and oil, as well as on T and w remain to be performed for such quaternary systems.

3. INTERMICELLAR EXCHANGES THROUGH FRAGMENTATION-COAGULATION PROCESSES

The first evidence for fragmentation-coagulation processes in aqueous micellar solutions was reported by KAHLWEIT et al./14/.These authors were led to postulate the existence of such processes between submicellar aggregates to explain the observed decrease of the relaxation time associated with the micelle formation-breakdown at high surfactant concentration and/or ionic strength.Fragmentation-coagulation can be viewed as reactions by which micelles form and break down much more efficiently (rapidly) than by a series of steps where one monomeric surfactant at a time asso-

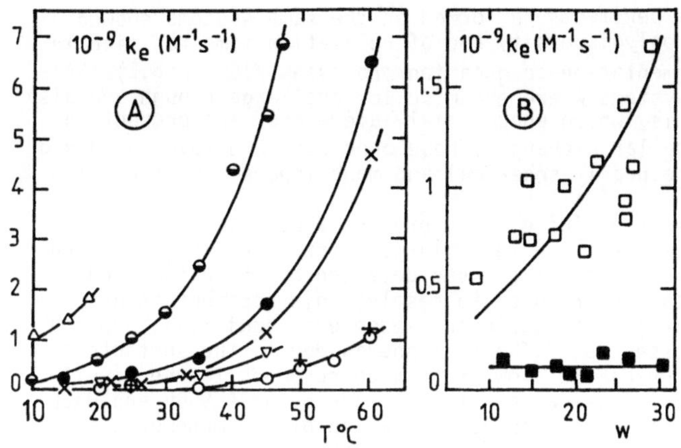

Figure 2. Variation of ke with temperature T and with w for AOT/water/n-alkane ternary systems. (A) w = 26.3: (+) n-hexane; (O) n-heptane; (▼) equimolar mixture of n-hexane + n-octane; (x) isooctane; (●) n-octane; (◒) n-decane; (▲) n-dodecane. (B) T = 25°C; (■) n-heptane; (◘) n-decane

Table 2. k_e values in various ternary and quaternary systems

System	$10^{-7}k_e(M^{-1}s^{-1})$	System	$10^{-7}k_e(M^{-1}s^{-1})$
Water/AOT/n-heptane		Water/KOl/1-hexanol/ n-dodecane	
w = 15 , T = 25°C	1.6		
w = 11 , T = 25°C	1.7	w = 20 , T = 25°C	130
Water/AOT/isooctane w = 15	19	Water/KOl/1-hexanol/ n-hexane w = 20 , T = 25°C	10
Water/AOT/n-dodecane w = 26.3 , T = 25°C	120	Water/KOl/1-pentanol/ n-hexadecane w = 16	10
Water/SDS/1-pentanol/ toluene (water volume fraction 0.04-0.12)	25	Water/DAP/cyclohexane w = 2.75	130

Abbreviations: AOT = sodium diethylhexylsulfosuccinate; KOl = potassium oleate; DAP = dodecylammonium propionate.

ciates with and dissociates from an aggregate /15/. In a more recent study /15/, alcohol additions were shown to induce or enhance fragmentation-coagulation processes, whereas additions of oil inhibit these processes. The time scale for micelle formation-breakdown through fragmentation-coagulation, as detected by T-jump /14, 16/, is 10^{-2}-10^{-5}s.

In recent time-resolved fluorescence probing studies of concentrated ionic micellar solutions, fast intermicellar exchanges with first order rate constants of 10^5-10^6 s^{-1} were observed with probe (pyrene) and quencher (cetylpyridinium ion, for example) which reside in micelles a time long compared to the fluorescence time scale which is shorter than 10^{-5} s /17/. Since the experimentally observed exchanges could not be explained in terms of processes A and B for the investigated systems, process C (Fig.1) was proposed for this purpose. It involves a full-sized micelle and a micellar fragment, at the difference of reactions between two submicellar aggregates invoked in the interpretation of the T-jump results /14/. As in T-jump studies /16/, additions of alcohol were found to induce or to strongly accelerate these processes, whereas alkane additions had the opposite effect /18/.

Quasi-elastic light scattering experiments performed on the same systems showed correlated increases of micelle polydispersity and of relaxation time for micelle formation-breakdown through fragmentation-coagulation processes /16/ (Fig.1).These processes are observed only in systems where the micelles are large enough and also sufficiently polydisperse, that is, where enough small aggregates are present to give rise to detectable intermicellar exchanges.These processes are never observed in systems where the micelles are nearly spherical and monodisperse in size.

Thus both T-jump and time-resolved fluorescence probing appear to detect fragmentation-coagulation processes.There are however large differences in the concentration ranges and the time scales of the processes detected by these two techniques.The following explanation can be proposed to resolve this apparent inconsistency.Recall that dilute micellar solutions, in the absence of additives, are characterized by two relation processes /6,7/.The fast one is due to the surfactant exchange (Process A in Fig.1) whereas the slow process corresponds to the micelle formation-breakdown under the now classical assumption made by ANIANSSON and WALL /15/, that is, through a series of association-dissociation of one monomeric surfactant at a time to/from aggregates.As noted above,in going to concentrated systems the role played by the monomeric surfactant is taken over by submicellar aggregates /14/.This results in a much more efficient (rapid) micelle formation-breakdown, through a series of fragmentation-coagulation reactions.In this regime, then, the elementary step for micelle formation-breakdown is no longer the monomeric surfactant exchange, but rather the fragment exchange detected by time-resolved fluorescence.This last process is characterized by a time $1/k_{\bar{F}}$ shorter than the relaxation time measured by T-jump,because the latter involves many such elementary steps.

The values found for $k_{\bar{F}}$ may at first sight look surprisingly large, and not sufficiently smaller than the rate constant of dissociation k^- of a monomeric surfactant.For instance for SDS at 25°C, $k^- \simeq 10^7 s^{-1}$ whereas $k_{\bar{F}}$ is of about $10^5 s^{-1}$.However, when a surfactant ion leaves a micelle, its whole alkyl chain becomes exposed to water.On the contrary when a micellar fragment leaves a micelle, the ten to twenty ions which may constitute this fragment pack together in the form of a sphere so as to minimize their contact with water. The free energy changes involved in the two processes may not be too different and so would be the rate constants which are determined by this free energy.

4. REFERENCES

1. R. Zana: J. Chim. Phys. 83, 603 (1986) and references therein
2. R. Zana, J. Lang: In Microemulsions: Structure and Dynamics, ed. by S. Friberg and P. Bothorel (CRC Press Inc., 1987) and references therein
3. R. Zana: In Surfactants in Solution, ed. by K.L. Mittal and P. Bothorel (Plenum Press, New York 1987) Vol.4, p.115 and references therein
4. J. Lang, R. Zana: In Surfactant Solutions: New Methods of Investigation, ed. by R. Zana, Surfactant Science Ser., Vol.22 (Marcel Dekker Inc., New York, Basel 1987) p.405
5. R. Zana: In Surfactant Solutions: New Methods of Investigation, ed. by R. Zana, Surfactant Science Ser., Vol.22 (Marcel Dekker Inc., New York, Basel 1987) p.241
6. H. Hoffmann: Ber. Bunsenges. Phys. Chem. 82, 988 (1978)
7. E.A.G. Aniansson, S.N. Wall, M. Almgren, H. Hoffmann, I. Kielmann, W. Ulbricht, R. Zana, J. Lang, C. Tondre: J. Phys. Chem. 80, 905 (1976)
8. A. Malliaris, J. Lang, R. Zana: J. Chem. Soc. Faraday Trans.1 82, 109 (1986)
9. C. Tondre, R. Zana: J. Dispersion Sci. Technol. 1, 179 (1980)
10. R. Zana, C. Weill: J. Phys. Lett. 46, L-953 (1985)
11. J. Lang, A. Malliaris, A. Jada: to be published
12. A. Jada, J. Lang, R. Zana: to be published
13. J.S. Huang, S.A. Safran, M.W. Kim, G.S. Grest, M. Kotlarchyk, N. Quirke: Phys. Rev. Lett. 53, 592 (1984) and references therein

14. E. Lessner, M. Teubner, M. Kahlweit: J. Phys. Chem. 85, 3167 (1981); M. Kahlweit: J. Colloid Interface Sci. 90, 92 (1982)
15. E.A.G. Aniansson, S. Wall: J. Phys. Chem. 78, 1024 (1974) and 79, 857 (1975)
16. J. Lang, R. Zana: J. Phys. Chem. 90, 5258 (1986)
17. A. Malliaris, J. Lang, R. Zana: J. Phys. Chem. 90, 655 (1986)
18. A. Malliaris, J. Lang, J. Sturm, R. Zana: J. Phys. Chem. 91, 1481 (1987)

Part VI

Microemulsions

A Statistical Mechanical Model for Microemulsions

J.C. Wheeler and T.P. Stockfisch

University of California, San Diego, La Jolla, CA 92093, USA

A microscopic statistical mechanical lattice model for microemulsions is presented whose energy parameters represent interactions among oil, water and surfactant molecules. Solving the model in the mean-field approximation allows phase diagrams and critical loci to be calculated. Features important in microemulsions are found in the model as well. These include ordinary critical, critical double point, critical endpoint, and tricritical loci, as well as three-phase equilibrium. The homogeneity assumption is checked against the possibility of a stable lamellar phase.

1. A Model Microemulsion

This paper introduces a lattice model for microemulsions. It is similar in purpose but distinct from the models of WIDOM [1] and ROBLEDO [2].

Each cell is assigned either an oil molecule (o) or a like volume of water molecules (w). Since cells are of molecular dimensions, we are defining a microscopic statistical mechanical model. This is in contrast to many earlier microemulsion theories (TALMON-PRAGER [3], JOUFFROY-LEVINSON-DEGENNES[4], WIDOM [5]) in which the cell size is chosen to give some desired coherence length, thereby building in certain microemulsion features from the start.

In addition to occupancy of the cells themselves, each face between two neighboring cells can be occupied by one surfactant molecule (s). In effect, there are two interpenetrating lattices, one containing oil and water, the other surfactant. We call the former the "cell" lattice and the latter the "bond" lattice. E_{ww}, E_{ow} and E_{oo} are different nearest-neighbor interaction energies between pairs of cells (with occupancy water-water, water-oil and oil-oil respectively) *not* separated by an intervening surfactant molecule. E_{wsw}, E_{osw}, and E_{oso} are the corresponding energies for cell pairs *with* intervening surfactant. Unlike the earlier microemulsion theories of TALMON-PRAGER, JOUFFROY-LEVINSON-DEGENNES, and WIDOM, the combinations *ow*, *wsw* and *oso* are allowed. Of course, we choose comparatively large values for the energies E_{wo}, E_{wsw} and E_{oso}.

There is an energy associated with bending of the surfactant surface. Because of the lattice, curvature of the surface is concentrated entirely at the edges between adjacent cell faces. Two surfactant-filled faces meeting at an angle have an energy E_{ss}. To account for the possibility that the surfactant may prefer to bend towards or away from water, there is an asymmetric bending energy E_b, defined as half the energy difference in bending towards oil instead of water.

Oil and water are represented by the occupation variables ν_i. They can take on the values one (representing oil occupancy) and zero (representing water occupancy), and reside on the cell lattice. The variables μ_{ij} also have allowed values one and zero, referring to the presence or absence, respectively, of surfactant on the face between cells i and j. The Hamiltonian is then

$$\begin{aligned} \mathbf{H} = {} & \Delta_{ow}\sum_i \nu_i + \Delta_s \sum_{<ij>} \mu_{ij} \\ & + E_{ow} \sum_{<ij>} (\nu_i + \nu_j - 2\nu_i\nu_j)(1-\mu_{ij}) \\ & + E_{osw} \sum_{<ij>} (\nu_i + \nu_j - 2\nu_i\nu_j)\mu_{ij} \\ & + E_{oo} \sum_{<ij>} \nu_i\nu_j(1-\mu_{ij}) \end{aligned}$$

$$+ E_{oso} \sum_{<ij>} \nu_i \nu_j \mu_{ij}$$
$$+ E_{ww} \sum_{<ij>} (1-\nu_i)(1-\nu_j)(1-\mu_{ij})$$
$$+ E_{wsw} \sum_{<ij>} (1-\nu_i)(1-\nu_j)\mu_{ij}$$
$$+ \sum_{<ijk>} \mu_{ij}\mu_{jk}[\, E_{ss} + E_b(2\nu_j-1)\,] \,, \tag{1}$$

where Δ_{ow} is the chemical potential difference between oil and water, Δ_s is the chemical potential of surfactant, and $\sum_{<ij>}$ and $\sum_{<ijk>}$ are nearest-neighbor sums on the cell and bond lattice, respectively.

We work with a linear transformation of the energy parameters:

$$w_0 \equiv \frac{1}{8}(2E_{ow}-E_{oo}-E_{ww}) \tag{2}$$
$$w_s \equiv \frac{1}{8}(2E_{osw}-E_{oso}-E_{wsw})$$
$$w_a \equiv \frac{1}{8}[E_{oso}-E_{oo}-(E_{wsw}-E_{ww})]$$
$$w_1 \equiv \frac{1}{8}(E_{oo}-E_{ww}+E_{oso}-E_{wsw})$$
$$w_2 \equiv \frac{1}{8}(2E_{osw}+E_{oso}+E_{wsw})$$
$$w_3 \equiv \frac{1}{8}(2E_{ow}+E_{oo}+E_{ww})$$
$$w_+ \equiv w_0+w_s$$
$$w_- \equiv w_0-w_s.$$

w_0 is the cost of mixing oil and water without benefit of surfactant. The adage "oil and water don't mix" implies that we should choose $w_0>0$.

w_s is the energy involved in moving surfactant from bulk oil or bulk water to an oil-water interface. Since surfactant is amphiphyllic, this is an energetically favorable operation and we require that $w_s<0$.

w_a (*a* for *asymmetric*) is proportional to the energy difference between putting a surfactant molecule in bulk oil as opposed to bulk water. If $w_a\neq 0$, oil and water don't enter the model on an equal footing and certain symmetries (discussed later) are broken.

Finally, since it costs surfaces energy to bend, we require that the quantities $(E_{ss}\pm E_b)$ be positive.

Certain symmetries become transparent if we translate the model into magnetic terminology. To this end we define spin variables

$$s_i \equiv 2\nu_i-1 \,, \tag{3}$$
$$\sigma_{ij} \equiv 2\mu_{ij}-1$$

and also magnetic field variables

$$H_c \equiv -\frac{\Delta_{ow}}{2} - qw_1 - \frac{qq'}{16}E_b \,, \tag{4}$$
$$H_b \equiv -\frac{\Delta_s}{2} - w_2 + w_3 - \frac{q'}{4}E_{ss}.$$

Terms in **H** not containing any spins can be ignored since they can't influence phase equilibrium. If we replace occupancies by spins, recombine terms, replace interaction energies by the linear transform (2), replace chemical potential differences by magnetic fields and ignore spinless terms, the Hamiltonian becomes

$$\begin{aligned}\mathbf{H} = &- H_c \sum_i s_i - H_b \sum_{<ij>} \sigma_{ij} \\ &+ \frac{1}{4} \sum_{<ijk>} \sigma_{ij}\sigma_{jk}(E_{ss}+s_j E_b) + \frac{q'}{4} E_b \sum_{<ij>} \sigma_{ij} s_j \\ &- w_0 \sum_{<ij>} s_i s_j (1-\sigma_{ij}) \\ &- w_s \sum_{<ij>} s_i s_j (1+\sigma_{ij}) \\ &+ w_a \sum_{<ij>} (s_i+s_j)\sigma_{ij} \ . \qquad (5)\end{aligned}$$

Note that w_1, w_2, and w_3 are subsumed in H_c and H_b. This form for **H** corresponds to a (hypothetical) magnetic system with two- and three-spin interaction terms, and two magnetic fields.

2. Phase Equilibrium

We have three degrees of freedom for the three components, but lose one due to the requirement that $\rho_o+\rho_w=1$ (recall that the cell lattice sites must contain either oil or water). This requirement can be interpreted as either an incompressible-fluid or an infinite-pressure limit. There are also five degrees of freedom from the energies $\overline{w}_0$, $\overline{w}_s$, $\overline{E}_{ss}$, $\overline{E}_b$ and $\overline{w}_a$ – seven degrees of freedom over all. If we settle on a particular choice of interaction energies, they can be replaced by one temperature variable, leaving three degrees of freedom. Various phase equilibria and critical loci reduce the degrees of freedom further, as summarized in Table 1.

Table 1. The **Global** column refers to the degrees of freedom for the full parameter space with all the interaction energies. For the **Particular** column the degrees of freedom are those remaining after w_o, w_s, E_{ss}, E_b, and w_a are set to a particular value.

Degrees of Freedom		
Feature	Global	Particular
single phase	7	3
two-phase equilibrium	6	2
three-phase equilibrium	5	1
ordinary critical locus	5	1
critical end point locus	4	0
tricritical locus	3	-

3. Results

We have solved the model in the mean field approximation. Minimization of the free energy results in the three equations

$$h_c = \tanh^{-1} m_c + q m_c(\overline{w}_- m_b - \overline{w}_+) + q \overline{w}_a m_b + \frac{qq'}{16} \overline{E}_b (2m_b + m_b^2) \ , \qquad (6)$$

$$h_b = \tanh^{-1} m_b + \frac{q'}{4} \overline{E}_{ss} m_b + \frac{q'}{4} \overline{E}_b m_c (1+m_b) + \overline{w}_- m_c^2 + 2\overline{w}_a m_c \ , \qquad (7)$$

$$\begin{aligned} f = &\ q m_c^2 \left(\frac{\overline{w}_+}{2} - \overline{w}_- m_b\right) - \frac{qq'}{16} \overline{E}_{ss} m_b^2 - \frac{qq'}{8} \overline{E}_b m_c m_b (1+m_b) + q \overline{w}_a m_c m_b \\ &+ \frac{1}{2} \log \frac{1-m_c^2}{4} + \frac{q}{4} \log \frac{1-m_b^2}{4} . \qquad (8)\end{aligned}$$

Here $h_c \equiv \frac{H_c}{kT}$, $h_b \equiv \frac{H_b}{kT}$, the bar over energy terms represents division by kT, q and q' are the coor-

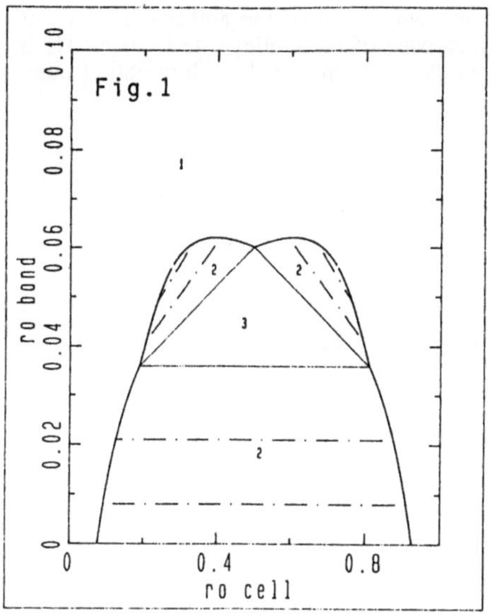

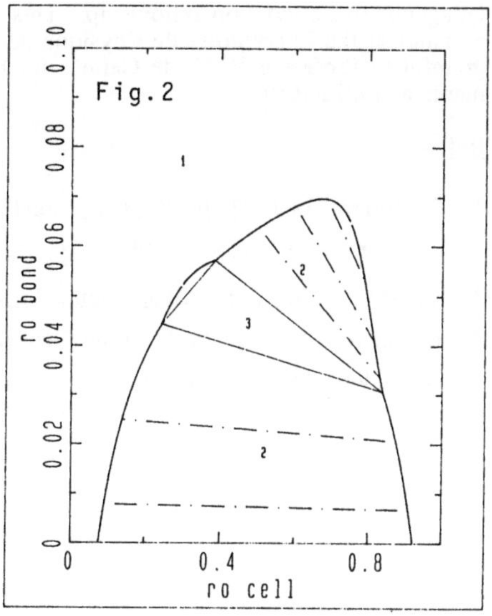

FIG. 1. Phase diagram for the energy assignments E_{ss}=0.0128, w_0=0.107, w_s=–0.504, $\overline{w}_a$=0, and $\frac{T}{T_t}$=0.873 (symmetric)

FIG. 2. Phase diagram for the energy assignments E_{ss}=0.0128, w_0=0.107, w_s=–0.504, $\overline{w}_a$=0.02, and $\frac{T}{T_{t,sym}}$=0.873 (asymmetric)

dination numbers for the cell and bond lattice (6 and 8 for simple cubic), m_c and m_b are per-site magnetizations, and f is the per-site free energy which is minimized at constant h_c and h_b.

Many phase diagrams for this model have been calculated, two of which are shown in Figs. 1-2. The triangle in Fig. 1 is symmetric, in the sense that oil may be interchanged with water. This occurs when $w_a \equiv E_b \equiv 0$, or when the effects of w_a and E_b happen to balance each other. Asymmetric triangles are also possible, as in Fig. 2.

Ordinary critical, critical double point, critical endpoint, and tricritical loci have also been calculated. The ordinary oil-water critical point occurs when $2q\overline{w}_0 \equiv 1$. For any physical choice of the energy parameters that satisfies $w_a = E_b = 0$ and $\overline{w}_0 > \frac{1}{2q}$, we have proven that there exists a unique, stable tricritical point. The three-phase equilibrium curve terminates in an upper critical endpoint or tricritical point. For a particular choice of energy parameters we do not expect (and have not found) a lower critical endpoint or tricritical point, but if we make w_a or E_b dependent on temperature, this is easily achieved.

As in any mean field calculation for a lattice model, we must check that a solution with sublattice ordering does not have a lower free energy. We have found that at either high surfactant concentrations or low temperature, lamellar phases are more stable. However, for a reasonably large neighborhood of the tricritical point, and small surfactant concentration, the homogeneous phase has been determined to be more stable.

Acknowledgment

Research supported by the National Science Foundation through NSF grant CHE 81-19247 and NSF grant INT 82-12577. One of the authors (JCW) acknowledges the support of a John Simon

Guggenheim Foundation fellowship. This work was started while one of the authors (JCW) was a visitor at the Laboratoire de Physique de la Matière Condensée of the College de France. He is grateful to Professor P. G. de Gennes for his kind hospitality there and for his illuminating comments and criticisms.

References

1. B. Widom: In J. Chem. Phys. 84 , 6943 (1986)
2. C. Varea, A. Robledo: In Phys. Rev. A 33 , 2760 (1986)
3. Y. Talmon, S. Prager: In J. Chem. Phys. 69 , 2984 (1978)
4. J. Jouffroy P. Levinson P. G. de Gennes, In J. Phys. (Paris) 43 , 1241 (1982)
5. B. Widom: In J. Chem. Phys. 81 , 1030 (1984)

Middle-Phase Microemulsions and Random Surfaces

S.A. Safran, D. Roux, S.T. Milner, M.E. Cates, and D. Andelman
Corporate Research, Exxon Research and Engineering Co.,
Annandale, NJ 08801, USA

I. Introduction

A characteristic feature of microemulsions, as opposed to simple liquid mixtures, is that the oil and water remain separated by surfactant monolayers with coherent domains, typically tens or hundreds of Angstroms in size [1]. The configuration of these domains varies with composition. For small fractions of oil in water or water in oil, the structure is that of globules [2] whose colloidal properties are well understood [3]. On the other hand, when the volume fractions of oil and water are comparable and the surfactant concentration is low, one expects random, bicontinuous [4] structures to form. The theoretical characterization of these structures and the predicted phase diagrams are a topic of current interest.

A common aspect of the phase diagrams of random microemulsions at low surfactant concentration (≤5%), is the presence of two- and three-phase regions [1, 5]. In the two-phase region, there is a coexistence between an almost pure phase (small amounts of either surfactant in oil or surfactant in water) and a microemulsion (lower- or upper-phase, respectively). In the three-phase region, a middle-phase microemulsion coexists simultaneously with almost pure water and almost pure oil. At higher surfactant concentrations, there is typically a first-order transition from the isotropic, disordered, microemulsion to an ordered, lamellar phase (or to other ordered phases).

DE GENNES and TAUPIN [6] were the first to suggest that at low surfactant concentration a random microemulsion phase may be favored over the ordered, lamellar phase. This is because (i) the random microemulsion has a greater entropy of mixing - first considered by TALMON and PRAGER [7], and (ii) the bending of the surfactant layer forced by the randomness of the oil and water domains occurs at the persistence length of the surfactant monolayer defined by

$$\xi_K = a\, e^{4\pi K/\alpha T}. \tag{1.1}$$

Here K is the bending constant which parameterizes the "splay energy" of a surfactant monolayer, a is a molecular length, T is the temperature, and α is a numerical constant which depends on the details of the calculation. (De Gennes and Taupin set α=2, but below we will find it convenient to use a different value.) However, this model did not predict three-phase equilibria, which is characteristic of these systems [8].

In Refs. [9] and [10] we proposed a simple model which took into account the thermal undulations of the fluctuating surfactant interface [6]. The calculated phase diagrams <u>did</u> show both two-and three- phase equilibria in qualitative agreement with experiment [5]. An alternate model, which also results in two- and three- phase equilibria was previously proposed by WIDOM [11] for <u>compressible</u> surfactant interfaces in microemulsions. In our model, the free energy of the microemulsion consists of the entropy of mixing of water and oil domains, and the bending energy of the surfactant film, which is assumed to be an <u>incompressible</u>

monolayer [2, 12, 13]. The effects of thermal fluctuations on length scales smaller than the domain size ξ are accounted for by calculating the curvature energy with a size-dependent bending constant, $K(\xi)$ [14, 15]. Our simple model leads to a phase diagram with both two- and three-phase regions; the "middle-phase" microemulsion, which coexists with both nearly pure oil and water, is characterized by a length scale $\xi \sim \xi_K$ and by a surfactant concentration, $\phi_s \sim 1/\xi_K$. At higher values of the surfactant concentration, the bending constant assumes its bare value; the random microemulsion is then unstable to an ordered lamellar phase due to the high energy cost of bending the surfactant monolayer in a random manner.

In this paper, we review our model and its major results, with an emphasis on the role of the renormalized bending constant, $K(\xi)$. We analyze the role of thermal fluctuations in the renormalization of the bending constant for a one-dimensional model at length scales $\xi<\xi_K$ as well as $\xi>\xi_K$. The analysis suggests that at large length scales, $\xi>\xi_K$, the free energy cost of a bend saturates at a value of order T. Finally, we discuss the connection between our microemulsion model and the general problem of the random surface.

II. Microemulsion Model

A. Free Energy of a Random Microemulsion

We consider microemulsions to be ternary mixtures of oil, water and surfactant. Space is divided into cubes of size ξ filled either with water or oil. The surfactant is constrained to stay at the water-oil interface; we divide it equally between the oil and water domains. Using the random mixing approximation, the probability ϕ for a cube to contain water is

$$\phi = \phi_W + \phi_S/2 \, ,$$

where ϕ_W and ϕ_S are respectively the volume fractions of water and surfactant. The probability for a cube to contain oil is $1-\phi$. The constraint for the surfactant to fill the water-oil interface allows us to relate the volume fractions of the components and the domain size ξ within the random mixing approximation:

$$\phi_s \Sigma_0 = z \, v_s \frac{\phi(1-\phi)}{\xi} \, , \tag{2.1}$$

where z=6 is the coordination number of the cubic lattice; v_S is the molecular volume of the surfactant and Σ_0 is the surface area per surfactant molecule, which is fixed in our model. Within this approximation relation (2.1) gives the domain size ξ at each point of the phase diagram (for fixed ϕ, ϕ_S)

$$\frac{\xi}{a} = z \, \frac{\phi(1-\phi)}{\phi_s} \, , \tag{2.2}$$

where, for convenience, $a=v_S/\Sigma_0$ is chosen to be equal to the lower cutoff in Eq. (1.1).

Since the area per surfactant is kept fixed, the free energy per unit volume, f, has only two terms: the entropy of mixing of the water and oil domains, f_S, and the energy of curvature of the interface, f_C. The first term is calculated using the random mixing approximation

$$f_s = \frac{T}{\xi^3} \, [\phi \log(\phi) + (1-\phi)\log(1-\phi)] \, . \tag{2.3}$$

The second term f_C is calculated as follows. First we associate the bending energy E_C of a cube of water (oil) of size ξ surrounded by oil (water) with that of a sphere of diameter ξ

$$E_C = 8 \pi K(\xi) (1 - \xi / \rho_0)^2,$$

where ρ_0 is twice the spontaneous radius of curvature and is defined to be positive for curvature towards the water and negative for curvature towards the oil. In our model the probability of having a bend is related to the probability of having an edge. However, we take the radius of curvature to be comparable to ξ, rather than presuming the interface to have a sharp edge. Thus, the total energy of curvature per unit volume is given by

$$f_c = \frac{8\pi K(\xi)}{\xi^3} \phi(1-\phi) [1 - 2\xi (1-2\phi)/\rho_0] + \dots . \tag{2.4}$$

We have explicitly incorporated into our model the renormalization of the bending constant by thermal fluctuations, since $K(\xi)$ is a function of the lattice size, ξ. In our calculation of the phase diagram (see Refs. [9] and [10] for details), we use the expression first derived by HELFRICH [14] and later by perturbation theory [15]

$$K(\xi) \cong K_0 [1- \tau \log(\xi/a)] + \dots , \tag{2.5}$$

where $K_0 \equiv K(a)$ is the bare bending constant, a is the molecular size, and $\tau = \alpha T/(4\pi K_0)$. The downward renormalization of K indicates that it becomes relatively easy to bend a sheet of size $\xi \gtrsim \xi_K$, since such a sheet is already spontaneously wrinkled by thermal fluctuations. The result presented in Eq.(2.5) is correct for small values of ξ; the extension to larger values of ξ is discussed in Sec. III.

B. Phase Behavior and Transition to Lamellar Structure

The phase diagram for the case of zero spontaneous curvature ($\rho_0 \to \infty$) is shown in Fig. 1. It exhibits a one-phase region (the random microemulsion) and three polyphasic regions at low surfactant concentration: two two-phase regions where a microemulsion phase is in equilibrium with a very dilute phase of surfactant in either water or oil, and a three-phase region where a middle-phase microemulsion is in equilibrium with both dilute phases. The length scale ξ in the middle phase is proportional to ξ_K. Moreover, along the two-phase coexistence curve, ξ remains on the order of ξ_K even far from the middle phase. The concentration of surfactant in the middle phase scales as $1/\xi_K \sim \exp(-\tau)$; the phase diagram is a strong function of the persistence length and hence of the bare curvature modulus, K_0.

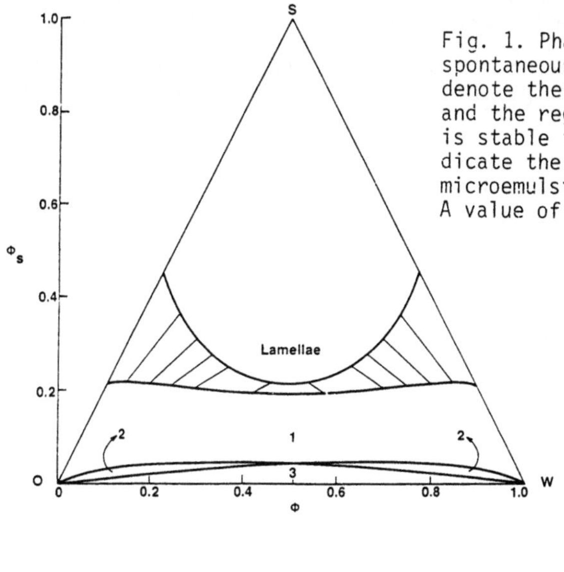

Fig. 1. Phase diagram for the case of no spontaneous curvature ($\rho_0 \to \infty$). The numbers denote the number of coexisting phases and the region where the lamellar phase is stable is indicated. The tie-lines indicate the coexistence of a disordered microemulsion phase with a lamellar phase. A value of τ=0.2 was chosen.

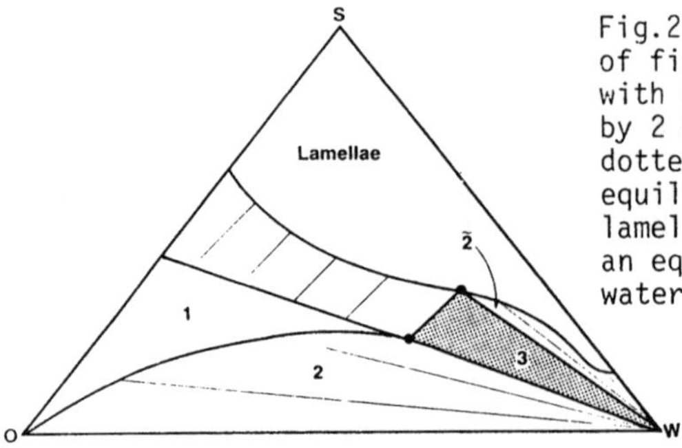

Fig.2 Schematic phase diagram for the case of finite spontaneous curvature ($\rho_0=\xi_K/5$), with a value of τ=0.15. The region denoted by 2 has been calculated in Ref. [10]. The dotted three-phase region consists of an equilibrium between the microemulsion, lamellar and water phases, while $\tilde{2}$ indicates an equilibrium between a lamellar phase and water.

For finite spontaneous curvature (e.g. $\rho_0>0$ favoring bending towards the water), there is an asymmetry in the extent of the two two-phase regions. For small enough values of ρ_0/ξ_K, the three-phase region disappears, as shown schematically in Fig. 2 where $\rho_0=\xi_K/5$. In that case, the length scale ξ along the two-phase coexistence curve bounding region 2, scales not with the persistence length ξ_K, but with the spontaneous radius of curvature, ρ_0; for small values of ϕ and ϕ_S, phase separation occurs when $\xi\sim\rho_0$. This is an indication of the emulsification failure instability which precludes the formation of globules with a size larger than ρ_0 [3]. At large values of ϕ and ϕ_S, a single phase of droplets is unstable to coexistence with a lamellar phase.

Returning the case of the symmetric microemulsion ($\rho_0\to\infty$), we note that the renormalization of the bending constant to values on the order of T, allows the energy cost of the random systems to be compensated by the entropy gain from the random mixing of the water and oil regions. However, at large values of ϕ_S, the length scale ξ decreases [see Eq. (2.2)], and $K(\xi)$ approaches its bare value K_0. The energy cost of the curved interfaces in the random microemulsion is no longer compensated by the entropy of mixing. Instead, the lamellar phase dominates, as shown in Fig. 1.

III. Renormalization of Bending Constant and Random Surfaces

Since the size dependence of the bending constant plays a crucial role in our model of microemulsions, it is important to go beyond the first-order perturbation formula for $K(\xi)$ [Eq. (2.5)], which is used to calculate the free energy cost of the bends imposed by the random mixing of the oil and water regions [Eq. (2.4)]. This free-energy cost, ΔF, can be easily calculated for a one-dimensional, semi-flexible (worm-like [16]) rod of length L. Here we sketch a calculation of $\Delta F(L)$ in a model which includes the bending energy and do not make any a priori assumptions about the persistence length. The details of this calculation will be presented elswhere.

The free energy ΔF, is calculated by taking the difference of the free energies of a chain constrained to bend and one which is free. The bending constraint is expressed as

$$< \vec{t}_1 > = \vec{\beta} < | \vec{R}_0 | > / L , \tag{3.1}$$

where $\vec{t}_1$ is the n=1 Fourier mode of the tangent vector $\vec{t}(s)$. The coordinate s is measured along the arclength of the chain.

$$\vec{t}(s) = \vec{R}_0/L + \sum_{m=1}^{N} \vec{t}_m \cos(m\pi s/L) \tag{3.2}$$

with N=L/a. In (3.1) $\vec{\beta}$ is a constant vector which determines the angle and direction of the bend, $\vec{R}_0$ is the end-to-end vector of the chain, and <...> represents a statistical average with the weight function $P[\vec{t}(s)]$, to be determined below. This choice for $\vec{t}_1$ ensures that the spatial displacement of the bend is of the same order as the mean end-to-end distance $< | \vec{R}_0 | >$; the bend is visible on scales of order $< | \vec{R}_0 | >$.

The free energy F is calculated variationally; F is minimized with respect to the statistical weight, $P[\vec{t}(s)]$ where,

$$F = T < \log P > + < H_b > . \tag{3.3}$$

Here, the first term is the entropy and $H_b = (x_0/2) \int ds\, (d\vec{t}/ds)^2$, is the bending energy, where x_0 has the units of energy × length. The incompressibility of the chain, which forces $\vec{t}^2(s)=1$, is approximated as a global constraint, which is imposed by a Lagrange multiplier conjugate to $\tau = \int ds\, \vec{t}^2(s)$. With these constraints, we determine $P[\vec{t}(s)]$ from $\delta G/\delta P=0$, where

$$G = F - \mu < \tau > - \bar{\mu} < \vec{t}_1 > . \tag{3.4}$$

The Lagrange multiplier $\bar{\mu}$ is chosen so that (3.1) is satified. The parameterization of the chain by its arclength allows the calculation of ΔF in both the stiff ($R_0 \sim L$) and floppy ($R_0 \sim L^{1/2}$) limits for chains without self-avoidance. Excluded volume effects can be included at the end of the calculation by a Flory argument.

The variational calculation yields an end-to-end distance, $R_0 = (\lambda/2z)^{1/2} L$. The Lagrange multiplier $z \equiv \mu L\lambda/2T$ with $\lambda = (4LT/\pi^2 x_0)$ is determined by the incompressibility constraint

$$\sum_{m=1}^{N} \frac{1}{m^2 + z} = \frac{4}{3\lambda} \left(1 - (1+\beta^2/2) \frac{R_0^2}{L^2} \right) . \tag{3.5}$$

For small values of λ, $R_0 \sim L$, and the chain is stiff. The free energy per unit length due to bending,

$$\Delta F(L)/L \approx \beta^2 (\pi^2/4)\, x(L)/L^2 ,$$

where the effective bending constant is

$$x(L) \approx x_0 (1 - \gamma (TL/x_0) + \ldots)$$

for $\beta \ll 1$, where γ is a numerical constant of order unity. The bending constant is renormalized by L and not log(L), due to the stronger effect of the one-dimensional fluctuations. As the length scale increases, $x(L)$ decreases. In the limit $\lambda \to \infty$, the chain is floppy, leading to $R_0 \sim L^{1/2}$ in the absence of excluded volume, and

$$\Delta F/L \approx T \beta^2 (1 - \bar{\gamma} \beta^2 (x_0/TL))/2L ,$$

where $\bar{\gamma}$ is a numerical constant. The free energy is higher for the bent chain; the change in free energy due to bending is on the order of kT per bend. An examination of the higher order terms in $\Delta F/L$ shows that they are positive and proportional to $x_0{}^2/L^3$.

A similar treatment of the two-dimensional interface (a surface) in a three-dimensional system can be carried out. An estimate of ΔF for the surface can be made by assuming that only the phase space in the Fourier decomposition [Eq. (3.4)] changes in going from the chain to the surface (this also implies that for the surface, where the bending constant K_0 has the units of energy, the parameter λ in (3.5) is proportional to T/K_0, independent of L). One predicts that for small values of L, the free energy per unit area due to the bend,

$$\Delta F/L^2 \approx K(L)/L^2,$$

with

$$K(L) \approx K_0 \, (1 - \tilde{\gamma} \, (T/K_0) \log(L/a) + \ldots) ,$$

where $\tilde{\gamma}$ is a numerical constant. This is in agreement with the functional form predicted by the perturbation theories [14, 15]. For large values of $L \gg \xi_K$, the free energy per unit area,

$$\Delta F/L^2 \approx T\beta^2/L^2 + \ldots ,$$

again representing the loss of entropy of one degree of freedom per surface.

While the small L limit of the stiff interface relates to recent treatments of membranes, the large L limit of the floppy interface may be relevant to the random surface discussed by Kantor et al. [17]. When L is much larger than the relevant persistence length, a large number of microscopic models may converge in a single universality class. The surface then executes the equivalent of the random walk of a polymer chain. Ref. [17] suggests that excluded volume interactions are extremely crucial in this regime; they change the expected radius of gyration from a weak logarithmic dependence on L to a power law.

The application of the statistics of a random surface to the microemulsion problem requires some care. In our microemulsion model, the length scale of the water/oil domains never grows larger than the persistence length ξ_K, because the entropy of mixing favors small ξ. If ξ is increased beyond $\sim\xi_K$, the system phase separates, since the microemulsion with $\xi \sim \xi_K$ is a local minimum (for fixed ϕ) of the free energy. Thus, for length scales $\xi \lesssim \xi_K$, one may use calculations of the effective free energy of bending such as those discussed above, as inputs to microemulsion thermodynamics. However, for length scales larger than ξ, the random walk of the surfactant interface as well as its excluded volume interaction are taken into account by the lattice construction. The random microemulsion is a random surface that is different from that considered in the previous discussion or by Ref. [17] in that (i) it has a sideness with respect to the oil/water and (ii) it is not one continuous surface; in random mixing, the characteristic surface size is of order ξ (iii) the coordination number is not fixed as in the tethered surface; coordination defects can lead to additional fluctuations which may increase the size of the surface even in the absence of excluded volume effects [18]. The random microemulsion is perhaps more analogous to the problem of equilibrium, branched polymerization. Nonetheless, the physics of the random surface and the analogy to the worm-like chain play an important role in our understanding of the free energy and the renormalization of the bending constant of microemulsions.

Acknowledgments

We are grateful for discussions with P. Pincus, W. Helfrich, and Y. Kantor.

References

(a) Present address: Department of Chemistry, UCLA, Los Angeles, CA 90024. Permanent address: Centre Paul Pascal, CNRS, Domaine Universitaire, 33405, Talence Cedex, France.

(b) Present address: Institute for Theoretical Physics, University of California, Santa Barbara, CA 93106.

(c) Also at Laboratoire de Physique de la matière condensée, Collège de France, 75231 Paris Cedex 05, France.

1. For a general survey see (a) Surfactants in Solution, ed. K. Mittal and B. Lindman, (Plenum, N.Y., 1984), and ibid 1986 (in press); (b) Physics of Complex and Supermolecular Fluids, ed. S.A. Safran and N.A. Clark (Wiley, N.Y., in press); (c) other papers in this volume.
2. A. Calje, W.G.M. Agerof, A. Vrij in Micellization, Solubilization, and Microemulsions, ed. K. Mittal, (Plenum, N.Y.) 1977, p. 779; R. Ober and C. Taupin, J. Phys. Chem. **84**, 2418 (1980); A.M. Cazabat and D. Langevin, J. Chem. Phys. **74**, 3148 (1981); D. Roux, A.M. Bellocq, P. Bothorel in Ref. 1a, p. 1843; J.S. Huang, S.A. Safran, M.W. Kim, G.S. Grest, M. Kotlarchyk, N. Quirke, Phys. Rev. Lett. **53**, 592 (1983); M. Kotlarchyk, S. H. Chen, J.S. Huang, M.W. Kim, Phys. Rev. A **29**, 2054 (1984).
3. C. Huh, J. Coll. Interface Sci. **97**, 201 (1984) and **71** (1979); S.A. Safran and L.A. Turkevich, Phys. Rev. Lett. **50**, 1930 (1983); S.A. Safran, L.A. Turkevich, P.A. Pincus, J. Phys. (Paris) Lett. **45**, L69 (1984).
4. L.E. Scriven, in Micellization, Solubilization, and Microemulsions, ed. K. Mittal, (Plenum, N.Y., 1977), p.877.
5. For example, see M. Kahlweit and R. Strey, J. Phys. Chem. **90**, 5239 (1986) and references therein.
6. P. G. de Gennes and C. Taupin, J. Phys. Chem. **86**, 2294 (1982).
7. Y. Talmon and S. Prager, J. Chem. Phys. **69**, 2984 (1978) and **76**, 1535 (1982).
8. J. Jouffroy, P. Levinson, P.G. de Gennes, J. Phys. (Paris) **43**, 1241 (1982).
9. S.A. Safran, D. Roux, M.E. Cates, D. Andelman, Phys. Rev. Lett. **57**, 491 (1986) and in Surfactants in Solution: Modern Aspects, ed. K. Mittal, (Plenum, N.Y., in press).
10. D. Andelman, M.E. Cates, D. Roux, and S.A. Safran, J. Chem. Phys., submitted.
11. B. Widom, J. Chem. Phys. **81**, 1030 (1984).
12. L. Auvray, J.P. Cotton, R. Ober, C. Taupin, J. Phys. (Paris) **45**, 913 (1984) and in Ref. 1.
13. Insoluble Monolayers at Liquid-Gas Interfaces, G.L. Gaines, (Wiley, N.Y., 1966).
14. W. Helfrich, J. Phys. (Paris) **46**, 1263 (1985).
15. L. Peliti and S. Leibler, Phys. Rev. Lett. **54**, 1690 (1985); D. Forster, Phys. Lett. **114A**, 115 (1986); W. Kleinert, Phys. Lett. **114A**, 263 (1986).
16. G. Ronca and D.Y. Yoon, J. Chem. Phys. **76**, 3295 (1982).
17. Y. Kantor, M. Kardar, and D.R. Nelson, Phys. Rev. Lett. **57**, 791 (1986) and Phys. Rev. A (in press).
18. J. M. Drouffe, G. Parisi, and N. Sourlas, Nucl. Phys. B **161**, 397 (1980); M. Karowski and H. J. Thun, Phys. Rev. Lett. **54**, 2556 (1985); M.E. Cates, Phys. Lett. **161 B**, 363 (1985).

Intimations of Bicontinuity in Microemulsion Theory

L.A. Turkevich

The Standard Oil Company, Corporate Research Center,
4440 Warrensville Center Road, Cleveland, OH 44128, USA

> Now while the birds thus sing a joyous song,
> The cataracts blow their trumpets from the steep;
> I hear the Echoes through the mountains throng,
> The Winds come to me from the fields of sleep.
>
> W. Wordsworth

1. INTRODUCTION

Microemulsions [1] are complicated multicomponent liquid mixtures with complex phase diagrams and multiphase equilibria [2]. We restrict our discussion to an idealised three-component liquid containing water, an immiscible nonaqueous "oil", and surfactant molecules, which have hydrophilic polar heads and hydrophobic hydrocarbon tails. Due to the bifunctional character of the surfactant molecules, the resulting complex fluid is not a purely random mixture of the three types of molecules but instead possesses a domain structure, typically on length scales ~ 100 Å. Our treatment emphasises the strong interplay between interglobular interactions and the intraglobular self-organisation.

In section 2 we study possible structures of microemulsion domains and develop a simple theory [3,4] of globule self-organisation, based on internal bending (splay and saddle-splay) energy considerations. We find generically three stable self-organised microemulsion states: spheres, cylinders, lamellae. As the topology of the structures changes, the role and importance of these large "internal" fluctuations also changes. The large internal fluctuations of these latter extended structures are reminiscent of the random domains that have been posited [5] for the structure of "middle-phase" microemulsions.

Self-organisation is conceptually clearest in the dilute regime, where the internal energy and possible structural changes are decoupled from any interaction effects. Similarly, the effect of interactions between microemulsion globules can be easily understood if the structures are "frozen". Sufficiently strong attractive interactions lead to a compressibility instability, and the single phase microemulsion phase separates (coacervation [6]) into a high-density "liquid"-like phase and a low-density "gas"-like phase. In section 3 we outline a mean-field phenomenological theory [7] of phase separation within this "colloidal" approximation.

In Section 4, we utilise the single-particle (emulsification failure) instability and the cooperative (liquid-gas) instability to describe the topology of microemulsion phase diagrams. We see that the phase behaviour is governed by the spontaneous radius of curvature ρ_0 and is relatively insensitive to the energetics of the model.

The decoupling of the self-organisation of microemulsion globules from their mutual interaction is only possible in the dilute regime, where, in fact, no phase separation is possible. Since microscopically, the forces responsible for the phase separation are identical with those responsible for the self-organisation, one cannot consider interacting microemulsion globules without also considering the possible effect of those interactions on the shapes and structure of the globules themselves. While we understand the phenomenology of the decoupled case (sections 2 and 3), the coupled case is considerably more difficult. In section 5, we re-examine the validity of the colloidal (integrity-preserving) limit, where the interactions had been essentially treated perturbatively. We discuss the simplest possible feedback effect of interactions on shape near the liquid-gas phase separation, namely the shape of clusters of spherical globules, interacting via a strong short-range interaction [8,9].

As the energy scales are similar, it is not surprising that the globules deform on interaction. We argue that microemulsions appear to be in the strongly interacting regime. A natural biproduct of this analysis is that for three-component water-in-oil microemulsions, v_w/v_s is conserved between clusters, in accord with the experimental suggestion [10] of this as a field variable. For reasonable interaction strengths, the spherical clusters severely deform when they aggregate. Beyond dimers, the spherical clusters aggregate into rods, and, as the volume fraction is increased, into polymeric chains with a persistence length (determined by the interaction energy and internal bending energy), analogous to the aggregation of ferromagnetic colloids [11]. The excluded volume of the linear clusters lowers the percolation threshold [12], with a resulting conductivity increase for water-in-oil microemulsions [13]. The semi-dilute regime, where the polymeric clusters have grown beyond the overlap concentration, resembles proposed bicontinuous structures [14], although this semi-dilute regime is highly disordered. Thus the high-density "liquid"-like microemulsion phase possesses considerable internal structure.

In section 6, we address the issue of whether aggregated microemulsion globules retain their integrity or whether they "merge" upon aggregation. We argue that for $K < T$, the interfaces between globules are permeated by pores, permitting the exchange of internal phase between droplets. Thus, in this regime, a large aggregate (as would obtain as the result of the cooperative liquid-gas instability) would be bicontinuous.

2. PHENOMENOLGICAL THEORY OF SELF-ORGANISATION OF MICROEMULSION GLOBULES

2.1 The Ginzburg-Landau Model Free Energy

Our theory of self-organisation of microemulsion globules recognises a hierarchy of microscopic molecular interactions in these systems. Dominant among these is the hydrophobic interaction. This expresses the relative attraction of water molecules for an aqueous environment and of oil molecules for a nonaqueous environment. The hydrophobic interaction is responsible for the familiar immiscibility of oil and water and also expresses the bifunctionality of the surfactant molecules, namely the hydrophilicity of the polar head and the hydrophobicity of the hydrocarbon tail. We incorporate the hydrophobic interaction into our model by precluding any naked oil-water interface--all potential oil-water interfaces must be saturated with surfactant. We also preclude any monomeric solubilisation of surfactant in either oil or water, as this would entail contact of polar head with oil or of hydrocarbon tail with water. In principle, surfactant may still be solubilised in aqueous solution as micelles or in nonaqueous solution as inverse micelles, but for simplicity we neglect these additional structures. Thus the hydrophobic interaction partitions the fluid into regions of oil and water separated by a surfactant layer.

We next focus on the properties of the surfactant layer. By analogy with liquid crystals, this layer behaves like a two-dimensional incompressible liquid. Namely, to distort the liquid away from some optimal area per molecule, either by compression or rarefaction, costs a large free energy, i.e. at fixed thermodynamic parameters (e.g. temperature T, pressure p, salinity s,...) the area per surfactant molecule is constant.

There remains the freedom to splay the surfactant interface without changing the overall area per molecule, but relatively changing the area per head and the area per tail. With any point $\vec{r}$ on the interface we associate two orthogonal curvature fields $\rho_1(\vec{r})$, $\rho_2(\vec{r})$. These curvature fields completely describe the shape of the interface. We now write down [3,4,15] the elastic energy associated with deviations of these curvature fields from some preferred values. With a change of coordinate system, these fields change. However we can construct two invariants, the mean curvature $2H = 1/\rho_1 + 1/\rho_2$ and the Gaussian curvature $Q = 1/\rho_1\rho_2$. The free energy per globule is written as a Ginzburg-Landau harmonic expansion in terms of these invariants

$$F = K_{spl}\int (H - H_o)^2\, dS + K_{ss} \int Q\, dS + \dots , \qquad (1)$$

where the integral is over the globule surface S. The magnitude of the elastic constants may be estimated [15] K_{spl}, $K_{ss} \sim .1$ eV from the analogous liquid crystal systems. K_{spl} represents the energy cost to splay the surface from some preferred mean curvature $H_o = 2/\rho_o$, while K_{ss} represents the energy cost associated with the formation of surface saddles. The preferred mean curvature H_o describes the inherent tendency of the interface to bend towards either the water ($H_o > 0$) or the oil ($H_o < 0$) side of the interface. The Gauss-Bonnet theorem guarantees the second term in (1) to be constant for all structures of the same topological genus, but since we are precisely interested in comparing structures of different topologies, this term acts as a topological chemical potential. It is simple algebra to rewrite (1) in the form

$$F = K/2 \int (1/\rho_1 + 1/\rho_2 - 2/\rho_o)^2 + \bar{K}/2 \int (1/\rho_1 - 1/\rho_2)^2\, dS, \qquad (2)$$

where $K,\bar{K}$ are trivially related to K_{spl}, K_{ss}.

While the above expansion is thoroughly general, it may be motivated by a simple model [4,16] for the elasticity of a "directed" surfactant layer. This model generalises that of Marčelja [17] for condensed paraffins and has been used to study the elasticity of lipid bilayers [18] and vesicles [19]. We idealize the surfactant molecule as a wedge with an area A_h per polar head and an area A_c per hydrocarbon chain, where these areas are measured within fictitious layers located at distances l_h and l_c from the "center of mass" of the molecule (Fig. 1). As one layer is "outside" the cm layer and the other is "inside", by convention $l_h l_c < 0$ (e.g. for water-in-oil microemulsions $l_h < 0$, $l_c > 0$). Within the harmonic approximation, the elastic energy of the head and tail layers may be written [17]

$$F = k_h/2\, (A_h/A_h^o - 1)^2 + k_c/2\, (A_c/A_c^o - 1)^2, \qquad (3)$$

where k_h and k_c are the elastic constants of the heads and chains respectively. The variables A_h and A_c are the actual areas per polar head and hydrocarbon chain respectively, while A_h^o and A_c^o are the optimal areas (the equilibrium areas in the absence of any geometric constraints).

For a curved interface, the areas are determined by geometry

$$A_h = (1 + l_h H + l_h^2\, Q)\, A , \qquad A_c = (1 + l_c H + l_c^2\, Q)\, A , \qquad (4)$$

where A is the area per surfactant molecule measured within the center of mass

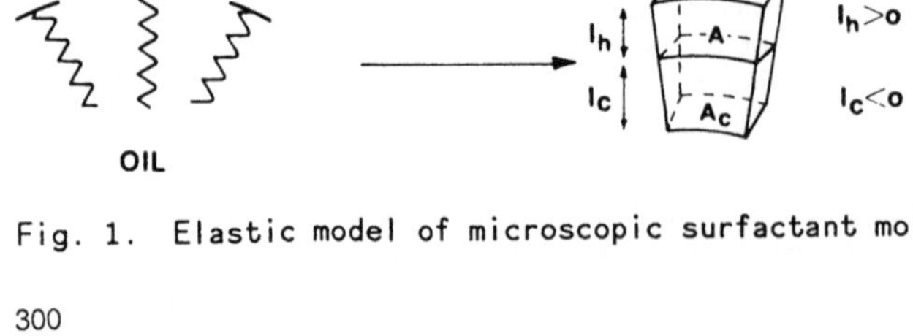

Fig. 1. Elastic model of microscopic surfactant molecules at an interface.

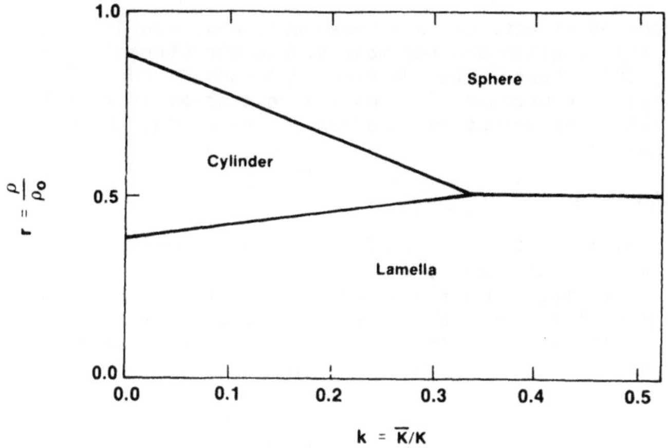

Fig. 2. Shape phase diagram for microemulsion globules.

layer. The free energy (3) for a single globule may be expressed as a function of the fields A,H,Q, and, minimizing with respect to the area A per surfactant molecule, yields (2). The parameters K, $\bar{K}$, ρ_o are thus derivable from the microscopic parameters A_h, A_c, l_h, l_c. Within this model,

$$k = \bar{K}/K = (l_c + l_h)/\rho_o \qquad (5)$$

so that a large spontaneous radius of curvature ρ_o necessitates a small saddle-splay constant $\bar{K}$. In fact, since l_h, l_c are microscopic lengths whereas ρ_o is macroscopic, $k = \bar{K}/K \ll 1$.

2.2 Single-Particle Shape Analysis

We now discuss the results of the above free energy in the dilute limit, where we may neglect interactions between globules. The mean-field phase diagram which obtains is shown in Fig. 2. The surfactant concentration, v_s, plays a major role in determining the structure of the domains. At low v_s, it becomes increasingly difficult to solubilise all of the internal phase. In fact, at $v_s = 0$, the oil and water merely phase separate, and a remnant of this situation, "emulsification failure", persists until a reasonable v_s is achieved. By contrast, at high v_s, it becomes increasingly difficult to accommodate all of the surfactant on the surfaces of closed globular domains, and the ensuing structures open up (lamellae and cylinders).

Consider the case of spherical microemulsion globules of radius ρ. The volume fractions x, v_s of globules and surfactant respectively are given by

$$x \sim n\, 4\pi/3\, \rho^3 , \qquad v_s \sim n\, 4\pi\, \rho^2\, \delta , \qquad (6)$$

where n is the number of drops per unit volume, and where δ is the microscopic surfactant length. Thus, given the geometry of the drops, the volume fractions of surfactant and internal phase completely determine the drop size ρ and the drop density n. The free energy per unit volume is given by

$$F_{sph} \sim n\, 16\, \pi\, K\, (1 - \rho/\rho_o)^2 . \qquad (7)$$

As the drop size ρ is increased to ρ_o, the internal free energy is decreased. When $3\delta x/v_s > \rho_o$, it is energetically more favorable for the fluid to self-organise into drops of the optimal radius ρ_o and phase separate ("emulsification failure") excess internal fluid rather than self-organise drops of radius $\rho > \rho_o$. By contrast, at high v_s, it is not possible to self-organise drops of radius $\rho < \delta/2$--the self-organised structures must "open up", and indeed, from Fig. 2 we see that cylinders and lamellae are the more stable structures. For large saddle-splay ($k > 1/3$), only shapes with identical orthogonal radii of curvature (spheres and lamellae) are present; for small saddle-splay ($k < 1/3$), there is little cost in generating an anisotropic radius of curvature, thereby permitting a stable cylindrical phase.

2.3 Finite Temperature Effects

Two entropic effects result from the introduction of finite temperature. The extended structures, which are so topologically distinct at $T = 0$, acquire defects, making them resemble more the compact spheres; the infinite cylinders pinch off after a certain degree of polymerisation, while the lamellae fold over into discs of a characteristic domain size. The $T = 0$ free energies must then be augmented by entropies of mixing, which we include within a mean-field approximation. With finite size extended structures, the first-order boundaries of the mean-field phase diagram (Fig. 2) metamorphise into cross-over regions, which, however, still have meaning for the fluctuation behaviour.

Finite temperature also introduces internal entropy to the structures. We stress that such fluctuations of size and shape occur in the ensemble of globules and not necessarily as a dynamical distortion of any one globule. One mechanism whereby such ensemble fluctuations occur is via collisions in which the globules may exchange surfactant and/or dispersed phase molecules, and, while not included in our model, there is also present in the physical microemulsions a reservoir of monomeric surfactant in the two fluids.

For spheres, thermal fluctuations lead mainly to polydispersity of the drops although there are fluctuations in the shapes (which may be analysed by a spherical harmonic decomposition) [20]. Of especial interest is the divergence in polydispersity (the $\lambda = 0$ fluctuation mode) at $\rho = 3/2\ \rho_0$. Thus, if one were able to "turn off" emulsification failure (at $\rho = \rho_0$) and access this larger droplet radius, the droplet would go unstable. While not originally recognized as such, this is essentially a phase inversion instability.

The fluctuations of lamellae, however, are always large [21]. The correlation function, $\langle\theta(\vec{r})\theta(0)\rangle \sim T/K \log(r/a)$, where $\theta(\vec{r})$ is the angle the normal to the lamella makes with some laboratory axis. Thus the individual lamellae are rigid only over a persistence length, $\xi \sim a \exp(2\pi K/T)$. There is a qualitative difference between the two regimes $R < \xi$, in which case the discs appear rigid over the entire domain, and $R > \xi$, in which case the discs are highly floppy and only appear locally lamellar. For higher concentrations, the rigid discs should order as a discotic liquid crystal, while the floppy highly fluctuating lamellae may be used to construct [21] a bicontinuous object with local lamellar structure.

For infinite cylindrical microemulsions, these fluctuations are even more severe [3,4]. The correlation function, $\langle[\vec{t}(s)-\vec{t}(0)]^2\rangle \sim T/K\ s/\rho$, w $\vec{t}(s)$ is the tangent to the cylinder at distance s along its arc length. Thus the cylinder is rigid only on length scales smaller than the persistence length, $\xi \sim K/T\ \rho$. Again there is a qualitative difference between the two regimes $N\rho < \xi$, in which case the cylinders apppear rigid over the entire domain, and $N\rho > \xi$, in which case the cylinders undergo random walk and only appear locally cylindrical. For higher concentrations, the rigid rods should order as a nematic liquid crystalline phase, while the highly fluctuating cylinders again resemble a bicontinuous phase.

3. COOPERATIVE EFFECTS: PHENOMENOLOGY OF THE LIQUID-GAS INSTABILITY

We now turn to cooperative effects among microemulsion globules, treated within the colloidal virial approximation [7,8], namely we take the microemulsion globular structure as fixed and unchanged by mutual interaction. For specificity we consider the case of microemulsion spheres. The internal energy of the noninteracting globules (2) is augmented by the entropy of mixing of hard spheres. Within a virial expansion, the interglobular interaction is written

$$F_{int} = -1/2 \quad 3/4\pi\rho^3 \; T \; A(\rho,T) \; x^2 , \tag{8}$$

where the dimensionless virial coefficient

$$A(\rho,T) \sim \int[\exp(-U(r,\rho)/T) - 1] \; d^3r \tag{9}$$

depends explicitly on the radius ρ of the droplet. This arises from the dependence of the drop-drop interaction $U(r,\rho)$ not only on the separation r of the drops but also on the droplet radius ρ. Because of the drop size dependence of the virial coefficient, as the volume fractions are varied, not only is the concentration of globules being changed but also the strength of their interaction through $\rho \sim v_w/v_s$.

The competition between the attractive interactions and the entropy of mixing permits a collective liquid/gas-like instability wherein the system separates from a single phase of spheres into two coexisting phases of spheres immersed in the same continuous phase. Such a phase separation, with a critical point, occurs in the AOT/water/decane system [22]. For $K \gg T$, the two phases consist of globules of the same radius, but with different numbers of globules per unit volume (i.e. $x_l \neq x_g$). Neutron scattering has verified this picture for the AOT system [23]. The critical point occurs at $x_l = x_g = x_c = .13$.

At fixed temperature, the phase separation can be driven by increasing the strength of the interaction; this can be achieved by varying the radius ρ of the globules, i.e. by varying the v_w/v_s. When the radius attains a critical value ρ_c, such that the virial coefficient $A(\rho_c,T) \sim 21$, the interactions are strong enough so that the microemulsion separates into "liquid" (high number density of globules) and "gas" (low number density of globules).

Of course, $\rho < \rho_o$, at which the single-particle emulsification failure instability intervenes. Thus if $\rho_o > \rho_c$ emulsification failure occurs in addition to the liquid-gas collective instability: as more water is added, the system phase separates out the excess water from two already coexisting microemulsion phases (one of which possesses a small volume fraction of microemulsion and hence looks like "oil")--i.e. a three-phase coexistence is achieved with microemulsion coexisting with excess water and "oil". This is illustrated in Fig. 3, where the solid line (liquid-gas coexistence curve) demarcates the collective instability between the single phase (1) and two-phase (2) microemulsions, and where the dashed line indicates the emulsification failure instability, either from single phase (1) to excess internal phase (2') (outside the liquid-gas coexistence curve), or from two-phase (2) to three-phase (3) inside the liquid-gas coexistence curve). If, however, $\rho_c < \rho_o$, the spheres can never grow past ρ_o, and the interactions between globules never become strong enough to drive the collective phase separation; in this case only the emulsification failure instability is achievable (Fig. 3 inset).

4. MICROEMULSION PHASE SEPARATION

We now utilise the above considerations to discuss the topology of microemulsion phase diagrams as occurs in the Winsor systems [2]. In those systems, with a variation in thermodynamic parameter (often salinity s), an upper phase microemulsion coexisting with brine evolves into a middle-phase microemulsion

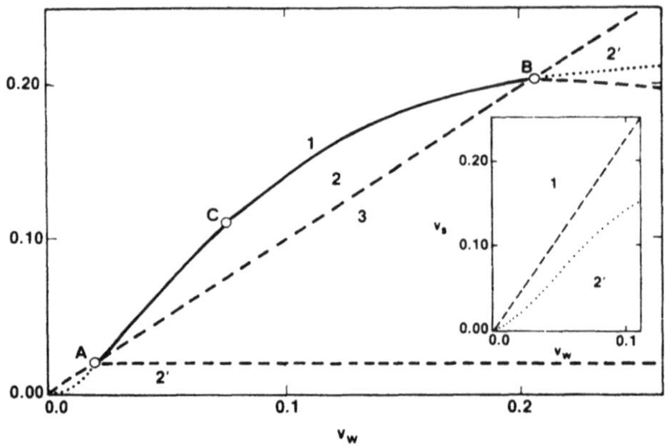

Fig. 3. Phase diagram for interacting, spherical microemulsion drops.

coexisting with excess oil and brine. Further variation in the thermodynamic parameter causes the system to evolve back into a two-phase coexistence, a lower phase microemulsion coexisting with excess oil. While the microemulsion/external phase interface vanishes with appropriate critical exponents at the liquid-gas instability (1 → 2), since the emulsification failure instability (1 → 2') is purely single-particle, the disappearance of the microemulsion/internal phase interface is not critical. Thus the simple "colloidal" model can achieve three-phase coexistence, but in its present form it cannot effect the progression 2 phase → 3 phase → 2 phase. In order to see such a progression, the model must be generalized to account for the phase inversion of the microemulsion (e.g. to effect the inversion from a water-continuous microemulsion to an oil-continuous microemulsion). This is the thrust behind the proposals [14] of a bicontinuous middle phase. Such a phase inversion cannot be achieved within a rigid colloidal treatment of the globules. Again, this decoupling of the interacting globules from their self-organisation must break down at these higher concentration regimes.

At low globule volume fraction, x, the combination of the emulsification failure (at $\rho = \rho_0$) and liquid-gas ($\rho = \rho_c$) instabilities can lead to a three-phase coexistence, of middle-phase microemulsion, coexisting with both excess internal and external phase. For x > .6, the globules perforce must touch and hence be deformed; the globule free energies are thus higher than in the dilute limit. Thus the emulsification failure instability, where excess internal phase is expelled, no longer occurs at $\rho = \rho_0$, since the chemical potential of internal phase at $\rho = \rho_0$ is no longer equal to that of the bulk internal phase ($\rho = \infty$); the coexistence must occur with globules of finite globule radius. But such globules are unstable, as seen in the spherical fluctuation calculation. Thus, for x > .6, it is possible to access the phase inversion instability--with its critical point.

Finally, as $\rho_0^{-1} \sim 0$, the spontaneous curvature flips, with the surfactant preferentially solubilising the "external" phase inside a globule in a solvent of "internal" phase. Thus changing ρ_0, generates the desired 2 → 3 → 2 phase progression. We thus see that the phase behaviour is governed by ρ_0, and that the phase behaviour is relatively insensitive to the energetics, namely K. We argue below that the parameter T/K governs the microstructure rather than the phase behaviour.

Projecting the ternary phase diagram at constant v_s, the above variation in the ternary phase diagram is evinced in a phase diagram with v_o/v_w and ρ_0^{-1} axes. Depending on whether the cut is taken above or below the three-phase body, one sees either a Schreinemaker groove or a Shinoda cut in the v_o/v_w-ρ_0^{-1} phase diagram. This is suggestive, as these topologies are well known to occur in

v_o/v_w-T phase diagrams [24]. Similarly, projecting the ternary phase diagram at constant v_o/v_w, the above variation in the ternary phase diagram is evinced as a "Kahlweit fish" topology in the v_s-ρ_o^{-1} phase diagram. Again this is suggestive, as this topology is well known to occur in a v_s-T phase diagram [24]. As ρ_o is expected to be temperature dependent, the experimental phase diagrams can thus be qualitatively understood.

5. COMPETITION BETWEEN INTERACTIONS AND SELF-ORGANISATION: DEFORMABLE INTERACTING DROPS

5.1 Dimers

We now combine interactions and self-organisation. In particular, we allow the particles to deform upon interaction. As the liquid-gas instability is approached, clusters of larger and larger size are generated; we are interested in the shape of those clusters. We first investigate the shape of dimers, bound by the strong short-range attractive interaction.

We approximate the strong short-range attractive interaction by a contact interaction. The attractive interaction tends to cause the monomers comprising the dimer to deform in order to maximise their area of contact. This deformation occurs at the expense of internal bending energy. We parameterise the deformed spherical monomers by the angle ϕ subtended at the spheres' centers (Fig. 4).

Dynamic equilibrium between monomers and dimers must conserve surface phase (surfactant) and internal phase (for specificity, water). If V_1 and S_1 are respectively the volume and area of a monomer sphere, then the number of water and surfactant molecules contained in a monomer are

$$n_w^1 = V_1/v, \qquad n_s^1 = S_1/a, \tag{10}$$

where v is the volume of a water molecule and a is the area per surfactant molecule (both of which are assumed constant--incompressibility assumption). Similarly if V_2 and S_2 are respectively the volume and area of a dimer (with no assumption as to its shape), then the number of water and surfactant molecules contained in a dimer are

$$n_w^2 = V_2/v, \qquad n_s^2 = S_2/a. \tag{11}$$

Dynamic equilibrium, N [monomers] $\rightleftarrows$ M [dimers], is achieved by conservation of water and surfactant molecules

$$N\, n_w^1 = M\, n_w^2, \qquad N\, n_s^1 = M\, n_s^2. \tag{12}$$

Combining (12) with (11) yields

$$V_2/S_2 = V_1/S_1. \tag{13}$$

Since the volume fractions of water and surfactant in the subsystem of monomers is

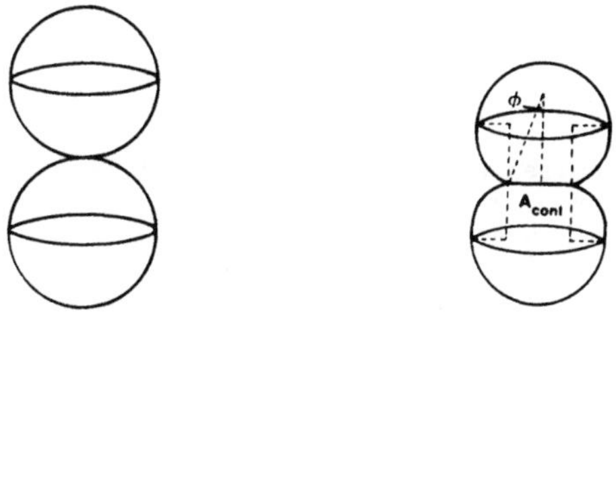

Fig. 4. A deformed dimer cluster.

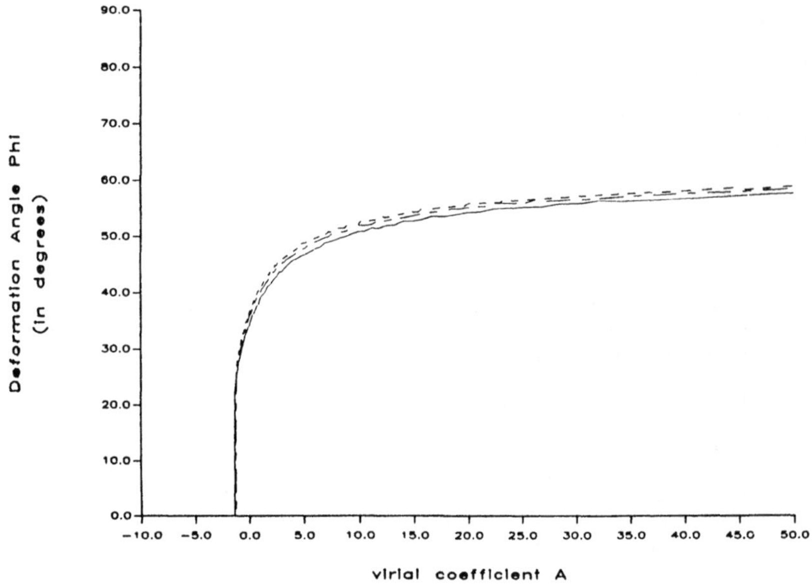

Fig. 5. Deformation angle ϕ as a function of second virial coefficient A. Upper curve ρ/ρ_o =0.6, middle curve ρ/ρ_o =0.8, lower curve ρ/ρ_o = 1.0.

$$v_w^1 = n_2 V_2 , \qquad v_s^1 = n_1 S_1 \delta , \tag{14}$$

where n_1 is the number density of monomers, and similarly since the volume fractions of water and surfactant in the subsystem of dimers is

$$v_w^2 = n_2 V_2 , \qquad v_s^2 = n_2 S_2 \delta , \tag{15}$$

where n_2 is the number density of dimers, (13-15) imply that

$$v_w^2/v_s^2 = v_w^1/v_s^1 , \tag{16}$$

i.e. v_w/v_s is conserved between the subsystems of dimers and monomers. The above argument clearly generalises to clusters of arbitrary size. Thus the appearance of v_w/v_s as a field variable [10] is a natural consequence of our model of self-organising globules in dynamic equilibrium.

Using the geometry of Fig. 4, we have variationally calculated [25] the free energy change of a dimer, subject to the constraint (13). Minimising the free energy with respect to ϕ, we obtain the shape of the dimers as a function of interaction strength (Fig. 5), as measured by the virial coefficient A; the deformation is relatively insensitive to ρ/ρ_o. We remark that globule deformation more efficiently utilises short-ranged interactions; this permits reasonable strength attractive interactions to drive coacervation.

5.2 Larger Clusters

Having calculated the deformation of the globules when they coalesce into a dimer, one can proceed to calculate the shape of a trimer. This will be a cluster of deformed spheres, with the central sphere possessing two flats. A linear trimer is shown in Fig. 6a. As in the dimer, these flats (of infinite radius of curvature) will relax to the radius of curvature that the sphere

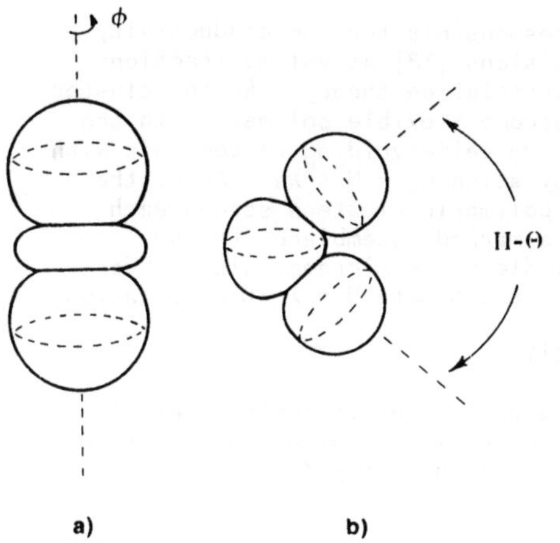

Fig. 6. Deformed trimer clusters:
a) linear trimer; b) trimer bent by Θ.

would normally possess over a "healing length" of order the drop size. The free energy of the trimer is a function of the angle Θ between the two flats (Fig. 6b), and because of the healing of the radius of curvature from infinite to the drop size, the bending energy is minimised when the flats are furthest apart. The details of this calculation will be published elsewhere [25]. For small angles Θ, the change in free energy for the off-axis trimer (Fig. 6b) is given by $\Delta F = \alpha\, K \sin^2\Theta$, where $\alpha(\rho/\rho_o)$ is a numerical constant of order unity, and the difference between conifigurations 6a and 6b is also of order K. The lower energy of the linear configuration mitigates towards the clusters lining up in chains. This chaining may be made quantitative by pursuing the strong analogy with ferrofluid clusters.[11]

In ferrofluids, the interaction among magnetically alined ferromagnetic colloids is given by $W = \mu^2/r^3\ (1 - \cos^2\theta)$, where μ is the magnetic moment of the colloid, r is the distance between particles, and θ is the angle $\vec{r}$ makes with the external magnetic field (which is the axis of the moments $\vec{\mu}$) and is a minimum along the alignment axis. The dimensionless coupling constant of the (liquid-gas) phase separation is $\lambda = \mu^2/Ta^3$, where a is the radius of the colloids. Similarly, the dimensionless coupling constant in the microemulsion problem is $\lambda = 2\alpha/3\ K/T$.

For the ferrofluid, in the high magnetic field (alined) limit, the condensing
ferromagnetic clusters tend to line up in chains with an anisotropy in the correlation length, $\xi_\parallel^2/\xi_\perp^2 \sim 6\,\lambda$. In the microemulsion problem, global alinement may be effected by an externally applied flow. The anisotropy in correlation length will be reflected by a pronounced flow birefringence.

For the ferrofluid, in the absence of an externally applied field, the chaining persists, the length of the chains (degree of polymerisation) being given by

$$N \sim [\ 1 - 2n/\lambda\ |A_\infty|\ ]^{-1}, \tag{17}$$

where A_∞ is the second virial coefficient of the aligned interacting colloidal fluid. The chain is rigid over a persistence length λa, and for $N > \lambda$ the chain behaves like a random-walking polymer.

In the microemulsion problem, we can expect the following behaviour. As the liquid-gas phase separation is approached by increasing the volume fraction of globules, the cluster size ξ grows. For $N < \lambda$, the clusters behave as rigid rods, whose excluded volume interaction will lower [12] the percolation threshold to p_c/λ^2. This mechanism may be responsible for the conductivity anomalies seen in some water-in-oil microemulsions [13] at volume fractions considerably below that predicted (15%) by percolation theory. As the cluster size is increased past $N > \lambda$, the clusters become flexible polymers. In the dilute regime, these polymeric clusters undergo self-avoiding random walk with persistence length λa, and hence radius of gyration $R_G \sim N^{3/5}\lambda a$. Above the overlap concentration $c^* \sim (\lambda a)^{-3}N^{-4/5}$, the polymeric clusters screen each other, and the structure which ensues bears a marked resemblance to the bicontinuous structures proposed [14] for middle-phase microemulsions. For realistic physical microemuslion parameters, we estimate $N \sim \lambda$ at percolation.

6. PORE DEFECTS AND THE ORIGIN OF BICONTINUITY

We discuss whether aggregated microemulsion globules retain their integrity upon aggregation or whether the droplets open up to permit the exchange of internal phase. Globule integrity is governed by the bending energy K.

6.1 Pore Defects in Surfactant Bilayers

The free energy of a lamellar bilayer between, say, two aqueous regions, is given by

$$F_{bilayer} = K/2 \int (1/\rho_1 + 1/\rho_2 - 2/\rho_o)^2 \, dS \; + \; 2K\, A_{bilayer}/\rho_o^2 \,, \qquad (18)$$

where the area of the bilayer is $A_{bilayer} = \pi\rho^2$. We now consider the possible penetration of the bilayer by a pore. For a bilayer of separation $2\Delta\rho$, we model the pore (of radius R) by the interior of a torus of revolution. The bending energy cost associated with such a model pore is given by

$$\Delta F_{pore} \sim 2\pi K \{-4 + x^2(x^2-1)^{-1/2} [\pi - 2\arctan([(x-1)/(x+1)]^{1/2})]\}, \qquad (19)$$

where $x = R/\Delta\rho$. The free energy associated with nucleating this pore is the pore bending energy minus the associated entropy (the log of the number of places to put the pore), $\log (\rho/x\Delta\rho)^2$. For large enough temperatures, the free energy change may be driven negative by the entropic term, thereby defining a pore nucleation temperature, T_N, above which the bilayer is no longer intact. For $T < T_N$, the droplets retain their integrity; for $T > T_N$, pores penetrate the bilayer interface, and the droplets no longer retain their integrity.

6.2 The Origin of Bicontinuous Microemulsions

The above considerations permit a qualitative phase diagram for microemulsions as a function of the two relevant energies, the bending energy K and the interaction strength, as measured by the dimensionless virial coefficient, A. For $A < 20$, the interdomain interactions are sufficiently weak, and we have a single phase of droplets. For $A > 20$, the interdomain interactions are strong enough to drive a liquid-gas instability, resulting in a phase separation into a high and low globule-density microemulsion phases. For $T < K$, the globules are well defined, and the globules retain their integrity, whether in the single-phase microemulsion fluid, or in the phase-separated microemulsion liquid and gas. However, for $T > K$, the aggregated globules do not retain their identity, with pores penetrating their mutual interfaces. The high globule-density microemulsion liquid is now bicontinuous. Similarly, the lower globule-density microemulsion fluid can percolate, unlike colloids which retain their integrity.

7. CONCLUSIONS

Microemulsions constitute complex fluids in which globular domains on intermediate length scales are organised by forces of the same magnitude as those which are responsible for macroscopic phase separations. We have discussed, within a Helfrich elasticity theory, the possible self-organised structures and the phase behaviour that then obtains, the phase separations being inextricably linked to possible changes in structure. The interdomain interactions drive a liquid-gas instability; steric interactions between globules and the intraglobular internal fluctuations drive a phase inversion instability. Qualitatively, the spontaneous radius of curvature, ρ_0^{-1}, controls the phase behaviour, with the 2 → 3 → 2 phase progression occurring as ρ_0^{-1} changes sign, i.e. the interesting middle-phase occurs for no preferred bias of the surfactant films to curve towards either oil or water domains. This middle-phase phase behaviour, however, is independent of the energetics; the energetics instead controls the microstructure of the globules. For $T < K$, the globules retain their integrity, and the middle-phase retains a cellular structure; for $T > K$, the flat interfaces between aggregated globules are unstable to pore formation, and the globules open up, and the middle-phase becomes bicontinuous. In addition, the deformation of the globules permits a more efficient utilisation of any attractive interaction, which gives rise to an attraction enhancement. The cluster anisotropy which results gives rise to chaining in the dense microemulsion liquid, which should be seen in both flow birefringence and in a percolation renormalisation (i.e. conductivity anomalies occurring at low volume fraction).

REFERENCES

1. For general surveys see Surfactants in Solution, ed. by K. Mittal and B. Lindman (Plenum, N.Y., 1984), and Micellization, Solubilization and Microemulsions, ed. by K.L. Mittal (Plenum, N.Y., 1977).
2. P.A. Winsor, Chem. Rev. 68, 1 (1968).
3. S.A. Safran, L.A. Turkevich, P.A. Pincus, J. Phys. (Paris) 45, L69 (1984).
4. L.A. Turkevich, S.A. Safran, P.A. Pincus, in Proc. 5th Symp. "Surfactants in Solution", ed. by K.L. Mittal, P. Bothorel (Plenum, N.Y., in press).
5. Y. Talmon, S. Prager, J. Chem. Phys. 69, 2984 (1978); J. Jouffroy, P. Levinson, P.G. de Gennes, J. Phys. (Paris) 43, 1241 (1982); B. Widom, J. Chem. Phys. 81, 1030 (1984); L. Auvray, J.P. Cotton, R. Obert, C. Taupin, J. Phys. (Paris) 45, 713 (1984).
6. C.A. Miller, et al, J. Colloid Int Sci., 61, 554 (1977).
7. S.A. Safran, L.A. Turkevich, Phys. Rev. Lett. 50, 1930 (1984).
8. S.A. Safran, L.A. Turkevich, J.S. Huang, in Proc. 5th Symp. "Surfactants in Solution", ed. by K.L. Mittal, P. Bothorel (Plenum, N.Y., in press).
9. J.S. Huang et al, Phys. Rev. Lett. 53, 592 (1984).
10. D. Roux, A.M. Bellocq, Phys. Rev. Lett. 52, 1895 (1984), and in Proc. 5th Symp. "Surfactants in Solution", ed. by K.L. Mittal, P. Bothorel (Plenum, N.Y., in press).
11. P.G. de Gennes, P.A. Pincus, Phys. Kondens. Mater. 11, 189 (1970).
12. I. Balberg, C.H. Anderson, S. Alexander, N. Wagner, Phys. Rev. B 30, 3933 (1984).
13. A.M. Cazabat et al, J. Phys. (Paris) 41, L441 (1980), 43, L89 (1982).
14. L.E. Scriven, Nature 263, 123 (1976), and in Micellization, Solubilization, and Microemulsions, ed by K.L. Mittal (Plenum, N.Y., 1977), page 877.
15. W. Helfrich, Z. Naturforsch., 28, 6693 (1973).
16. L.A. Turkevich, S.A. Safran, to be published.
17. S. Marčelja, Biochim. Biophys. Acta 367, 165 (1974).
18. A.G. Petrov,A. Derzhanski, J. Phys. (Paris) Colloq. 37, C3-155 (1976); A.G. Petrov, M.D. Mitov, A. Derzhanski, Phys. Letts. 65A, 374 (1978).
19. W. Helfrich, Z. Naturforsch. 28C, 693 (1973); F.C. Frank, Discuss. Faraday Soc. 25, 19 (1958).
20. S.A. Safran, J. Chem. Phys. 78, 2073 (1983), and in ref. 1, p. 1781.
21. P. G. de Gennes, C. Taupin, J. Phys. Chem. 86, 2294 (1982).
22. J.S. Huang, M.W. Kim, Phys. Rev. Lett. 47, 1462 (1981); M. Kotlarchyk, S.H. Chen, J.S. Huang, Phys. Rev. A 28, 508 (1983).
23. M. Kotlarchyk, S.H. Chen, J.S. Huang, J. Phys. Chem. 86, 3273 (1982), M. Kotlarchyk, S.H. Chen, J.S. Huang, M.W. Kim, Phys. Rev. A 29, 2054 (1984).
24. M. Kahlweit, R. Strey. Angew. Chem. Int. Ed. 24, 654 (1985), M. Kahlweit, R. Strey, P. Firman, J. Phys. Chem. 90, 671 (1986).
25. L.A. Turkevich, to be published.

Microemulsions and Their Precursors

H.T. Davis, J.F. Bodet, L.E. Scriven, and W.G. Miller

Dept. of Chemical Engineering and Materials Science and of Chemistry, University of Minnesota, 421 Washington Avenue S.E., Minneapolis, MN 55455, USA

Introduction

Amphiphiles are generally defined with reference to water. An amphiphilic molecule possesses a hydrophilic (water soluble) moiety and a hydrophobic (water insoluble) moiety separable by a mathematical surface. In this paper the only hydrophobic moieties considered are those composed of hydrocarbon chains (as opposed, for example, to fluorocarbon chains). According to IUPAC [1], surfactants are substances which lower the surface tension of the medium in which they are dissolved and/or the interfacial tension with other phases. This definition is too broad as it would include substances such as benzene which lower the surface tension when dissolved in water but have none of the other properties expected of a surfactant. A generally accepted restriction of the definition is that surfactant molecules must be amphiphilic [2-4]. We adopt this restriction and, furthermore, share the view of Laughlin [5] that those properties associated with surfactancy (e.g., detergency, co-solubilization of oil and water, emulsification, foaming, and the like) are not common to all amphiphiles. He reserves the term "surfactant" for molecules which form association colloids such as micelles and liquid crystals. Thus, amphiphiles such as long-chained fatty acid salts (soaps), sodium alkyl sulfates and sodium alkyl sulfonates (detergents), and lecithins are clearly surfactants, whereas monohydric straight-chained alcohols and amines are mere amphiphiles. A special property of the association colloids formed by amphiphiles is that the amphiphiles associate into monolayer or sheet-like structures with the water soluble moieties on one side of the sheet and the water insoluble moieties on the other side (a bilayer is a pair of opposed monolayers). In the spirit of Laughlin, ***we define surfactants as amphiphiles which form association colloids distinguished by sheet-like surfactant microstructure.***

By alcohol titration of an emulsion of oil and water stabilized by a surfactant, Schulman and coworkers obtained isotropic, transparent fluids [6-9]. Hoar and Schulman [6] described these fluids as what would in the current language of the field be called an aqueous solution of oil-swollen micelles and an oil solution of water-swollen inverted micelles. Noting that small spherical objects (of the order of 5 to 100 nm in diameter) were observed by chemical staining electron microscopy, Schulman, Stoeckenius and Prince [9] called these fluids "micro emulsions" and described them as "optically isotropic, fluid, transparent oil and water dispersions, consisting of uniform spherical droplets of either oil or water in the appropriate continuous phase."

In 1976, Scriven [10,11] put forward the idea that microemulsions containing comparable amounts of oil and water could be bicontinuous, with irregular sample spanning water-rich regions separated by surfactant sheet-like zones from irregular sample spanning oil-rich regions. As in two-component association colloids, the surfactant sheet-like zone admits a dividing surface on one side of which are the water-soluble moieties of the surfactant molecules and on the other side are the water-insoluble moieties. Thus, the surfactant sheet-like zone provides a topological ordering of the water-rich and oil-rich regions of the microemulsion, water on one side of the surfactant sheet and oil on the other. To include the possibility of such a microstructured fluid in the classification of microemulsions we prefer the following definition: ***A microemulsion is a thermodynamically stable, isotropic, topologically ordered microstructured phase containing at least surfactant,*** *hydrocarbon and water*. With the interpretation that the topological order is that imposed by sheet-like surfactant zones, the above definition of microemulsion admits the possibility of dispersed droplet structures (swollen micellar or inverted micellar solutions) as well as bicontinuous ones. Often salt, alcohol, or a second surfactant is added to the surfactant-hydrocarbon-water mixture to obtain an isotropic microemulsion with the desired phase behavior. In many of the microemulsions of practical importance the surfactant and hydrocarbon materials are mixtures whose compositions are often not very well characterized.

In Figure 1, we present an idealized ternary phase diagram in which are illustrated fluid microstructures that have been identified in solutions of surfactants with water and/or oil [12-14]. The existence of solutions of spherical and cylindrical micelles or inverted micelles is generally accepted, as are lyotropic liquid crystalline phases with lamellar, hexagonal and cubic symmetries. However, the detailed local shapes of the surfactant sheet-like zones, the role and importance of molecular fluctuations and bending and stretching motions of the sheet-like zones in micellar solutions and liquid crystals are issues not entirely resolved and are the subjects of much current research. A convincing body of data now exists supporting the existence of bicontinuous microemulsions, but their detailed microstructure has not been established.

One of the goals of the research program at the University of Minnesota is to understand the relationship between molecular structure, fluid compositions and the properties of microemulsions. A logical step toward attaining this goal is to distinguish solutions of brine, hydrocarbon and mere amphiphile from microemulsions of brine, hydrocarbon and surfactant. To this end we have been carrying out a systematic study of the microstructure of a sequence of solutions of straight-chained ethoxylated alcohols with hydrocarbon and brine [15-22]. The sequence begins with propanol and includes $H(CH_2)_n(OCH_2CH_2)_mOH$ (denoted as C_nE_m) with n ranging from 3 to 12 and m from 0 to 7. We have made a concerted effort to investigate the low end of this sequence to try to

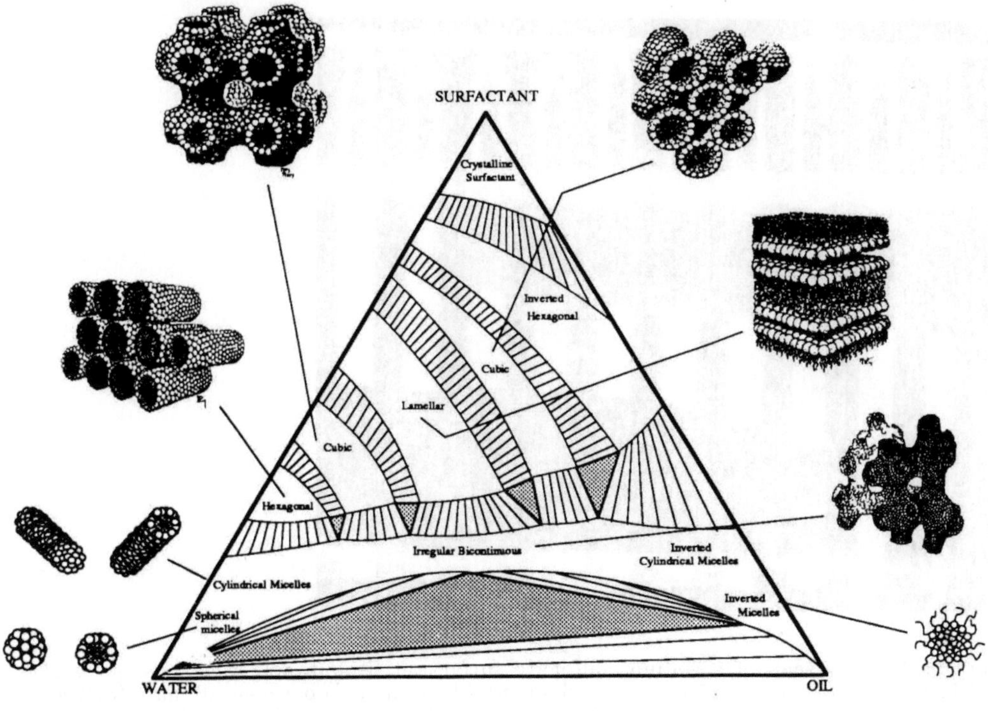

Figure 1. Schematic oil-water-surfactant phase diagram with microstructures depicted. Ref. 12.

capture the pre-surfactant trends and pre-microemulsion patterns of amphiphiles that do not qualify as surfactants. Our work is complementary to the extensive studies of microemulsions of C_nE_m carried out by Kahlweit and coworkers [23-27], Lindman and coworkers [28,29], and Shinoda and coworkers [30,31].

We report here the status of our studies of C_3E_0, C_4E_0, C_4E_1 and C_6E_2 — hydrocarbon-brine systems. Before giving these results, however, we present in the next section the established behavior of microemulsions with which we contrast that of these simple systems.

Patterns of Microemulsion Behavior

Because of possible applications in enhancing oil recovery, the phase and interfacial behavior of microemulsions has received particular attention during the last decade. The special interest is to find the thermodynamic conditions for which a microemulsion phase has ultralow tensions against coexisting oil and water-rich phases. Under these conditions microemulsions injected into an oil reservoir will surround oil blobs trapped by capillary forces in the pores of the rock and will lower the overall tension sufficiently to mobilize the trapped blobs into a recoverable oil-bank.

For a given oil-water-surfactant system, the most frequently used method to adjust the phase splits and tensions in a microemulsion system is to add an alcohol and a salt, often NaCl. An example [32] of the phase behavior in a salinity scan of a low tension microemulsion system is shown in Fig. 2. The system is formed by mixing a given volume of brine solution with an equal volume of a mixture of alcohol, surfactant, and hydrocarbon. The type and amount of alcohol have been chosen so that in the three phase region the microemulsion (middle phase) will have ultralow tension against both the oil-rich (top) and water-rich (bottom) phases. In the salinity scan the phase splits obey

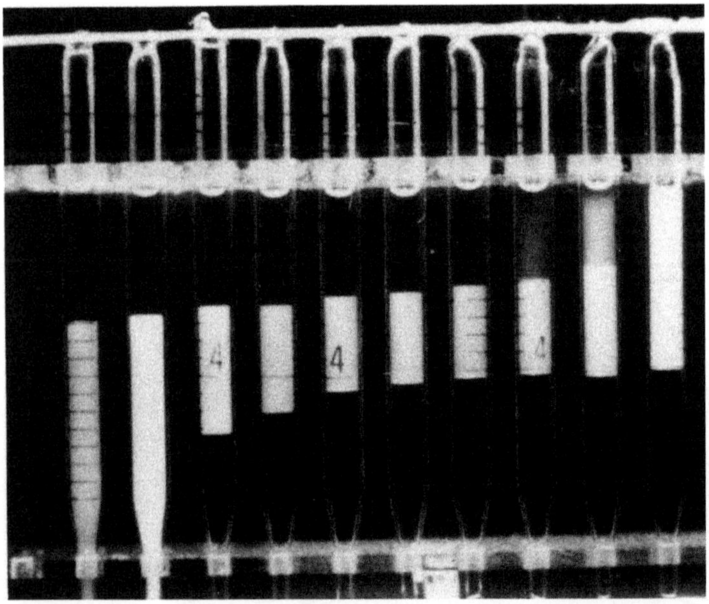

Figure 2. Salinity scan of a sodium sulfonate surfactant, isobutanol, n-hexadecane, and NaCl brine mixture. The surfactant is sodium 4-(1'-heptylnonyl) benzenesulfonate (SHBS). Mixtures were in the volume percent of about 5% surfactant plus alcohol, 45% hydrocarbon and 50% brine. The brine ranged in salinity from about 0.2 to 2 wt% NaCl. Ref. 32.

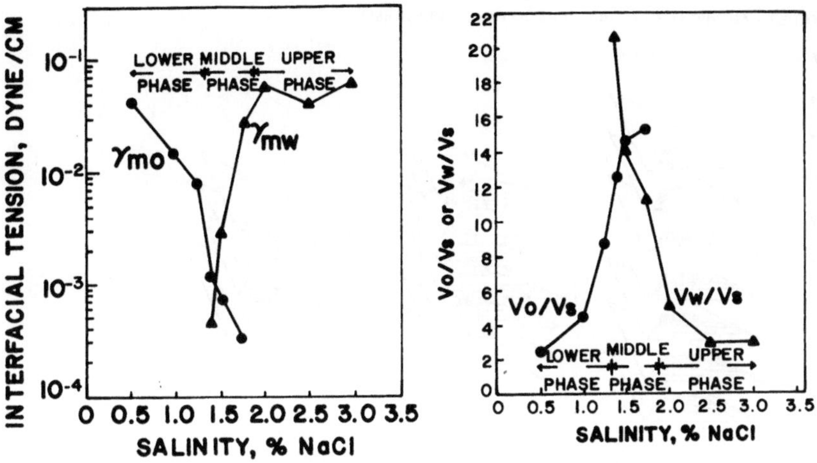

Figure 3. Interfacial tensions between microemulsion and oil-rich and water-rich phases. On the right is shown the ratio of volume of water or oil to volume of surfactant in the microemulsion phase. The system is a mixture of oil, NaCl brine, tertiary amyl alcohol, monoethanol amine salt of an alkylorthoxylene sulfonic acid. Composition is 4% surfactant and alcohol, 48% oil and 48% brine. Ref. 34.

the $\underline{2}3\overline{2}$ progression characteristic of low tension microemulsions. Following Knickerbocker *et al.* [15,16], we denote by $\underline{2}$ a two-phase system in which the microemulsion is the lower, water-rich phase; by 3 a three-phase system in which the microemulsion is the middle phase and contains appreciable amounts of oil and water; and by $\overline{2}$ a two-phase system in which the microemulsion is the upper, oil-rich phase.

The pattern of interfacial tension behavior of a low-tension salinity scan is given in Fig. 3. The salinity at which the microemulsion interfacial tensions against the coexisting oil-rich phase and water-rich phase are equal, i.e., $\gamma_{om} = \gamma_{wm}$, was designated as "optimal" by Reed and Healy [33,34] because in this state the microemulsion is most favorable for coating and mobilizing capillary-trapped oil blobs. The ratio of the volumes of oil and water in the microemulsion phase oil is approximately unity in the optimal microemulsion state shown in Fig. 3. Reed and Healy noted this and found that it is generally true for low-tension microemulsions. Besides being a fascinating fact, the correlation of equal oil-water uptake with optimal tension lets one replace tension measurements by the easier volume uptake measurement in scanning candidate formulations for ultralow tension microemulsions.

As illustrated by the example given above, the search for a surfactant formulation for enhancing oil recovery has generally involved the scanning of at least two field variables to find conditions under which the microemulsion will have ultralow tension against coexisting oil-rich and water-rich phases. A field variable is defined as one which is the same in all phases in thermodynamic equilibrium. Common choices of field variables in microemulsion studies are the activity of added salt, alcohol or cosurfactant, temperature, pressure, hydrocarbon chain length, and chain length of the hydrocarbon moiety of the surfactant.

The phase and tension behavior that has been observed can be summarized in terms of the generic patterns shown in Fig. 4. In Fig. 4, as Field Variable 1 increases from a low value, a middle or third phase appears, the 3-phase tie triangle springing from a critical endpoint tie line of the 2-phase system. As Field Variable 1 continues to increase, the middle phase progresses from water-rich to increasing oil-rich until the tie triangle collapses into a critical endpoint tie line of the 2-phase system. As one of the critical endpoints (CEP) is approached, the tension between middle phase and its near-critical partner of course approaches zero. As Field Variable 2 increases the two CEP's move

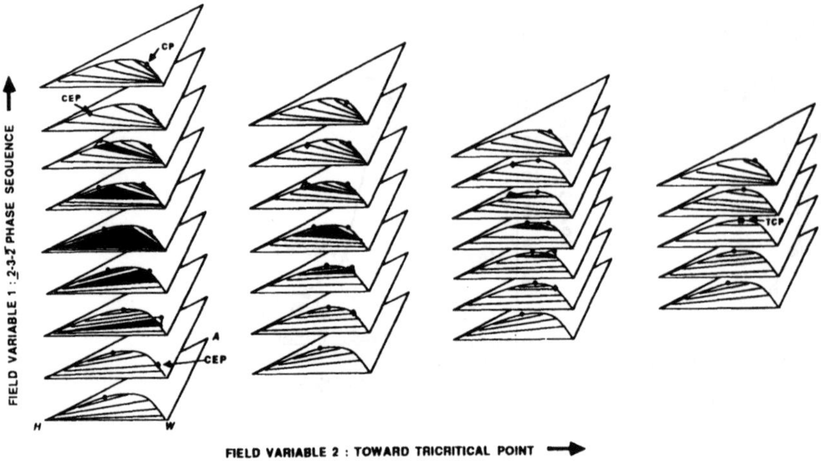

Figure 4. Generic progression of amphiphile, hydrocarbon, water ternary phase diagrams. Ref. 20.

together, i.e., the range of Field Variable 1 needed for a $\underline{2}$ to 3 to $\overline{2}$ scan is reduced. At sufficiently large Field Variable 2 the two CEP's collapse into a tricritical point (TCP). Since all three phases of the three-phase system become identical at the TCP, it follows that the tension at optimal salinities will decrease as Field Variable 2 is increased towards the TCP. Furthermore, as the three phases tend to become increasingly more alike it is reasonable to expect oil and water to be equally welcome in the middle phase, i.e., to expect the oil-to-water ratio to tend towards unity as the TCP is approached. In the example shown in Fig. 2, Field Variable 1 is salinity (which should be converted to salt activity to be a proper field variable). Field Variable 2 is the alcohol which was added to advance the microemulsion towards the TCP and obtain ultra-low tension at optimal. The patterns represented in Fig. 2 have been verified for many field variables. For example, Bennett *et al.* [35] have observed the trends displayed in Fig. 4 with both a pure and a commercial alkyl aryl sodium sulfonate surfactant with salinity as one field variable and carbon numbers as the other field variable (in hydrocarbon mixtures average carbon numbers have been found to function as field variables [36,37]. Kahlweit and coworkers [23-27] have demonstrated the patterns of Fig. 4 for ethoxylated alcohol-alkane-water mixtures with temperature and alkane carbon number as field variables. Kilpatrick *et al.* [19,20] have found them for alkane carbon number and ethoxylated alcohol homologous series.

To model microemulsion as an isotropic fluid with water-rich and oil-rich regions separated by sheet-like surfactant zones, Talmon and Prager [38,39] introduced the following statistical theory. They used the Voronoi polyhedral construction, Fig. 5, to randomly intersperse oil and water regions. They assumed that the surfactant lies entirely on the polyhedral faces separating the oil and water regions. The theory yields the purely entropic free energy density

$$f = ckT\left[\phi_0 ln\left(\frac{\omega c}{e}\phi_o\right) + \phi_w\, ln\left(\frac{\omega c}{e}\phi_w\right)\right] , \tag{1}$$

where k is Boltzmann's constant, T the absolute temperature, ϕ_o and ϕ_w the oil and water volume fractions, c is the number density of Voronoi polyhedra, ω a volume parameter characteristic of the size scale of the microstructure of the microemulsion, and e the base of natural logarithms. The surfactant per unit volume c_s given by the formula

$$\begin{aligned} c_s &= 5.82\alpha_1 c^{1/3}\, \phi_o\phi_w - 17.52\alpha_2 c^{2/3}\, \phi_0\phi_w \\ &= [Ac^{1/3} - Bc^{2/3}]\phi_o\phi_w \ . \end{aligned} \tag{2}$$

In this expression, $5.82c^{1/3}\phi_0\phi_w$ is the mean area of oil-water boundary per unit volume and $17.52c^{2/3}\phi_0\phi_w$ the mean edge length per unit volume of those Voronoi polyhedral faces that separate oil and water. The parameter α_1 gives the number of molecules per unit surfactant internal oil-water area. The amount of surfactant that can reside between oil and water regions depends on the curvature of the surfaces dividing these regions. It is a mathematical artifact of the Voronoi tesselation that all the curvature lies along the polyhedral edges. Thus, the parameter α_2 represents the effect of curvature on the inventory of surfactant in the sheet-like surfactant zones.

In the Talmon-Prager theory $A = 5.82\alpha_1$ and $B = 17.5\alpha_2$ are adjustable parameters. Davis and Scriven [40] showed that the $23\overline{2}$ phase sequence is predicted by the theory if it is assumed that $B = B_1 + B_2\,\phi_0$, where B_2 advances as field variable (see Fig. 6). If salinity is the field variable, the interpretation of Fig. 6 is that an increase in salinity increases the tendency of the sheet-like surfactant zones to curve into the water-rich regions.

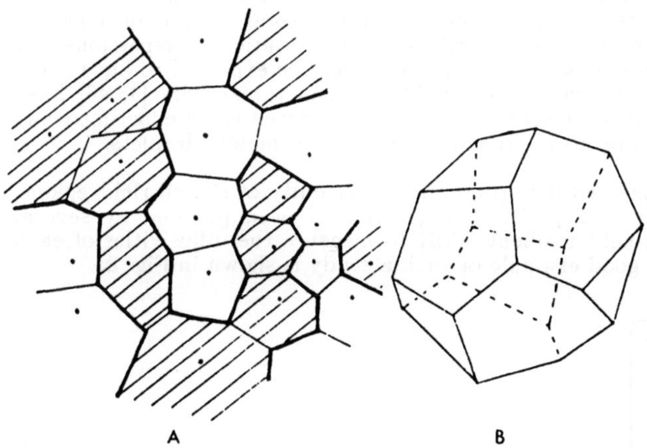

Figure 5. A Voronoi representation of a microemulsion. A) Two dimensional example. Shaded regions are occupied by oil-rich fluid, unshaded by water-rich material; heavy lines indicate the surfactant sheets. B) Typical Voronoi polyhedron. Ref. 38.

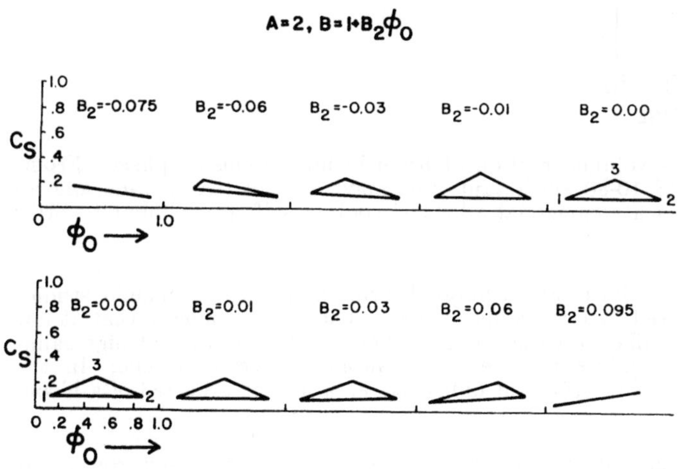

Figure 6. Sequence of ternary phase diagrams, opening and closing at critical end points (CEP), predicted by the Talmon-Prager model. Ref. 40.

Variations of the Talmon-Prager theory have been developed by de Gennes and coworkers [41] and by Widom [42]. de Gennes *et al.* [41] add to the theory the Schulman concept of zero internal film tension. In Widom's improvement of the theory, the characteristic size scale ω is obtained as a consequence of minimization of the free energy.

If oil and water-rich regions are chaotically interspersed, it is a general prediction of percolation theory that there exists a volume fraction of water, ϕ_{wc} below which the oil-rich fluid is continuous, a volume fraction of oil, ϕ_{oc}, below which the water-rich fluid is continuous, and intermediate range of volume fractions in which the system is bicontinuous. The critical volume fractions are percolation thresholds, and, for the Voronoi tessellation, Monte Carlo simulations yield the estimate $\phi_{wc} = \phi_{oc} = 0.145$ [43]. The implication of the theory is that the electrical conductivity of a microemulsion should decrease dramatically as the water/oil ratio decreases to the value corresponding to percolation threshold. This expected behavior has been verified experimentally for many microemulsions [13,32,44-46]. In Fig. 7 we compare experimental conductivity results with those calculated in a Monte Carlo simulation [32,43] of a random oil-water interspersion on a Voronoi tessellation. Similar agreement has been obtained for several microemulsions. It should be noted that the percolation theory refers to the static interspersions. A microemulsion is a dynamic system and so the conductivity may be expected to have small but not zero value at the static percolation threshold. French researchers [45,46] have shown that dynamic effects are indeed present and have investigated these experimentally and explained them theoretically in terms of stirred percolation or mobile droplets.

Electrical conductivity cannot be used for probing the continuity of the oil-rich regions. However, self-diffusion provides such a probe and has indeed been pursued by several researchers using spin-echo pulsed-field gradient NMR to measure the diffusivities of each component of a microemulsion. A good example of such a study is shown in Fig. 8.

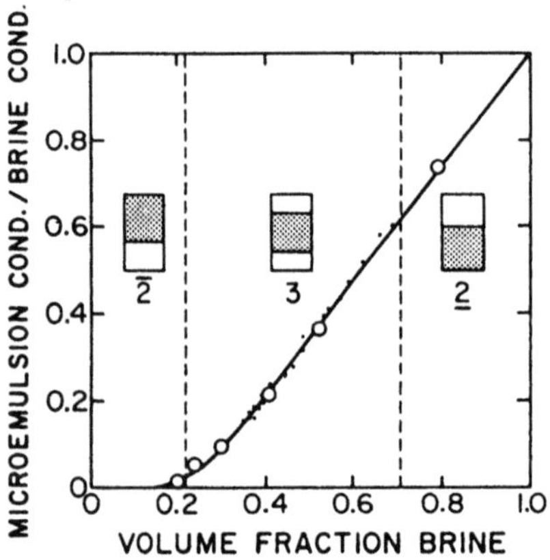

Figure 7. Conductivity versus volume fraction of brine in microemulsion phase. From a salinity scan of a Witco TRS 10-80 surfactant, tert-amyl alcohol (tAA), n-decane and NaCl brine. The surfactant is a commercial alkylaryl sodium sulfonate similar to SHBS. Ref. 32.

The system is sodium dodecycl sulfate (SDS), butanol, brine and toluene. Similar studies have been independently reported by two groups [47,48]. The results indicate that at low salinity the oil and surfactant diffuse together as a swollen micelle and that at high salinity the water and surfactant diffuse together as a swollen inverted micelle. In the midrange we see a maximum in the surfactant diffusivity as would be expected of a bicontinuous microemulsion.

The data in Fig. 8 are consistent with the picture that as salinity is increased from a low value there is a transition from a water-continuous swollen micellar solution to a bicontinuous microemulsion in an intermediate salinity range at the end of which there is a transition to an oil-continuous swollen inverted micellar solution.

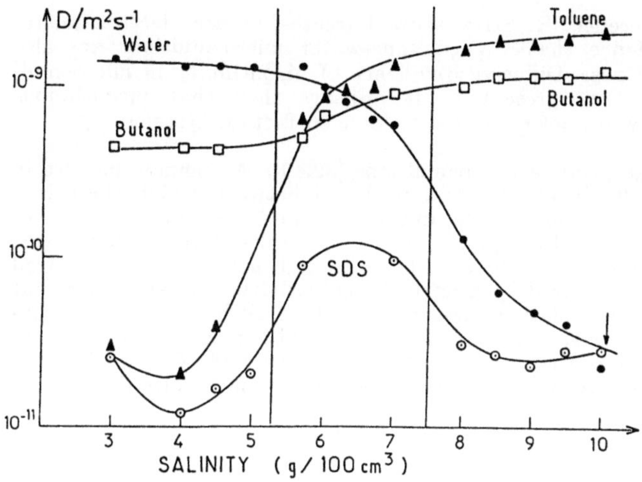

Figure 8. Diffusivities of components of SDS, butanol, toluene and NaCl brine mixture. Vertical lines denote $\underline{2}3$ and $3\overline{2}$ phase transitions. Ref. 47.

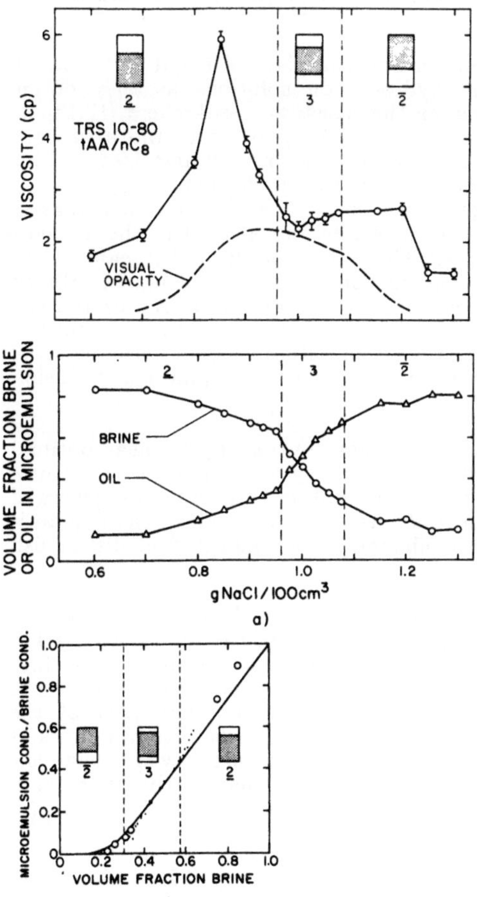

Figure 9. Microemulsion properties in a TRS 10-80/tAA/n-octane salinity scan. Ref. 32.

In a presentation to this conference, R. Strey showed freeze-fracture electron micrographs that are convincing evidence that $C_{12}E_5$-octane-water microemulsions are also bicontinuous under certain conditions. Diffusivity evidence of bicontinuity in microemulsions involving $C_{12}E_5$ has also been presented. It appears then that bicontinuous microemulsions exist and that they are not restricted to ionic surfactant systems.

Viscosity is another special fingerprint of microemulsions [32,49]. A common pattern in a salinity scan is shown in Fig. 9. There are two peaks in between which there is a minimum. The minimum typically occurs at the optimal salinity, i.e., at the point of equal oil/water uptake in the microemulsion phase. In the several sodium sulfonate surfactant systems studied at Minnesota it appeared that the high salinity peak occurred near the percolation threshold as measured by electrical conductivity. We speculate that the other peak corresponds to the percolation threshold at which oil becomes the discontinuous component. Near the viscosity peaks the microemulsion is somewhat non-Newtonian, shear thinning by a factor of about 2 to 5 at shear rates of 1000 sec^{-1}. Elsewhere the microemulsion is approximately Newtonian, although it is shear birefringent.

Alcohol–Brine–Hydrocarbon Solutions

The patterns of phase behavior shown in Fig. 4 have been observed for many monohydric alcohol solutions with brine and hydrocarbons [15,16,50,51]. Thus, the pattern is not specific to microemulsion systems, but rather is common to amphiphile, brine and hydrocarbon mixtures.

A natural question then is whether optimal tension and the state of equal water to oil uptake volumes coincide in alcohol, brine, and hydrocarbon solutions as they do in ultralow tension microemulsions (Fig. 3). To answer the question, researchers [17,18] in our laboratory studied the phase and tension behavior of solutions of brine and hydrocarbons with n-propanol and t-butanol. The results for n-propanol are summarized in Fig. 10. Similar results were obtained for the t-butanol solutions. In the figure γ_{ao} and γ_{aw} denote the tension of the alcohol-rich phase against the oil-rich and water-rich phases, respectively. V_o, V_w and V_a denote the volumes of oil, water and alcohol in the alcohol-rich phase. The hydrocarbons were normal alkanes. Thus, the field variables scanned are salinity and alkane carbon number. With decreasing carbon number the range of salinity needed to accomplish the $\underline{2}3\overline{2}$ phase scan decreases, the optimal tension decreases and the salinities of optimal tension and equal oil/water uptake volumes approach one another. From this we conclude that, qualitatively, solutions of brine, oil and monohydric alcohols, which are amphiphiles but not surfactants, exhibit the same trends of phase *and* tension behavior as do microemulsion systems.

There are, however, quantitative differences. The volume uptake ratios near optimal salinity are one or two orders of magnitude smaller in alcohol systems than in microemulsions. Furthermore, the tensions of microemulsions at optimal are typically two orders of magnitude smaller than optimal tensions of the alcohol systems. We believe these quantitative differences arise from the microstructure of microemulsions and in fact distinguish them from solutions containing mere amphiphiles.

According to the thermodynamic theory implied by Fig. 4, ultralow tensions are achieved at optimal salinity by adjusting field variables to move in the direction of a tricritical point (TCP). In the light of the near-critical point interpretation, it is at first surprising that the ultralow tensions (0.01-0.001 dyn/cm) can be achieved even when the compositions of the oil-rich and water-rich phases differ markedly from that of the microemulsion phase. As pointed out in the preceding paragraph, these ultralow tensions are in contrast to the low tension of water-hydrocarbon-alcohol mixtures, even though the phase compositions in the alcohol system are closer to each other than in the microemulsion systems.

This quantitative difference can be understood in terms of the special microstructure of microemulsions. In the near-critical regime, the Talmon-Prager theory [39] predicts that the interfacial tension obeys the asymptotic formula

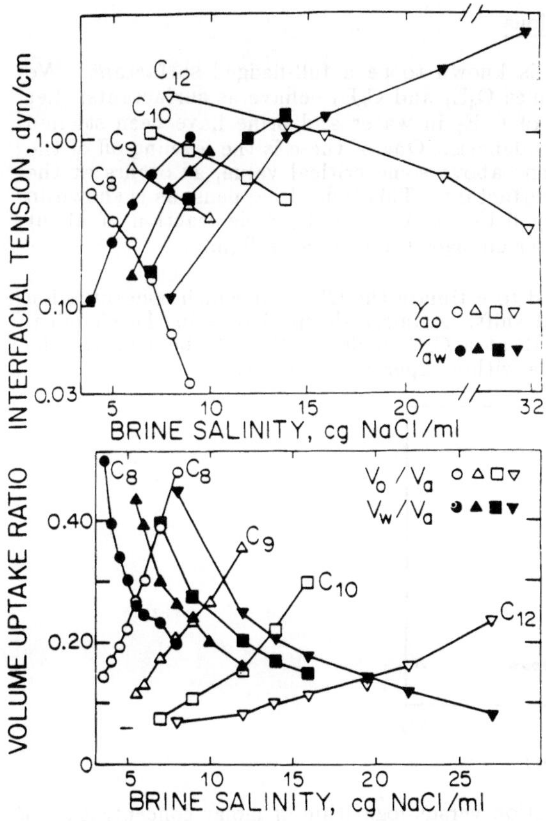

Figure 10. Salinity scans of interfacial tensions and volume uptake ratios versus brine salinity as a function of alkane chain length for mixtures of equal volumes of n-propanol hydrocarbon and brine. Ref. 18.

$$\gamma \sim 0.175\mathrm{kT}|\phi_w^2 - \phi_w^1|^3/\xi^2\,, \tag{3}$$

where T is temperature. k is Boltzmann's constant, ϕ_w^i is the volume fraction of water in phase i, and $\xi = \omega^{1/3}$ is a length scale characteristic of the microstructure of the solution.

Typically in light scattering or small angle x-ray or neutron scattering experiments values of the order of 10 nm are deduced in microemulsions [44, 52]. On the other hand,in molecular solutions of oil, water and alcohol there is no appreciable aggregation and so ξ is of the order of molecular sizes, say 1 nm. Thus, from Eq. (3) it follows that at the same distance from a critical point, as measured by the value of $(\phi_w^1 - \phi_w^2)$, the tension of a microemulsion system is about two orders of magnitude smaller than that of a simple alcohol-water-oil system, in agreement with the observed differences between alcohol and microemulsion systems.

That the values of the volume uptake of water and oil in microemulsion systems are one to two orders of magnitude higher than in alcohol systems is the result of the "packaging" ability of the sheet-like surfactant structures present in microemulsions but absent in alcohol solutions.

C_4E_1 and C_6E_2 —Brine—Hydrocarbon Solutions

If n and m are sufficiently large, C_nE_m is known to be a full-fledged surfactant. We want to know to what degree the amphiphiles C_4E_1 and C_6E_2 behave as surfactants. Let us first examine C_4E_1. Several properties of C_4E_1 in water and brine have been studied with the aim of discovering micellization tendencies. One of these is the leveling off of the surface tension at amphiphile concentrations above some critical value, as occurs at the critical micelle concentration (CMC) of a surfactant. This indeed happens, as is shown in Fig. 11. The surface tension changes very little beyond a C_4E_1 mole fraction of about 0.02. Substituting 0.2M NaCl brine for water changes the results negligibly.

The ^{13}C NMR chemical shift is often used to estimate the CMC, at which concentration there is an abrupt change in the chemical shift. A fairly sharp change in the chemical shift is evident in the data shown in Fig. 12 at a C_4E_1 mole fraction of about 0.02. The water and brine results are indistinguishable within experimental error.

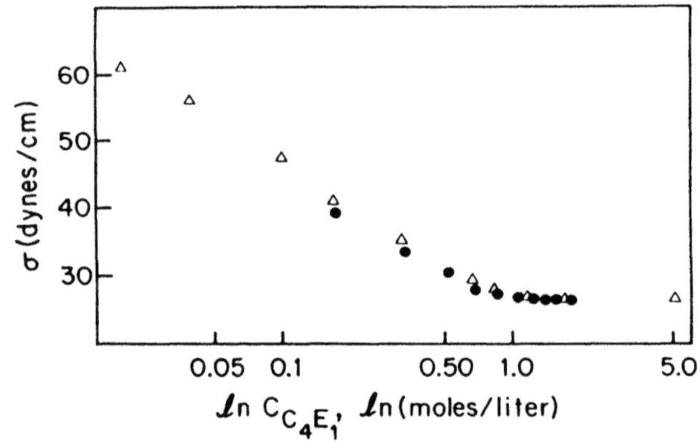

Figure 11. Surface tension of aqueous solution versus logarithm of molar concentration of C_4E_1 in water (open triangles) and 0.2M NaCl brine(filled circles) at 25 ° C. Ref. 21.

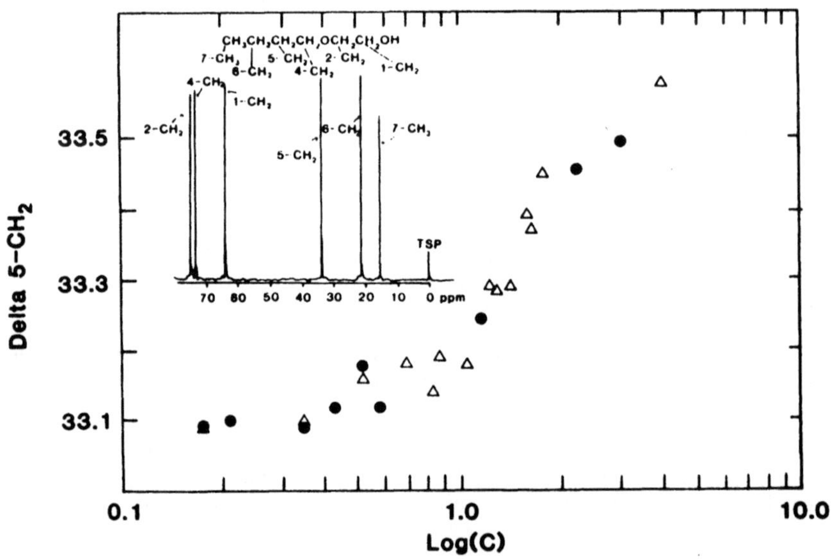

Figure 12. ^{13}C NMR chemical shift of 5-CH_2 carbon versus logarithm of molar concentration of C_4E_1 in water (open triangles) and 0.2 M NaCl brine (filled circles) solutions. Ref. 21.

Table 1. Physical properties changing abruptly at a C_4E_1 mole fraction of about 0.02 in water.

PROPERTY	TECHNIQUE	REFERENCE
Partial molar volume	Densitometry	Refs. 53, 54
Partial molar heat capacity	Microcalorimetry	Refs. 53, 54
Relaxation frequency	Ultrasonic absorption	Refs. 55
Diffusion coefficient	Quasielastic light scattering	Ref. 56
Raman spectra	Raman spectroscopy	Ref. 57
Partial molar refractive index	Refractometry	Ref. 21
Surface tension	Ring tensiometry	Ref. 21
Paramagnetic shielding	C^{13}NMR chemical shift	Ref. 21

As is summarized in Table 1, several physical properties of aqueous solutions of C_4E_1 have been observed to change abruptly with C_4E_1 concentration at a mole fraction of about 0.02. Thus, on the basis of these properties it appears that there is significant molecular aggregation suddenly occurring when a critical mole fraction is reached, similarly to what happens in micellization.

Next consider the behavior of solutions of C_4E_1, water or brine, and hydrocarbons. Kahlweit *et al.* [25,26] have shown that C_4E_1-water-decane mixtures exhibit the $\underset{\sim}{2}3\overline{2}$ sequence as a function of temperature, the three-phase triangle opening at a water-rich critical endpoint (CEP) at about 23 ° C and closing at an oil-rich CEP at about 50 ° C. The trend with carbon number is shown in Fig. 13. The $\underset{\sim}{2}3\overline{2}$ phase sequence is again followed

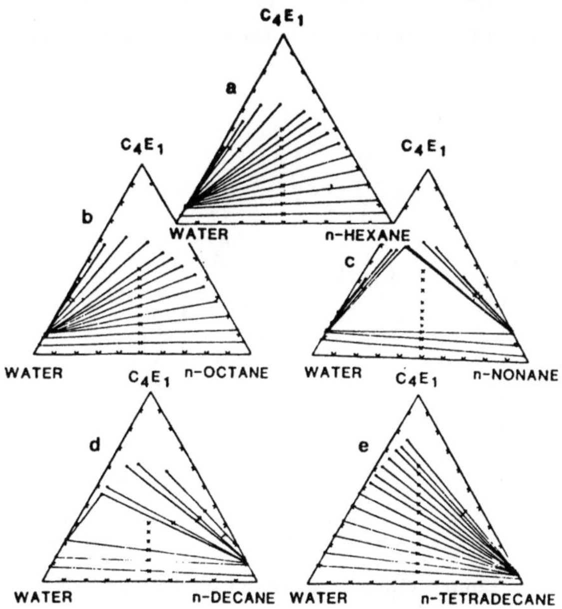

Figure 13. Ternary phase diagrams of C_4E_1, n-alkane, water solutions at 25 ° C. Compositions are in wt%, x's denote mixpoints, and ends of straight lines or vertices of triangles denote two or three phase compositions. Ref. 20.

(with decreasing carbon number) except, of course, the discreteness of the carbon number variation does not allow a continuous scan through the CEP's.

We saw earlier that the electrical conductivity of a microemulsion passes through a percolation threshold if the water fraction drops to a sufficiently low value. In Fig. 14 the equivalent conductance is plotted versus weight percent brine along a composition path passing through the points 2,3,...,6 in Fig. 15. The equivalent conductance becomes very small below a brine fraction of about 2% by weight. This is qualitatively similar to microemulsion percolation behavior, although microemulsion percolation thresholds are usually about 10 or 15%.

Another characteristic property of microemulsions is that quasielastic light scattering (QLS) indicates large diffusive microdomains (of the order of 10nm) of refractive index heterogeneities, arising presumably from the microstructure of the fluid. In Table 2 are given diffusivities of the microdomains deduced by QLS for the compositions 1-6 shown in Fig. 15. They are quite low in view of the viscosities (Table 3) of the solution. The implication is that the microdomains are rather large compared to molecular dimensions.

Table 2. QLS results for 0.2M NaCl brine, C_4E_1, and normal decane mixtures. Ref. 21.

Solution #	D $Dx10^7$	R_H (Å)
1	10	5.4
2	2.8	27
3	5.6	15
4	4.9	20
5	5.9	20
6	4.0	31

If we assume that the domains obey the Stokes-Einstein law, then the radius R_H of a domain can be estimated from

$$R_H = \frac{kT}{6\pi\eta D} , \tag{4}$$

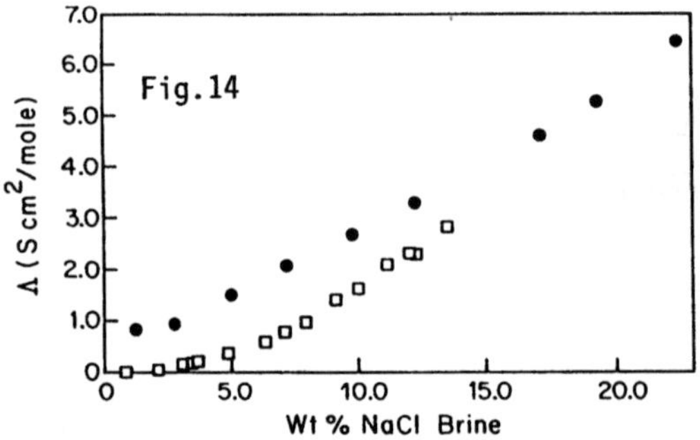

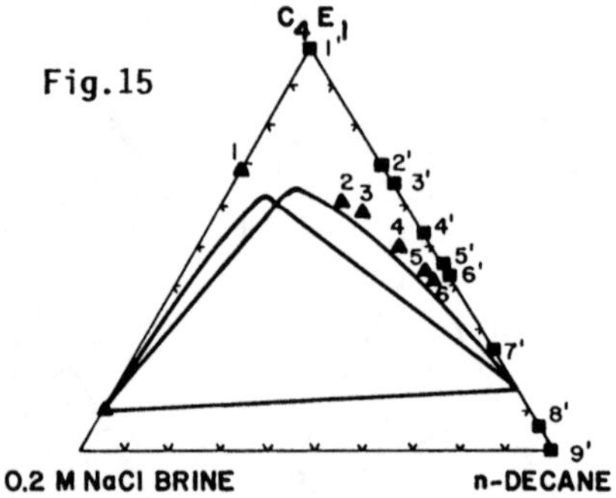

Figure 14. NaCl equivalent conductance of binary C_4E_1-brine (filled circles) and ternary C_4E_1-n-decane-brine mixtures (open squares). Ref. 21.

Figure 15. Pseudoternary phase diagram at 25 °C. Compositions are given in weight percent. Ref. 20.

Table 3. Compositions and viscosities of 0.2M NaCl brine, C_4E_1 and normal decane mixtures. Refs. 21,22.

Solution #	Composition (Weight %)			Viscosity (cP)
	Brine	Decane	C_4E_1	
1	30.92	0	69.08	4.13
2	13.14	26.60	60.26	2.84
3	12.06	30.76	57.18	2.67
4	8.68	41.96	49.30	2.29
5	5.82	51.41	42.77	1.89
6	5.17	42.01	52.82	1.75
1′	0	100	0	2.83
2′	0	30.45	69.55	1.69
3′	0	35.02	64.97	1.67
4′	0	45.99	54.01	1.43
5′	0	53.96	46.04	1.28
6′	0	55.56	44.44	1.22
7′	0	74.97	25.03	1.02
8′	0	4.98	95.02	0.85
9′	0	100	0	0.84

where η is the viscosity and D is the diffusivity. The radius so derived is large compared to the diameters of the fluid molecules, even in the decane-free case (#1).

If we were to stop at this point in the examination of the properties of the C_4E_1 solutions, it would be reasonable to designate C_4E_1 as a surfactant owing to (1) the implications of spontaneous aggregation implied by the measurements summarized in Table 1, (2) the patterns of ternary phase behavior, (3) the percolation threshold of the conductivity and (4) the microdomain sizes deduced from QLS. However, in a QLS experiment, scattering objects are tracked for microseconds only. In the micellar or swollen micellar (or inverted micellar) regimes of microemulsions, the self-diffusivities of the surfactant and the component dissolved in the micelle (or inverted micelle) are approximately the same and are similar to the QLS diffusivity. This is not the case for the C_4E_1 system. Self-diffusivities determined by spin echo, pulsed field gradient NMR (SEPFG NMR) are given in Table 4 for the composition path 1-6 and along the water-free path 1′ - 9′ shown in Fig. 15. The SEPFG NMR measurements were made for two diffusion times, 40 msec and 300 msec, and the results were the same to within the 1% error in the reproducibility of the experiments.

The Stokes radii for all the components are consistent with simple molecular diffusion. In fact the Stokes radius of C_4E_1 varies very little with solution composition, including the cases of pure C_4E_1 and its binary solutions with oil and water. Thus, the molecular association responsible for the results of Table 1 and the microdomains responsible for the light scattering have too short a life time or are too small in population to affect significantly the diffusive migration of the molecules of the solution. This is in sharp contrast to micellar solutions and microemulsions. On this basis then we judge that at $n = 4$ and $m = 1$, C_nE_m has not quite made it to the status of surfactant and is not quite capable of forming a microemulsion. Nevertheless, it does exhibit many of the patterns associated with surfactants and microemulsions.

A similar, but not as detailed analysis has been made for solutions of C_6E_2 with brine and dodecane. The QLS and SEPFG NMR results are given in Tables 5-7 for the composition paths labelled in Fig. 16. Again, although the QLS indicates microdomains larger than molecular sizes we see that, as in the case of the C_4E_1 solutions, these domains do not appreciatively retard molecular diffusivity.

Table 4. SEPFG NMR results for 0.2M NaCl brine C_4E_1, and normal decane mixtures. Ref. 22.

	H_2O		n-decane		C_4E_1	
Solution #	Dx10^7 (cm^2/sec)	R_H (A)	Dx10^7 (cm^2/sec)	R_H (A)	Dx10^7 (cm^2/sec)	R_H (A)
1	44	1.2	-	-	26	2.0
2	53	1.5	52	1.5	37	2.1
3	48	1.7	57	1.4	39	2.1
4	47	2.0	71	1.3	45	2.1
5	-	-	84	1.3	49	2.3
6	-	-	87	1.4	53	2.3
1′	--		-	-	39	2.0
2′	-	-	-	-	63	2.1
3′	-	-	-	-	65	2.0
4′	-	-	-	-	81	1.9
5′	-	-	-	-	83	2.0
6′	-	-	-	-	88	2.0
7′	-	-	-	-	110	1.9
8′	-	-	-	-	142	1.8
9′	-	-	159	-	-	-

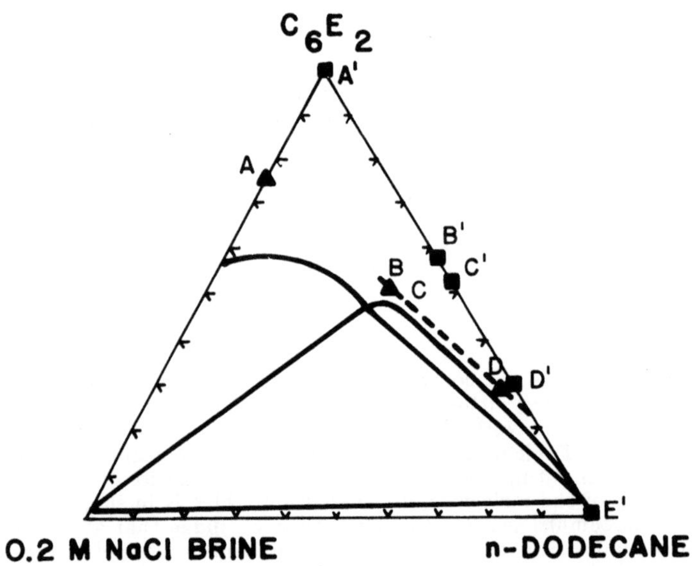

Figure 16. Pseudoternary phase diagram at 25 ° C. Compositions are given in weight percent. Ref. 20.

Table 5. Compositions and viscosities of 0.2M NaCl brine, C_6E_2 and normal dedecane mixtures. Ref. 21,22.

Solution #	Composition (Weight %)			Viscosity (cP)
	Brine	Decane	C_6E_2	
A	24.61	0	75.39	9.31
B	13.71	36.56	49.73	5.66
C	12.74	40.80	46.46	5.23
D	3.80	68.16	28.04	2.45
A′	0	0	100	6.8
B′	0	43.01	56.99	2.97
C′	0	47.28	52.72	2.76
D′	0	71.86	28.14	1.87
E′	0	100	0	1.34

Table 6. QLS results for 0.2M NaCl brine, C_6E_2 and normal dodecane mixtures. Ref. 21.

Solution #	D $Dx10^7$	R_H (Å)
A	6.2	3.8
B	5.3	7.1
C	5.2	7.4
D	3.6	25.7

Table 7. SEPFG NMR results for 0.2M NaCl brine, C_6E_2 and normal dodecane mixtures. Ref. 22.

	H_20		n-decane		C_6E_1	
Solution #	$Dx10^7$ (cm^2/sec)	R_H (Å)	$Dx10^7$ (cm^2/sec)	R_H (Å)	$Dx10^7$ (cm^2/sec)	R_H (Å)
A	51	0.5	-	-	11.8	2.0
B	31	0.5	44	0.9	16.5	2.3
C	25	1.7	51	0.8	18.2	2.3
D	-	-	117	0.8	40	2.2
A′	-	-	-	-	15	2.1
B′	-	-	-	-	31	2.4
C′	-	-	-	-	35	2.3
D′	-	-	-	-	45	2.6

References Cited

1. IUPAC Manual of Symbols and Terminology, Pure and Appl. Chem. *31*, No. 4, 577 (1972).
2. K. Shinoda, T. Nakagawa, B. I. Tamamushi and T. Isemura, *Colloidal Surfactants. Some Physicochemical Properties* (Academic Press, New York 1963).
3. M. J. Rosen, *Surfactants and Interfacial Phenomena* (Wiley-Interscience, New York 1978).
4. B. Lindman and H. Wennerstrom, Topics in Current Chemistry **87**, 3 (1980).
5. R. G. Laughlin: *In advances in Liquid Crystals*, Vol. 3, ed. by G. H. Brown (Academic Press, New York 1978) p. 41.
6. T. P. Hoar and J. H. Schulman, Nature **152**, 102 (1943).
7. J. H. Schulman and D. P. Riley, J. Colloid Sci. **3**, 383 (1948).
8. J. H. Schulman and J. A. Friend, J. Colloid Sci. **4**, 497 (1949).
9. J. H. Schulman, W. Stoeckenius and L. M. Prince, J. Phys. Chem. **63**, 167 (1959).
10. L. E. Scriven, Nature **263**, 123 (1976).
11. L. E. Scriven: In *Micellization, Solubilization and Microemulsions*, ed. by K. L. Mittal (Plenum, New York 1977) p. 877.
12. Composed by W. F. Michels, Ph.D. Thesis, University of Minnesota, 1987, using a schematic phase diagram published in Ref. 11, and combining cartoons of microstructures published in Ref. 13 and 14 with some newly drawn ones.
13. J. C. Hatfield, Ph.D. Thesis, University of Minnesota, 1978.
14. G. H. Brown, J. W. Doane and V. D. Neff, *A Review of the Structure and Physical Properties of Liquid Crystals*, (CRC Press, Cleveland, Ohio 1971).
15. B. M. Knickerbocker, C. V. Pesheck, L. E. Scriven and H. T. Davis, J. Phys. Chem. **83**, 1984 (1979).
16. B. M. Knickerbocker, C. V. Pesheck, H. T. Davis and L. E. Scriven, J. Phys. Chem. **86**, 393 (1982).
17. J. E. Puig, Ph.D. Thesis, University of Minnesota, 1982.
18. J. E. Puig, D. C. Hemker, A. Gupta, H. T. Davis, and L. E. Scriven, J. Phys. Chem. **91**, 1137 (1987).
19. P. K. Kilpatrick, Ph.D. Thesis, University of Minnesota, 1983.
20. P. K. Kilpatrick, C. A. Gorman, H. T. Davis, L. E. Scriven and W. G. Miller, J. Phys., Chem. **90**, 5292 (1986).
21. P. K. Kilpatrick, H. T. Davis, L. E. Scriven, and W. G. Miller, J. Colloid and Interface Sci. (to appear 1987).
22. J. F. Bodet, H. T. Davis, L. E. Scriven and W. G. Miller, J. Colloid and Interface Sci. (submitted).
23. M. Kahlweit, E. Lessner and R. Strey, J. Phys. Chem. **87**, 5032 (1983).
24. M. Kahlweit, E. Lessner and R. Strey, J. Phys. Chem. **88**, 1937 (1984).
25. M. Kahlweit, R. Strey and D. Haase, J. Phys. Chem. **89**, 163 (1985).
26. M. Kahlweit, R. Strey, P. Firman and D. Haase, Langmuir **1**, 281 (1985).
27. M. Kahlweit, R. Strey, P. Firman, J. Phys. Chem., **90**, 671 (1986).
28. P. G. Nilsson and B. Lindman, J. Phys. Chem. **88**, 4764 (1984).
29. U. Olsson, K. Shinoda and B. Lindman, J. Phys. Chem. **90**, 4083 (1986).
30. K. Shinoda and H. Saito, J. Colloid and Interface Sci. **26**, 70 (1968).
31. K. Shinoda, Prog. Colloid Polym. Sci. **68**, 3 (1983).
32. K. E. Bennett, J. C. Hatfield, H. T. Davis, C. W. Macosko and L. E. Scriven: In *Microemulsions*, ed. by I. D. Robb (Plenum, New York 1982), p. 65; K. E. Bennett, Ph.D. Thesis, University of Minnesota, 1985.
33. R. L. Reed and R. N. Healy: In *Improved Oil Recovery by Surfactant and Polymer Flooding*, ed. by D. O. Shah and R. S. Schechter (Academic Press, New York 1977). p. 383.
34. R. N. Healy, R. L. Reed and D. C. Stenmark, Soc. Pet. Eng. J. **16**, 147 (1976).
35. K. E. Bennett, C. H. K. Phelps, H. T. Davis and L. E. Scriven, Soc. Pet. Eng. J. **21**, 747 (1981).
36. J. L. Salager, J. C. Morgan, R. S. Schechter, W. H. Wade and E. Vazquez, Soc. Pet. Eng. J. **19**, 107 (1979).
37. J. L. Salager, M. Bourrel, R. S. Schechter and W. H. Wade, Soc. Pet. Eng. J. **19**, 271 (1979).
38. Y. Talmon and S. Prager, J. Chem. Phys. **69**, 2984 (1978).
39. Y. Talmon and S. Prager, J. Chem. Phys. **76**, 1535 (1982).

40. H. T. Davis and L. E. Scriven, Soc. Pet. Eng. Reprint # 9278, 55th Annual Fall Technical Conference and Exhibition of the SPE of AIME, Dallas, TX, 1980.
41. P. G. de Gennes and C. Taupin, J. Phys. Chem. **86**, 2294 (1982); J. Jouffroy, P. Levinson and P. G. de Gennes, J. Physique **43**, 1241 (1982).
42. B. Widom, J. Chem. Phys. **81**, 1030 (1984).
43. G. R. Jerauld, L. E. Scriven and H. T. Davis, J. Phys. Chem. **17**, 3429 (1984).
44. E. W. Kaler, K. E. Bennett, H. T. Davis and L. E. Scriven, J. Chem. Phys., **79**, 5673 (1983); J. Chem. Phys. **79**, 5685 (1983).
45. M. Lagues, R. Ober and C. Taupin, J. Physique Lett. **39**, L487 (1978); M. Lagues, J. Physique Lett. **40**, L331 (1979).
46. B. Lagourette, J. Peyrelasse, C. Boned and M. Clausse, Nature **281**, 60 (1979).
47. P. Guering and B. Lindman, Langmuir **1**, 464 (1985); D. Dhatenay, P. Guering, W. Urbach, A. M. Cazabat, D. Langevin, J. Meunier, L. Leger, B. Lindman, 5th Int. Symp. on Surfactants in Solution, Bordeaux, France, July, 1984.
48. M. T. Clarkson, D. Beaglehole and P. T. Callaghan, Phys. Rev. Lett. **54**, 1722 (1985).
49. A. T. Papaioannou, L. E. Scriven and H. T. Davis: In *Proceedings of 5th International Symposium on Surfactants in Solution,* 1984, ed. by K. Mittal and P. Botherel.
50. J. C. Lang and B. Widom, Physica A **81**A, 190 (1975); J. C. Lang, P. K. Lim and B. Widom, J. Phys. Chem, **80**, 1719 (1976).
51. M. Kahlweit and R. Strey, Angew. Chem. Int. Ed: Engl. **24**, 654 (1985).
52. N. J. Chang, J. F. Billman and R. A. Lickliden: In *Statistical Thermodynamics of Micellar and Microemulsion Systems,*, ed. by S. H. Chen (Springer-Verlag 1986).
53. G. Roux, G. Perron and J. E. Desnoyers, J. Phys. Chem. **82**, 966 (1978).
54. G. Roux, G. Perron, and J. E. Desnoyers, J. Soln. Chem. **7**, 639 (1978).
55. S. Nishikawa, M. Tanaka, and M. Mashima, J. Phys. Chem. **85**, 686 (1981).
56. N. Ito, K. Saito, T. Kato, and T. Fujiyama, Bull. Chem. Soc. Jpn. **54**, 991 (1981).
57. N. Ito, T. Fujiyama, and Y. Udagawa, Bull. Chem. Soc. Jpn. **56**, 379 (1983).

Three-Component Microemulsion Structure: Curvature and Geometric Constraints

V. Chen[1], *G.G. Warr*[1], *D.F. Evans*[1], *and F.G. Prendergast*[2]

[1]Dept. of Chemical Engineering and Material Science, University of Minnesota, Minneapolis, MN 55455, USA

[2]Dept. of Pharmacology, Mayo Foundation, Rochester, MN 55905, USA

1. Abstract

Steady state anisotropy measurements using an amphiphilic fluorescence probe, trimethylammoniumdiphenylhexatriene (TMA-DPH) are reported for three-component microemulsions, employing didodecyldimethylammonium bromide (DDAB) as the surfactant, simple alkanes, and water. The anisotropies of TMA-DPH are almost constant when oil is added to the microemulsions but decrease upon addition of water. The results are interpreted in terms of a structural model based on geometric packing constraints of surfactant coated cylinders and spheres. The anisotropy measures local curvature at the oil-surfactant-water interface which in turn can be related to global structure.

2. Introduction

Since Schulman's work four decades ago, a major question in colloid science concerns the relationship between microemulsion structure and measurable physical properties [1]. It has persisted because microemulsions are inherently complex systems. At present, the most thoroughly characterized systems consist of four or five components, utilize surfactants or cosurfactants which are soluble in oil and water, and possess labile structures which rearrange in response to small variations in composition or temperature.

In an attempt to circumvent some of these difficulties, we have characterized a simple, three-component microemulsion system. It employs as surfactant didodecyldimethylammonium bromide which is only sparingly soluble in either oil or water. From conductance, viscosity, and NMR diffusion coefficient measurements, and observations on oil and counterion specificities, we have made the case that microemulsion structure can be directly related to curvature of the surfactant film at the oil-water interface [2]. This is determined by a delicate balance between headgroup repulsion and oil penetration into the surfactant chains.

In this paper, we employ fluorescence anisotropy measurements using a cationic, amphiphilic probe in order to characterize in more detail the properties of the surfactant-oil-water interface. We interpret changes in anisotropy in terms of changes in local interfacial cuvature and describe in more detail a model for microemulsions which relates curvature and structure.

3. Experimental

The fluorescent probe trimethylammoniumdiphenylhexatriene p-toluenesulfonate (TMA-DPH) was purchased from Molecular Probes Inc. (Eugene, Oregon) and was used without further purification. The preparation of the probe and microemulsion samples for fluorescence anisotropy measurements are discussed in reference 3. Ratios of fluorophore to surfactant were $\leq 4 \times 10^{-6}$. Didodecyldimethylammonium bromide (DDAB) was purchased from Eastman Kodak and recrystallized at least

twice from diethyl ether. Reagent grade alkanes were purchased from Aldrich or Fisher, and the distilled water was deionized and purified with a Millipore® system.
excitation wavelength of 360 nm. Emission intensities were taken above 418 nm. The anisotropy is given by $(I_\perp - I_\parallel) / (I_\perp + 2 I_\parallel)$, where $I_\perp$ and $I_\parallel$ represent intensities of fluorescence emission polarized perpendicular and parallel to the excitation polarization, respectively [4].

4. Results

The amphiliphic probe, TMA-DPH, is tethered to the interface with its hydrophobic tail parallel to the bilayer and its motion restricted to a cone-like region [5]. Thus anisotropy measures the ability of the TMA-DPH to move within the surfactant layer and provides information about the steric constraints imposed upon the surfactants. For TMA-DPH in phospholipid bilayers, the anisotropy ranges from 0.32 and 0.12 from gel state to the fluid state [4]. We first examined the anisotropy behavior of TMA-DPH at fixed water to surfactant ratios (w/s) along oil dilution paths for various alkanes (Figure 1 shows the results for decane.). Over the bulk of the one-phase region, the anisotropy remains relatively constant, but as the compositions approach the oil corner, the anisotropies decrease slightly.

In contrast, measurements along constant s/o for DDAB microemulsions give anisotropies which decrease substantially with increasing water. For a given oil, the measurements were taken at several s/o ratios. Figure 2 shows anisotropy values which are almost independent of the oil volume fraction and suggest a single curve when plotted against molar water to surfactant ratio (w/s). Because each curve consists of data from widely varying s/o ratios, substantial scatter about the curve is expected.

At low w/s ratios, the anisotropy is highest for all of the oils. The anisotropy decrease with added water is not linear but falls rapidly at first and finally reaches a plateau. The onset of the plateau region occurs at the lowest w/s for the most highly penetrating oil and is not readily discernible for tetradecane [2]. Lifetime measurements were checked for these various microemulsion systems and suggest that the anisotropy data are comparable between them for our analysis [3].

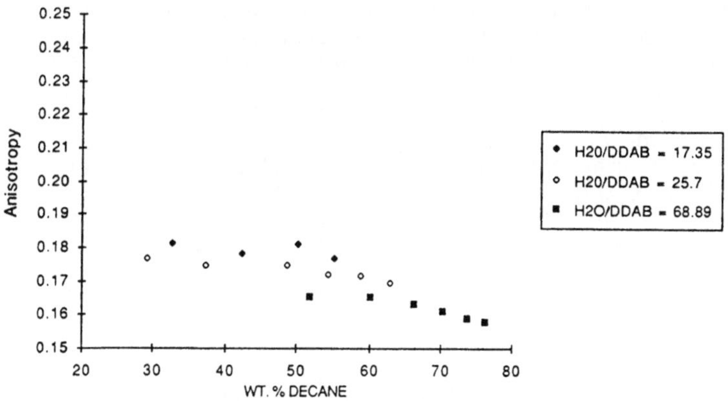

Fig. 1 Steady state anisotropies for decane, DDAB, and water microemulsions at constant H_2O to DDAB ratios (by mole)

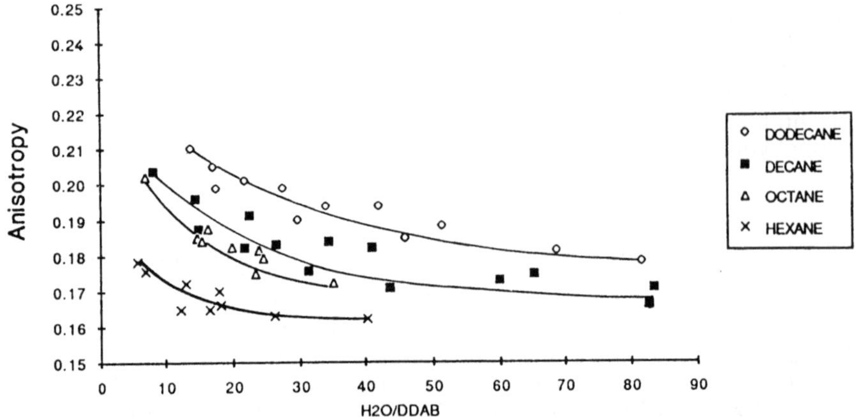

Fig. 2 Steady state anisotropies for alkanes, DDAB, and water microemulsions plotted as function of H_2O to DDAB ratios (by mole) for various surfactant to oil ratios (s/o)

5. Discussion

In previous papers, we have suggested that the steric constraints are the result of interfacial curvature imposed by a balance between headgroup repulsion and oil penetration but have not developed these suggestions in detail. These ideas are made more explicit by first summarizing the important features of the DDAB microemulsions, developing a model based on packing constraints, and discussing the anisotropy data in light of this model.

5.1 Curvature and Properties of the DDAB Microemulsions

Important properties of the surfactant are: (1) In the absence of oil, DDAB forms two lamellar phases with headgroup areas (a_0) of 60 [2] and 68 $Å^2$ [6]. The volume of the hydrocarbon tail (V) is estimated to be 704 $Å^3$ [7], that of the headgroup 120 $Å^3$ [8], and chain length (l_c) of 11.5 Å which is 80% of the fully extended length [2]. This gives surfactant parameters ($V/l_c a_0$) of 1.0 and 0.9 in each lamellar phase. (2) The surfactant is only sparingly soluble in water and oil so that the interfacial area is fixed by the surfactant concentration and the effective headgroup area (a_0) of the surfactant. (3) The surfactant chains are fluid at room temperature so there is flexibility in how they pack and allow penetration of oil. In this situation, the surfactant parameter becomes $V_{eff}/a_0 l_c$, where V_{eff} includes both the volume of the surfactant chains and the volume of the oil taken up by the chains. When $V_{eff}/l_c a_0$ becomes greater than unity, inverted structures form. From studies on bilayers, oil uptake by amphiphilic molecules is greatest for small alkanes [9]. When the chain length of the alkane exceeds that of the surfactant, oil penetration is greatly diminished [9]. (4) The role of electrostatic headgroup interactions is complex, as indicated by the very large changes in the microemulsions in going from iodide to bromide to chloride counterions. The systematic counterion effects on structures in these microemulsions are discussed in reference 10.

The salient features of the microemulsion are as follows: (1) The single-phase region of the DDAB microemulsions resembles roughly a wedge with two well defined sides [11]. These correspond to the minimum (A-B) and maximum water content. (2) Upon addition of water (as one moves from the A-B line towards the water corner along constant s/o), the conductance decreases and for all the oils except tetradecane, a transition to a nonconducting liquid is observed. This occurs as a locus of percolation points that is a constant w/s line lying just above the high water content phase boundary. Self-diffusion coefficients, D, by NMR measurements give D_{oil} and D_{DDAB} which are almost constant throughout the single-phase region [12]. Those for oil are about 1/2 of that of bulk oil. Upon addition of water, D_{H2O} decreases by a factor of twenty which substantiates the conductance measurements. Thus all of the DDAB microemulsions are bicontinuous along the A-B line until close to the oil corner and, upon addition of water, transform into w/o microemulsions near the high water content phase boundary (except for tetradecane which remains bicontinuous throughout its one-phase region).

Based on these observations, we have suggested that a range of interfacial curvature is allowed for each system as set by the balance of headgroup repulsion (which favors diminished curvature) [10] and oil penetration (which favors increased curvature) [13]. This in turn determines the structure. We make these ideas more explicit by describing a geometric model below.

5.2 A Geometric Model for Three-Component Microemulsions

The relationship between microemulsion composition, curvature, and allowed structures can be considered by using a geometric model that includes the steric constraints of packing surfactants at the interface and packing of surfactant coated water cylinders and spheres. By using simple cylinders and spheres to represent microemulsion structure, we can calculate the curvature of the water core from the volume of the aqueous phase and the surface area provided by the surfactant with a fixed headgroup area. The details of these calculations are described elsewhere [3], but the main features can be summarized as follows: (a) If simple cylinders and spheres are used, the lines of constant curvature are constant water to surfactant ratios. This is because the surface area and internal volume are set only by DDAB and water concentrations respectively. (b) Geometric constraints of importance are: (1) The volume fraction that the impenetrable surfactant coated cylinders and spheres can occupy is limited to 90.7% and 74.05% [2]. (2) The volume of the surfactant tails cannot exceed that of the annulus of l_c beyond the water core. For $a_o = 60$ Å^2, this restriction is not very severe, but for $a_o = 50$ Å^2, almost half of the ternary diagram is disallowed [3]. (c) There is a minimum (R_{min}) and a maximum (R_{max}) radius of curvature that defines the limits of the one-phase microemulsion region. The structures that achieve R_{min} at minimum water content are cylinders. When water is added, the cylinders can swell until R_{max} is exceeded, and then spheres must form if more water is to be accommodated. This line we determined from the locus of percolation points. Once the spheres exceed R_{max} as more water is added, the one-phase boundary occurs. In reality, the transition from bicontinuous to discrete water structures occurs continuously as water is added. R_{min} and R_{max} can be determined experimentally from the A point (minimum water content) and the locus of percolation line as shown in Fig. 3 [3].

These features of the geometric model yield a wedge-like shape of allowable microemulsion region as set by a range of interfacial curvature and packing constraints as well as structural behavior consistent with experimental observation. The geometric model also provides a basis for understanding the meaning of the anisotropy measurements. The constant w/s lines are loci of constant curvature along which the anisotropy is almost invariant. The constant s/o lines sweep across changing curvature, and the anisotropy changes in a corresponding way. The magnitude and change in the anisotropy in fact mirror changes in the radius of curvature which dictates global microemulsion structure.

The notion of packing constraints associated with impenetrable cylinders and spheres explains many of the salient properties of DDAB microemulsions; however, we are still faced with some unresolved issues. (1) a_o is assumed to be fixed across the microemulsion phase although it is likely to increase slightly upon

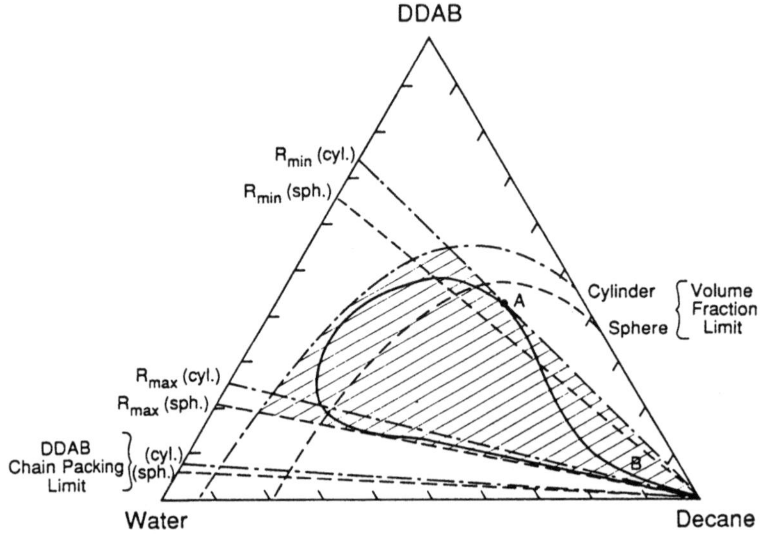

Fig. 3 Phase boundaries calculated from geometric constraints for decane and a_o = 60 $Å^2$ with experimental R_{min} and R_{max} determined from actual one-phase region for DDAB-decane-water microemulsions (solid line)

addition of water. (2) Headgroup repulsion and oil penetration are coupled, thus making the use of experimental R_{min} and R_{max} necessary to characterize each system. (3) Only steric (repulsive) interactions are included so far. However, the dependence of the surface tension between DDAB microemulsions and oil along the A-B line can account for this in terms of extraordinarily long-range attractive forces between water-filled cylinders (but not spheres) in an oil continuum [8]. These attractive interactions increase as one goes from hexane to tetradecane and account in part for the departure of the initial A-B line from a constant w/s line as in our model.

6. Conclusion

(1) The anisotropy of the amphiphilic fluorescence probe, TMA-DPH, has been determined in the DDAB three-component microemulsion as a function of oil and water-to-surfactant ratios. Along constant w/s lines in the triangular phase diagrams, the anisotropy is almost constant, but it decreases as one moves along water dilution lines toward the water corner. Since this probe is located in the oil-surfactant -water interface, it samples the steric constraints imposed by local curvature. Upon addition of water, changes in curvature are mirrored by changes in anisotropy.

(2) A model based on the packing constraints is used to define the phase regions where conduits and spheres can exist. When combined with experimentally determined minimum and maximum water lines, phase diagrams which approximate that in Figure 3 are obtained. The major discrepancy occurs along the A-B line and is discussed in terms of the large attractive forces associated with water-filled conduits in an oil continuum. This model permits local curvature to be related to global structure as well as to the results from anisotropy measurements.

Acknowledgement

V. C. thanks Hercules Corporation for support as a predoctoral fellow. D.F.E. acknowledges support by the U.S. Army (Contract No. DDAG 29-85-K-0169).

References

1. L. M. Prince: Microemulsions (Academic Press Inc. , New York 1977)
2. D. F. Evans, D. J. Mitchell, B. W. Ninham: J. Phys. Chem. 90, 2817 (1986)
3. V. Chen, G. G. Warr, D.F. Evans, F. G. Prendergast: J. Phys.Chem., submitted
4. F. G. Prendergast, R. P. Haugland, P. J. Callahan: Biochemistry 20, 7333 (1981)
5. L. W. Engel and F. G. Prendergast: Biochemistry 20, 7338 (1981)
6. K. Fontell, A. Ceglie, B. Lindman, B. W. Ninham: Acta Chemica Scandinavica A 40, 247 (1986)
7. David W. R. Gruen: J. Phys. Chem. 89, 146 (1985)
8. Martin Allen, D. F. Evans, D. J. Mitchell, B. W. Ninham: J. Phys. Chem., in press
9. D. W. R. Gruen, D. A. Haydon: Pure and Appl. Chem. 52, 1229 (1980)
10. V. Chen, D. F. Evans, B. W. Ninham: J. Phys. Chem. 91,1823 (1987)
11. S. J. Chen, D. F. Evans, B. W. Ninham: J. Phys. Chem. 88, 1631(1984)
12. S. J. Chen, D. F. Evans, B. W. Ninham, D. J. Mitchell, F. D. Blum, S.Pickup: J. Phys. Chem. 90, 842 (1986)
13. F. D. Blum, S. Pickup, B. Ninham, S. J. Chen, D. F. Evans: J. Phys. Chem. 89, 711 (1985)

Transition of Rodlike to Globular Micelles by the Solubilization of Additives

H. Hoffmann and W. Ulbricht

Lehrstuhl für Physikalische Chemie I der Universität Bayreuth, Universitätsstraße 30, D-8580 Bayreuth, Fed. Rep. of Germany

1. ABSTRACT

The solubilization of hydrocarbons in surfactant solutions containing rodlike micelles was studied using static and dynamic light scattering techniques. It was found that these systems have a rather high solubilization capacity for aliphatic hydrocarbons; this solubilization leads to a shrinkage of the rods, and above a characteristic concentration of the hydrocarbon to a rod to sphere transition. The spheres can then accommodate more hydrocarbon and grow to microemulsion droplets. On the other hand the solubilization capacity of rodlike micelles for aromatic hydrocarbons is much smaller and these hydrocarbons even stabilize the rods. A model is proposed which can explain the experimental data on the basis of simple geometrical considerations and the packing parameter of the surfactants.

2. INTRODUCTION

The solubilization capacity of surfactants for additives like pharmaceutical drugs, dye molecules or simple hydrocarbons varies greatly with the physical conditions of the surfactant system /1/. For a given surfactant ion the capacity is dependent on the counterion, the temperature, the excess salt and cosurfactant concentration. It was recently shown by us that the solubilization capacity is particularly large for single chain hydrocarbons when rodlike micelles exist in the isotropic surfactant solution /2/. We can therefore also state that whenever we have rodlike micelles in the isotropic solution we can expect to find a rather large solubilization capacity for aliphatic hydrocarbons. It is thus possible to make predictions on the solubilization capacity of a surfactant solution on the basis of the determination of micellar structures. As we will show, this is actually not surprising, because both the capacity and the structures are largely determined by the packing parameter of a surfactant molecule at an interphase. Very often rodlike micelles are formed in binary ionic surfactant systems with globular micelles by increasing the excess salt concentration.

Systems under such conditions for both cationic and anionic surfactants have been studied in detail by many groups /3,4,5,6/. When the ionic surfactant is combined with hydrophilic counterions, usually large salt concentrations are necessary for the formation of rodlike micelles. As a consequence of the high ionic strength the Debye length in these solutions is rather short (less than 20 Å). This has consequences on the flexibility of the micellar rods and their persistence lengths. Both PORTE et al and IKEDA et al have reported persistence lengths for rodlike micelles of the order of several hundred Å while the contour lengths were much longer. Rodlike micelles are therefore usually referred to as being rather flexible.

In our previous studies on rodlike micelles, transition from globular to rodlike micelles was enforced by using strongly binding counterions /7/. In a way these counterions act like cosurfactants. The resulting rodlike micelles are only weakly dissociated and the ionic strength in the solutions can be rather low. As a consequence these rods are usually much less flexible and have persistence lengths exceeding 1500 Å.

The rodlike micelles in these solutions can be very large and therefore easy to detect by different techniques. We had observed recently that these long rods are transformed to globular structures when aliphatic hydrocarbons are solubilized into the micelles /2/.

In this paper we shall present more data to demonstrate the general validity of this behaviour for rodlike systems and emphasize the basic principles for the transformation of the rods to globular particles. We also show that the transformation has many consequences in surfactant chemistry, in particular for the phase diagrams of ternary systems.

3. RESULTS

The transformation of the rodlike micelles to globular ones can easily be monitored by various techniques. The transition shows up in the disappearance of electric and flow birefringence, in the decrease of the viscosity, in the decrease of the intensity of scattered light, of the radius of gyration and of the hydrodynamic radius of the micelles. All such measurements have been carried out and the data usually are very consistent /2,8/. Here we will mainly present light scattering data; most of these data will be presented for the first time. They will be supplemented by a few results which have been published before to demonstrate the general validity of the phenomenon.

Typical light scattering data are shown in Fig. 1 for the systems Tetradecyltrimethylammoniumsalicylate (TTMASal), Tetradecylpyridini-

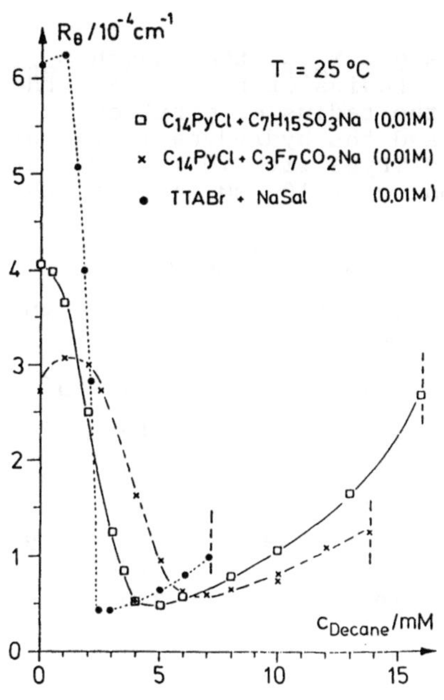

Figure 1. Plot of the Rayleigh ratio R_θ for solutions of 10 mM TTMABr + 10 mM NaSal, 10 mM TPyCl + 10 mM NaPFB and 10 mM TPyCl + 10 mM NaHS against the concentration of added n-Decane at 25°C

umperfluorobutyrate (TPyPFB) and Tetradecylpyridiniumheptanesulfonate (TPyHS). Both systems form already large rodlike micelles in a 10 mM solution without excess salt. In the binary system the scattering intensity is however rather small due to the intermicellar repulsion of the rods. The structure factor S is below 0,1 under these conditions.

In order to increase the structure factor and hence to increase the scattering intensity excess salt in equivalent amounts to the surfactant concentration was therefore added to the system. Detailed investigations on the systems have shown that for these conditions the structure factor is close to unity /9/. Table 1 summarizes micellar parameters for the experimental conditions. The table also gives the parameters at the overlap concentration c* for the micellar rods.

Increasing amounts of Decane were solubilized in these solutions and the scattering intensity was measured. The amount of hydrocarbon in the saturated solution was determined by refractive index increment measurements.

All systems show a drastic decrease of the scattering intensity at a rather low hydrocarbon : surfactant ratio and then a slow increase with further addition of Decane. Qualitatively the drop of the intensity signals the transition of the rods to the globules. The molecular weight of the spheres at the transition is more than ten times smaller than the one of the rods.

The transition concentration depends on the chain lengths of the solubilized hydrocarbon, as is shown in Fig. 2. With increasing chain length both the transition and the saturation concentrations shift to smaller values. As will be shown in the theoretical part, a certain absolute amount of a hydrocarbon is necessary for the transition. As a consequence the hydrocarbons with more than 16 C-atoms are not soluble enough to enforce the transition in the TTMASal-system. It is likely that the critical number of C-atoms depends on the chain length of the surfactant and shifts with shorter surfactant

Table 1. Values for the effective molar weight M_{eff}, the length L_{eff} of the rodlike micelles assuming a short radius of r = 20 Å, the mean distance d between the aggregates, the radius of gyration R_G, the effective diffusion coefficient D_{eff} and the hydrodynamic radius R_H of the rods in solutions of TTMASal and TPyHS with 10 mM NaCl at the experimental surfactant concentration c = 10 mM and at the overlap concentration c* at 25°C

	TTMASal + 10 mM NaCl		TPyHS + 10 mM NaCl	
	c = 10 mM	c^* = 4 mM	c = 10 mM	c^* = 6 mM
M_{eff} (g/mol)	$8,24 \cdot 10^5$	$2,33 \cdot 10^6$	$7,13 \cdot 10^5$	$8,94 \cdot 10^5$
L_{eff} (Å)	1210	3420	950	1190
d (Å)	720	1430	650	840
R_G (Å)	360	1200	720	890
D_{eff} (cm^2/s)	$9,89 \cdot 10^{-8}$	$5,32 \cdot 10^{-8}$	$1,15 \cdot 10^{-9}$	$1,09 \cdot 10^{-9}$
R_H (Å)	220	410	190	200

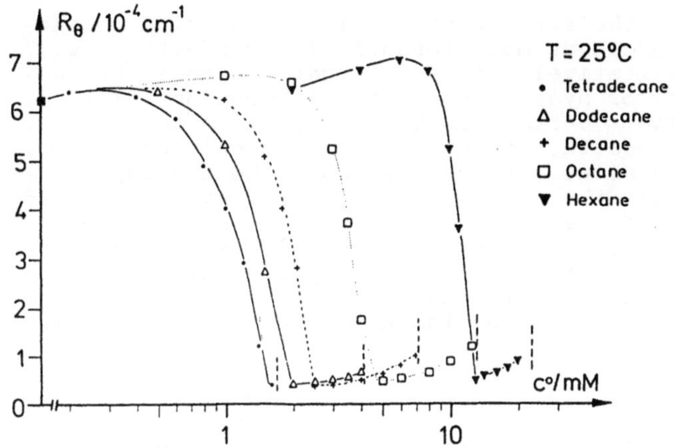

Figure 2. Plot of the Rayleigh ratio R_θ for solutions of 10 mM TTMABr + 10 mM NaSal against the logarithm of the concentration of added n-alcanes with different chain lengths at 25°C

chains to lower numbers. For a given surfactant the transition concentration also depends to some degree on the used counterions, as Figure 1 demonstrates. It should also be noted that the required amount of hydrocarbon for the transition depends also on the temperature. With increasing temperature less hydrocarbon is required for the transition of ionic surfactant systems. In these situations both temperature and hydrocarbon have the same tendency.

In Fig. 3 the chainlength of the surfactant was varied between 12 and 16 for the solubilization of the same hydrocarbon. The concentration of the C_{14}- and C_{16}-surfactant was the same while for the C_{12}-system it was 20% higher in order to have the same amount of surfactant in the micellar form, because the C_{12}-surfactant has a cmc of about 2 mM. The results show that both the transition and the saturation concentrations shift to higher decane concentrations with

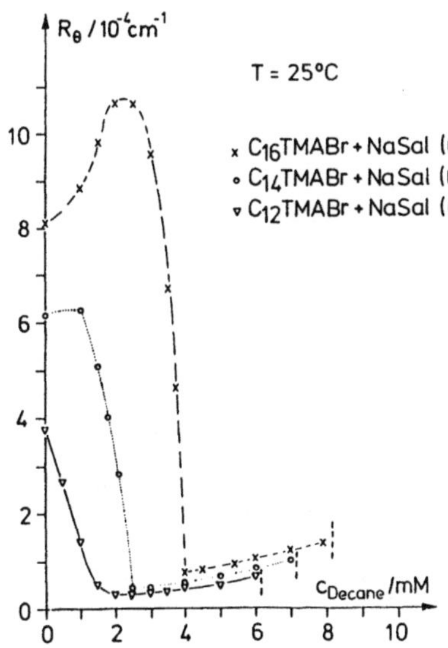

Figure 3. Plot of the Rayleigh ratio R_θ for solutions of 10 mM CTMABr + 10 mM NaSal, 10 mM TTMABr + 10 mM NaSal and 13 mM DTMABr + 13 mM NaSal against the concentration of added n-Decane at 25°C

increasing chain length of the surfactant. It is furthermore noticable that the transition proceeds more abruptly for surfactants with higher chain lengths. Qualitatively, it seems that the rods can accommodate a certain amount of hydrocarbon without becoming smaller. Only when the amount is above a certain value do they begin to shrink. In this respect the light scattering data could be deceiving. In our starting conditions we are about a factor 2 above the overlap concentration, which means that the scattering intensity was suppressed already with increasing surfactant concentration. It is thus conceivable that the size of the rods begins to become smaller in the range where the light scattering intensity is increasing with the hydrocarbon concentration. There could indeed be a decrease of the lengths of the rods right from the beginning of the solubilization experiment.

In a previous investigation we observed that hydrocarbons like Benzene, Toluene and Cyclohexane did not cause a rod to sphere transition upon solubilization /8/. These hydrocarbons actually made the rods more stable. It can be expected that the influence of Toluene should be reversed when the length of the alkyl sidechain on the molecule is increased. Measurements to test this idea are presented in Fig. 4 where solubilization data for Toluene, Ethylbenzene and n-Butylbenzene in TTMASal-solutions are shown. While Ethylbenzene still behaves in the same way as Toluene, Butylbenzene can already induce the rod to sphere transition. In the globular state the micelles can then solubilize more Butylbenzene so that the total solubilization capacity of the surfactant solution is higher for Butylbenzene than for Toluene /2/.

Figure 5 shows the influence of various solubilized pentanols on the rodlike micelles. There are remarkable differences for the isomeric alcohols. For small concentrations all alcohols behave very similar. The rods become smaller with increasing alcohol concentration. Even the absolute values of the scattering intensities for a given alcohol are very similar. For higher alcohol concentrations,

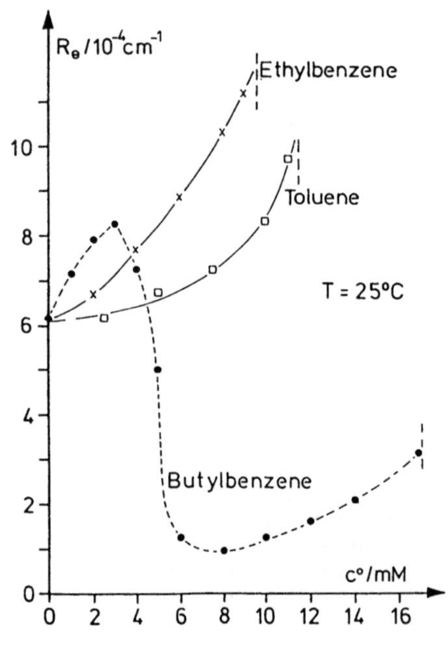

Figure 4. Plot of the Rayleigh ratio R_θ for solutions of 10 mM TTMABr + 10 mM NaSal against the concentration of added n-Alkylbenzenes with different lengths of the side chain at 25°C

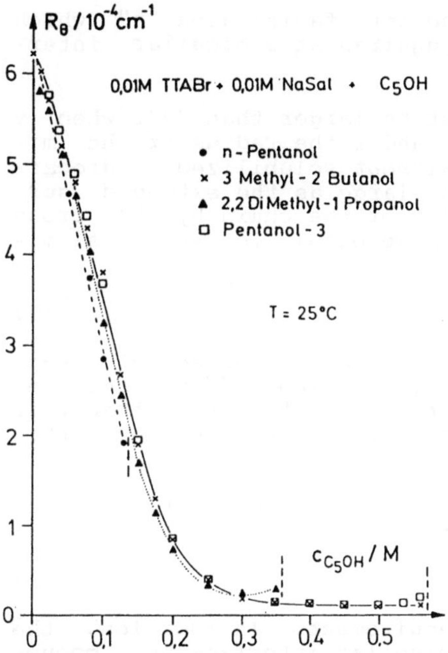

Figure 5. Plot of the Rayleigh ratio R_θ for solutions of 10 mM TTMABr + 10 mM NaSal against the concentration of added isomeric Pentanols at 25°C

however, there are significant and interesting differences. For n-Pentanol a phase boundary is reached at a concentration at which still large anisometric micelles exist in the solution. The coexisting phase in equilibrium with the micellar solution is a lamellar phase. This shows that the n-Pentanol is not capable to suppress the anisometric micelles completely. All the other branched Pentanols are more soluble in water and with increasing alcohol concentration solutions are obtained in which only globular micelles exist. It is however noteworthy that the scattering intensity levels off at rather small intensities, which shows that the aggregates which exist in solution must be fairly small, that is much smaller than the aggregates which are obtained with the solubilization of hydrocarbons.

These results in dilute solutions provide already insight into the understanding of ternary phase diagrams between water, surfactant and alcohols. When to an isotropic micellar solution n-Pentanol is added in increasing amounts, usually a liquid crystalline lamellar phase is observed and it is not possible to move on an isotropic pathway from the binary surfactant base line of the triangle diagram to the corner of the alcohol. With isomeric Pentanols the situation is different. Due to their hydrophobic nature these alcohols are also surface active and are incorporated into the micellar interior. But due to their bulkiness these compounds destroy any liquid crystalline order and prevent the formation of liquid crystalline phases. As a consequence much larger single-phase areas are observed in such systems than in systems with n-alcohols. This is also the situation when hydrocarbons are added and microemulsions are formed. We can predict already from the results of the dilute range the influence which the additives have in the concentrated region.

4. MODEL AND THEORETICAL CONSIDERATIONS

The observed phenomenon can be explained on a simple model which is based on geometrical equations for the packing of the chains. The essential parameter in this model is the interfacial area "a" which a headgroup of a surfactant molecule requires at a micellar interface.

For globular micelles this area must be larger than 3v/r when v is the volume of the hydrocarbon chain and r the radius of the micelle. For pure surfactant solutions without solubilized hydrocarbons the maximum radius can only be as large as the extended surfactant chain. By expressing the volume v of the chain by its cross section a_o and the length r of the chain we obtain for spherical micelles

$$a_s = 3a_o. \tag{1}$$

For a globular micelle the available area a per surfactant headgroup is three times as large as its cross section. If the required area is because of repulsive or steric requirements larger than this value, the radius of the micelles will even be smaller than this largest possible value.

For rodlike micelles the headgroups can pack more tightly and by geometrical constraints we obtain for rods

$$a_r = 2a_o. \tag{2}$$

All systems used fulfilled this requirement. In addition, the counterions are strongly bound to the micellar interface and occupy also interfacial area. The situation at the micellar interface of the rodlike micelles corresponds therefore to a situation which is schematically drawn in Fig. 6. The hydrophobic core of the micelles is more or less completely covered by the headgroups and counterions and there is no area left at which bulk water is directly in contact with the hydrophobic core. In these systems we have therefore a situation which is quite different from the one for micelles with hydrophilic counterions where there is considerable contact between water and the alkyl groups.

With the globular micelles we can theoretically also reduce the required area per headgroup by expanding the size of the sphere above the normally possible value of the length of the extended alkyl chain. In real surfactant solutions this is not possible because the interior of the micelles cannot be filled up by surfactants with polar headgroups. In the presence of hydrocarbons however this becomes possible and is what happens. With our model it is thus straightforward to calculate a radius R for the microemulsion droplets for

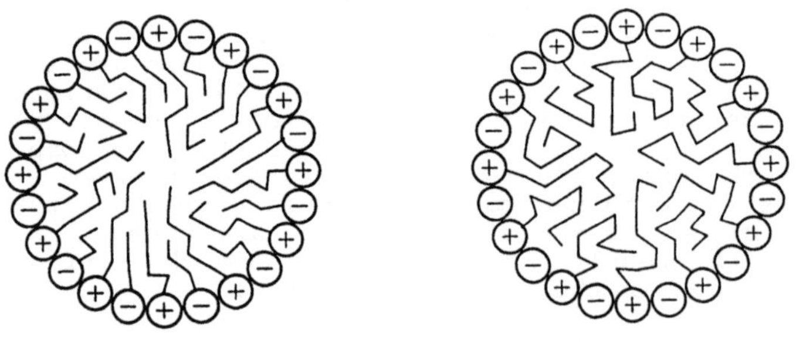

Figure 6. Schemes of the cross-section of rodlike micelles of a cationic surfactant with strongly binding or hydrophobic counterions

which the surfactant ion has the same area available as on the rodlike micelle. We obtain for microemulsion droplets

$$a_r = 2v/r = 2a_o = a_{ME} = 3v/R = 3v/x \cdot r \tag{3}$$

and hence

$$x = 1.5 \text{ and } R = 1.5r. \tag{4}$$

Equation 4 shows that given the conditions that the interior of the micellar droplets is filled up with hydrocarbon and that the surfactant remains in the palisade layer without mixing between the chains of the surfactant and the hydrocarbon core, the smallest droplets in which the headgroups are in the same state as in the rodlike micelles will have a radius of 1.5 times the length of an extended alkyl chain.

The model furthermore predicts that the rod to sphere transition should occur at a volume fraction of $(0.5/1.5)^3$ hydrocarbon to surfactant chain. The observed volume fractions are always somewhat larger. The reason is very likely that in the real situation some mixing between chains of the solubilized hydrocarbons and the surfactant takes place. This effect seems to gain in importance with decreasing chain length of the solubilized hydrocarbon. As a consequence the transition concentration shifts to higher concentration.

The model can also be used to predict the change of the scattering intensity at the transition. If we neglect the change in the volume concentration the scattering ratio should be given by the ratio of the volumes of the micelles before and after the transition and we obtain

$$I_{rod}/I_{ME} = r^2\pi L/(4/3)\pi(1.5r)^3 = L/4.5r. \tag{5}$$

Equation 5 shows that the rods should have an axial ratio of larger than 5 in order to obtain a drastic change of the scattering intensity. For ionic micelles the length L is of the order of 100 r and the scattering ratio is of the order of 10.

Because the configuration and the surrounding of the surfactant remains constant during the transformation it is likely that the heat of transformation is negligible. The entropy is increasing however because of the increase of the number density of the particles. We can assume therefore that the whole process is driven by the associated entropy gain.

The model also predicts a radius R for the growing microemulsion droplets from the volume fraction of the hydrocarbon. Under the assumption that the hydrocarbon is restricted to the inner core we obtain the following equations for the volume V and the surface O of the droplet

$$V = (4/3)\pi R^3 = v_T \cdot n + v_{HC} \cdot m \tag{6}$$

and

$$O = 4\pi R^2 = a \cdot n. \tag{7}$$

The indices T and HC stand for surfactant and hydrocarbon, n and m mean the number of surfactant and hydrocarbon molecules per micelle. From equations 6 and 7 we derive

$$R = 3(v_T \cdot n + v_{HC} \cdot m)/a \cdot n = 3(v_T + v_{HC})/a. \quad (8)$$

With $v_{HC}/v_T = x$ and $a = 2v_T/r$ we obtain

$$R = 1.5r(1 + x). \quad (9)$$

Figure 7 shows evaluated data for the system TTMASal which had some n-Pentanol added in order to increase the solubilization capacity for hydrocarbons even further. The figure shows a comparison of the radii of the droplets which are calculated from the theoretical model, and R-values which are calculated from the molecular weight of the droplets assuming a density of $\rho = 0.9$ g/cm^3. In addition, the hydrodynamic radii from dynamic light scattering data are shown. The agreement of the results with the simple model is good. In particular the slope and the intercept for x = 0 are about the same.

For our model we assume that the interfacial area per surfactant molecule remains constant during the solubilization and independent of the size of the microemulsion droplet. It should be realized that this requires the thickness of the surfactant palisade layer to become smaller with increasing droplet size. The thickness of the surfactant layer has to decrease from a value l for the smallest droplet size to l/2 when R approaches infinity.

While it is possible with our model to calculate the amount of hydrocarbon which must be solubilized for the transformation to take

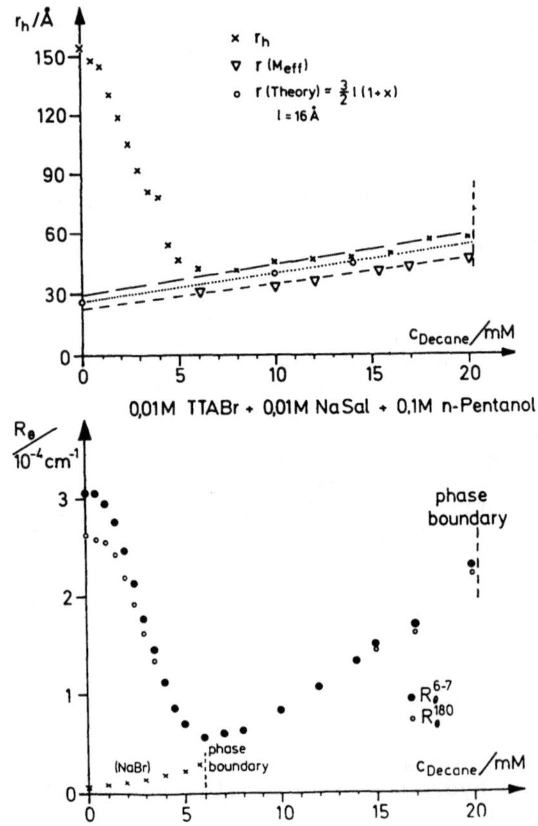

Figure 7. Plot of the hydrodynamic radii R_H, the radii r of microemulsion droplets calculated from the effective molar weight or from the theory, respectively, and the Rayleigh ratio R_θ for solutions of 10 mM TTMABr + 10 mM NaSal + 100 mM n-Pentanol against the concentration of added n-Decane at 25°C

place, it does not predict how much hydrocarbon can be solubilized in the surfactant solution.

The transformation can also be predicted with a somewhat different approach. Lately the importance of the radius of curvature for micellar structures has been realized /10/. The mean radius of curvature ξ for the rodlike micelles is determined by the two radii r and r' of the rod according to

$$1/\xi = 0.5(1/r + 1/r'), \qquad (10)$$

from which follows that $\xi = 2r$ for a long rod. We could thus expect that rodlike micelles should at least solubilize enough hydrocarbon so that microemulsion droplets with $R = 2r$ can form. From simple algebra follows that such a droplet would be formed at $x = 1/3$. For most of the solubilized hydrocarbons this is indeed the case, as the data show.

5. GENERAL CONSEQUENCES OF THE ROD TO SPHERE TRANSITION FOR SURFACTANT CHEMISTRY

The fact that rodlike micelles can be transformed into globular ones has many consequences in various aspects of surfactant chemistry and can be of considerable interest in practical applications of surfactants. Surfactant systems with rodlike micelles can have high viscosities or can even be viscoelastic. If the rods are destroyed in these solutions, the viscosity will decrease to the range of the solvent viscosity and the elasticity will disappear. The solubilization of aliphatic hydrocarbons in surfactant solutions with an undesired high viscosity is therefore one possibility to get solutions with low viscosities, even up to surfactant concentrations of the order of 20%.

While in the binary systems usually a highly viscous birefringent liquid crystalline phase is formed when the surfactant concentration is increased, the situation is different when aliphatic hydrocarbon is solubilized. In this case the system forms a liquid crystalline phase which is optically isotropic.

These phases which have volume fractions of water in the range of 60 - 70% are highly viscous and elastic. They have usually very peculiar features. The damping coefficient of propagating waves is rather low and consequently a wave oscillates many times before it dies out. Because of the high elasticity of the systems the frequencies of such waves can be in the audible range; for this reason the phases are referred to as "ringing gels" or "Brummgele". Such systems play a role in pharmaceutical preparations for the solubilization of drugs. They are usually prepared from Oleylpolyglycolethers, Paraffinoil and water; they can, however, also be formed from other surfactants. Phase diagrams of such systems have been established by NÜRNBERG et al. /11/ and MÜLLER-GOYMANN /12/. In some of these gels the radii of the droplets could be increased up to 150 Å by increasing the hydrocarbon:surfactant ratio.

There is a well-described phenomenon regarding the cloud point of nonionic surfactants in the literature which is also probably related to the rod to sphere transition. The phenomenon has been described by SHINODA et al /13/ but not well explained.

In a very fundamental paper on the cloud point of nonionics the effect of different additives on the cloud point was studied /13/. It was observed for solutions of Triton X 100 that solubilization of hydrocarbons into the systems the concentrations of which were close to the critical concentration led to an increase of the clouding temperature. Long chain alcohols on the other hand lowered the cloud point.

In nonionic systems the critical volume fraction at the lowest clouding temperature is very often around 1%. Since this phase separation is in physical terms a condensation of the micelles in the gaseous state to the liquid state, the structures present in these systems must have a rather high axial ratio when the volume fraction is low. However, when these long rods are transformed to globular micelles by the solubilization of hydrocarbons the attractive forces between the micelles become smaller and consequently the cloud point must increase. With a rod to sphere transition it is thus straightforward to explain the rather unusual phenomenon of the increase of the cloud point by solubilization. On the other hand, alcohols stabilize the rods and for this reason lower the cloud point.

Generally it can be said that all phenomena in surfactant chemistry which depend on the presence of micellar rods disappear with the rod to sphere transition. A very subtle phenomenon is the occurence of shear-induced states in dilute micellar solutions. Conditions for such encounters are found for sol-gel-transitions in surfactant solutions /14/.

In these solutions small rodlike micelles are already present but because of their low axial ratio and number density the rotational volumes of the rods do not yet overlap. If such solutions are sheared, viscoelastic states are induced if the shear rate is increased above a threshold value which is characteristic for the system.

The shear-induced structures can easily be monitored by their induced flow birefringence. With increasing shear rate the flow birefringence reaches saturation. Upon solubilization of small amounts of hydrocarbon into the system the shear-induced states disappear completely. These states are due to the formation of strings of small rods under shear. With the solubilization of the aliphatic hydrocarbons the rods disappear and the strings are no longer formed. This is remarkable because the packing of the surfactants in the globules should be very similar to that for the rods.

6. REFERENCES

1. M. E. L. Mc Bain and E. Hutchinson, "Solubilization and Related Phenomena", Academic Press, New York 1955
   P. H. Elworthy, A. T. Florence and C. B. Mc Farlane, "Solubilization by Surface Active Agents and Its Application in Chemistry and the Biological Sciences", Chapman and Hall, London 1968
   P. Mukerjee, in "Solution Chemistry of Surfactants", vol. 1, p. 153, ed. by K. L. Mittal, Plenum Publishing Corporation, New York 1979
2. H. Hoffmann and W. Ulbricht, Tenside Detergents 24, 23 (1987)
3. P. J. Missel, N. A. Mazer, G. B. Benedek and M. C. Carey, J. Phys. Chem. 87, 1264 (1983)
4. S. Ozeki and S. Ikeda, J. Phys. Chem. 89, 5088 (1985)
   T. Imae, R. Kamiya and S. Ikeda, J. Colloid Interface Sci. 108, 215 (1985)
5. G. Porte, J. Appell and Y. Poggi, J. Phys. Chem. 84, 3105 (1980)
6. P. Lianos and R. Zana, J. Phys. Chem. 84, 3339 (1980)
7. H. Hoffmann, G. Platz, H. Rehage, W. Schorr and W. Ulbricht, Ber. Bunsenges. phys. Chem. 85, 255 (1981)
   H. Hoffmann, H. Rehage, G. Platz, W. Schorr, H. Thurn and W. Ulbricht, Colloid Polymer Sci. 260, 1042 (1982)

8. O. Bayer, H. Hoffmann, H. Thurn and W. Ulbricht, Adv. Colloid Interface Sci. 26, 177 (1986)
9. H. Hoffmann, G. Platz, H. Rehage and W. Schorr, Adv. Colloid Interface Sci. 17, 275 (1982)
10. S. A. Safran, L. A. Turkevich and P. Pincus, J. Phys. Lett. 45, 69 (1984)
11. E. Nürnberg and W. Pohler, Progr. Colloid Polymer Sci. 69, 64 (1984)
12. C. Müller-Goymann, Seifen-Öle-Fette-Wachse 110, 395 (1984)
13. K. Shinoda, T. Nakagawa, B. Tamamushi and T. Isemura, "Colloidal Surfactants", Academic Press, New York 1963
14. H. Hoffmann, H. Rehage and I. Wunderlich, Progr. Colloid Polymer Sci. 72, 51 (1986)

A Study of Dynamics of Microemulsion Droplets by Neutron Spin Echo Spectroscopy

J.S. Huang[1], *S.T. Milner*[1], *B. Farago*[2], *and D. Richter*[2]

[1]Exxon Research and Engineering Co., Annandale, NJ 08801, USA
[2]Institut Laue-Langevin, F-38042 Grenoble, France

We have employed neutron spin-echo techniques to study the dynamics of the shape fluctuations of microemulsion droplets. It is found that the scattering spectrum of a droplet system with a labelled layer of surfactants exhibits a sharp peak in the relaxation frequency spectrum. We have determined that the fluctuations are driven mainly by the elastic modes of the surface, and the height of the peak allows us to estimate the surface elastic bending coefficient to be of the order of 5 kT.

Microemulsions[1] are homogeneous mixtures of oil, water and surfactants (and co-surfactants). One of the most striking properties of the microemulsion is that they can form a so-called middle phase in equilibrium with both a pure oil phase and a pure brine phase simultaneously. The interfaces that separate the middle phase microemulsion with the excess pure phases generally have extremely low surface tensions (10^{-3} dyne/cm or less.) The origin of this ultralow interface tension has been of interest for many people[2,3]. It is known that microemulsions generally contain large internal surface areas that separate the hydrocarbon domains from the aqueous domains. For instance, for a microemulsion containing 10% surfactants, a typical value[4] for the interfacial surface area is of the order of 100 m^2 /cc. It is thus expected that the property and structure of microemulsions will be dependent on the nature of this surfactant layer. When microemulsions were first formulated, it was recognized that the surfactants must reside mainly on the interface, and it was believed that the surface tension of the interfaces (of the microscopic droplets) must be extremely low[5,6] so that the free energy of formation of the dispersed droplets would not be unduly penalized by a large surface energy contribution. Alternatively, it could also be argued[7] that the formation of the microemulsion droplets was mainly due to the natural bending tendency of the surfactant layer, the minimization of the bending elastic energy resulted in the observed microemulsion structure. For surfactant layers that have strong natural bending tendencies, a droplet phase often forms, while for systems with neutral bending tendencies, a random bicontinuous phase[8,9,10,11] becomes more favorable. Recent experiments using a "double--contrast variation" method [12,13] in small angle neutron scattering have discovered the existence of a droplet phase in the AOT system (at equal water-oil ratio) and a random bicontinuous structure for other systems that usually contain cosurfactant such as alcohol. We decided that in order to understand the roles of surface tension and the bending energy in the microemulsion formation, one must study the dynamics of surfactant layer.

In the case of a 3-component microemulsion containing a surfactant called AOT (sodium di-2-ethyl-hexyl sulfosuccinate), water (or heavy water) and decane (or per-deuterated decane), it is well established that a droplet structure exists in the rather extensive single phase region [12,14]. It is also known for this system that the mean radius of the water droplets is linearly dependent on the surfactant-to-water ratio[14,15,16].

The basic idea of the experiment is described as follows: thermal fluctuations of the surfactant film distort the droplet from its average (presumably spherical) form, and the amplitude of these fluctuations are expected to be small as indicated by the moderate effective polydispersity determined through small angle neutron scattering measurements[14,16]. In order to measure the relaxation frequency of these small fluctuations it is necessary to contrast match the internal droplet phase neutron scattering length density with that of the external continuum so that the scattering is given by a shell of surfactant layer. The distortion of the shell from its mean spherical shape can be expanded in spherical harmonics. As we shall see later, the scattering produced by the lowest mode of the fluctuations (the $l = 2$ mode) is effectively coupled to j_2, the 2nd order spherical Bessel function. It turns out that the first maximum of j_2 occurs in the vicinity of the first zero of the j_0, which couples to the scattering of the surfactant shell of the mean spherical droplet. This near orthogonality of the Bessel functions is what allows us to measure the fundamental frequency of the fluctuation with small amplitudes. From the dispersion relation of the relaxation frequency, we can ascertain whether interfacial tension or the elastic bending energy drives the dynamics of the shape fluctuations.

The dynamic experiment is carried out on the quasi-elastic neutron spin echo spectrometer (IN11) at Institut Laue-Langevin in Grenoble, France. The energy change of the scattered neutron due to the scattering process is measured directly through the Larmor precessions of the neutron spin in an external guide field. The intermediate scattering function F(Q,t) is directly given by the final polarization of the scattered neutron[17].

The five microemulsion samples and one micellar solution were used for this study. The water-to-AOT molar ratios ranges from 0 to 40 giving a mean droplet radius (center to the outer edge of surfactant tails) from 70 to 17 Å. The dispersed phase volume for all the samples is a constant 5%. The continuous phase is a per-deuterated decane which matches the neutron scattering length density of the internal D_2O phase almost exactly so the surfactant coated droplet appears to be only an empty shell to the scattering beam. The scattering intensity I(Q,t) from a droplet system can be formally expressed as

$$I(\vec{Q},t) = \frac{1}{v}\sum_{i,j} F_i(\vec{Q},t)F_j(\vec{Q},t)\exp\left(i\vec{Q}\cdot(\vec{R}_i-\vec{R}_j)\right), \quad (1)$$

where $F_i(Q,t)$ is the form factor of the ith droplet at time t measured at scattering wave vector Q ($Q = 4\pi B/\lambda \sin(\theta/2)$ is given by the wavelength of the incident neutron λ and the scattering angle θ):

$$F_i(\vec{Q},t) = \int d\vec{r}\left(\rho_i(\vec{r},t)-\rho_s\right)\exp(i\vec{Q}\cdot\vec{r}). \quad (2)$$

Here ρ is the mean scattering length density of the surfactant shall and $\rho(r)$ is the scattering length density profile. Assuming the mean radius of the surfactant shell is r_0, the shape of the surface is given by the following spherical harmonic expansion[18]:

$$r(\theta,\phi,t) = r_0 + \sum_{l,m} a_{lm}(t)Y_{lm}(\theta,\phi), \quad (3)$$

where a_{lm} describes the small amplitude of the fluctuation from the mean sphere. Y_{lm}'s are the spherical harmonics. If we further assume that the fluctuations of different eigenstates (l,m) are uncorrelated, then the scattering intensity can be written down as

$$S(q,t)\propto\exp(-Dq^2t)\left[(4\pi j_0(qr))^2+\sum 4\pi f_\ell(qr)\langle a_{\ell 0}(t)a_{\ell 0}(0)\rangle\right], \quad (4)$$

where $f_l(Qr_0) = [(l+2)\, j_l(Qr_0) - (Qr_0)j_{l+1}(Qr_0)]^2$, and V_s is the volume of the surfactant shell, j_l is the spherical Bessel function of order l, and D is the translational diffusion constant. The time dependent amplitudes $<a_l(0)\, a_l(t)>$ $e^{-t/\tau}$ are expected to be overdamped[19]. We can see this by considering the magnitude of the Reynolds number $R_e = \rho vL/\eta$. Here L is distance over which the velocity field decays, so L must be of the order of 100 Å, the diameter of the microemulsion droplets, v must be of the order of L/τ, τ is the characteristic time, which is roughly 10^{-7} s[20]. This yields a $R_e \sim 0.001$ for this system, so the motion is completely dominated by viscous force, and we expect to see overdamped motions only.

As we stated earlier, the lowest order fluctuations (l = 2 mode) coupled mainly to $j_2(QR)$, which reaches a maximum near the 1st zero of j_0, so if we plot the measured relaxation frequency (inverse decay time)/Q^2 as a function of Q, we expect to see a peak at the scattering vector corresponding to $QR \sim \pi$. The decay of the echo signals is shown in Figure 1. The solid line represents the first cumulant fit to the data. In Figures 2 and 3 we have plotted the effective diffusion constant Γ/Q^2 as a function of Q for two droplet radii. In Figure 2 we show the results obtained for samples for which the mean droplet radius of the microemulsion is 70 Å and in Figure 3 the mean radius is 50 Å. In fact, a pronounced peak is observed for each of the samples, and the position of the peak occurs at $QR = 3.2 \pm 0.1$ for all the samples measured (Table I). It is also found that as $Q \to 0$, Γ/Q^2 approaches the value for the center of mass diffusion estimated for each of the droplet sizes. Both the low Q limit and the occurrence of a peak in our spectra are qualitatively understood, and the arguments do not depend on the models for the dynamics. We can distinguish between the bending- and, alternatively, the surface-tension-driven modes by the dependence of Γ at the peak given by the droplet radius. Both the center of mass diffusion and the bending modes scale as r_0^{-3}, while (over-damped) surface tension modes scale as r_0^{-1}. Measurements performed on our samples with radii from 70 to 30Å give Γ ($Qr_0 = \pi$) consistent with r_0^{-3} scaling. As we have mentioned earlier, the surfactant packing

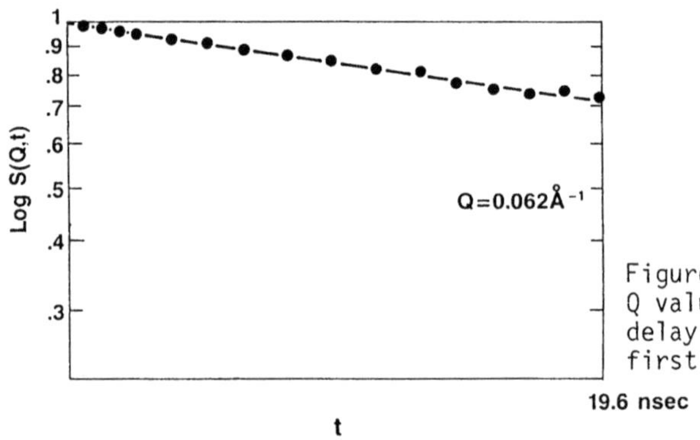

Figure 1: The spin-echo signal at several Q values is plotted as a function of the delay time. The solid curves are the first cumulant fits for the data.

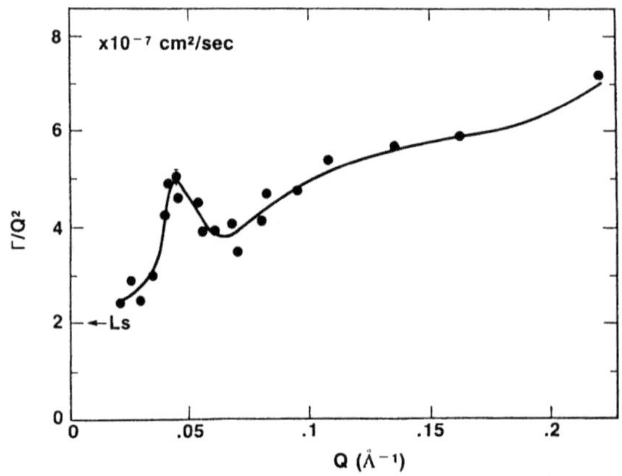

Figure 2: The effective constant Γ/Q^2 is plotted versus Q for the microemulsion with 70 Å droplet mean radius.

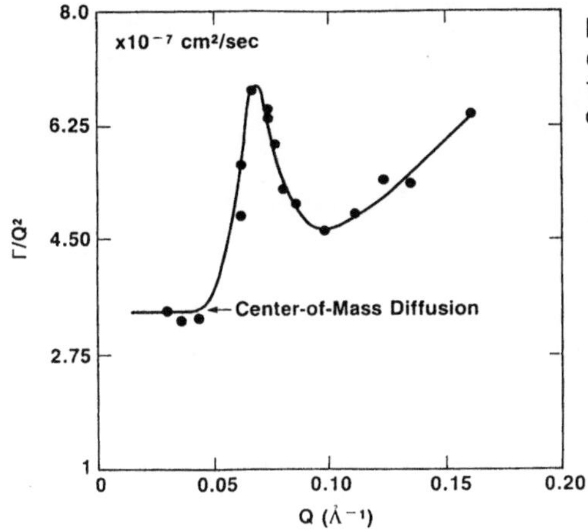

Figure 3: The effective diffusion constant Γ/Q^2 is plotted versus Q for the microemulsion with 50 Å droplet mean radius.

TABLE I

$[D_2O]$	Radius r_o (Å)	Q_mR_o	Γ/DQ_m^2
40.8	70	3.15	2.4
32.6	59.2	3.2	2.3
24.5	48.7	3.30	2.3
16.3	38.3	3.16	2.0
8.2	27.5	3.16	2.2

The molar ratio of D_2O to AOT in this study is listed in Column 1. The mean radius is listed in Column 2; the peak position in Column 3 and the peak height in Column 4.

density on the interface is found to be independent of the radius of the droplets[15], but only a weak function of the temperature. In Figure 4 we show the mean radius of the microemulsion containing different amounts of water (the internal phase) for a given amount of surfactant in the system. We see a linear dependence of the size on the internal phase volume. This suggests that the total surface area is determined by the surfactant concentration alone, so that the interfacial layer is virtually incompressible, and so we may expect the fluctuations to be driven mainly by the elastic modes of the interface.

At higher Q, the behavior of Γ is observed to be roughly $\Gamma \sim Q^3$, which at first sight is further evidence for bending modes, since the characteristic frequency of a bending mode with wavenumber Q is K_cQ^3/η where η is the viscosity. However, the limit of Equ. 4 for Q>>1 is not dominated by scattering off the lth mode such that $Q = 2\pi l/r_0$; scaling arguments cannot tell us how r_0 appears in the function $\Gamma(Qr_0)$. Further analysis is necessary[21].

In order to be sure the existence of the peak is not due to some artifacts, we have also measured the scattering from a 5 % AOT micellar solution. Here no internal dynamics in the time resolution of the spin echo measurement is expected. The result is shown in Figure 4. The plateau value is that expected from the center-of-mass diffusion of micelles with 15 Å radius. The upturn at lower values of Q is the "de Gennes narrowing" due to the excluded volume depression in the

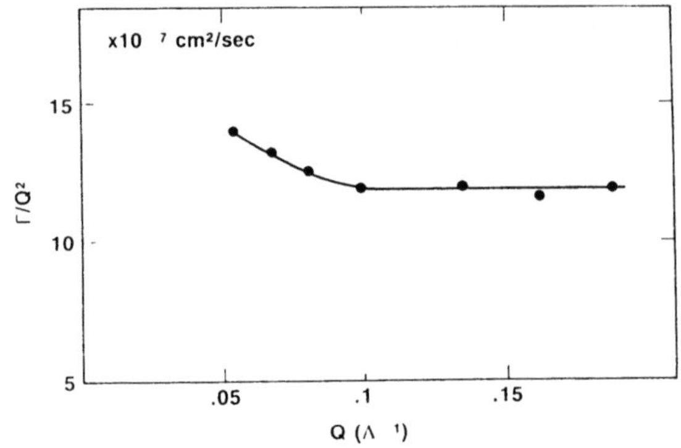

Figure 4: The effective diffusion constant Γ/Q^2 is plotted versus Q for AOT micellar solution at 5 vol. % concentration. The upsweep at small Q is due to the excluded volume interaction[22].

structure factor S(Q) at small Q. A short range interaction as determined by HUANG et al.[22] can account for the 20% rise in the apparent diffusion constant.

To estimate the peak height, we extend and modify the the work of SCHNEIDER et al.[23]. Their expression for relaxation rates is extended to nonzero spontaneous curvature. The "apparent surface tension" d_0 as described in Ref.23 is determined by requiring the mean-square fluctuations (given by equipartition with $E_{eff} = E_{bend} - d_0 A$, where A is the area of the droplet) to give rise to a mean excess area in agreement with the equilibrium predictions of SAFRAN[24]. (The excess area is the difference between the droplet area and that of the sphere of equivalent volume.) The expression for the damping frequency of the lth mode is

$$\Gamma_1 = \frac{K_c}{\eta r^3} \frac{(\ell+3)(\ell-2) + 4w}{Z(\ell)} \tag{5}$$

and the amplitude of the lth mode is

$$\langle | a_{\ell m} |^2 \rangle = \frac{kT}{K_c} \Big[(\ell+2)(\ell-1) \big(\ell(\ell+1) - 6 + 4w \big) \Big] \tag{6}$$

where kT is the thermal energy at absolute temperature T, and $Z(l) = (2l+1)(2l^2 + 2l-1)/[l(l+1)(l+2)(l-1)]$ is given in ref. 23, and $w = r_0/r_s$. At finite r_s, a small, temperature-dependent, entropy of mixing term can be neglected in the above expressions. The experimentally measured decay time is then given by the initial slope of I(Q,t), the mean decay rate of the modes being weighted by their amplitudes and the form factors:

$$\overline{\Gamma}(q) \equiv \frac{1}{S(q,0)} \frac{\partial S(q,t)}{\partial t} \Big|_{t=0}$$

$$= Dq^2 + \Big[4\pi \sum_{\ell>1} \omega_\ell f_\ell(qr) \langle | a_{\ell m} |^2 \rangle \Big]$$

$$\times \Big[(4\pi j_0(qr))^2 + 4\pi \sum_{\ell>1} f_\ell(qr) \langle | a_{\ell m} |^2 \rangle \Big]^{-1} .$$

For the range of accessible Qr, the denominator (the static structure factor) is dominated by the first two terms (the undistorted spherical shell and the l=2 mode.) In the numerator the l=2 mode accounts for about half of the sum, assuming the summation cuts off at $l \sim r_0/l_0 \sim 5$ where l_0 is the surfactant size.

We take into account polydispersity in the expression for $\Gamma(Q)$ by averaging numerator and denominator over the different radii in the sample. Near $Qr_0 = \pi$, the main effect is to replace $j_0(\pi)^2=0$ by $\langle\delta R^2\rangle/R^2$; this variance is an equilibrium property which may be either measured or predicted [16,25]:

$$\langle \bar{\Gamma}(q)\rangle_r \sim Dq^2 + \frac{K_c}{\eta r^3} \frac{(.2)(qr)}{4\pi K_c/kT(qr)^{-2} + (0.48)} . \tag{7}$$

A simple semiquantitative result for $\Gamma(Qr_0=\pi)$ follows:

$$\bar{\Gamma}(qr=\pi) \sim D(\pi/r)^2 + \frac{(0.53)K_c}{\eta r^3}\left(\frac{\pi/2}{3/2 - w} + \frac{(0.15)}{w}\right)^{-1} . \tag{8}$$

For $w = r_0/r_s$ intermediate between 0 and 3/2 (which are the limits of stability of the droplet phase), the r-dependence of the denominator is weak, giving $\Gamma \sim r^{-3}$. Using Equa. 8 we can estimate the value for the bending elastic constant K_c:

$$K_c/kT \doteq (\pi/0.9)[\Gamma/(DQ_m^2) - 1].$$

For $Qr_0 = \pi$, $\Gamma/DQ^2 = 1.5$, and $r/r_s = 0.5$, we have $K_c = 5$ kT, an eminently reasonable value.[25]

We have benefitted greatly from the many discussions with S. Safran. Experimental assistance from J. Sung is also acknowledged.

References

1. For a general survey, see for instance Surfactants in Solutions edited by K. Mittal and B. Lindman (Plenum, New York, 1984).
2. A. Pouchelon, D. Chatennay, J. Meunier, and D. Langevin, J. Coll. Int. Sci., 82, 418, (1981).
3. C. A. Miller, R. Hwan, W. J. Benton, and T. Fort, J. Coll. Int. Sci., 61, 554 (1977)
4. This is estimated by the surfactant packing area/molecule for a AOT/Water/Decane system. It is known that the surfactant (AOT) resides almost exclusively on the interfacial regime and the packing density is ~65 Å^2 /molecule.
5 J. H. Schulman, W. Stoeckenius and L. Prince, J. Phys. Chem., 63, 1677 (1959).
6. C. E. Cooke, Jr. and J. H. Schulman, Surface Chem. pp231-251 (1965)
7. S. A. and L. A. Turkevich, Phys. Rev. Lett., 50, 1930 (1982).
8. L. E. Scriven, Nature (London) 263, 123 (1976) and L. E. Scriven, Micellization, Solubilization, and Microemulsions, Edited by K. Mittal (Plenum Press NY) 877, (1977).
9. J. Jouffray, P . Levinson, and P. G. DeGennes, J. Phys. (Paris) 43, 1241 (1982)
10. Y. Talman and S. Prager, J. Chem. Phys., 76, 1535 (1982).
11. B. Widom, J. Chem. Phys., 81, 1030 (1984).
12. J. S. Huang and M. Kotlarchyk, Phys. Rev. Lett., 57, 2587 (1986).
13. L. Auvray, T. P. Cotton, R. Ober, and C. Taupin, J. Phys. Chem. 88, 4382 (1985).
14. M. Kotlarchyk, S. H. Chen, J. S. Huang and M. W. Kim, Phys. Rev. A29, 2054 (1984).
15. M. Kotlarchyk, S. H. Chen and J. S. Huang, J. Phys. Chem. 86, 3273 (1982).
16. M. Kotlarchyk, R. B. Stephens, and J. S. Huang, J. Chem. Phys. to be published (1987).
17. F. Mezei, Neutron Spin Echo, Lecture Notes in Physics, Vol. 128, Springer Verlag 1979.
18. Heinrich B. Stuhrmann, Acta Cryst. A26, 297 (1969).

19. For overdamped capillary waves, also see J. S. Huang and W. W. Webb, Phys. Rev. Lett. 23, 160 (1969).
20. A rough estimate of the characteristic time for the elastic mode is $\eta r^3/K_c$ 10^{-7}. For a surface tension driven mode the estimate is $r\eta/\sigma \sim 10^{-7}$.
21. S. Milner and S. A. Safran, to be published (1987).
22. J. S. Huang, S. A. Safran, M. W. Kim, G. S. Grest, M. Kotlarchyk, and N. Quirke, Phys. Rev. Lett., 53, 592 (1984)
23. M. B. Schneider, J. T. Jenkins and W. W. Webb, J. Phys. 45, 1457 (1984)
24. S. A. Safran, J. Chem. Phys. 78, 2073 (1983).
25. C. Huh, J. Coll. Interface Sci., 71, 408 (1979).

Images of Bicontinuous Microemulsions by Freeze Fracture Electron Microscopy

W. Jahn and R. Strey

Max Planck Institut für Biophysikalische Chemie,
Postfach 2841, D-3400 Göttingen, Fed. Rep. of Germany

1. INTRODUCTION

Direct images could be of help to clarify the microstructure of microemulsions. Due to the dynamic nature and the smallness of the objects this has never been achieved. In structural research on biological tissues freeze fracture electron microscopy is a well-established method /1/. Attempts to apply this method to microemulsions have not yet been successful /2/. Nevertheless, we have started a systematic study using this method paying particular attention to a number of important points. First of all, we studied systematically, how to prepare microemulsions with only three components, namely water, oil and a single amphiphile /3,4/ at the desired working temperature. Secondly, we selected an amphiphile that promised sufficiently large structures to be easily resolved by the method. The third, but not last, argument for the system were experimental results /5/ from other, complementary techniques. Among these SAXS, SANS, NMR-selfdiffusion and electrical conductivity proved to be most useful, drawing conclusions on the characteristic sizes and connectivity of the water and oil rich domains. In the following we give a brief description of the method and images of the fracture faces of rapidly frozen microemulsions.

2. EXPERIMENTAL

We studied the system H_2O-n-octane-$C_{12}E_5$. All solutions contained 7wt% of $C_{12}E_5$, fixing the size scale (the repeat distance /6/) to about 50 to 80 nm as determined by SAXS and SANS /5/. The water-to-oil ratio is defined by α=100(n-octane)/((H_2O)+(n-octane))wt%. Details of the freeze fracture technique will be described elsewhere /7/. Briefly, the microemulsion is thermostated at a temperature in the one-phase channel /3-5/. A small copper sandwich is assembled comprising a copper disc (3 mm diam.), a TEM grid and a U-shaped, sandbeamed copper foil. The sandwich is immersed into the microemulsion, then, it is, using a mechanical plunging device /7/, rapidly (<20 ms) transferred to liquid propane at 80 K and stored under liquid nitrogen. The achievable cooling rate /8/ is of the order of 10^4 K/s. After fracturing in vacuum at 173 K a Ta-W shadow is cast onto the fracture face for contrast and a carbon replica taken. The shadow material, originally chosen to give a fine grain, shows a decoration effect on the oil fracture face, permitting to distinguish between oil-and water-rich domains.

3. RESULTS

Figure 1 shows the replica of a bicontinous microemulsion composed of equal volumes water and n-octane (α=40, T=32.3°C). The fracture through the three-dimensional bicontinuous network is not planar, for which reason the carbon film may rupture (white portions) or crumple to appear

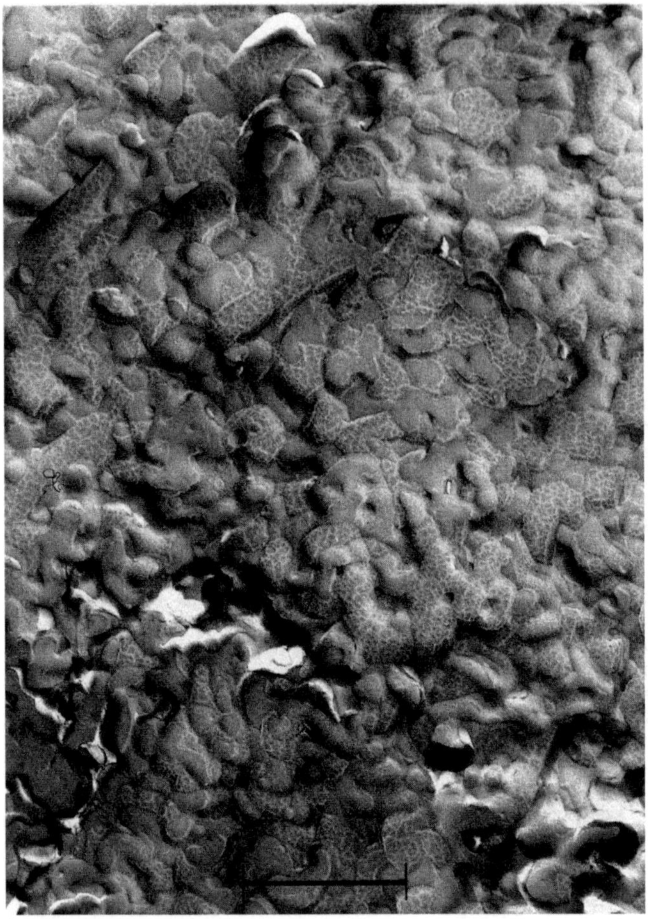

Fig. 1. Bicontinuous microemulsion containing equal volumes water and oil. The shadow material decorates the oil-rich parts of the fracture face. bar: 500 nm

black. The undamaged parts, however, disclose that both water and oil domains are mutually intertwined. Also both signs of curvature are visible. Moreover, the characteristic repeat distances observed appear to be close to the 80 nm measured by SAXS and SANS /5/. Figure 2 shows the result for an oil-rich microemulsion (α=70,T=35.5°C) known to be still bicontinuous by NMR-selfdiffusion and electrical conductivity. Correspondingly, the image shows larger oil domains in which the water-rich domains are imbedded like branched "veins". The fracture cuts through these veins and the cross-section appears as gray as pure water. The diameter of these often round cross-sections is again of expected order of magnitude. Apparently the fracture quite often occurs along the hydrocarbon side of the amphiphilic layer, as is known from biological specimens /1/. Although there is little doubt that one can indeed discriminate water-rich and oil-rich domains and even the amphiphilic layer, one cannot be sure that the freezing process has not changed the shape of the aggregates. Figure 3 shows two examples of specimens located in the two-phase regions above and below the one-phase channel (α=60). From the phase behavior one expects a water-rich microemulsion

Fig. 2. Oil-rich microemulsion, water domains are imbedded in oil-like veins. Fracture often occurs along the amphiphilic layer. bar: 500 nm

in equilibrium with excess oil at low temperatures, whereas at high temperatures one expects an oil-rich microemulsion with excess water. Indeed, one finds at T=29.0°C ,Fig. 3 bottom, large oil droplets that have been fractured. On the other hand, at T=37.5°C one finds, Fig. 3 top, spheres, presumably water, imbedded in an oil-rich matrix. It is interesting that in both cases the droplet interface seems to be structured.

4. CONCLUSIONS

We have shown that, by carefully choosing the system, one can obtain electron micrographs of replicas of rapidly frozen microemulsions. The structures seen show both water- and oil-rich domains mutually interwoven in an apparently bicontinuous network. The fact that the pictures agree with our knowledge of the system from other techniques, in particular with respect to characteristic sizes, indicates that freezing artefacts are of minor influence. The application of this method to less well-known systems is straightforward. The work is in progress.

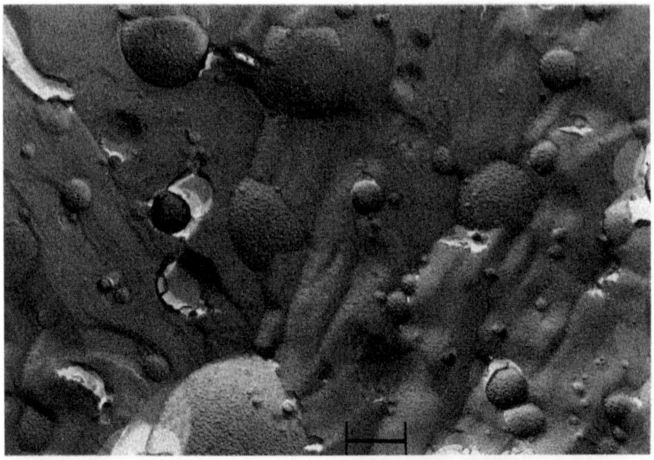

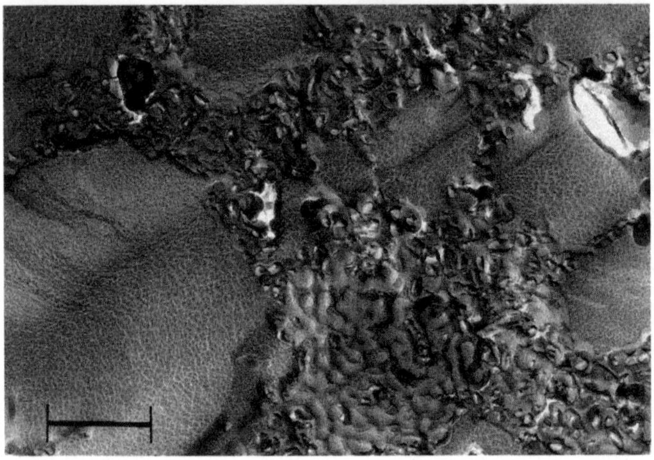

Fig. 3. Top, excess water emulsion droplets imbedded in oil-rich matrix; bottom, excess oil emulsion in the two-phase region. bar: 500 nm

ACKNOWLEDGEMENT: This work has been performed in the department of Prof. M. Kahlweit. We wish to thank him for his continuing support and interest in the work.

5. REFERENCES

1. J.H.M. Willison, A.J. Rowe: In Practical Methods in Electron Microscopy ed. by A.M. Glauert, Vol. 8, (North-Holland, Amsterdam 1980)
2. J. Biais, M. Mercier, P. Bothorel, B. Clin, P. Lalanne, B. Lemanceau: J. Microscopy 121, 169 (1981)
3. M. Kahlweit, R. Strey: Angew. Chem. 24, 654 (1985)
4. M. Kahlweit, R. Strey, P. Firman: J. Phys. Chem. 90, 671 (1986)
5. M. Kahlweit et.al.: J. Colloid Interface Sci. to appear (1987)
6. F. Lichterfeld, T. Schmeling, R. Strey: J. Phys. Chem. 90, 5762 (1986)
7. W. Jahn, R. Strey: in preparation
8. M.J. Costello, R. Fetter, J.M. Corless: In Science of Biological Specimen Preparation for Microscopy and Microanalysis ed. by J.R. Revel, SEM Inc., 105 (1984)

Molecular Self-Diffusion and Microemulsion Bicontinuity

B. Lindman

Physical Chemistry 1, Chemical Center, Lund University,
P.O. Box 124, S-22100 Lund, Sweden

1. INTRODUCTION

Microemulsions are thermodynamically stable solutions of oil, water (or other polar solvent), surfactant and often salt and/or cosurfactant. Most intriguing from many points of view are microemulsions of three-component systems in which small amounts of surfactant can mix equal volumes of water and oil. The high volume ratio solvents-to-surfactant immediately suggests these solutions to be microstructured and surfactant molecules are generally assumed to form monomolecular layers between oil and water domains. This point is directly verified in different types of experiments, but to provide a structural description is quite demanding and many experimental approaches are difficult to apply under conditions of highly dynamic situations with about equal volumes of the two solvents.

Bicontinuity of microemulsions has been proposed for quite a long time /1-4/, partly with reference to liquid crystalline phases. (We note that most l.c. phases - formed in the same or analogous systems under somewhat different conditions - are bicontinuous).

2. GENERAL ASPECTS

It emerged naturally that a simple way of establishing if a microemulsion is bicontinuous or not would be by considering the molecular self-diffusion coefficients of oil and water. In the first studies in collaboration with the Montpellier group we used a combination of the capillary tube method with radioactive labelling and NMR spin-echo with pulsed gradients to study some ionic and nonionic surfactant microemulsion systems /5,6/. From the observation that oil and water diffusion are simultaneously of the same order of magnitude as diffusion in the neat liquids it was concluded that the microemulsions are bicontinuous. The experimental techniques used in the early studies have severe disadvantages for microemulsions - very time-consuming experiments, the need of isotope-labelling - so the development of the FT version of the NMR spin-echo method /7/ was a major step forward for this type of study.

Noting that all these experiments monitor molecular displacements over macroscopic distances (typically microns and upwards), it is clear that complete confinement of the molecules of one component in closed domains (of any shape) results in very slow diffusion (with a radius of a spherical droplet of 50Å, the Stokes-Einstein equation gives a droplet D value of ca. $5 \cdot 10^{-11} m^2 s^{-1}$ or below). Molecules of components forming domains which extend over macroscopic distances diffuse very rapidly; except for obstruction /8/ and solvation effects they will diffuse as in the neat liquids. These have self-diffusion coefficients of the order of $10^{-9} m^2 s^{-1}$ (for example, $1.9 \cdot 10^{-9} m^2 s^{-1}$ for water $1.4 \cdot 10^{-9} m^2 s^{-1}$ for n-decane and $2.4 \cdot 10^{-9} m^2 s^{-1}$ for toluene all at $25^{\circ}C$) and are suitable reference system in discussions. The value given for water is for HDO in heavy water, generally used for experimental reasons. For normal water D is $2.2 \cdot 10^{-9} m^2 s^{-1}$. Surfactant diffusion is expected to be considerably more rapid in a bicontinuous structure than in droplet structures where the surfactant molecules are located at the surfaces, in the same way as surfactant diffusion along the bilayer in a lamellar l.c. phase is more rapid than in a cubic l.c. phase where the surfactant occurs in closed aggregates /9/. For systems without marked microstructure where surfactant molecules occur dominantly as single non-associated molecules, surfactant self-diffusion will be rapid; suitable reference states are here the D values at infinite dilution in the solvents.

Based on these simple arguments, we can evidently envisage very different self-diffusion behaviour depending on whether the solutions are microstructured or not, and in case of microstructure whether we have closed domains of one of the solvents or we have a bicontinuous structure. All these four limiting cases have indeed been experimentally documented, as have also intermediate situations and continuous transitions between the limiting structures (as a function of oil/water ratio, temperature, salinity etc.).

In the present account we will merely give a few examples of the limiting cases and give a list of studies in the field. For a fuller account of the principles and more complete references to the literature see refs. 10 and 11.

3. OIL-IN-WATER DROPLET STRUCTURE

For systems of nonionic surfactant, hydrocarbon and water at low temperatures D_{oil} may be well below $10^{-11} m^2 s^{-1}$ while D_{water} is above $10^{-9} m^2 s^{-1}$ (cf. figure 1) /12/. Another example is the system of Aerosol OT, brine and isooctane where at isooctane weight fractions of 0.1-0.2 and temperatures of 55-60°C, D/D_o is of the order of 1 for water but may be below 10^{-3} for oil /13/. For five-component microemulsions of ionic surfactant, cosurfactant, brine and oil, water diffusion is above $10^{-9} m^2 s^{-1}$ at low salinities while oil and surfactant diffusion have similar and orders of magnitude lower values (fig. 2) /14/.

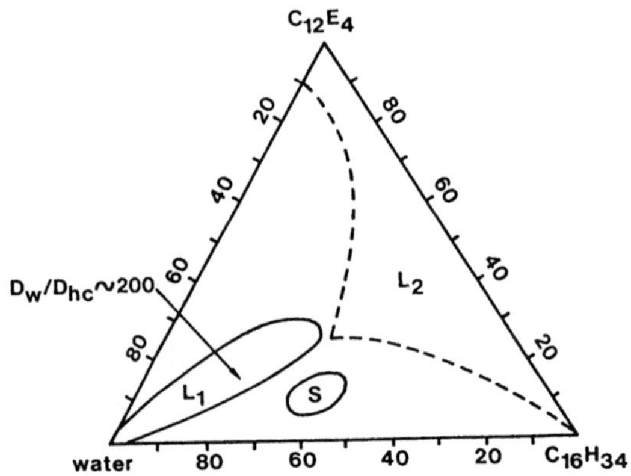

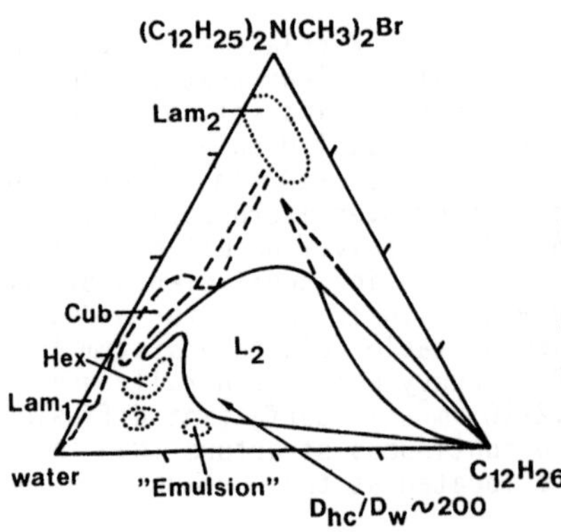

Fig. 1. The self-diffusion behaviour of three-component microemulsions may be very different for different surfactants. Thus at similar compositions $C_{12}H_{25}(OCH_2CH_2)_4OH$ (data from Ref. 12) gives O/W droplets and $(C_{12}H_{25})_2 N(CH_3)_2Br$ (data from Ref. 16) W/O droplets.

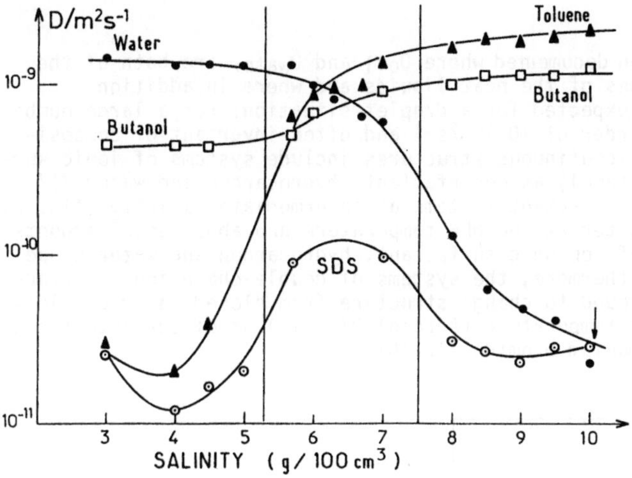

Fig. 2. In the system composed of sodium dodecylsulphate, butanol, toluene, water and sodium chloride the self-diffusion behaviour and hence microemulsion structure is strongly dependent on salinity. (From Ref. 14)

4. WATER-IN-OIL DROPLET STRUCTURE

Classical examples are those of double-chain ionic surfactant, hydrocarbon and water with didodecyldimethylammonium bromide (fig. 1) /15,16/ and Aerosol OT (fig. 3) /17/ as examples. In both cases D_{oil}/D_{water} may be of the order of 100. Other examples are provided by systems of nonionic oligo (ethylene oxide) surfactant, hydrocarbon and water at temperatures above the HLB temperature /18/ and by five-component ionic surfactant systems at very high salinities where now $D_{surfactant}$ is close to D_{water} (fig. 2) /14/.

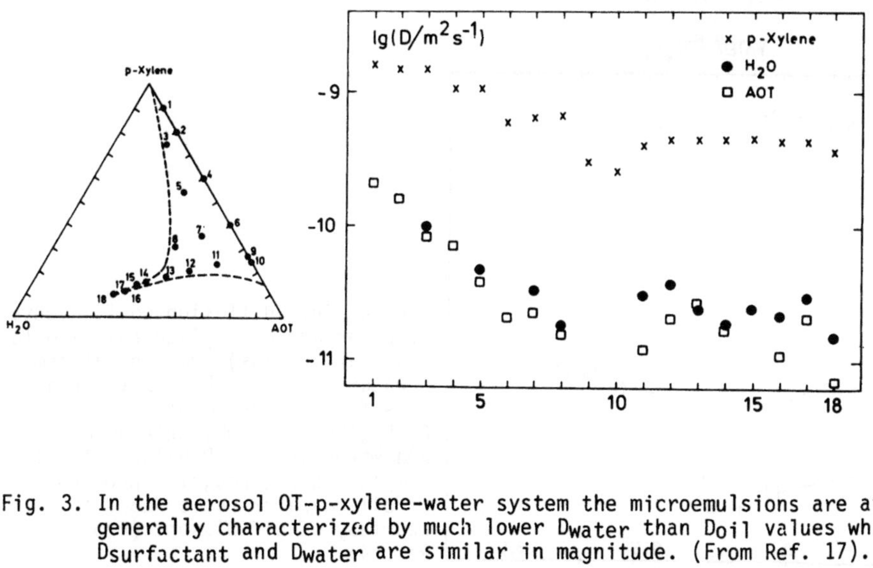

Fig. 3. In the aerosol OT-p-xylene-water system the microemulsions are at 25°C generally characterized by much lower D_{water} than D_{oil} values while $D_{surfactant}$ and D_{water} are similar in magnitude. (From Ref. 17).

5. BICONTINUOUS STRUCTURES

A large number of cases have been documented where D_{oil} and D_{water} are both of the order of magnitude of the D values of the neat liquids and where in addition $D_{surfactant}$ is much higher than expected for a droplet situation. For a large number of cases $D_{surfactant}$ is of the order of $10^{-10} m^2 s^{-1}$ and often invariant of composition in a wide range. Cases of bicontinuous structures include systems of ionic surfactant, short-chain alcohol (butanol) as cosurfactant, hydrocarbon and water /17, 19,20/; of five-component ionic surfactant systems at intermediate salinity (fig. 2) /14/; of Aerosol OT, brine and water at the HLB temperature and about equal amounts of water and hydrocarbon /13/; of nonionic surfactant, hydrocarbon and water close to the HLB temperature /18/. Furthermore, the systems of double-chain ionic surfactant mentioned above have been found to change structure from closed water droplets to bicontinuous with increase of temperature (Aerosol OT /21/) or of surfactant concentration (didodecyldimethylammonium bromide /15,16/).

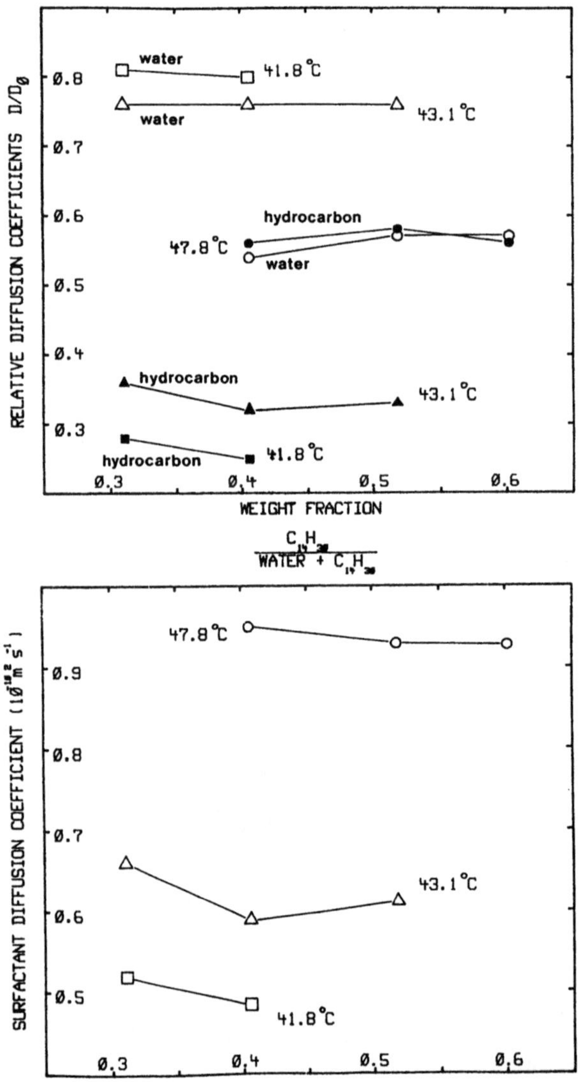

Fig. 4. Self-diffusion coefficients of water and oil (given relative to the neat liquids) are high at temperatures around the HLB temperature for the system of $C_{12}H_{25}(OCH_2CH_2)_5OH$, tetradecane and water and the D values of all components are roughly independent of oil-to-water ratio /18/. (By courtesy of Ulf Olsson. (From Ref. 4).

Especially striking are those cases where at equal fractions of water and oil, D/D_0 of both solvents are above 0.5 and where $D_{surfactant}$ is of the order of $10^{-10} m^2 s^{-1}$ (fig. 4). As there must always be some geometric obstruction effect /8/ these solvent diffusion values are not far from the maximum values conceivable.

While bicontinuous structures have thus been identified for a number of cases it remains to closer characterize the structure. Certainly, more than one bicontinuous structure is possible depending on system. It appears reasonable to assume that, for example, the bicontinuous structures of the double-chain cationic surfactant and of the nonionic surfactant systems mentioned are different in nature /4/. In the latter case we have argued that a highly defect layered structure is consistent with the self-diffusion characteristics /18/ and such a structure is also consistent with a current multi-field NMR relaxation study in Lund. (The ^2H and ^{13}C NMR relaxation work on microemulsions is due to O. SÖDERMAN, U. OLSSON, M. JONSTRÖMER and others). X-ray scattering and electron microscopy results have been taken to argue against this structure /22,23/ but the work of the Göttingen Group has been performed for a slightly different system and under different temperature conditions. Furthermore, it is not obvious that the X-ray data for the range of oil volume fraction of the solvents between 0.35 and 0.65 are inconsistent with a defect layered structure. It is hoped that joint efforts using different techniques under identical conditions can help to resolve these problems.

6. SYSTEMS WITHOUT MICROSTRUCTURE

For simple (molecule-disperse) solutions the self-diffusion coefficients of the components show a weak dependence of the composition in the entire range. For most mixtures of simple liquids the variations are within a factor of ca. 2 from the values of the neat liquids. For systems where one of the components is water (or another strongly hydrogen-bonded compound) the variations are slightly larger (Fig.5). As an example, in ternary mixtures of p-xylene, 1-pentanol and formamide, the relative D values, D/D_0, are over wide concentration ranges 0.7-1 for formamide, 0.3-1 for p-xylene and 1-1.6 for pentanol /24/.

For systems of a very slightly amphiphilic compound, water and hydrocarbon, the D values of all components are high and thus the solutions ("microemulsion" may not be an appropriate term in this case) are structureless (A. CEGLIE et al., unpublished study). A nice example of this behaviour is given by the system of C_4E_1, decane and water (H.T. DAVIS, this meeting). On increasing the amphiphilicity there is a continuous changeover from structureless solutions to microstructured ones.

Some non-aqueous surfactant systems have recently been investigated. For the system of glycerol, hexanol and SDS, with and without hydrocarbon, the self-diffusion behaviour showed the microemulsions to be much less structured than for the corres-

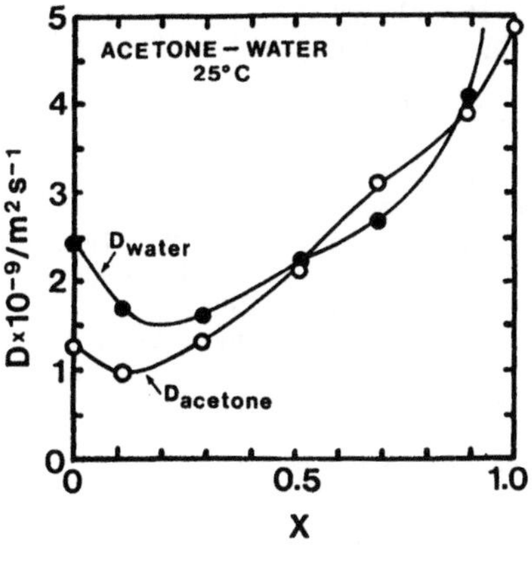

Fig. 5. Self-diffusion coefficients in the acetone-water system at 25°C as a function of the mole fraction of acetone. (Modified after Ref. 26).

ponding aqueous systems and to be closer to a structureless solution /25/. For microemulsions of formamide, hydrocarbon, alcohol (butanol to octanol) and SDS, the self-diffusion coefficients at 25°C are all high and the behaviour is again very different from the aqueous case /24/. For example, over wide composition ranges of the system SDS-pentanol-formamide-p-xylene the D values relative to those of the neat liquids are 0.1-1 for p-xylene, 0.4-1 for formamide and 0.7-1.6 for pentanol and relative to the surfactant-less ternary mixture these factors are considerably closer to unity. $D_{surfactant}$ is within a factor of 2 from the value for infinite dilution in formamide. While it seems that replacing water with another high dielectric solvent often leads to a closely structureless situation, situations have also been encountered which show a marked microstructure. (Surfactant type and temperature appear to play a great role).

CONCLUSION

It appears that multi-component molecular self-diffusion coefficients, now easily accessible using the FT NMR technique, allow a particularly convenient characterization of solution microstructure. The approach easily distinguishes between microstructured and structureless microemulsions and also between different types of microstructure.

ACKNOWLEDGEMENTS

The fruitful collaboration over the years with the Montpellier group, with P. Stilbs, with D. Langevin's group (P. Guéring), with K. Shinoda and with others (see reference list), in particular in Lund, is gratefully acknowledged.

LITERATURE

1. L.E. Scriven: In Micellization, Solubilization and Microemulsions, ed. by K.L. Mittal, Vol. 2 (Plenum, New York, 1977) p. 877
2. K. Shinoda: Progr. Colloid Polym. Sci. 68, 1 (1983)
3. S.J. Chen, D.F. Evans, B.W. Ninham, D.J. Mitchell, F.D. Blum, S. Pickup: J. Phys. Chem. 90, 842 (1986)
4. K. Shinoda, B. Lindman: Langmuir, in press
5. B. Lindman, N. Kamenka, T-M. Kathopoulis, B. Brun, P-G. Nilsson: J. Phys. Chem. 84, 2485 (1980)
6. B. Lindman, N. Kamenka, B. Brun, P-G. Nilsson: In Microemulsions, ed. by I.D. Robb, (Plenum, New York, 1982) p. 115
7. P. Stilbs: Progress in NMR Spectroscopy 19, 1 (1987)
8. B. Jönsson, H. Wennerström, P-G. Nilsson, P. Linse: Colloid Polym. Sci. 264, 77 (1986)
9. G. Lindblom, H. Wennerström: Biophys. Chem. 6, 167 (1976)
10. B. Lindman, P. Stilbs: In Microemulsions, ed. by S. Friberg and P. Bothorel, (CRC Press)
11. B. Lindman, P. Stilbs: In Microemulsions, ed. by H. Rosano, M. Clausse (Marcel Dekker, New York)
12. P-G. Nilsson, B. Lindman: J. Phys. Chem. 86, 271 (1982)
13. J.O. Carnali, A. Ceglie, B. Lindman, K. Shinoda, Langmuir 2, 417 (1986)
14. D. Dhatenay, P. Guéring, W. Urbach, A.M. Cazabat, D. Langevin, J. Meunier, L. Léger, B. Lindman, 5th Int. Symp. on Surfactants in Solution, Bordeaux, France, July, 1984; P. Guéring, B. Lindman: Langmuir 1, 464 (1985)
15. F.D. Blum, S. Pickup, B.W. Ninham, S.J. Chen, D.F. Evans: J. Phys. Chem. 89, 711 (1985)
16. K. Fontell, A. Ceglie, B. Lindman, B.W. Ninham: Acta Chem. Scand. A40, 247 (1986)
17. P. Stilbs, B. Lindman: Progr. Colloid & Polymer Sci. 69, 39 (1983)
18. U. Olsson, K. Shinoda, B. Lindman: J. Phys. Chem. 90, 4083 (1986)

19. B. Lindman, T. Ahlnäs, O. Söderman, H. Walderhaug, K. Rapacki, P. Stilbs: Faraday Discuss Chem. Soc. 76, 317 (1983)
20. P. Stilbs, K. Rapacki, B. Lindman: J. Colloid Interface Sci. 95, 583 (1983)
21. S. Geiger, H.F. Eicke: J. Colloid Interface Sci. 110, 181 (1986)
22. F. Lichtenfeld, T. Schmeling, R. Strey: J. Phys. Chem. 90, 5762 (1986)
23. R. Strey, this workshop
24. K.P. Das, A. Ceglie, B. Lindman: J. Phys. Chem., in press
25. K.P. Das, A. Ceglie, B. Lindman, S. Friberg: J. Colloid Interface Sci., in press.
26. D.W. McCall, D.S. Douglass, J. Phys. Chem. 71, 987 (1967)

Scattering Probes of Microemulsion Structure

E.W. Kaler

Dept. of Chemical Engineering, BF-10, University of Washington, Seattle, WA 98195, USA

Small-angle neutron (SANS) and quasielastic light scattering (QLS) techniques have been used to study a family of microemulsions formed in a salinity scan with sodium 4-(1'-heptylnonyl benzene sulfonate) (SHBS), iso-butanol, NaCl brines and dodecane. The growth of microemulsion "droplets" and the evolution of bicontinuous structures have been examined, and a model of the dynamics of microemulsions that contain comparable amounts of both water and oil is reviewed. SANS measurements reveal the decrease in surfactant head group area from ca.200 Å^2 to 125 Å^2 as the salinity increases and the composition varies from oil-in-water to water-in-oil.

1. INTRODUCTION

Microemulsions are isotropic dispersions of oil and water stabilized by surfactant molecules located at the interfaces between oil-rich and water-rich domains. Microemulsions dilute in either oil or water usually exist with the minority component dispersed in the continuous one as spherical or quasi-spherical droplets [1]. Much less well understood is the nature of the continuous structural evolution of microemulsions (as temperature or salinity changes) from those dilute in water to those dilute in oil. Usually, but not always, these "mid-range" microemulsions coexist with both an excess water-rich phase and an excess oil-rich phase, and they therefore have been loosely termed "middle-phase" microemulsions.

There is experimental evidence and theoretical support of the idea that middle-phase microemulsions contain bicontinuous structures. The most convincing bits of evidence for the existence of a bicontinuous structure are electrical conductivity and self-diffusion coefficient measurements that suggest that the percolation thresholds in several microemulsion systems occur at approximately 20% of the minority component [e.g. 2]. The first theoretical model of bicontinuous microemulsion structure was proposed by TALMON AND PRAGER [3]. In their model, the structure is generated using a Voronoi tessellation to form many microscopic cells which are filled at random with oil and water, while surfactant molecules are confined at the internal interfaces between oil and water. This model has been modified and extended by others, but the key idea that bicontinuous microemulsions contain a random dispersion of surfactant-rich sheets separating oil-rich and water-rich domains remains unaltered.

Current efforts to improve the predictions of phenomenological models of bicontinuous microemulsions are focussed on refining the representation of the surfactant-rich interfacial sheets. In WIDOM'S [4] model, the film is assumed to behave as a two-dimensional gas of surfactant molecules, in which the area per surfactant head-group Σ can vary over a substantial range. In contrast, the model of SAFRAN and coworkers [5] begins with the assumption that the surfactant film is an incompressible layer with zero surface tension. Σ is thus fixed at a value Σ_o. These authors identify the bending energy of the film as the crucial element in the free energy expression and find that renormalization of the "bending constant" by thermal undulations of the film is responsible for the stability of the middle phase.

Despite the relatively successful use of these models to predict microemulsion phase behavior, there is as yet no general quantitative model of the structure of bicontinuous microemulsions. In particular, there is no model of the continuous evolution of the structure from a dispersion of oil-rich droplets in water (or vice-versa) as the oil/water ratio approaches one, nor is there a model of the changes in the fluctuations of the microemulsion as its composition varies. The only previous attempt to model the structure of the middle phase was a calculation of the density-density correlation functions of the Voronoi model [6]. The calculation is based on the assumption that the oil and water cells in the model are filled at random. This assumption is not consistent with the ordering of the droplets seen in the oil-in-water and water-in-oil microemulsions from which bicontinuous ones evolve. Remnants of that structure manifest themselves in the scattering curves measured for bicontinuous structures, so the scattering predicted from the Voronoi model does not match the experimental curves well. The time-dependent properties of a dynamic Voronoi tessellation were also calculated, and although the results of that work are in rough agreement with the experimental results then available, the model is flawed by an incorrect fluid mechanical treatment of the fluctuating tessellation.

The lack of a quantitative model useful for the interpretation of scattering data from the middle phase is in contrast to the situation for data measured from microemulsions that are dilute in oil or water (i.e., those with compositions below the percolation thresholds). In this case it is possible to apply the tools developed for the treatment of scattering from colloidal suspensions to determine with good accuracy droplet sizes and polydispersity, as well as estimate the mean-field potential of interaction between the droplets [7-9].

The structure of the microemulsion is controlled by the configuration of the surfactant-rich sheet. To learn about the average and the time-dependent state of the surfactant film, we have probed the structure of microemulsions with both small-angle neutron (SANS) and quasielastic light scattering (QLS) [7,8]. Analysis of these measurements yields direct information on the properties of the film, such as the surfactant head group area Σ, and in terms of an ad-hoc model [10] suggests the existence of capillary waves damped by the surface tension of the film. The results are reviewed and extended here.

2. EXPERIMENTAL

The microemulsions studied here are formed with purified sodium 4-(1'-heptylnonyl)benzenesulfonate (SHBS), and reagent grade iso-butyl alcohol and dodecane with NaCl brines containing different ratios of H_2O and D_2O. As the salinity increases, the microemulsion phase behavior follows the usual Winsor I-III-II pattern. Sample preparation is described elsewhere [7].

The equipment used for light scattering is described elsewhere [7]. The sample temperature was maintained at 25±0.02°C and the sample cells used for measurement of dilute microemulsion samples were 2 cm diameter ultra-precision glass cylindrical cells (NMR tubes). For measurement of visually turbid microemulsions, the sample was placed in a cylindrical cell (either 0.4, 0.2, or 0.1 cm inside diameter) that was then immersed in a large vat containing toluene, which served as an index-matching fluid. The magnitude of scattering vector is $q = (4\pi n/\lambda_o)\sin(\theta/2)$ and was varied by changing the scattering angle θ from 30° to 150°. Here n is the index of refraction of the sample and λ_o is the wavelength of light in vacuum, which was 4880 Å for all measurements.

The normalized intensity autocorrelation function was fitted to a two-cumulant expansion using a non-linear least-squares fitting routine. The first cumulant is equal to $2q^2D_{app}(q)$. In a solution containing particles, D_{app} is the z-averaged diffusion coefficient, and its q-dependence is a result of inter-particle and hydrodynamic interactions. In a solution that only contains random microstructured phase domains rather than distinct particles, D_{app} is a measure of the relaxation of fluctuations of the phase domains.

Small-angle neutron scattering measurements were made with the 30-m camera at the National Center for Small-Angle Scattering Research at Oak Ridge National Laboratory. The neutron wavelength was 4.75 Å and q ranged from 0.005 to 0.1 $Å^{-1}$. The data were corrected and placed on an absolute scale using standard procedures.

3. THEORY

3.1 SMALL-ANGLE NEUTRON SCATTERING

Small-angle neutron scattering from microemulsions containing swollen micelles of oil in water or inverted swollen micelles of water in oil can be usefully modeled in terms of a particle form factor P(q) and a solution structure factor S(q) [9]. The form factor depends on the size and shape of the droplets, while the structure factor, which is essentially the Fourier transform of the radial distribution function, depends on the relative locations of the individual droplets. In the absence of significant polydispersity, the scattered intensity I(q) is proportional to the product P(q)S(q). This treatment is particularly useful because of the limiting forms of P(q) and S(q). In general, in the limit of q=0, P(0) = 1.0 and S(0) is proportional to the isothermal compressibility of the solution. In the limit $q \to \infty$, S(q) approaches 1.0 and the scattering then depends entirely on the form of the particle.

It is clear that the formalism developed for understanding the behavior of concentrated colloidal dispersions is not an appropriate tool to use to understand the scattering of neutrons from bicontinuous microemulsions. To model a bicontinuous structure quantitatively, one can introduce the density correlation function $\gamma(r) = \langle\rho(o)\,\rho(r)\rangle$, which is essentially the Fourier transform of the scattered intensity. Further analysis depends on the choice of a model for $\gamma(r)$. In the case of bicontinuous microemulsions, a useful model may be that of a distorted lamellar structure. VONK [11] has treated a similar problem, and the use of such a model for microemulsions will be reported elsewhere. Note that this correlation function treatment is not particularly useful when applied to the scattering from samples that contain droplets because $\gamma(r)$ contains contributions from both inter and intra-particle scattering. The relative influence of the two is difficult to separate.

Finally, there are fundamental structural properties obtainable from the scattering curve without assumptions. The one of particular interest here is the specific surface S_p. The scattering intensity has the asymptotic form

$$I(q) = S_p q^{-4} + B$$

where B is a background term. The neutron scattering contrast in a microemulsion can be adjusted such that the scattering arises only from the internal oil-water interfaces. In that case, the specific surface measured is equal to Σ times the surfactant concentration.

3.2. QUASIELASTIC LIGHT SCATTERING

The analysis of quasielastic light scattering measurements of concentrated colloidal suspensions has developed to the point where the influence of both direct (thermodynamic) and hydrodynamic interactions on the apparent diffusion coefficient found from the first cumulant may be treated riqorously to first order in the volume fraction $\emptyset$ of the particles or droplets [12]. The theory allows the intrinsic diffusion coefficient D_o to be extracted from measurements of the q-dependence of D_{app}^{-1} in terms of functions of the potential of interaction and the flow fields set up by the diffusing particles. D_o is related to the hydrodynamic radius of the droplet R_h via the Stokes-Einstein relation, so that D_o^{-1} is proportional to R_h.

Once again, it is clear that the formalism developed for understanding the behavior of dispersions is not appropriate for describing the sources of the fluctuations of dielectric constant that scatter light in bicontinuous microemulsions. We have developed a model of the fluctuations of the surfactant film in a bicontinuous microemulsion that incorporates an assumption of two independent modes of film motion[10]. Only a concise description of the calculation of the first cumulant result is given here.

The first mode results from the assumption that the surfactant film has a large extent and has a low - but finite - interfacial tension. As a result, there are capillary waves on the interface, and these waves scatter light. We have solved the Navier-Stokes equations that govern the motion of the interface and find that if the interfacial elasticity is small, the first cumulant D_w of the spectrum of scattered light is given by

$$D_w = \sigma/4\eta q$$

where σ is the interfacial tension of the internal oil-water interface and η is the average of the viscosity of oil and water. Essentially the same result was found by HUANG [13].

The q-dependence of this capillary wave mode is an important feature in that D_w^{-1} will increase with q. This behavior is much distinct from that of a bending mode, for which D_{app}^{-1} goes as q^{-1}

(HUANG, this volume), or for a diffusive mode for which D_{app} is independent of q.

A second mode is assumed because as the droplet volume fraction increases, the droplets that are present in dilute microemulsions begin to link and fuse into an extended structure. The droplets sometimes remain as isolated droplets, and other times they merge with neighboring droplets to form a flexible coil-like "macromolecule". This extended structure fluctuates , and the diffusion of such an aggregate is approximately described by a Rouse polymer-like model [14]. The chains are assumed to grow in proportion to the volume fraction of the minority phase. The first cumulant expression for this mode involves a estimate of the dimension of the coil and the relaxation times of its normal modes [10]. The dimension of the coil is estimated by assuming a linear growth law above the percolation threshold, while the bending mode frequency depends on the stiffness of the surfactant sheet. A useful feature of this mode is that the diffusive relaxation depends on the viscosity of the oil when water is the minority phase and on the viscosity of water when oil is the minority phase.

The basic premise of the model of microemulsion dynamics proposed here is that fluctuations of the surfactant sheets separating oil and water are controlled by the motion of the bulk oil and water domains, with the decoration of surface waves. An important feature of the model is the relative contribution of the two modes to the total scattering. The full theory shows that the magnitude of the surface wave term is inversely proportional to the interfacial tension on the film. As a result, the surface scattering becomes increasingly important as σ decreases, which occurs as the composition approaches that of equal uptake of oil and water. When the volume fractions of oil and water are equal, light is scattered nearly entirely from surface wave fluctuations; thus there is a smooth transition from water-rich microemulsions to oil-rich ones.

4. RESULTS AND DISCUSSION

4.1 SMALL-ANGLE NEUTRON SCATTERING

The small-angle I(q) curves measured from a variety of microemulsion systems show the characteristic progression of the building of a scattering peak and the movement of the peak to smaller angles as the ratio of oil to water approaches one (Fig.1). The scattering curves for samples containing less that approximately 20 % water in oil can be well represented by models of polydisperse hard-spheres in which the interactions are accounted for in the Percus-Yevick approximation. The scattering from samples with less than about 20% oil dispersed in water can be modeled well as a monodisperse population of ellipsoidal particles interacting with the usual DLVO potential. The droplet sizes obtained from the SANS data agree well with those measured with QLS [8].

The specific surface measured for the SHBS microemulsions is shown in Fig.2. The salinity of the brine used to make the microemulsion varies from 0.8 gm NaCl/100 ml water when the volume fraction of water is 0.16 to 0.33 gm NaCl/100 ml water when the volume fraction of water is 0.80. It is to be expected that the surface area per molecule of an ionic surfactant would vary strongly with salinity when cosurfactant is present in the interfacial film.

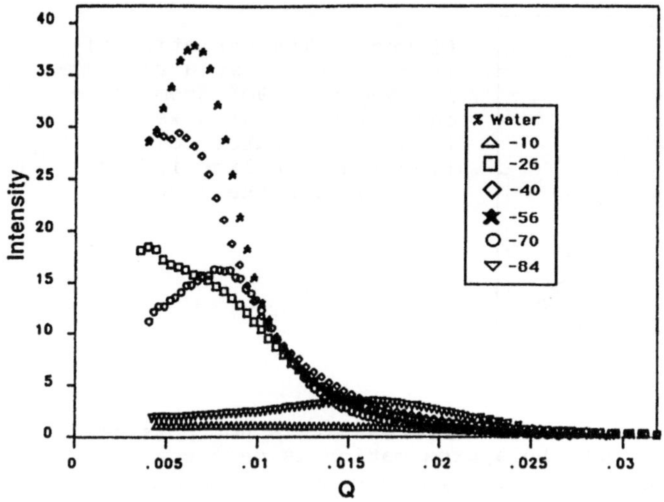

Figure 1. The evolution of scattered intensity (SANS) as the volume fraction of water increases

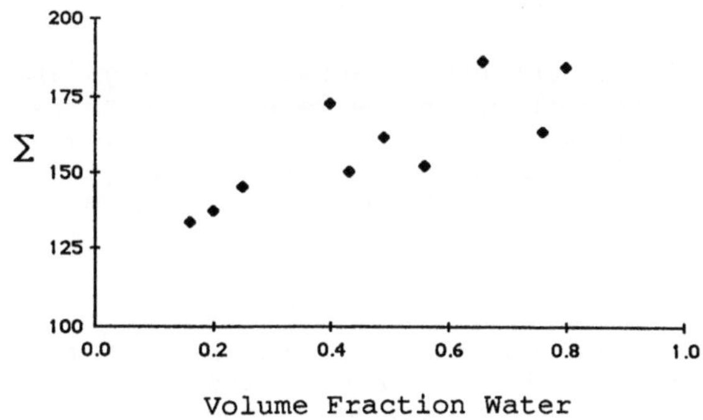

Figure 2. The area per surfactant head group as a function of water volume fraction. Σ has units of $Å^2$

4.2 QUASIELASTIC LIGHT SCATTERING

Measurements of the q-dependent diffusion coefficient of both oil-in-water and water-in-oil SHBS microemulsions have been made and the results indicate that the droplets grow in proportion to their volume fraction until the percolation thresholds are reached [7]. When the apparent diffusion coefficients, measured at $\theta=90^\circ$, of microemulsions over the entire range of composition are examined (Fig.3), this growth accounts for the nearly linear increase of D_{app}^{-1} with water volume fraction $Ø_w$ up to 0.2 and for its decrease with $Ø_w$ for $Ø_w$ above 0.8. These values of $Ø_w$ correspond approximately to the percolation thresholds for oil and water, respectively. In the region where bicontinuous structures probably exist, the behavior of the apparent diffusion coefficient is more complicated.

Quantitative comparison of the experiment with the theoretical results outlined above requires accurate values of the interfacial

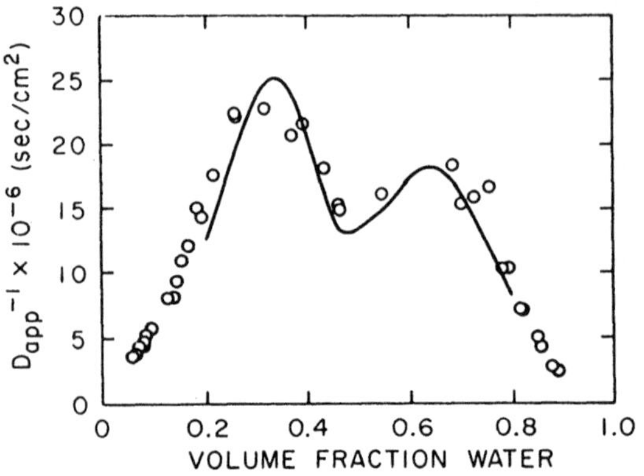

Figure 3. The variation of measured D_{app}^{-1} at a scattering angle of 90° from dodecane microemulsion as a function of water volume fraction. The line is calculated with the model

tension of the internal oil-water interfaces. It is not possible to measure that tension, so here the approximation is made that $\sigma_{ow}^{internal} = \sigma_{me,o} + \sigma_{me,w}$. $\sigma_{me,o}$ and $\sigma_{me,w}$ are, respectively, the interfacial tensions measured between the macroscopic microemulsion phase and the excess oil-rich and water-rich phases. σ_{ow} varies from 10^{-1} to 10^{-3} dynes/cm. Data for this system are taken from the literature [15].

The expression for D_{app} is fit to the experimental data (Fig. 3) with essentially one parameter that determines the rate of growth of the coils [10]. The ratio of the maxima in the plot of D_{app}^{-1} are related to the ratio of water viscosity to oil viscosity and change dramatically when another hydrocarbon is substituted for dodecane [10].

5. SUMMARY

The results of SANS and QLS have been combined to characterize the droplet structures in dilute microemulsions with good accuracy. The structure of bicontinuous microemulsions is less well understood, but the results of QLS measurements can be fitted well with the model described here. The major features of the model are the consideration of scattering from surface-wave fluctuations as well as from the motion of oil-rich and water-rich domains. The change in surfactant head group area with salinity and composition can be measured, and that information should be used in the continuing refinement of theory.

6. ACKNOWLEDGEMENTS

This work was supported by the National Science Foundation (PYIA-8351179). The National Center for Small-Angle Scattering is funded by the NSF(DMR-77-244-59) through Interagency Agreement No. 40-636-77 with the US Department of Energy under contract number DE-AL05-840R21400 with Martin Marietta Energy Systems Inc. I acknowledge useful discussions with J.M. Schurr, S. Mildner, and J.S. Huang and the important contributions of J.F. Billman and N.J. Chang.

7. REFERENCES

1. A.M. Cazabat, D. Langevin, and A.Pouchelon, J. Colloid Interface Sci. 73 1 (1980)
2. M.T. Clarkson, D. Beaglehole, and P.T. Callaghan, Phys. Rev. Lett.54 1722 (1985)

3. Y. Talmon and S. Prager, J. Chem. Phys. 69, 2984 (1978)
4. B. Widom, J. Chem. Phys. 81, 1030 (1984)
5. S.A. Safran, D. Roux, M.E. Cates, and D. Andelman, to be published in Surfactants in Solutions, ed. by K. Mittal (Plenum Press, 1987)
6. E.W. Kaler and S. Prager, J. Colloid Interface Sci. 86 359 (1982)
7. N.J. Chang and E.W. Kaler, Langmuir, 2 184(1986)
8. N.J. Chang, J.F. Billman, R.A. Licklider, and E.W. Kaler, in Statistical Thermodynamics of Micellar and Microemulsion Systems, ed. by S.-H. Chen (Springer-Verlag, 1987)
9. S.-H. Chen, Ann. Rev. Phys. Chem. 37 351 (1986)
10. N.J. Chang and E.W. Kaler, J. Chem. Phys. (submitted).
11. C.G. Vonk, J. Appl. Cryst. 6 81 (1973)
12. H.M. Fijnaut, J. Chem. Phys. 74 6857 (1981)
13. J.S. Huang, Ph.D. Thesis, Cornell U. (1969)
14. R. Pecora, J. Chem. Phys. 49, 1032 (1968)
15. D. Guest and D. Langevin, J. Coll. Interface Sci. 112, 208 (1986

Small-Angle Scattering on Microemulsion Systems. Evidence for an Ordered Bicontinuous Structure

A. de Geyer and J. Tabony*

Departement de Physico-Chimie, CEN-Saclay,
F-91191 Gif-sur-Yvette Cedex, France

1. INTRODUCTION

Microemulsions are oil and water dispersions stabilised by the presence of a surfactant and a co-surfactant /1/. Often, the co-surfactant is a short chain alcohol such as butanol or pentanol. Microemulsions frequently exist over a wide range of oil and water volume fractions. At low volume fractions of either oil or water, the structure of the optically isotropic phase has been shown to consist of a dispersion of droplets of oil or water surrounded by an interfacial layer of surfactant and co-surfactant /2,3/. The size of these droplets may vary from about 50 /Å/ to several hundred /Å/. In many cases it is possible, by changing the relative amounts of oil and water, to effect a continuous change from an oil-in-water microemulsion to a water-in-oil microemulsion. Moreover this transition occurs in a gradual manner without any abrupt change in the properties or the behavior of the microemulsion. This raises the question of the structure of the microemulsion when there are equal amounts of oil and water and the mechanism of the structural inversion.

Numerous investigations including self-diffusion and conductivity measurements /4-6/ have led to the proposal that the structure in the inversion domain may be one of intertwined oil and water bicontinuous zones /7/. Although the structure has often been stated to be random /8-10/, a more microscopic approach is required to obtain further insight into these systems. Neutron or X-ray small-angle scattering is a technique which can probe the structures over distances usually found in microemulsion systems. Some authors have concluded from X-ray or neutron investigations in favour of a random structure for the bicontinuous microemulsions found in the WINSOR III region /11-14/. In contrast to this, small-angle neutron scattering results obtained on similar systems suggest that the structure of these bicontinuous microemulsions is considerably more ordered /15-17/. We give in this paper more quantitative arguments in favour of a locally ordered structure.

Ordered bicontinuous structures may be generated by infinite periodic minimal surfaces /7/. Such surfaces have been studied in the past by SCHWARZ /18/ and NEOVIUS /19/ and more recently by SCHOEN /20/. Among the seventeen infinite periodic minimal surfaces described by SCHOEN, there are eight of cubic symmetry. Among these cubic minimal surfaces, SCHWARZ's primitive and diamond minimal surfaces constitute suitable candidates for optically isotropic bicontinuous microemulsions existing at the middle of the path connecting oil-in-water and water-in-oil droplets structures, as is schematically suggested in Fig. 1.

(*) Present address: Centre de Recherches Rhône-Poulenc, 14 rue des Gardinoux, 93308 Aubervilliers, France.

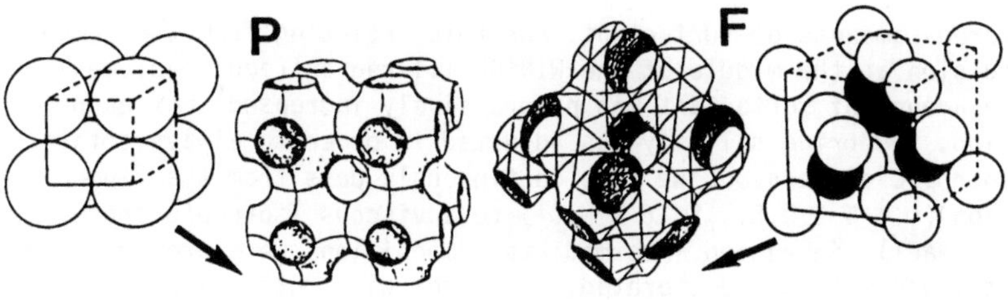

Fig.1 SCHWARZ's primitive (P) and diamond (F) infinite periodic minimal surfaces may be obtained by fusion of droplets on a primitive or a diamond lattice

One of the interests of neutrons over X-rays comes from the fact that using the difference in the coherent scattering length density between protonated and deuterated components, different parts of a structure can be selectively seen. However for concentrated systems containing equal volume fractions of oil and water, the oil and water zones cannot be separately observed. The structural information obtained when either oil or water is deuterated results from an average over both the oil and water zones. In contrast to this, when both oil and water are deuterated and contrast matched, it is the interfacial layer which dominates the scattering. We have simulated the scattering profiles for the SCHWARZ's primitive and diamond minimal surfaces /17/. Spectra obtained at different contrasts for the WINSOR III microemulsion are in agreement with the latter. SCHWARZ's diamond surface separates continuous oil and water sub-volumes into two identical and interpenetrating diamond lattices. Microemulsions are normally fluid mixtures and this is an obvious limitation to the presence of highly extended ordered structures in such systems. However, as a local order is seen to exist in this bicontinuous microemulsion, it may be possible in some circumstances to considerably increase the long-range translational order. This is what is observed when the surfactant concentration in the microemulsion is increased and diffraction patterns which index onto cubic crystal symmetries are observed.

2. MATERIALS and EXPERIMENTAL METHODS

The system investigated was composed of toluene (47.2 % by weight), brine (46.8 %), sodium dodecyl sulfate (2.0 %) and butanol (4.0 %). The brine was made with NaCl. This system, which has been studied by other methods /21/, is convenient because by changing the water salinity it is possible to pass in a gradual manner from an oil-in-water microemulsion to a water-in-oil microemulsion. At low water salinities ($<$ 5.4 % NaCl), the mixture separates into two coexisting phases. The lower phase is an oil-in-water microemulsion, whilst the upper phase is almost pure toluene. At high salinities ($>$7.4 % NaCl) the opposite occurs; an upper water-in-oil microemulsion coexisting with brine. At intermediate salinities, a three-phase equilibrium exists where a microemulsion of unknown structure is sandwiched between toluene and brine. Using the nomenclature introduced by WINSOR /22/, the microemulsions formed with increasing salinity are of the types I, III and II. In the WINSOR I microemulsion, the volume fraction of oil is low, whereas in the WINSOR II microemulsion it is the volume fraction of water which is low. In contrast to this, the WINSOR III microemulsion contains approximately equal volumes of oil and water.

In order to selectively highlight the scattering from different structural parts of the microemulsions, mixtures were prepared for different isotopic compositions using either D_2O or toluene-d as deuterated components. Microemulsions were also prepared containing different amounts of surfactant. For this, starting from the composition of the microemulsion at the middle of the WINSOR III region (equal volumes of oil and water), the amount of surfactant was progressively increased. For each surfactant concentration, the brine salinity was adjusted in order to always obtain a single optically isotropic dispersion. We thus continuously pass from the fluid WINSOR III microemulsion (∿ 4 % S.D.S., S∿ 6 % NaCl) to a viscous isotropic phase (∿ 20 % S.D.S., S ∿ 2 % NaCl). Seventeen intermediate compositions were prepared for two different contrasts; one with oil deuterated, the other with both oil and water deuterated.

Neutron small-angle scattering measurements were made on the D11, D17 and D16 instruments at the Institut Laue-Langevin (Grenoble, France) and on the PACE spectrometer at the Orphee reactor at Saclay. The range of momentum transfer, Q, obtained using the different instruments was from 2×10^{-3} /Å^{-1}/ to 0.6 /Å^{-1}/ ($Q=(4\pi/\lambda)\sin\theta$ where 2θ is the scattering angle and λ the wavelength, $\Delta\lambda/\lambda \sim 10$ %). Samples were contained in 1 or 0.5 /mm/ pathlength optical quartz cells maintained at 20 /°C/.

The information contained in a scattering process can be given in reciprocal space by the function I(Q) or in the direct space by the function $\gamma(r)$. For isotropic scatterers these functions are connected by the integral transformation

$$\gamma(r) = 1/(2\pi^2)\int_0^\infty Q^2\ I(Q)\ \sin(Qr)/(Qr)\ dQ.$$

The information in the r-space is often represented as $r\gamma(r)$ or $r^2\gamma(r)$. The latter ($p(r)=4\pi r^2\gamma(r)$) corresponds to the distance distribution function /23/. One of the advantages of this transformation is that distances and geometries that contribute to the scattering pattern are represented in a visually convenient manner /24/. However, one of the constraints of the method is that it is necessary to measure the scattering profil over a large Q-range. In practice a window in Q-space from Q_1 to Q_2 corresponds to a finite range of distances of π/Q_1 to π/Q_2. Experimental spectra were measured from 2×10^{-3} /Å^{-1}/ to 0.6 /Å^{-1}/. These experimental conditions correspond to a spatial resolution of about 5 /Å/ and permit to probe structures up to distances of about 1000 /Å/.

3. RESULTS

Figure 2 shows the spectra obtained when the oil is the only deuterated component in the different domains WINSOR II, WINSOR III and WINSOR I. At high salinities

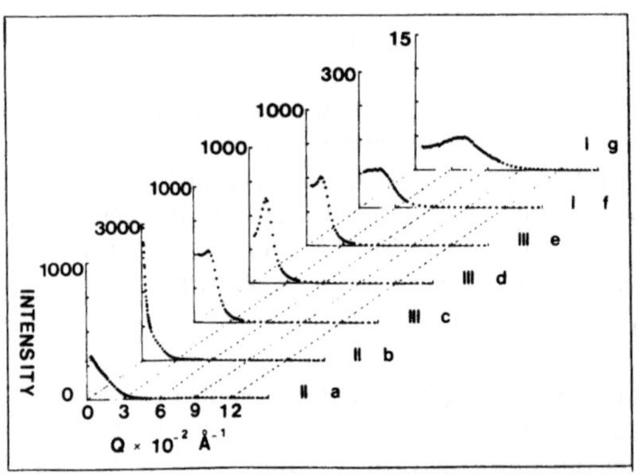

Fig.2 Spectra obtained from the WINSOR microemulsions (marked I, II, III) formed at different salinities (% NaCl). (a) 8.8 %, (b) 7.5 %, (c) 6.8 %, (d) 6.4 %, (e) 5.8 %, (f) 4.1 %, (g) 1.5 %. The intensities are in units of /cm^{-1}/. The only deuterated component was toluene

(WINSOR II) the scattering curves resemble those for polydisperse spheres. Close to the WINSOR II → WINSOR III transition, intense scattering occurs at small angles. This is in keeping with the light scattering results of CAZABAT et al. /21/ who attributed this effect to critical fluctuations. Assuming this to be the case, we obtained correlation lengths in agreement with these authors. Once the WINSOR III phase is entered, a pronounced maximum appears whose position is independent of the salinity. Close to the WINSOR III → WINSOR I phase transition, small-angle scattering once again occurs, but it is not as intense as at the WINSOR II → WINSOR III transition. For the WINSOR I microemulsions, a maximum also appears, but its shape and position vary with salinity, becoming less pronounced and moving to larger values of Q as the salinity and the volume fraction of oil diminish.

For the opposite case, where water is the only deuterated component, the spectra have the same behavior as described above. However, when both oil and water are deuterated, the spectra are qualitatively different. In the WINSOR II and I phases, distinct secondary maxima occur at high Q. In contrast to this, for the WINSOR III microemulsion these maxima are not seen and the intense low-angle maximum seen with either D_2O or toluene-d disappears.

The distance distribution functions for the microemulsions in the different WINSOR domains are compared in Fig. 4. These functions for the WINSOR I and WINSOR II microemulsions are in agreement with that for polydisperse spheres. We first consider the case of the WINSOR II microemulsion. The function p(r) shows a maximum for a distance corresponding to about half the largest distance occurring in the distribution. This is in agreement with spherical objects. The largest distance in the p(r) corresponds to a diameter and the position of the maximum corresponds, in a first approximation, to the radius of droplets. In the case of the WINSOR I microemulsion, a negative part occurs in the p(r). This arises from spatial correlations between the droplets. The diameter of the objects is given by the minimum in this part of the curve. It should be noted that for the WINSOR I and WINSOR II microemulsions, the distance distribution functions when either oil or water are deuterated are translated from about

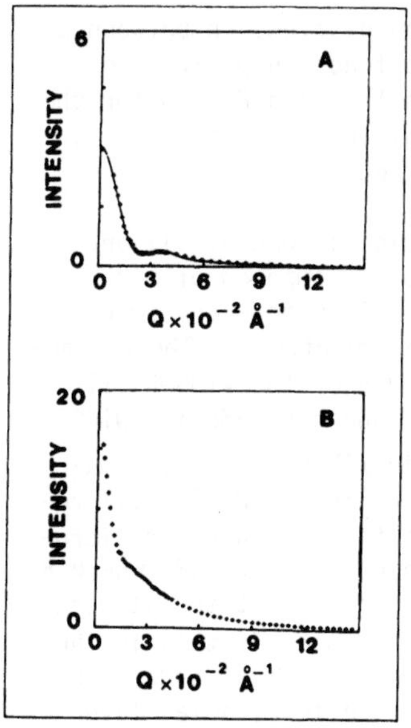

Fig.3 Spectra from the interfacial layer of the WINSOR II (A) (S=9.0% NaCl) and WINSOR III (B) (S=6.4% NaCl) microemulsions. The intensities are in units of /cm^{-1}/. The scattering length densities in units of $10^{10}/cm^{-2}$/ are the following. (A) ρ_w=2.2, ρ_i=-0.4, ρ_o=1.9. (B) ρ_w=4.5, ρ_i=-0.4, ρ_o=4.5. (w : aqueous phase, i : interfacial layer, o : hydrocarbon phase). The full line in A is a simulation from a collection of closed shells with an outer radius of 95 /Å/, an interfacial thickness of 10 /Å/ and a gaussian size polydispersity with σ/R =28%

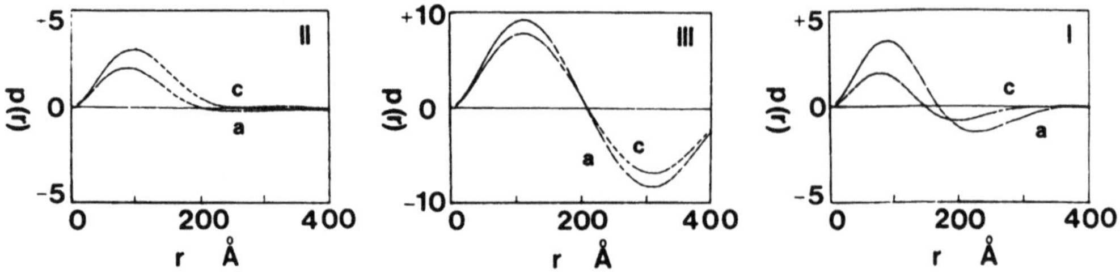

Fig.4 Distance distribution functions for the WINSOR II (S=8.8% NaCl), WINSOR III (S=6.4% NaCl) and WINSOR I (S=4.1% NaCl) microemulsions. (a) only the water is deuterated. (c) only the oil is deuterated

10 /Å/. This arises from the thickness of the interfacial layer and is in agreement with a dispersion of water-in-oil and oil-in-water droplets in the WINSOR II and WINSOR I regions respectively. The function p(r) for the WINSOR III microemulsion has a profile qualitatively similar to that for the WINSOR I or WINSOR II microemulsions. Thus, oil and water parts in the WINSOR III microemulsion can be described, as in the WINSOR I and WINSOR II regions, on the basis of spheroidal units.

However, evidence that the structure of the WINSOR III microemulsion is not one of disconnected droplets comes mainly from two observations. For a dispersion of one phase in another as droplets, the radius is related to the volume fraction of the dispersed phase by the expression $R=3\Phi/(\Sigma C_S)$. Σ represents the surfactant area per polar head ($\Sigma \sim 60$ /Å^2/ for the sodium dodecyl sulfate) and C_S is the number concentration of surfactant molecules. For the WINSOR III microemulsion, this expression gives a radius of curvature of 240 /Å/, which corresponds to about twice the value for the oil and water domains as determined previously. This suggests that a transformation of the interfacial layer involving some area annihilation has occurred. This may, for instance, result from fusion at points of contact between droplets. That a bicontinuous structure is formed is also supported by the fact that the scattering curve for the interfacial layer of the WINSOR III microemulsion differs from that for closed shells, as is shown Fig. 3. Another manner to distinguish between a layer forming a spherical shell and other structures is the function p(r)/r. For spherical shells, this function shows a horizontal part as is shown Fig. 5 for the amphiphilic layer of a WINSOR I and WINSOR II microemulsion. In contrast to this, for the WINSOR III microemulsion a linear decrease is observed.

So, although the structure of the WINSOR III microemulsion is one of oil and water bicontinuous zones, oil and water volumes are seen to be composed of micro-domains of well-defined size and geometry. Moreover, correlations between these micro-domains cannot be satisfactorily described by excluded-volume effects. The characteristic size of the dispersed phase is seen to be unchanged over the WINSOR III region and its volume fraction varies from about 25% at the phase boundaries WINSOR I → WINSOR III or WINSOR II → WINSOR III to 50% for the middle WINSOR III microemulsion. Then the fact that the correlation peak shows no change in the WINSOR III region, combined with the fact that the correlation length in the concentrated microemulsion is about twice the characteristic size of the dispersion ($2\pi/Q \sim 4R$, where R represents the radius of curvature of the micro-domains) suggests that the structure is not random and that a locally ordered structure might be formed. If this is the case, the single layer of amphiphiles is expected to show a topological order. This is what is observed. The distance distribution function for the interfacial layer

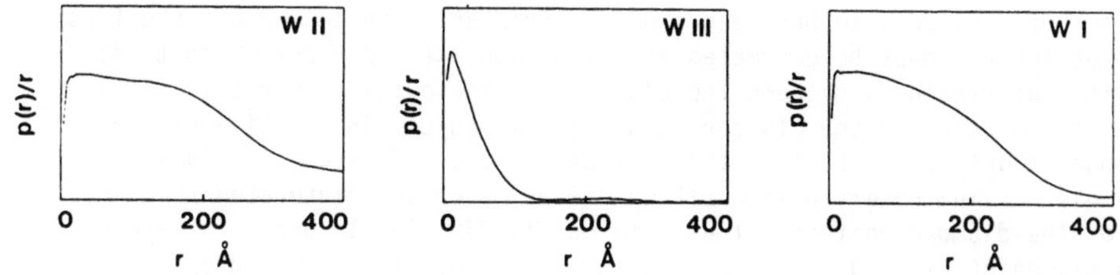

Fig.5 Functions p(r)/r for the amphiphilic layer of the different WINSOR microemulsions (marked I, II, III). (a) near W II → W III transition (S=7.5% NaCl), (b) middle WINSOR III microemulsion (S=6.4% NaCl), (c) near W I → W III transition (S=5.3% NaCl)

of the WINSOR III microemulsion is represented Fig. 8. This function shows three characteristic distances with a maximum for about 40 /Å/ and two other maxima at larger distances. We will see that this profile is in agreement with SCHWARZ's diamond minimal surface.

4. INTERPRETATION and DISCUSSION

SCHWARZ's diamond minimal surface separates infinitely connected volumes forming two interpenetrating diamond lattices. These volumes can be described, in a first approximation, by a packing with spherical droplets. Figure 6 shows a simulation of the scattering by droplets on a diamond lattice. The intensity is simulated by the expression :

$$I(Q) = F^2(Q) \sum_{ij} \sin(Qr_{ij})/(Qr_{ij}) \exp-(D_0 r_{ij} Q^2) + I_c(Q). \qquad (1)$$

$F^2(Q)$ is the scattering function of the unit droplet ($F(Q)=3(\sin(QR)-QR\cos(QR))/(QR)^3$) and r_{ij} represents the distances between the droplets. The factor $\exp-(D_0 r_{ij} Q^2)$ takes into account the long-range disorder in the liquid structure ($\sigma_{ij}{}^2=2D_0 r_{ij}$ in the Debye-Waller factor $\exp-(Q^2\sigma_{ij}{}^2/2)$). The summation in the expression (1) is of course performed on a limited assembly of droplets. The size of this domain is physically related to the extension length of the structure. In order to take into account

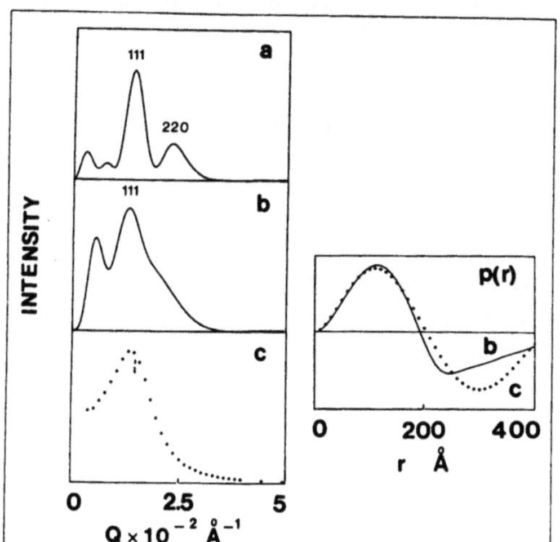

Fig.6 Full lines are simulations from droplets on a diamond lattice (a and b). In points (c), the experimental curve for the WINSOR III microemulsion (oil is deuterated). The simulations are obtained for a cluster of 29 spheres having radii 120 /Å/ with a lattice distance 740 /Å/. See the text for the expression of I(Q). (a) $D_0=0$, (b) $D_0=10$ /Å/. Also shown are the functions p(r) derived from spectra (b) and (c)

that a cluster of limited size has been isolated from an infinite medium, the first term in expression (1) must be corrected by a function $I_c(Q)$ /25,26/ which takes account of the interferences between the cluster and the outside. For this, it is assumed that the outside of the cluster is a uniform medium. The lattice parameter of the diamond structure can be adjusted so that the 111 reflexion coincides with the small-angle maximum observed when either oil or water are deuterated (Fig. 6). The size A of the diamond unit cell thus obtained is 740 /Å/. In order to be consistent with a bicontinuous structure, there must exist contacts between droplets. This implies that the diameter of the micro-domains must correspond to about 320 /Å/ ($D = A\sqrt{3}/4$). This is what is found to be the case. As is shown Fig. 6, the distance distribution function for the WINSOR III microemulsion contains a minimum for 320 /Å/. This clearly shows that a diamond-like lattice for both the oil and water volumes is compatible with the bicontinuity of the microemulsion. In contrast to this, for a simple cubic arrangement, the diameter of the droplets needs to be 480 /Å/. This value is of course inconsistent with the radius of curvature 120 /Å/ of the micro-domains.

Droplets on a diamond lattice, once fusion at points of contact between droplets is allowed for, lead to SCHWARZ's diamond minimal surface. Infinite periodic minimal surfaces are obtained by translation of a unit element. For the diamond surface, this element is shown Fig. 7. The unit element contained in a cube of size ξ is composed of six identical elements similar to a hyperbolic paraboloid of side $\xi/\sqrt{2}$. We have simulated the scattering from a hyperbolic paraboloid and have shown that it behaves in the same manner as is observed for the amphiphilic layer of the WINSOR III microemulsion /17/: the function $p(r)/r$ for a hyperbolic paraboloid shows a linear decrease. To go further, we have simulated the scattering for a more extended part of the surface. The function $p(r)$ derived from simulated intensities is shown in Fig. 8. Maxima in the experimental $p(r)$ are correctly described by the simulations.

These results, obtained at different contrasts, for the WINSOR III microemulsion are clearly in favour of an ordered bicontinuous structure as delimited by the SCHWARZ diamond minimal surface. In this case, as the solution is fluid, there is no extensive long-range order. However, it is possible to induce more long-range order in the dispersion. As already mentioned, it is possible to pass continuously from the fluid WINSOR III microemulsion to a viscous isotropic phase by increasing the surfactant concentration. Since the dispersion is always optically isotropic and there is no phase separation, it is reasonable to assume that the local structure remains unchanged and that the increase in the viscosity is related to the extension of the three-dimensional order in the bicontinuous microemulsion. To verify these points we consider spectra obtained when varying the surfactant concentration.

The SCHWARZ surface of symmetry Pn3 separates two interpenetrating diamond lattices. When oil is deuterated it is a diamond structure which is seen. The size A of

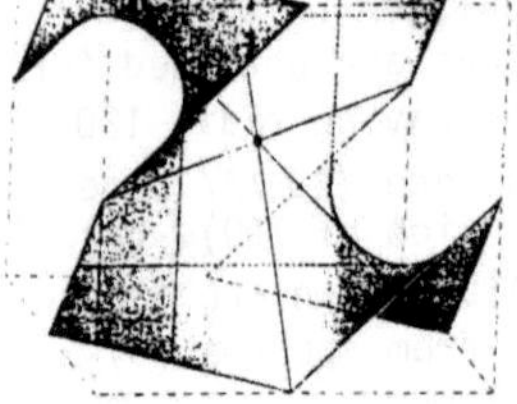

Fig.7 SCHWARZ's original drawing of a basis element of the diamond infinite periodic minimal surface

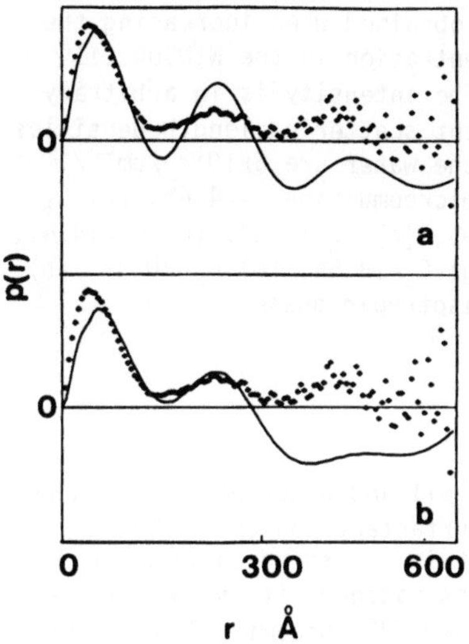

Fig.8 Full lines are p(r) simulations for SCHWARZ's diamond minimal surface. For this, I(Q) has first been calculated (see the expression 1 in the text) and then Fourier transformed. Simulations were obtained for a cluster of four of the unit elements represented Fig. 7 with, as discretising elements, spheres of radii 10 /Å/. (a) D_0=6 /Å/, (b) D_0=8 /Å/. Points are the p(r) derived from the scattering curve (Fig. 3-B) of the interfacial layer of the WINSOR III microemulsion

the diamond unit cell is 2ξ where ξ represents the size of the unit element of the surface. ξ is related to the microemulsion composition by the expression $\xi=\alpha/(\Sigma C_S)$. α has no dimension and its value is characteristic of the surface ($\alpha=S/V^{2/3}$). If the local structure is unchanged as the surfactant concentration increases, A^{-1} vs C_S gives a straight line which goes through the origin. This is what is observed. Figure 9 shows the spectra obtained for samples containing deuterated oil when increasing the surfactant concentration. The low-angle peak becomes sharper and moves to higher values of Q as C_S increases. Also represented on this figure is the curve A^{-1} vs C_S. Points give a straight line which passes through the origin. The slope of this line is proportional to Σ/α. For the SCHWARZ diamond surface α=2.42 /20/. This leads to Σ=58 /Å^2/, in agreement with values known for the sodium dodecyl sulfate (S.D.S.).

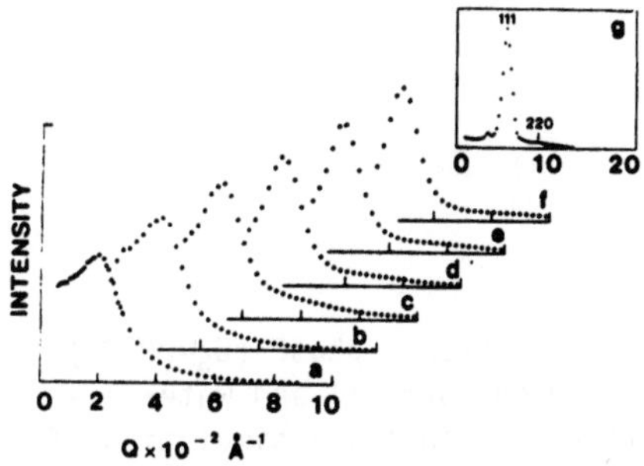

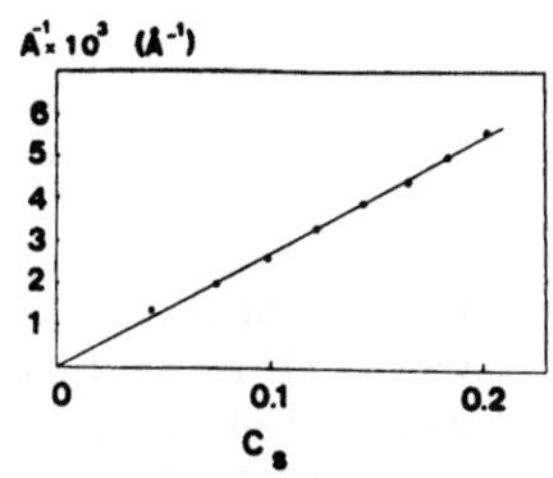

Fig.9 Small-angle part of the curves obtained when increasing the surfactant concentration in the WINSOR III microemulsion. The oil is deuterated. (a) C_S=7.5%, (b) C_S= 9.9%, (c) C_S=12.2%, (d) C_S=14.4%, (e) C_S=16.5%, (f) C_S=18.5%, (g) C_S=20.3%. (% surfactant are given in volume). (g) is the viscous isotropic phase. From the position of the low-angle maximum, the size A of the diamond unit cell can be determined. On the right part of this figure, the curve A^{-1} vs C_S

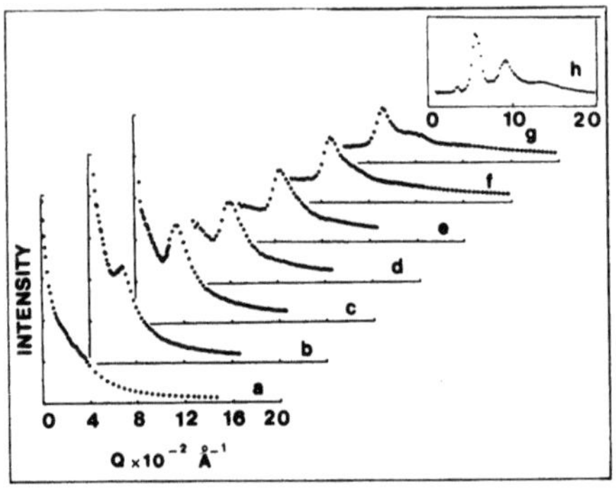

Fig.10 Spectra obtained when increasing the surfactant concentration in the WINSOR III microemulsion. The intensity is in arbitrary units. The coherent scattering length densities of the oil and the water are $5x10^{10}$ /cm^{-2}/. (a) WINSOR III microemulsion C_S=4.4%, (b) C_S=7.5%, (c) C_S=9.9%, (d) C_S=12.2%, (e) C_S=14.4%, (f) C_S=16.5%, (g) C_S=18.5%, (h) C_S=20.3%. (h) is the viscous isotropic phase

Figure 10 shows spectra obtained from samples where oil and water have same coherent scattering length densities with increasing the surfactant concentration. As already noted, when only oil or water are deuterated it is a diamond lattice which is seen. Now, when both oil and water parts are contrast-matched, it is the interfacial layer which is seen. Thus the symmetry is no longer Fd3m becoming Pn3 and the dimension of the unit cell is half that of the diamond lattice. Consequently, in the case of the diamond surface, there is a change in the symmetry and in the size of the unit cell with isotopic substitution. The experimental spectra are consistent with these remarks (Fig. 11). When oil is deuterated only two diffraction peaks are clearly seen. These peaks can be indexed on the reflexions 111 and 220. Spectrum obtained when both oil and water are deuterated differs from this one in that other intense maxima occur at higher values of Q. Also shown fig. 11 is a simulation of the diffraction spectrum from the SCHWARZ diamond surface. The value of ξ introduced in the simulation is 94 /Å/. This value is derived from the position of the 111 reflexion when the oil is deuterated ($2\xi=2\pi\sqrt{3}/Q_{111}$). We see that the 110 reflexion for the simulation occurs at the same angle as the maximum which appears when both oil and water are deuterated. More, this reflexion coincides very well with the 220 reflexion

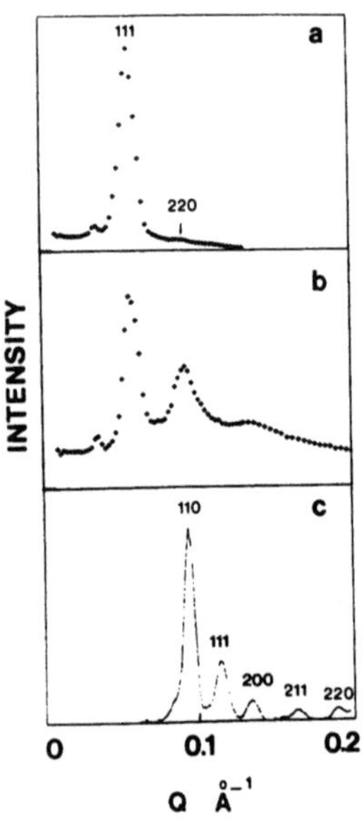

Fig.11 Spectra for the viscous isotropic phase (C_S=20.3%) when oil is deuterated (a) and when both oil and water are deuterated: $\rho_O=\rho_W=5x10^{10}$ /cm^{-2}/ (b). In (c) a simulation of a diffraction spectrum for the SCHWARZ's diamond minimal surface. This simulation is obtained from expression (1) in the text for a cluster of 7^3 of the elements represented in Fig. 7. Discretising elements are spheres having radii 10 /Å/. The size of the Pn3 unit cell introduced in the simulation is 94 /Å/ (half the value of the unit cell of the diamond structure obtained from spectrum (a))

observed when either oil or water is deuterated. These observations are in agreement with the model of minimal surface proposed. However it may be argued that for the viscous isotropic phase the 111 reflexion related to the diamond structure is not completely extinguished when oil and water are deuterated. This can presumably be attributed to a contrast effect: the aqueous and hydrocarbon parts may be, in that case, slightly mismatched. It should be noticed that, in contrast to this, for the WINSOR III microemulsion where aqueous and hydrocarbon phases are exactly contrast matched, the intense low-angle maximum which appears when either oil or water are deuterated disappears for the interfacial layer, as predicted by this model of bicontinuous structure.

5. CONCLUSION

Spectra obtained by small-angle neutron scattering on microemulsion systems can be satisfactory interpreted in terms of an ordered bicontinuous structure. Amphiphilic molecules form a single layer which can be described by SCHWARZ's minimal periodic surface of diamond symmetry. In the case of the fluid WINSOR III microemulsion the three-dimensional order extends over only limited distances. However, with increasing surfactant concentration a viscous isotropic phase is formed which has a considerable amount of long-range three-dimensional order.

ACKNOWLEDGEMENT

We would like to thank the Institut Laue-Langevin and the Laboratoire Léon Brillouin for providing the neutron beam facilities, also P. Jeener for drawing the view of the diamond surface included in Fig. 1.

REFERENCES

1. L.M. Prince: Microemulsions, theory and practice (Academic Press, New York 1977)
2. D.J. Cebula, L. Harding, R.H. Ottewill and P.N. Pusey: Colloid Polymer Sci. 258, 973 (1980)
3. M. Dvolaitzky, M. Lagues, J.P. Le Pesant, R. Ober, C. Sauterey, C. Taupin: J. Phys. Chem. 84, 1532 (1980)
4. F. Larche, J. Rouviere, P. Delord, B. Brun, J.L.Dussossoy: J. Physique Lettres 41, L-437 (1980)
5. B. Lindman, N. Kamenka, T.M. Kathopoulis, B. Brun, P.G. Nilsson: J. Phys. Chem. 84, 2485 (1980)
6. M. Clausse, P. Peyrelasse, J. Heil, C. Boned, B. Lagourette: Nature 293, 636 (1981)
7. L.E. Scriven: Nature 263, 123 (1976)
8. Y. Talmon, S. Prager: J. Chem. Phys. 69, 2984 (1978)
9. J. Jouffroy, P. Levinson, P.G. de Gennes: J. de Physique 43, 1241 (1982)
10. B. Widom: J. Chem. Phys. 81, 1030 (1984)
11. E.W. Kaler, K.E. Bennett, H.T. Davis, L.E. Scriven: J. Chem. Phys. 79, 5673 (1983)
12. E.W. Kaler, H.T. Davis, L.E. Scriven: J. Chem. Phys. 79, 5685 (1983)
13. L. Auvray, J.P. Cotton, R. Ober, C. Taupin: J. Physique 45, 913 (1984)
14. L. Auvray, J.P. Cotton, R. Ober, C. Taupin: J. Phys. Chem. 88, 4586 (1984)
15. A. de Geyer, J. Tabony: Chem. Phys. Lett. 113, 83 (1985)

16. A. de Geyer, J. Tabony: Chem. Phys. Lett. 124, 357 (1986)
17. A. de Geyer: thesis, Univ. Paris VI (1987)
18. M.A. Schwarz: Gesammelte Mathematische Abbandung Vol.1 (Springer, Berlin 1890)
19. E.R. Neovius: Bestimmung zweier speziellen periodischen minimalflachen (J.C. Frenkel, Helsingfors 1883)
20. A.H. Schoen: Nasa Technical Notes (1970) NASA TN D-5541
21. A.M. Cazabat, D. Langevin, J. Meunier, A. Pouchelon: J. Physique Lettres 43, L-89 (1982)
22. P.A. Winsor: Trans. Faraday Soc. 44, 376 (1948)
23. O. Kratky, G. Porod: Acta Phys. Austriaca 2, 133 (1948)
24. O. Glatter: J. Appl. Cryst. 12, 166 (1979)
25. B.E. Warren: The Physical Review 45, 657 (1934)
26. K. Nishikawa, Y. Murata: Bulletin of the Chemical Society of Japan 52, 293 (1979)

Relation Between Elasticity of Surfactant Layers and Characteristic Sizes in Microemulsions

O. Abillon and D. Langevin

Laboratoire de Spectroscopie Hertzienne de l'Ecole Normale Supérieure, 24 Rue Lhomond, F-75231 Paris Cedex 05, France

We discuss here the correlations between measured dispersion sizes, surfactant film curvature elasticity and interfacial tensions in model multiphase microemulsion systems. Sizes L are in agreement with simple random space filling models, the area per surfactant molecule in the film remaining constant. Bicontinuous structures are found when the spontaneous curvature of the film is small : the dispersion size is then also close to the persistence length of the film. Interfacial tensions correlate only approximately with kT/L^2.

I. INTRODUCTION

Microemulsions are dispersions of oil and water stabilized with surfactant molecules. Like emulsions, the dispersion is frequently composed of droplets either of oil or water, surrounded by a monolayer of surfactant molecules, in a continuous medium which is respectively water or oil. Because of the presence of surfactant molecules that spontaneously aggregate and form monolayers at the oil-water interfaces, ordered phases are also frequently encountered in these mixtures, especially at high surfactant concentration. We will be more concerned here with mixtures containing small amounts of surfactant.

The kind of dispersion obtained, i.e. oil in water (O/W) or water in oil (W/O), depends on the spontaneous curvature C_0 of the surfactant layer. Again, like for emulsions, o/w dispersions are obtained with surfactant molecules which are more soluble in water ($C_0 > 0$) and w/o dispersions with surfactant molecules which are more soluble in oil ($C_0 < 0$). The droplet size depends mainly on composition : because a small scale dispersion can only be obtained if the oil-water interfacial tension is ultralow or zero, the surface pressure of the surfactant film is high and the area per surfactant molecule Σ is fixed (or varies little) /1,2/. If ϕ_d is the volume fraction of the dispersed phase, c_S the number of surfactant molecules per unit volume, the droplet radius is simply :

$$R = 3 \frac{\phi_d}{c_S \Sigma} . \tag{1}$$

When the surfactant concentration is decreased, R increases until it reaches about the spontaneous radius of curvature $R_{max} = R_0 = C_0^{-1}$. Above this point, the dispersed phase that cannot be solubilized into larger droplets is rejected and forms an excess phase /2/. The corresponding phase equilibria are called Winsor I (o/w microemulsion in equilibrium with excess oil) and Winsor II (w/o microemulsion in equilibrium with excess water). Other mechanisms and other kinds of phase equilibria also exist (two coexisting microemulsions, microemulsion in equilibrium with a dilute lamellar phase, ...) but they won't be described here.

Although less frequent, a more interesting situation arises when the spontaneous curvature C_0 of the surfactant layer is almost zero. One would then expect to observe a lamellar structure in the dispersion. In fact many systems of this kind are fluid, optically isotropic, and there is much strong experimental evidence which shows that the structure is bicontinuous for both oil and water. Scriven proposed that the struc-

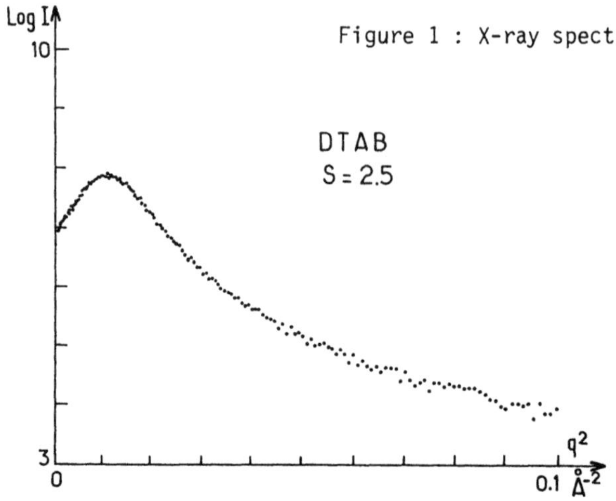

Figure 1 : X-ray spectrum of a DTAB microemulsion.

ture could be analogous to the bicontinuous cubic phases of lyotropic liquid crystals /3/. But the X-ray spectra of the microemulsions exhibit very broad peaks (figure 1) indicating that the structure is much more disordered. DE GENNES then proposed that the reason why the disorder is so important was the great flexibility of the surfactant films. He introduced the persistence length of the film ξ_K :

$$\xi_K = a \exp 2\pi K/kT , \qquad (2)$$

where a is a molecular length, K the bending elastic constant of the film, k the Boltzmann constant, T the temperature. If K is small enough ($K \sim kT$) ξ_K is microscopic and the lamellar structure is broken.

Using the fact that K depends on the length scale, SAFRAN et al. /4/ were recently able to show that stability of the microemulsions can be explained without introducing interfacial energies like DE GENNES and WIDOM /5/. Indeed, dispersion entropy favours mixing at a small scale, but bending energy (K decreases with scale) favours large scales. This model leads naturally to $\xi_{max} \sim \xi_K$. Like for droplets, the dispersion scale can be calculated if the area per surfactant molecule Σ is given. Then for different types of random bicontinuous structural models : Voronoi polyedra of TALMON and PRAGER, cubic cells of DE GENNES, one gets :

$$\xi \sim \frac{6\ \phi_o\ \phi_w}{c_S\ \Sigma} , \qquad (3)$$

where ϕ_o and ϕ_w are respectively the oil and water volume fractions (the surfactant volume fraction being assumed to be negligible). When c_S decreases, ξ increases until according to ref. 4 it reaches a value of the order of ξ_K. Then the system cannot accommodate more oil and water in the microemulsion and phase separates into a microemulsion and both excess oil and water. The corresponding phase equilibrium is called Winsor III.

When c_S increases, the model also explains the formation of lamellar phases. Indeed ξ decreases, then K and ξ_K both increase.

We have undertaken a study of this kind of systems, in order to test these ideas. We have studied sequences of Winsor equilibria Winsor I → Winsor III → Winsor II with ionic surfactants. The spontaneous curvature has been varied by adding salt to a mixture of constant composition in oil, water and surfactant. The electrostatic repulsion between surfactant ions is progressively screened by the salt and a continuous variation of C_0 is obtained.

For the microemulsions phases being in equilibrium with excess oil and/or water phases, the interfacial tensions γ between the different phases could also be measured. Like the free energy of the microemulsion phase, γ contains contributions of dispersion entropy and of curvature energy. DE GENNES /1/ showed that for microemulsion droplets in equilibrium with excess dispersed phase :

$$\gamma = \frac{2K}{R_0\, R} \,.$$

Many predictions were made for the entropy contribution /6/. They all lead to :

$$\gamma = \frac{kT}{L^2} \; f(\phi_O \, , \phi_W) ,$$

where f is a universal function of volume fraction and L = 2R or ξ.

In the following we will discuss the results obtained on one of those systems : structural data, interfacial tensions and bending elasticity, and compare them with previous results on other systems.

II. EXPERIMENTAL PROCEDURE

1. Sample composition

We have studied in a previous work two model systems. The first was a mixture of brine (47 wt%), toluene (47 wt%), butanol (4 wt%) and sodium dodecyl sulfate (SDS, 2 wt%). The brine is an aqueous solution of S wt% sodium chloride. Butanol is slightly soluble in oil and in water but acts mainly like a cosurfactant, and very importantly decreases the rigidity of the SDS layers. The salinities of the transitions between the different Winsor regimes were found to be S_1 = 5.4 and S_2 = 7.4 at 20°C /7/.

The second model system was a mixture of brine (water + NaCℓ, 56.83 wt%), dodecane (38.19 wt%), butanol (3.32 wt%) and sodium hexadecyl benzene sulfonate (SHBS, 1.66 wt%). This composition corresponds to equivalent volumes of oil and brine. S_1 = 0.52, S_2 = 0.61 at 20°C /8/.

Characteristic sizes were found larger in the second system, and correspondingly the interfacial tensions lower. More recently, we have undertaken the study of a new model system were the sizes were expected to be smaller, in order to test the theories in the widest possible range.

This system is a mixture of brine (water + NaBr, 47 wt%), toluene (47 wt%), butanol (4 wt%) and dodecyl trimethyl ammonium bromide (DTAB, 2 wt%). S was varied between 0 and 9 and S_1 = 1.98, S_2 = 5.1 at 20°C /9/.

2. Experimental techniques

The composition of the different phases was investigated by gas chromatography. It was found that the ratio of the number of butanol molecules to surfactant molecules in the amphiphilic layer is about one, like in the SDS system (it was 3 in the SHBS one).

The characteristic sizes were measured with small-angle X-ray scattering in collaboration with R. OBER at the Collège de France. A typical spectrum is shown in fig.1. A broad peak is clearly visible. From it we deduced $\xi \sim \Pi/q_{max}$ = 95 Å. At higher q values, the spectrum follow Porod's law $I(q) \propto q^{-4}$ and from this asymptotic behavior we deduced Σ = 72 Å^2. The departure ot the asymptotic regime in the intermediate q range leads to a minimum in the curve q I(q) versus q : $2\Pi/q_{min}$ = 86 Å is a typical radius of curvature of the film. The measured sizes for different salinities are presented in table 1.

Table 1 : Composition and characteristic sizes in the DTAB microemulsions. Theoretical values L_{th} are calculated from eq. 1 and 3 with $\Sigma = 70$ Å^2. We have incorporated in each ϕ_o and ϕ_w half the film volume fraction.

S	ϕ_o	ϕ_w	Π/q_{max}	R	L_{th}
1	0.10			64 Å	62 Å
2	0.37	0.63	120 Å		135
2.5	0.52	0.48	95		113
3	0.58	0.42	103		104
4	0.71	0.29	90		89
5	0.75	0.25	78		73
6		0.076		51	54
7		0.065		54	43
9		0.053		42	36

Interfacial tensions have been measured using surface light scattering techniques. Data are reported in table 2. Interfacial tensions between the microemulsions and respectively the excess oil and water phase γ_{mo} and γ_{mw} are equal at the salinity S^* called "optimal" (for oil recovery purposes). At this salinity $\phi_o = \phi_w$ and the spontaneous curvature of the surfactant layer is zero /10/. The corresponding value of the tension γ^* is therefore expected to arise from entropy contributions to the free energy only. This does not seem to be the case since $\gamma^* \xi^2/kT$ depends on the system (0.65 for the SDS system, 0.35 for SHBS and 0.38 for DTAB).

Table 2 : Interfacial tensions in the DTAB system. Measured and calculated values. Units are mN/m.

S	γ_{mo}	γ_{mw}	γ_c	γ_e
1	0.12		0.13	0.02
2.5 $\sim S^*$	1.7×10^{-2}	$1.7\ 10^{-2}$		
6		0.105	0.21	0.03
7		0.138	0.19	"
9		0.134	0.33	"

Ellipsometry has been used to study the interface between the excess phases in the three-phase region. The interfacial tension γ_{OW} is also very low because the interface is covered by a surfactant layer of high surface pressure and the phases in equilibria contain no microstructure. The ellipsometry signal is then dominated by the roughness of the interface due to thermal motion, and the measured ellipticity is proportional to $(\gamma_{OW}\ K)^{-1/2}$. Since K depends on the scale it was necessary at this stage to renormalize γ and K consistently /11/ in order to use these measurements to calculate the persistence length.

III. DISCUSSION

As in SDS and SHBS systems, the size of structural elements L is, within experimental error, proportional to the product $\phi_o\phi_w$ as predicted by random space filling models. The calculated value of ξ (eq. 3) is close to Π/q_{max} (see table 1) although it should be equal to $2\Pi/q_{max}$ (random objects of size ξ). This feature seems universal ; it has recently been found also in non ionic surfactant systems /12/. The interpretation of the discrepancy is not yet clear /13/.

As expected, the K values for DTAB system are lower than in SDS system. We find (scale of ellipsometry measurement) $K = 1.7 \times 10^{-14}$ erg at $S = S^{*}$ instead of $K = 3 \times 10^{-14}$ for the SDS sytem. With the appropriate renormalization treatment, this leads to :

$$\xi_K = 150\ \text{Å},$$

close to the measured ξ value (table 1).

Measurements of K in the SHBS system are in progress. Let us now briefly discuss the interfacial tension data and estimate the curvature and entropic contributions. From /6/ :

$$\gamma_C = \frac{2K}{R\ R_0} \qquad R = R_0 \left| 1 + \frac{kT}{8\pi K}\left(\ln\phi + \frac{1.5-\phi}{\phi}\ln(1-\phi)\right)\right|,$$

$$\gamma_e = \frac{kT}{4\pi R^2}\ln\phi$$

for droplet microemulsions.

The results are given in table 2. It is seen that since curvature energy is very small in these systems, the entropic contribution is no longer negligible as in the SDS and SHBS systems /6/. The large difference between γ and $\gamma_C + \gamma_e$ at high salinity could be due to an eventual salinity variation of the bending elastic constant K, which has only been measured in the Winsor III domain.

IV. CONCLUSION

The present data confirm that bending elastic constants are very small and of the order kT in microemulsion systems. The persistence length of the surfactant layer being microscopic, when the spontaneous curvature of the film is small, the lamellae break and give rise to a bicontinuous structure. The structural sizes are in good agreement with those calculated assuming that the area per surfactant is constant. Our experiments also show that at maximum swelling (in presence of excess oil and water) the size ξ is strongly dependent on the K value, and close to the persistence length. These facts strongly suggest that the interfacial tension at the microscopic oil-water interface is zero and that microemulsion stability is probably related to the scale variations of the elastic constant.

REFERENCES

1. P.G. de Gennes and C. Taupin; J. Phys. Chem. 86, 2294 (1982)
2. S. Safran and L.A. Turkevich; Phys. Rev. Lett. 50, 1930 (1983)
3. L.E. Scriven; Nature 263, 123 (1976)
4. S.A. Safran, D. Roux, M.E. Cates and D. Andelman; Phys. Rev. Lett. 57, 491 (1986)
5. B. Widom; J. Chem. Phys. 81, 1030 (1984)
6. D. Langevin, D. Guest and J. Meunier; Coll. and Surf. 19, 159 (1986)
7. A.M. Cazabat, D. Langevin, J. Meunier and A. Pouchelon; Adv. Coll. Int. Sci. 16, 175 (1982)
8. D. Guest, L. Auvray and D. Langevin; J. Phys. Lett. 46, L-1055 (1985); D. Guest and D. Langevin; J. Coll. Int. Sci. 112, 208 (1986)
9. O. Abillon, C. Otero, B.P. Binks, D. Langevin and R. Ober; in preparation
10. L. Auvray, J.P. Cotton, R. Ober and C. Taupin; Physica 136B, 281 (1986)
11. J. Meunier; These proceedings
12. L. Lichterfeld, T. Schmeling and R. Strey; J. Phys. Chem. 90, 5762 (1986)
13. A. de Geyer; These proceedings

Phase Behavior of Quinary Systems: The $\tilde{X}$ Surface

M. Kahlweit and R. Strey

Max Planck Institut für Biophysikalische Chemie,
Postfach 2841, D-3400 Göttingen, Fed. Rep. of Germany

Based on the features of the phase behavior of ternary and quaternary systems repoted on previously /1,2/, we have studied the phase behavior of quinary systems H_2O - oil - nonionic amphiphile - ionic amphiphile - lyotropic salt. We have shown /3/, that the phase behavior of quinary systems evolves continously from the phase behavior of the corresponding ternary and quaternary systems. It can systematically be studied by tracing the point $\tilde{X}$ /4/ as a function of the weight percentage δ of the ionic in the mixture of the two amphiphiles and the brine concentration ϵ. At constant weight percentage α of the oil in the mixture of oil and brine, $\tilde{X}$, being that point at which the three-phase body touches the homogeneous solution of the five components, is unambiguously defined by its temperature $\tilde{T}$ and the weight fraction $\tilde{\gamma}$ of the two amphiphiles in the mixture. Furthermore, we suggested /5/ to represent $\tilde{T}$ in T-δ-ϵ space, and $\tilde{\gamma}$ in γ-δ-ϵ space. The first representation gives the temperature at which to search for a three-phase body, the latter what efficiency of the amphiphile to expect. As an example, we show in the Figure (left) the $\tilde{T}$-surface created by uniform clockwise rotation of the $\tilde{T}_\delta$ trajectory from its position at δ=0, $\tilde{T}_N$, to that at δ=100, $\tilde{T}_I$, assuming the $\tilde{T}_\delta$ trajectories to be straight lines. For comparison, in the Figure (right) the actual $\tilde{T}$ surface for the system H_2O - n-decane - $C_{12}E_4$ - AOT - NaCl has been constructed from experimental $\tilde{X}$ data / 5 / in T-γ space as function of δ and ϵ at constant α=50 wt%.

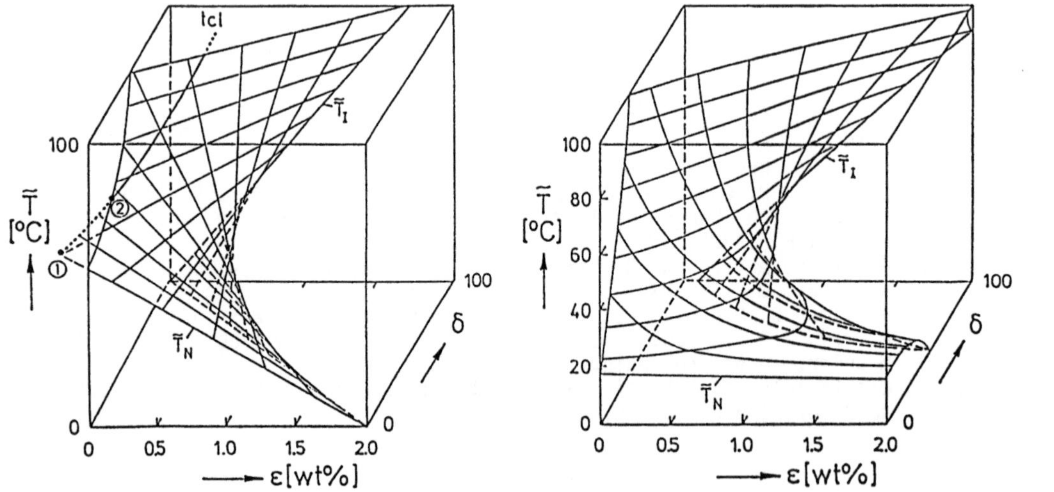

The antagonism of the temperature dependence of nonionic and ionic amphiphiles leads to hyperbolic $\tilde{T}_\epsilon$ trajectories along the $\tilde{T}(\delta,\epsilon)$ surface that diverge at a pole $0 \leq \delta^* \leq 100$ wt%. Accordingly, the trajectories for $\delta < \delta^*$ can be considered as nonionic, those for $\delta > \delta^*$ as ionic branches. Furthermore, we found that $\tilde{\gamma}$ can, in first approximation, be considered as straight interpolation between $\tilde{\gamma}_N$ of the nonionic and $\tilde{\gamma}_I$ of the ionic amphiphile at the respective temperature. We have compared this phenomenological description with measurements for a number of different systems and found that it permits qualitative predictions on the phase behavior of such quinary mixtures, as it depends on the chemical nature of the oil, the two amphiphiles, and the brine concentration.

References

1. M. Kahlweit, R. Strey: Angew. Chem. 24, 654 (1985)
2. M. Kahlweit, R. Strey, P. Firman: J. Phys. 90, 671 (1986)
3. M. Kahlweit, R. Strey: J. Phys. Chem. 90, 5239 (1986)
4. M. Kahlweit, R. Strey: J. Phys. Chem. 91, 1553 (1987)
5. M. Kahlweit, R. Strey: J. Phys. Chem. submitted

3. [illegible]
4. [illegible]
5. [illegible]

Part VII

Porous Media

Fractal Flow Patterns in Porous Media

A. Soucemarianadin[1], *R. Lenormand*[1], *G. Daccord*[1], *E. Touboul*[1], and *C. Zarcone*[2]

[1]Dowell Schlumberger, BP 90, F-42003 Saint Etienne Cedex 1, France
[2]Institut de Mécanique des Fluides, 2 Rue Camichel, F-31701 Toulouse Cedex, France

The displacement of one fluid by another fluid in a porous medium is of importance in many processes, especially petroleum recovery. The purpose of this paper is to describe three types of displacements where patterns are very ramified : i) when a non-wetting fluid is injected at very low flow rate (capillary fingering), ii) when a low viscosity fluid pushes a more viscous fluid (viscous fingering), iii) when the porous matrix is etched by the invading fluid (reactive flow).

The two first cases are studied in two-dimensional transparent networks of interconnected capillaries which simulate the porous medium and the third one in samples of plaster. For each case, the experimental results are presented and the technique for measuring the fractal dimension is described.

1. CAPILLARY FINGERING

In this case, viscous and gravity forces are assumed to be negligible and consequently all the displacement mechanisms are linked to capillary forces and randomness due to the different sizes of pores in a porous medium.

Generally speaking, when one fluid (say oil) is slowly displacing another immiscible fluid (say water) in a capillary tube of diameter D_o, the fluid for which the contact angle θ (between the tube and the meniscus) is smaller than $\pi/2$ is called the "wetting fluid" (W); the other one is the "non-wetting fluid" (NW). The pressure in the NW fluid exceeds the pressure in the wetting fluid by a value P, called capillary pressure and linked to the interfacial tension γ by Laplace's law:

$$P = 4\ \gamma\ \cos\theta / D_o\ . \tag{1}$$

A displacement where the NW fluid is pushing the wetting fluid is called *drainage*, the reverse is *imbibition*. We will describe here only the case of drainage (imbibition is more complex and some results can be found in previous publications /1/, /2/).

Capillary forces prevent the NW fluid from spontaneously entering a porous medium. It can only enter a duct (diameter D_o) when the pressure exceeds the capillary pressure. From a statistical point of view a duct with $D > D_o$ is an "active" or "conductive" bond and a duct with $D < D_o$ an inactive bond. The fraction p of active bonds can easily be deduced from the throat size distribution.

At a given pressure P, the injected fluid invades all the percolation clusters connected to the injection point; this mechanism has been called invasion percolation /3-5/. During the displacement, the wetting phase is trapped in the network when the invading NW fluid breaks the continuous path toward the exit.

The micromodels used in the experiments are made of transparent resin cast on a photographically etched mold (Fig. 1) /6/. The cross-section of each duct of the etched network is rectangular with a constant depth (x=1 mm) and a width d which varies from pore to pore with a given distribution and a random location. Various sizes of network are used for the experiments, the largest one containing more than 250,000 ducts. We will call *ducts* the cylindrical capillaries (bonds) and *pores* the volumes of the intersections (sites).

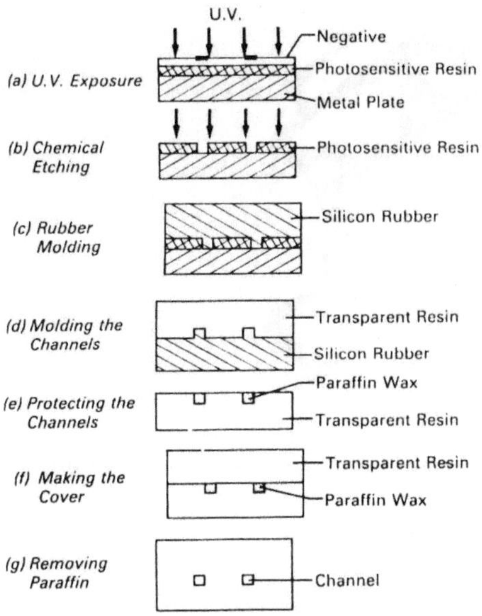

Figure 1. Procedure for the making of the etched network

For this study, air is displacing a very viscous oil (1000 cp) at very low flow rate. Figure 2 shows a close up of the situation of the non-wetting fluid (white) and the wetting fluid (black) in the ducts of the etched network. In order to obtain the geometrical properties of the invasion pattern, we must digitize the photographs taken during the displacement. For this purpose we developed an original technique: i) a negative copy of the image is made on a transparent high contrast film (Fig. 3a), ii) a positive copy of the micromodel before the displacement is made at the same scale (Fig. 3b), iii) the two films are then superimposed to eliminate the initial background due to the micromodel (Fig. 3c). The final image (Fig. 4a) is then digitized.

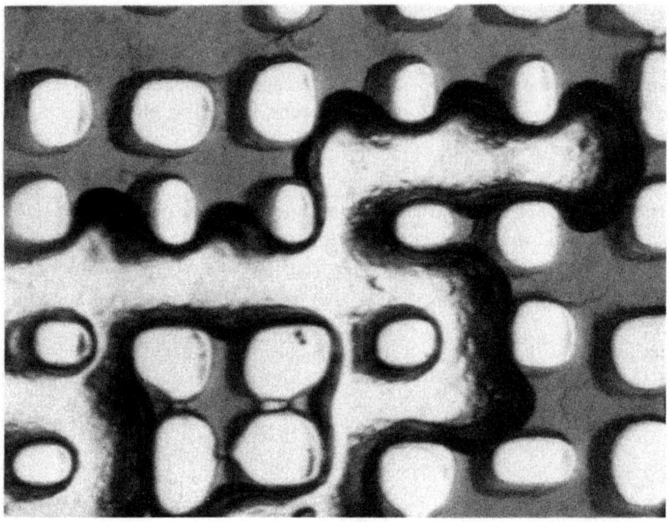

Figure 2. Close-up of the situation of the wetting fluid (black) and the non-wetting fluid (white) in the ducts of the etched network

Figure 3. Image subtraction (close-up): a) original image, b) back-ground, c) final image

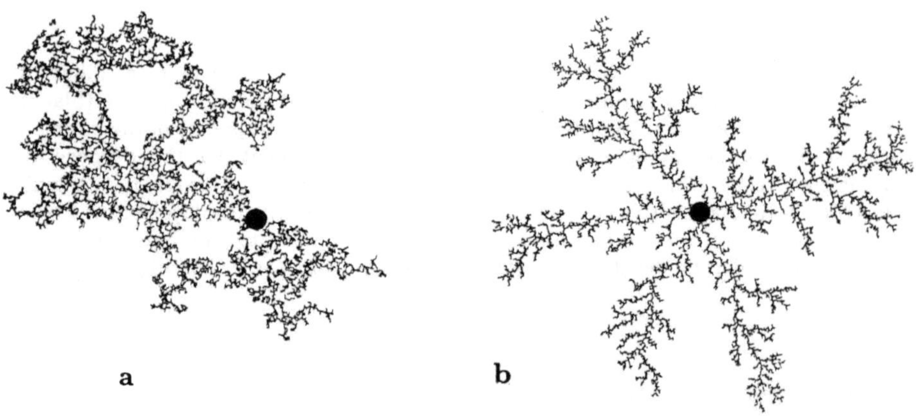

Figure 4. Air (black) displacing very viscous oil in a radial micromodel containing 250,000 ducts: a) capillary fingering (invasion percolation), b) viscous fingering (DLA)

For this kind of pattern, the more accurate method to obtain the fractal dimension consists in plotting the density-density correlation function $C(r)$ versus the distance r separating the bonds /7/:

$$C(r) = \sum_{r'} \rho(r')\ \rho(r' + r)\,. \tag{2}$$

The density $\rho(r)$ is defined to be 1 for the occupied bonds and 0 for the others. The plot in log-log scale is linear over more than two decades and leads to the value D = 1.80, in good agreement with the theoretical value of 1.82 /3/.

2. VISCOUS FINGERING

In this case, capillary forces are negligible compared to viscous forces. However, we need a small amount of capillary effects, otherwise the fluids mix together inside the ducts and the problem is more complex (miscible displacements).

Viscous displacements, either stable or unstable, are governed by the pressure field between the entrance and the exit. Consequently, even in the case of a stable displacement, a local model based on some rules at the interface cannot be used for modeling viscous displacements. A model, called *Gradient-Governed Growth* has been developed simultaneously by several authors /8/,/9/ to solve this problem, using both a continuum approach to calculate the pressure field and a discrete displacement of the interface which accounts for the granular structure of the porous medium.

If we assume that the injected fluid has negligible viscosity, the growing pattern can also be represented by a model known as *Diffusion-Limited Aggregation* /7/,/10/,/11/. A computation /12/ based on a network of *random* conductances, leads to similar results. The validity of this model is demonstrated by the similarity between computer simulations and experimental patterns. For instance, Fig. 4b shows the displacement of a very viscous oil by air in a radial geometry (same experimental technique as for drainage). The measured fractal dimension again is very close to the theoretical value D = 1.70.

3. REACTIVE FLOW

It has been known for a long time that injecting a reactive fluid into a soluble porous medium yields very ramified dissolution patterns. The physical phenomenon of the formation of dissolution patterns involves the flow of a liquid in a porous medium coupled with a chemical reaction.

We used the system plaster/water because its properties are well known and it is convenient for studying various geometries /13/, /14/. Good reproductions of the final dissolution patterns are obtained by injecting a low melting point alloy (Wood's metal) into the dried plaster. After cooling, the plaster is dissolved. Figure 5 shows the result in a two-dimensional radial cell, consisting of a thin disk of plaster (typically 1 mm thick and 200 mm diameter).

At first glance, this structure is very similar to other experimental growth patterns associated with the model of diffusion-limited aggregation (DLA): dielectric breakdown, electrodeposited metal leaves, viscous fingering ... This analogy can be understood by comparison with the above-described viscous fingering. The velocity u of a fluid of viscosity μ in a porous medium

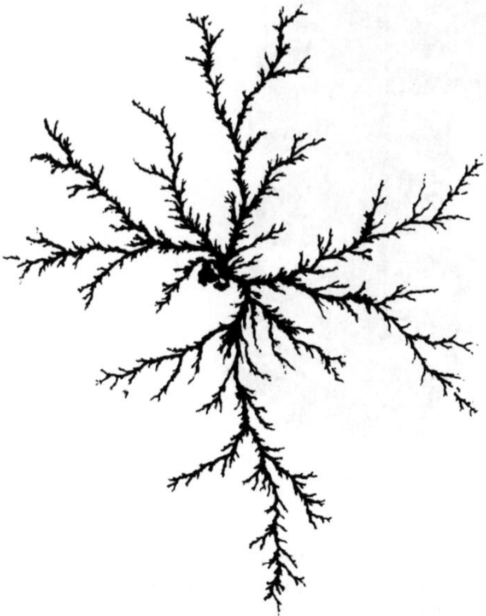

Figure 5. Photograph of a dissolution pattern obtained in a two-dimensional cell (diameter 200 mm)

of permeability k is proportional to the pressure gradient: $u = -M\text{grad}P$. Here $M = k/\mu$ is the mobility and P the pressure field. During the displacement of a viscous fluid by a non-viscous one, viscous fingering occurs due to the very sharp increase of mobility when a small perturbation appears at the interface between the fluids (the viscosity jumps from a finite value to a very small value). In the case of dissolution, the injected reactive fluid and the saturating non-reactive fluid have the same viscosity but, at the interface (the reactive front), the permeability jumps from a low value in the porous medium to a quasi-infinite value in the etched channels. Consequently, the effects on the mobility term are analogous to viscous fingering.

The two-dimensional patterns are quantified by digitizing the photographs and computing the density-density correlation functions. We obtained D = 1.6 ± 0.1 /13/, in reasonable agreement with results from DLA.

We also performed experiments with 3-dimensional samples. All the experiments produce very ramified, tree-like structures. Figure 6a is the moulding of a dissolution pattern obtained by radial injection through a cylindrical sample.

Now, the problem is to measure the geometrical properties of these 3-D dissolution patterns. So far, the experimental techniques available for measuring the fractal dimension of the surface of a 3D object are of two kinds /15/: i) the digitization of 2D sections, ii) the use of physical techniques which associate the change of a characteristic variable as a function of a tunable yardstick (neutron scattering, electrochemistry, etc...). The first of these techniques gives only 2D informations and destroys the fractal object, and the range of length scale of the second is always less than a micrometer. Thus none of these existing techniques is suitable for characterizing the large structures (Fig. 6a) obtained in the dissolution of a porous media by a reactive fluid. So, we developed an original technique based on capillary effects /16/.

The principle of this technique is to wrap the fractal object within a fluid "cocoon" (Fig. 6b). First, the object is immersed in a wetting fluid (Fig. 7a) and then slowly lowered into a

Figure 6. Experimental results: a) Photograph of the fractal object, b) Same object as in a) immersed in the non-wetting fluid (in white) and surrounded by the cocoon of wetting fluid (in grey) for a large value of the radius of curvature

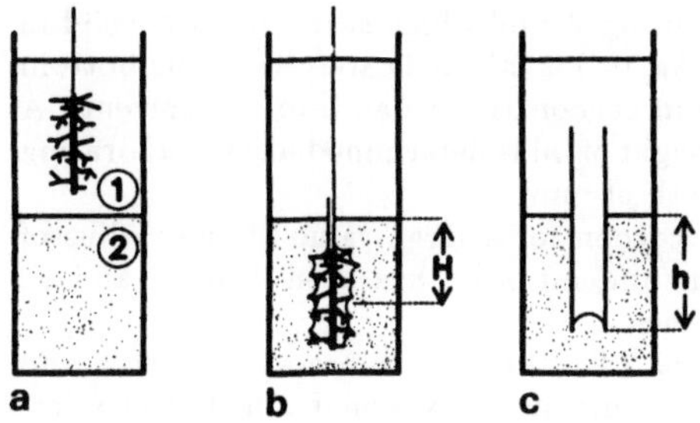

Figure 7. Principle of the method.
a) The object is immersed in the wetting fluid. b) The object is lowered in the non-wetting fluid and becomes surrounded by a cocoon of wetting fluid 1. The thin tube allows the continuity of fluid 1. c) Measurement of the radius of curvature R by a tube of radius r ($R = rh/H$)

non-wetting fluid (Fig. 7b). Due to capillary effects, a volume V of wetting fluid remains like a cocoon around the ramified object. The way the cocoon covers the object is ruled by the radius of curvature of the interface between the two fluids, the first fluid wetting the solid.

At each point located at a depth H in the non-wetting fluid, the radius of curvature R is determined by the balance between capillary and hydrostatic pressures (a thin tube makes the continuity between fluid 1 in the cocoon and the upper layer (Fig. 7c):

$$\frac{2\gamma}{R} = \Delta\rho\, g\, H, \tag{3}$$

where $\Delta\rho$ is the difference of density between the fluids and g the gravitational acceleration. This equation determines the "yardstick" R which can be tuned by changing the difference of density (we could also change the interfacial tension, however this in general alters the wettability). We obtain the relation $V(R)$ by measuring the volume V for different couples of fluids (different R). The difference of density can vary from 1 kg/m^3 to a few thousands of kg/m^3 (by using air as fluid 2), which leads theoretically to a range of R from 0.03 mm to a few tens of mm. (with $\gamma = 0.015$ N m^{-1} and H =0.10 m.).

If the object has a geometrical self-similarity, this capillary volume can be related to the fractal dimension D of the surface. For porous media, De GENNES /17/ found that $V(R)$ should scale as $R^{(3-D)}$ in the restrictive case of iterative pits and iterative flocs . If the self similarity is only statistical, such as for experimental objects, the way this capillary volume scales with the radius R is less trivial to derive. In a first approach, we can nevertheless assume that the scaling law will remain the same.

Although this principle seems simple, the experiments are difficult to perform. Nevertheless, we got reproducible results showing a power-law behavior over about one decade. The sensitivity to physico-chemical parameters (purity of fluids, diffusion of chemicals, evolution of interfacial tension with time...) forced us to make many tests before selecting appropriate couples of fluids and an experimental procedure which led to reproducible results. Fluid 1 is a silicon oil (methylphenyl polysiloxane), viscosity 0.2 Pa.s (200 cP) and density 1056 kg/m^3 at room temperature, chosen for its insolubility in water and its high density. Fluid 2 is a water solution of glucose at various concentrations (a saline solution would react with the Wood's metal making the sample). The pH of the solution is stabilized with 10% of a buffer and a few drops of bactericide to prevent the growth of bacteria.Furthermore, in order to obtain a well-defined wettability, the metallic object is coated with a very thin layer of polyvinylchloride before each experiment.

The yardstick R is determined directly by replacing the fractal object by a tube of known radius r, of the order of R. The measurement of the distance h of the meniscus in the tube from the interface between the fluids gives directly the value $R = r\, h/H$ (from the balance Eq. 3). This method gives more accurate results than the separate measurement of γ and $\Delta\rho$. During the experiment, we monitor the apparent weight of the object hung under a balance to determine when the equilibrium is reached (drainage of the wetting fluid through the cocoon and the tube). Then, the volume V is obtained by diluting the oil which surrounds the solid in a solvent: a known volume of chloroform ($\rho = 1.4\ 10^3$ kg/m^3) is injected carefully at the bottom of the cell and the object still surrounded by the fluid cocoon is lowered into the solvent. A given volume of solution is then withdrawn and the weight of oil is determined after evaporating the solvent. The accuracy of this method is a few tenth of mm^3.

Figure 6b shows the object surrounded by its oil cocoon for a large value of the radius of curvature R. For a small radius of curvature, a large amount of oil has been drained and oil remains only around the smallest details of the branches.

The preliminary results give a fractal dimension of 1.8. However, more experiments and a better understanding of the relationship between the measured exponent and the classical fractal dimension are needed.

REFERENCES

1. R. Lenormand, Physica **140A**, 114, (1986).
2. R. Lenormand, Soc. Petrol. Eng. paper **15390**, (1986).
3. D. Wilkinson and J. F. Willemsen, J. Phys. A **16**, 3365, (1983).
4. J. T. Chayes, L. Chayes and C. M. Newman, Commun. Math. Phys. **101**, 383, (1985).
5. R. Lenormand and C. Zarcone, Phys. Rev. Let. **54**, 2226, (1985).
6. R. Lenormand and Zarcone, Phys. Chem. Hydrodyn. **6**, 497 (1985).
7. T. A. Witten and Sander, Phys. Rev. B **27**, 5686, (1983).
8. M. King and H. Sher, Soc. Petrol. Eng. paper **14366**, (1985).
9. J. D. Sherwood and J. Nittmann, J. Physique **47**, 15, (1986).
10. L. Paterson, Phys. Rev. Lett. **52**, 1621, (1984).
11. K. J. Måløy, J. Feder and T. Jøssang, Phys. Rev. Lett. **55**, 1885, (1985).
12. J. D. Chen and D. Wilkinson, Phys. Rev. Lett. **55**, 1892, (1985).
13. G. Daccord and R. Lenormand, Nature, **325**, 41, (1987).
14. G. Daccord, Phys. Rev. Lett. **58**, 479, (1987).
15. F. Brochard, J. Physique (Paris), **46**, 2117, (1986).
16. R. Lenormand, A. Soucemarianadin, E. Touboul and G. Daccord, submitted to Phys. Rev. Lett. (1987).
17. P. G. de Gennes, in Physics of Disordered Materials, ed. D. Adler, H. Fritzsche and S. R. Ovshinsky, Plenum Publishing Corporation, (1985).

Index of Contributors

Springer Proceedings in Physics

Volume 12
J. D. Harvey, D. F. Walls (Eds.)

Quantum Optics IV

Proceedings of the Fourth International Symposium, Hamilton, New Zealand, February 10–15, 1986
1986. 110 figures. Approx. 310 pages.
ISBN 3-540-16838-9

Volume 13
M. J. Kelley, C. Weisbuch (Eds.)

The Physics and Fabrication of Microstructures and Microdevices

Proceedings of the Winter School, Les Houches, France, March 25–April 5, 1986
1986. 328 figures. XI, 469 pages.
ISBN 3-540-16898-2

Volume 14
L. M. Falicov, J. L. Moran-López (Eds.)

Magnetic Properties of Low-Dimensional Systems

Proceedings of an International Workshop, Texco, Mexico, January 6–9, 1986
1986. 126 figures. XI, 189 pages.
ISBN 3-540-16261-5

Volume 15
S. Rosenwaks (Ed.)

Gas Flow and Chemical Lasers

Proceedings of the 6th International Symposium, Jerusalem, September 8–12, 1986
1987. 479 figures. XIII, 579 pages.
ISBN 3-540-17481-8

Volume 16
N. K. Rahman, C. Guidotti, M. Allegrini (Eds.)

Photons and Continuum States of Atoms and Molecules

Proceedings of a Workshop, Cortona, Italy, June 16–20, 1986
1987. 144 figures. XI, 289 pages.
ISBN 3-540-17387-0

Volume 17
A. Heidemann, A. Magerl, M. Prager, D. Richter, T. Springer (Eds.)

Quantum Aspects of Molecular Motions in Solids

Proceedings of an ILL-IFF Workshop, Grenoble, France, September 24–26, 1986
1987. 117 figures. XII, 221 pages.
ISBN 3-540-17562-8

Volume 18
P. Günter (Ed.)

Electro-optic and Photorefractive Materials

Proceedings of the International School on Material Science and Technology, Erice, Italy, July 6–17, 1986
1987. 286 figures. IX, 385 pages.
ISBN 3-540-17550-4

Volume 19
R. Graham, A. Wunderlin (Eds.)

Lasers and Synergetics

A Colloquium on Coherence and Selforganization in Nature
1987. 114 figures. X, 271 pages.
ISBN 3-540-17940-2

Volume 20
T. Kobayashi

Primary Processes in Photobiology

Proceedings of the 12th Taniguchi Symposium, Fujiyoshida, Yamanashi Prefecture, Japan, December 7–12, 1986
1987. Approx. 173 figures. Approx. 260 pages.
ISBN 3-540-18068-0

MIX
Papier aus verantwortungsvollen Quellen
Paper from responsible sources
FSC® C105338

If you have any concerns about our products,
you can contact us on
ProductSafety@springernature.com

In case Publisher is established outside the EU,
the EU authorized representative is:
Springer Nature Customer Service Center GmbH
Europaplatz 3, 69115 Heidelberg, Germany

Printed by Libri Plureos GmbH
in Hamburg, Germany